Methods in Enzymology

Volume 347
PROTEIN SENSORS AND REACTIVE
OXYGEN SPECIES
Part A
Selenoproteins and Thioredoxin

METHODS IN ENZYMOLOGY

EDITORS-IN-CHIEF

John N. Abelson   Melvin I. Simon

DIVISION OF BIOLOGY
CALIFORNIA INSTITUTE OF TECHNOLOGY
PASADENA, CALIFORNIA

FOUNDING EDITORS

Sidney P. Colowick and Nathan O. Kaplan

Methods in Enzymology

Volume 347

Protein Sensors and Reactive Oxygen Species

Part A
Selenoproteins and Thioredoxin

EDITED BY

Helmut Sies

HEINRICH-HEINE-UNIVERSITÄT
DÜSSELDORF, GERMANY

Lester Packer

UNIVERSITY OF SOUTHERN CALIFORNIA
LOS ANGELES, CALIFORNIA

ACADEMIC PRESS

An Elsevier Science Imprint

San Diego   San Francisco   New York   Boston   London   Sydney   Tokyo

Academic Press
An Elsevier Science Imprint.
525 B Street, Suite 1900, San Diego, California 92101-4495, USA
http://www.academicpress.com

Academic Press
32 Jamestown Road, London NW1 7BY, UK
http://www.academicpress.com

International Standard Book Number: 0-12-182248-6

PRINTED IN THE UNITED STATES OF AMERICA
02 03 04 05 06 07 SB 9 8 7 6 5 4 3 2 1

Table of Contents

Section I. Selenoproteins

Section II. Thioredoxin

Contributors to Volume 347

Article numbers are in parentheses following the names of contributors.
Affiliations are current.

ELIAS S. J. ARNÉR (20), *Department of Medical Biochemistry and Biophysics, Medical Nobel Institute for Biochemistry, Karolinska Institute, SE-17177 Stockholm, Sweden*

GAVIN E. ARTEEL (11), *Laboratory of Hepatobiology and Toxicology, Department of Pharmacology, University of North Carolina, Chapel Hill, North Carolina 27599*

KAORI ASAMITSU (33), *Department of Molecular Genetics, Nagoya City University Medical School, Mizuho-ku, Nagoya 467-8601, Japan*

KATJA BECKER (13, 35, 36), *Interdisciplinary Research Center, Justus Liebig University, D-35392 Giessen, Germany*

JON BECKWITH (34), *Department of Microbiology and Molecular Genetics, Harvard Medical School, Boston, Massachusetts 02115*

MARLA J. BERRY (2), *Thyroid Division, Brigham and Women's Hospital, Harvard Institutes of Medicine, Harvard Medical School, Boston, Massachusetts 02115*

NEIL A. BERSANI (29), *Department of Biochemistry and Biophysics, Oregon State University, Corvallis, Oregon 97331*

DURGA K. BHUYAN (40), *Membrane Biochemistry Laboratory, Department of Ophthalmology, Columbia University, New York, New York 10032*

KAILASH C. BHUYAN (40), *Membrane Biochemistry Laboratory, Department of Ophthalmology, Columbia University, New York, New York 10032*

ALBERTO BINDOLI (28), *Study Center for Biomembranes, National Council of Research, I-35121 Padova, Italy*

AUGUST BÖCK (1, 17), *Institute of Genetics and Microbiology, University of Munich, D-80638 Munich, Germany*

REGINA BRIGELIUS-FLOHÉ (9), *Department of Vitamins and Atherosclerosis, German Institute of Human Nutrition, D-14558 Potsdam-Rehbrücke, Germany*

BOB B. BUCHANAN (32), *Department of Plant and Microbial Biology, University of California, Berkeley, California 94720*

DARIUS P. BUCHCYZK (11), *Institut für Physiologische Chemie I, Heinrich-Heine-Universität, D-40001 Düsseldorf, Germany*

BRADLEY A. CARLSON (3), *Section on Molecular Biology of Selenium, Basic Research Laboratory, Center for Cancer Research, National Cancer Institute, National Institutes of Health, Bethesda, Maryland 20892*

MYEONG-JE CHO (32), *Department of Plant and Microbial Biology, University of California, Berkeley, California 94720*

PAUL R. COPELAND (4), *Department of Cell Biology, Lerner Research Institute, Cleveland Clinic Foundation, Cleveland, Ohio 44195*

ALAN M. DIAMOND (15), *Department of Human Nutrition, University of Illinois, Chicago, Illinois 60612*

DONNA M. DRISCOLL (4), *Department of Cell Biology, Lerner Research Institute, Cleveland Clinic Foundation, Cleveland, Ohio 44195*

REGINA EBERT-DUEMIG (13), *Clinical Research Unit, Medizinische Poliklinik, University of Wuerzburg, D-97070 Wuerzburg, Germany*

NOBUYOSHI ESAKI (16), *Institute for Chemical Research, Kyoto University, Uji, Kyoto 611-0011, Japan*

ALEXANDRA FISCHER (24), *Institute of Animal Nutrition and Nutrition Physiology, Justus Liebig University, D-35392 Giessen, Germany*

LEOPOLD FLOHÉ (18, 22), *Department of Biochemistry, Technical University of Braunschweig, D-38124 Braunschweig, Germany*

DANIEL GAUTHERET (6), *ESIL-GBMA, Ecole Supérieure d'Ingénieurs de Luminy, Université de la Méditerranée, 13288 Marseille Cedex 09, France*

VADIM N. GLADYSHEV (8, 15, 43), *Department of Biochemistry, University of Nebraska, Lincoln, Nebraska 68588*

STEPHAN GROMER (36), *Biochemistry Center, Heidelberg University, D-69120 Heidelberg, Germany*

ELISABETH GRUNDNER-CULEMANN (2), *Thyroid Division, Brigham and Women's Hospital, Harvard Institutes of Medicine, Harvard Medical School, Boston, Massachusetts 02115*

JOHN W. HARNEY (2), *Thyroid Division, Brigham and Women's Hospital, Harvard Institutes of Medicine, Harvard Medical School, Boston, Massachusetts 02115*

DOLPH L. HATFIELD (3, 15), *Section on Molecular Biology of Selenium, Basic Research Laboratory, Center for Cancer Research, National Cancer Institute, National Institutes of Health, Bethesda, Maryland 20892*

HANS-JÜRGEN HECHT (22), *Department of Structure Research, German Center for Biotechnology (GBF), D-38124 Braunschweig, Germany*

BIRGIT HOFMANN (22), *Department of Biochemistry, Technical University of Braunschweig, D-38124 Braunschweig, Germany*

ARNE HOLMGREN (21, 26), *Department of Medical Biochemistry and Biophysics, Medical Nobel Institute for Biochemistry, Karolinska Institute, SE-17177 Stockholm, Sweden*

ROBERT J. HONDAL (7), *Department of Biochemistry, University of Wisconsin, Madison, Wisconsin 53706*

JEAN-PIERRE JACQUOT (39), *Biochimie et Biologie Moléculaire, Université de Nancy I, F-54506 Vandoeuvre Cedex, France*

FRANZ JAKOB (13), *Orthopedic Department, University of Wuerzburg, D-97074 Wuerzburg, Germany*

JUAN JURADO (42), *Departamento de Bioquímica y Biología Molecular, Campus de Rabanales Edificio C-6, Universidad de Córdoba, 14071 Córdoba, Spain*

STEFAN M. KANZOK (35), *Biochemistry Center, Heidelberg University, D-69120 Heidelberg, Germany*

LARS-OLIVER KLOTZ (11), *Institut für Physiologische Chemie I, Heinrich-Heine-Universität, D-40001 Düsseldorf, Germany*

JOSEF KÖHRLE (12), *Division of Molecular Internal Medicine, Medizinische Poliklinik, University of Wuerzburg, D-97070 Wuerzburg, Germany*

KONSTANTIN V. KOROTKOV (15), *Department of Biochemistry, University of Nebraska, Lincoln, Nebraska 68588*

R. LUISE KRAUTH-SIEGEL (23), *Biochemie-Zentrum Heidelberg, Universität Heidelberg, 69120 Heidelberg, Germany*

ALAIN KROL (6), *Unité Propre de Recherche 9002 du CNRS, Institut de Biologie Moléculaire et Cellulaire, 67084 Strasbourg Cedex, France*

GREGORY V. KRYUKOV (8), *Department of Biochemistry, University of Nebraska, Lincoln, Nebraska 68588*

EASWARI KUMARASWAMY (15), *Basic Research Laboratory, National Cancer Institute, National Institutes of Health, Bethesda, Maryland 20892*

YOUNG-MOO LEE (32), *Molecular Structure Facility, University of California, Davis, California 95616*

XIN GEN LEI (19), *Department of Animal Science, Cornell University, Ithaca, New York 14853*

STEPHANE D. LEMAIRE (39), *Institut de Biotechnologie des Plantes, Université Paris-Sud, F-91405 Orsay Cedex, France*

ALAIN LESCURE (6), *Unité Propre de Recherche 9002 du CNRS, Institut de Biologie Moléculaire et Cellulaire, 67084 Strasbourg Cedex, France*

ORVILLE A. LEVANDER (10), *Beltsville Human Nutrition Research Center, USDA/ARS, Beltsville, Maryland 20705*

NATHAN I. LOPEZ (29), *Department of Biochemistry and Biophysics, Oregon State University, Corvallis, Oregon 97331*

JUAN LÓPEZ-BAREA (42), *Departmento de Bioquímica y Biología Molecular, Campus de Rabanales Edificio C-6, Universidad de Córdoba, 14071 Córdoba, Spain*

MATILDE MAIORINO (18), *Department of Biological Chemistry, University of Padova, I-35121 Padova, Italy*

JOHN B. MANSELL (2), *Thyroid Division, Brigham and Women's Hospital, Harvard Institutes of Medicine, Harvard Medical School, Boston, Massachusetts 02115*

LYNNE E. MAQUAT (5), *Department of Biochemistry and Biophysics, School of Medicine and Dentistry, University of Rochester, Rochester, New York 14642*

GLOVER W. MARTIN III (2), *Harvard M.I.T. Division of Health Sciences and Technology, Massachusetts Institute of Technology, Cambridge, Massachusetts 02139*

HIROSHI MASUTANI (25), *Department of Biological Responses, Institute for Virus Research, Kyoto University, Sakyo-ku, Kyoto 606-8507, Japan*

JAMES M. MAY (30), *Departments of Medicine and Molecular Physiology and Biophysics, Vanderbilt University Medical Center, Nashville, Tennessee 37232*

HEIKO MERKLE (36), *Biochemistry Center, Heidelberg University, D-69120 Heidelberg, Germany*

GARY F. MERRILL (29), *Department of Biochemistry and Biophysics. Oregon State University, Corvallis, Oregon 97331*

JASON R. MERWIN (29), *Department of Biochemistry and Biophysics, Oregon State University, Corvallis, Oregon 97331*

YVES MEYER (37), *Laboratoire de Physiologie et Biologie Moléculaire des Plantes, Université UMR CNRS, 5096 Genome et Developpement des Plantes, 66860 Perpignan, France*

MYROSLAWA MIGINIAC-MASLOW (39), *Institut de Biotechnologie des Plantes, Université Paris-Sud, F-91405 Orsay Cedex, France*

HISAAKI MIHARA (16), *Institute for Chemical Research, Kyoto University, Uji, Kyoto 611-0011, Japan*

AKIRA MITSUI (41), *Pharmaceutical Research Laboratories, Ajinomoto Co., Inc., Kawasakiku, Kawasaki 210-8681, Japan*

FERNANDO MONJE-CASAS (42), *Departamento de Bioquímica y Biología Molecular, Campus de Rabanales Edificio C-6, Universidad de Córdoba, 14071 Córdoba, Spain*

NADYA MOROZOVA (2), *Thyroid Division, Brigham and Women's Hospital, Harvard Institutes of Medicine, Harvard Medical School, Boston, Massachusetts 02115*

CORDULA MÜLLER (9), *Department of Vitamins and Atherosclerosis, German Institute of Human Nutrition, D-14558 Bergholz-Rehbrücke, Germany*

HAJIME NAKAMURA (31, 41), *Department of Biological Responses, Institute for Virus Research, Kyoto University, Sakyo-ku, Kyoto 606-8507, Japan*

BERNHARD NEUHIERL (17), *Haematologikum der GSF, D-81377 Munich, Germany*

YUMIKO NISHINAKA (31), *Human Stress Signal Research Center, National Institute of Advanced Industrial Science and Technology, Ikeda, Osaka 563-8577, Japan*

TAKASHI OKAMOTO (33), *Department of Molecular Genetics, Nagoya City University Medical School, Mizuho-ku, Nagoya 467-8601, Japan*

ELLEN PAAR (13), *Clinical Research Unit, Medizinische Poliklinik, University of Wuerzburg, D-97070 Wuerzburg, Germany*

JOSEF PALLAUF (24), *Institute of Animal Nutrition and Nutrition Physiology, Justus Liebig University, D-35392 Giessen, Germany*

GEORGE D. PEARSON (29), *Department of Biochemistry and Biophysics, Oregon State University, Corvallis, Oregon 97331*

MARÍA-JOSÉ PRIETO-ÁLAMO (42), *Departamento de Bioquímica y Biología Molecular, Campus de Rabanales Edificio C-6, Universidad de Córdoba, 14071 Córdoba, Spain*

CARMEN PUEYO (42), *Departamento de Bioquímica y Biología Molecular, Campus de Rabanales Edificio C-6, Universidad de Córdoba, 14071 Córdoba, Spain*

STEFAN RAHLFS (35), *Interdisciplinary Research Center, Giessen University, D-35392 Giessen, Germany*

RONALD T. RAINES (7), *Departments of Biochemistry and Chemistry, University of Wisconsin, Madison, Wisconsin 53706*

PABBATHI G. REDDY (40), *Head and Neck Service, Laboratory of Epithelial Cancer Biology, Memorial Sloan-Kettering Cancer Center, New York, New York 10021*

JEAN PHILIPPE REICHHELD (37), *Laboratoire de Physiologie et Biologie Moléculaire des Plantes, Université UMR CNRS, 5096 Genome et Developpement des Plantes, 66860 Perpignan, France*

MARIA PIA RIGOBELLO (28), *Department of Biological Chemistry, University of Padova, I-35121 Padova, Italy*

GERALD RIMBACH (24), *School of Food Biosciences, Hugh Sinclair Human Nutrition Unit, University of Reading, Whiteknights, Reading RG6 6AP, United Kingdom*

DANIEL RITZ (34), *Department of Microbiology and Molecular Genetics, Harvard Medical School, Boston, Massachusetts 02115*

ANTONELLA ROVERI (18), *Department of Biological Chemistry, University of Padova, I-35121 Padova, Italy*

R. HEINER SCHIRMER (35, 36), *Biochemistry Center, Heidelberg University, D-69120 Heidelberg, Germany*

HEIDE SCHMIDT (23), *Biochemie-Zentrum Heidelberg, Universität Heidelberg, 69120 Heidelberg, Germany*

PETER SCHÜRMANN (38), *Laboratoire de Biochimie Végétale, Université de Neuchâtel, CH-2007 Neuchâtel, Switzerland*

NORBERT SCHÜTZE (13), *Orthopedic Department, University of Wuerzburg, D-97074 Wuerzburg, Germany*

HELMUT SIES (11), *Institut für Physiologische Chemie I, Heinrich-Heine-Universität, D-40001 Düsseldorf, Germany*

ALLEN D. SMITH (10), *Beltsville Human Nutrition Research Center, USDA/ARS, Beltsville, Maryland 20705*

THRESSA C. STADTMAN (27), *Laboratory of Biochemistry, National Heart, Lung and Blood Institute, National Institutes of Health, Bethesda, Maryland 20892*

PETER STEINERT (22), *Department of Biochemistry, Technical University of Braunschweig, D-38124 Braunschweig, Germany*

QI-AN SUN (43), *Department of Biochemistry, University of Nebraska, Lincoln, Nebraska 68588*

XIAOLEI SUN (5), *Department of Biochemistry and Biophysics, School of Medicine and Dentistry, University of Rochester, Rochester, New York 14642*

TAKASHI TAMURA (27), *Department of Bioresources Chemistry, Faculty of Agriculture, Okayama University, Tsushima, Okayama 700-8530, Japan*

TOSHIFUMI TETSUKA (33), *Department of Molecular Genetics, Nagoya City University Medical School, Mizuho-ku, Nagoya 467-8601, Japan*

MARTIN THANBICHLER (1), *Institute of Genetics and Microbiology, University of Munich, D-80638 Munich, Germany*

ROZA TUJEBAJEVA (2), *Thyroid Division, Brigham and Women's Hospital, Harvard Institutes of Medicine, Harvard Medical School, Boston, Massachusetts 02115*

FULVIO URSINI (18), *Department of Biological Chemistry, University of Padova, I-35121 Padova, Italy*

FLORENCE VIGNOLS (37), *Laboratoire de Physiologie et Biologie Moléculaire des Plantes, Université UMR CNRS, 5096 Genome et Developpement des Plantes, 66860 Perpignan, France*

ALEXIOS VLAMIS-GARDIKAS (26), *Department of Medical Biochemistry and Biophysics, Medical Nobel Institute for Biochemistry, Karolinska Institute, SE-17177 Stockholm, Sweden*

PHILIP D. WHANGER (14), *Department of Environmental and Molecular Toxicology, Oregon State University, Corvallis, Oregon 97331*

KIRSTIN WINGLER (9), *Rudolf Buchheim Institute of Pharmacology, University of Giessen, D-35392 Giessen, Germany*

JOSHUA H. WONG (32), *Department of Plant and Microbial Biology, University of California, Berkeley, California 94720*

HIROYUKI YANO (32), *Hokuriku National Agricultural Experiment Station, Joetsu, Niigata 943-0193, Japan*

JUNJI YODOI (25, 31, 41), *Department of Biological Responses, Institute for Virus Research, Kyoto University, Sakyo-ku, Kyoto 606-8507, Japan*

LIANGWEI ZHONG (21), *Department of Medical Biochemistry and Biophysics, Medical Nobel Institute for Biochemistry, Karolinska Institute, SE-17177 Stockholm, Sweden*

Preface

In this era of genomics and proteomics, control of the functioning of expressed proteins becomes a key issue. Protein Sensors and Reactive Oxygen Species, Parts A and B (*Methods in Enzymology,* Volumes 347 and 348) present the current state of knowledge in this fast-growing field of proteins as sensors and transmitters of redox signals. A fundamental means of mediating and regulating activity of enzymes is through oxidation-reduction, with cysteine and selenocysteine residues in the protein as key players. Redox potentials in different cellular compartments are reflected by low-molecular-weight redox systems, notably glutathione/glutathione disulfide and related components which interact with the sensing proteins. Proteins sense the redox equilibria by a few types of posttranslational modifications. These include the formation of protein mixed disulfides, protein disulfides, and *S*-nitrosylated proteins. Selenocysteine, recently named the twenty-first amino acid, as well as selenomethionine occur in a number of proteins.

The first section of Part A: Selenoproteins and Thioredoxin, *Methods in Enzymology,* Volume 347, is concerned with the rapidly developing field of selenoproteins synthesis and its related molecular genetics, including novel selenoproteins identified from genomic sequence data using bioinformatics as well as current knowledge on glutathione peroxidases, selenoprotein P, iodothyronine deiodinases, and thioredoxin reductases. In the second section of this volume thioredoxin, glutaredoxin, and peroxiredoxin systems and their related enzymology are covered. These systems afford specificity and are connected to important cell biological processes, such as gene transcription, cytokine action, inflammatory response, and cell signaling in general. There are structural and functional differences in these proteins in mammals and parasites, which can potentially be exploited pharmacologically.

In Part B: Thiol Enzymes and Proteins, *Methods in Enzymology,* Volume 348, direct sensing of reactive oxygen species and related free radicals by thiol enzymes and proteins is addressed. A prime example is the role of the tyrosyl radical and of thiols in the reaction mechanism and control of ribonucleotide reductase. There are several enzymes catalyzing protein disulfide formation, e.g., protein disulfide isomerase, flavin-containing monooxygenase, and sulfhydryl oxidase. The role of protein *S*-glutathionylation, in particular, has attracted much interest in recent years. This process is a means of regulating protein function comparable to that of protein phosphorylation. The enzymology and significance of protein glutathionylation regarding several processes and specific enzyme reactions are part of the scope of this volume.

HELMUT SIES
LESTER PACKER

METHODS IN ENZYMOLOGY

VOLUME 73. Immunochemical Techniques (Part B)
Edited by JOHN J. LANGONE AND HELEN VAN VUNAKIS

VOLUME 74. Immunochemical Techniques (Part C)
Edited by JOHN J. LANGONE AND HELEN VAN VUNAKIS

VOLUME 75. Cumulative Subject Index Volumes XXXI, XXXII, XXXIV–LX
Edited by EDWARD A. DENNIS AND MARTHA G. DENNIS

VOLUME 76. Hemoglobins
Edited by ERALDO ANTONINI, LUIGI ROSSI-BERNARDI, AND EMILIA CHIANCONE

VOLUME 77. Detoxication and Drug Metabolism
Edited by WILLIAM B. JAKOBY

VOLUME 78. Interferons (Part A)
Edited by SIDNEY PESTKA

VOLUME 79. Interferons (Part B)
Edited by SIDNEY PESTKA

VOLUME 80. Proteolytic Enzymes (Part C)
Edited by LASZLO LORAND

VOLUME 81. Biomembranes (Part H: Visual Pigments and Purple Membranes, I)
Edited by LESTER PACKER

VOLUME 82. Structural and Contractile Proteins (Part A: Extracellular Matrix)
Edited by LEON W. CUNNINGHAM AND DIXIE W. FREDERIKSEN

VOLUME 83. Complex Carbohydrates (Part D)
Edited by VICTOR GINSBURG

VOLUME 84. Immunochemical Techniques (Part D: Selected Immunoassays)
Edited by JOHN J. LANGONE AND HELEN VAN VUNAKIS

VOLUME 85. Structural and Contractile Proteins (Part B: The Contractile Apparatus and the Cytoskeleton)
Edited by DIXIE W. FREDERIKSEN AND LEON W. CUNNINGHAM

VOLUME 86. Prostaglandins and Arachidonate Metabolites
Edited by WILLIAM E. M. LANDS AND WILLIAM L. SMITH

VOLUME 87. Enzyme Kinetics and Mechanism (Part C: Intermediates, Stereochemistry, and Rate Studies)
Edited by DANIEL L. PURICH

VOLUME 88. Biomembranes (Part I: Visual Pigments and Purple Membranes, II)
Edited by LESTER PACKER

VOLUME 89. Carbohydrate Metabolism (Part D)
Edited by WILLIS A. WOOD

VOLUME 90. Carbohydrate Metabolism (Part E)
Edited by WILLIS A. WOOD

Section I

Selenoproteins

[1] Selenoprotein Biosynthesis: Purification and Assay of Components Involved in Selenocysteine Biosynthesis and Insertion in *Escherichia coli*

By Martin Thanbichler and August Böck

Introduction

In 1954, selenium was discovered to be an essential nutrient for certain enteric bacteria,[1] and in the following years, the same finding was reported for mammals[2] and birds.[3] The first information about the role of selenium in the metabolism of these organisms then came from experiments showing that this trace element is closely associated with proteins.[4–7] However, its chemical form remained a matter of speculation until the first selenoprotein was sequenced[8] and shown to contain the hitherto unknown amino acid selenocysteine (Sec). This discovery raised the issue of how selenocysteine incorporation into proteins takes place. The initially favored hypothesis, that selenocysteine may be introduced into the polypeptide chain via posttranslational modification of a certain amino acid side chain, was soon replaced by the concept of a cotranslational incorporation mechanism, when the isolation and sequencing of the first selenoprotein genes showed that selenocysteine residues were represented by a UGA codon.[9,10]

The system responsible for selenoprotein biosynthesis has been characterized in detail for *Escherichia coli,* but considerable progress has been made in the elucidation of this process in archaeal organisms (reviewed by Rother *et al.*[11]) and eukaryal organisms (see [2]–[4] in this volume[12–14]). A selenocysteine-specific tRNA

[1] J. Pinsent, *Biochem. J.* **57**, 10 (1954).

[2] K. Schwarz and C. M. Foltz, *J. Am. Chem. Soc.* **79**, 3292 (1957).

[3] E. Patterson, R. Milstrey, and E. Stockstad, *Proc. Soc. Exp. Biol. Med.* **95**, 617 (1957).

[4] J. R. Andreesen and L. G. Ljungdahl, *J. Bacteriol.* **116**, 867 (1973).

[5] D. C. Turner and T. C. Stadtman, *Arch. Biochem. Biophys.* **154**, 366 (1973).

[6] L. Flohé, W. A. Günzler, and H. H. Schock, *FEBS Lett.* **32**, 132 (1973).

[7] J. T. Rotruck, A. L. Pope, H. E. Ganther, A. B. Swanson, D. G. Hafeman, and W. G. Hoekstra, *Science* **179**, 588 (1973).

[8] W. Günzler, G. Steffens, A. Grossmann, S. Kim, F. Ötting, A. Wendel, and L. Flohé, *Hoppe-Seyler's Z. Physiol. Chem.* **365**, 195 (1984).

[9] I. Chambers, J. Frampton, P. Goldfarb, N. Affara, W. McBain, and P. R. Harriason, *EMBO J.* **5**, 1221 (1986).

[10] F. Zinoni, A. Birkmann, W. Leinfelder, and A. Böck, *Proc. Natl. Acad. Sci. U.S.A.* **84**, 3156 (1987).

[11] M. Rother, A. Resch, R. Wilting, and A. Böck, *Biofactors* **14**, 75 (2001).

[12] M. J. Berry, G. W. Martin III, R. Tujebajeva, E. Grundner-Culemann, J. B. Marsell, N. Marozova, and J. W. Harney, *Methods Enzymol.* **347**, [2], 2002 (this volume).

[13] P. R. Copeland and D. M. Driscoll, *Methods Enzymol.* **347**, [4], 2002 (this volume).

[14] B. A. Carlson and D. L. Hatfield, *Methods Enzymol.* **347**, [3], 2002 (this volume).

(tRNA^{Sec}), which is encoded by the *selC* gene in *E. coli*,[15] is first aminoacylated with L-serine by seryl-tRNA synthetase (SerS) (Fig. 1A). Because of its unusual structural features, tRNA^{Sec} is aminoacylated with low efficiency compared with the charging efficiency of cognate serine isoacceptors.[16] Seryl-tRNA^{Sec} is then converted to selenocysteyl-tRNA^{Sec} (Sec-tRNA^{Sec}) by the action of selenocysteine synthase (SelA).[17] In this reaction, the amino group of the seryl residue forms a Schiff base with the carbonyl of the pyridoxal 5-phosphate cofactor of the enzyme. After dehydration yielding an aminoacrylyl intermediate, selenide is introduced into the molecule and selenocysteyl-tRNA^{Sec} is released.[18] The selenium donor for this reaction is selenophosphate,[19] which is synthesized from selenide and ATP by selenophosphate synthetase (SelD).[20,21] Sec-tRNA^{Sec} is bound by the specialized elongation factor (EF) SelB,[22] which shows partial homology to EF-Tu (for a review see Thanbichler and Böck[23]), but differs from it in having a carboxy-terminal extension of 272 amino acids containing an mRNA-binding domain.[24] The SelB · GTP · Sec-tRNA^{Sec} ternary complex is tethered to an mRNA secondary structure [selenocysteine insertion sequence (SECIS) element] positioned immediately downstream of the UGA triplet encoding selenocysteine.[25,26] After formation of this quaternary complex (Fig. 1B), SelB adopts a conformation enabling it to interact with the approaching ribosome.[27] It delivers Sec-tRNA^{Sec} to the ribosomal A site containing the UGA codon, which in turn triggers GTP hydrolysis and the subsequent release of SelB · GDP. Spontaneous exchange of GDP by GTP then confers on SelB the ability to enter the next reaction cycle. This intricate mechanism of selenocysteine synthesis and incorporation ensures that suppression of normal UGA stop codons is prevented.

In the domains Archaea and Eukarya, selenoprotein biosynthesis is mediated by a system similar to that described for *E. coli*. SelC homologs have been identified in both domains, but they differ from bacterial SelC variants in some structural features.[14] SelA activity has been measured in crude extracts from

[15] W. Leinfelder, E. Zehelein, M. Mandrand-Berthelot, and A. Böck, *Nature* (*London*) **331**, 723 (1988).
[16] C. Baron and A. Böck, *J. Biol. Chem.* **266**, 20375 (1991).
[17] K. Forchhammer, W. Leinfelder, K. Boesmiller, B. Veprek, and A. Böck, *J. Biol. Chem.* **266**, 6318 (1991).
[18] K. Forchhammer and A. Böck, *J. Biol. Chem.* **266**, 6324 (1991).
[19] R. Glass, W. Singh, W. Jung, Z. Veres, T. Scholz, and T. Stadtman, *Biochemistry* **32**, 12555 (1993).
[20] A. Ehrenreich, K. Forchhammer, P. Tormay, B. Veprek, and A. Böck, *Eur. J. Biochem.* **206**, 767 (1992).
[21] Z. Veres, I. Y. Kim, T. D. Scholz, and T. C. Stadtman, *J. Biol. Chem.* **269**, 10597 (1994).
[22] K. Forchhammer, W. Leinfelder, and A. Böck, *Nature* (*London*) **342**, 453 (1989).
[23] M. Thanbichler and A. Böck, *Biofactors* **14**, 53 (2001).
[24] M. Kromayer, R. Wilting, P. Tormay, and A. Böck, *J. Mol. Biol.* **262**, 413 (1996).
[25] F. Zinoni, J. Heider, and A. Böck, *Proc. Natl. Acad. Sci. U.S.A.* **87**, 4660 (1990).
[26] A. Hüttenhofer, E. Westhof, and A. Böck, *RNA* **2**, 354 (1996).
[27] A. Hüttenhofer and A. Böck, *Biochemistry* **37**, 885 (1998).

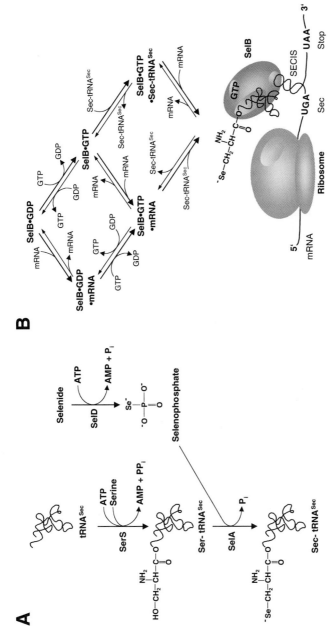

FIG. 1. Reactions involved in the biosynthesis of selenocysteine and its insertion into proteins. (A) Biosynthesis of selenocysteine in *E. coli.* (B) Interactions involved in the assembly of a functional quaternary SelB · GTP · mRNA · Sec-tRNA^{Sec} complex. The directions of reactions that are favored under physiological conditions are shown with large arrowheads.

archaeal[28] and eukaryal[29] organisms synthesizing selenoproteins, but until present, all attempts to purify the enzyme have failed. Moreover, homologs for the bacterial *selA* gene have not been found in the databases thus far. SelD homologs have been identified in archaeal and eukaryal genome sequences, but no functional protein has been purified to homogeneity up to now.[30] SelB homologs have also been described both for archaeal and eukaryal organisms.[31,32] They differ from the *E. coli* protein in lacking an mRNA-binding domain. Binding of the SECIS elements is, rather, mediated by an additional protein, designated SECIS-binding protein 2 (SBP2).[13] In contrast to the situation seen in the domain Bacteria, SECIS elements are located in the 3' untranslated region of selenoprotein mRNAs in the domains Archaea and Eukarya,[12] which enables the insertion of more than one selenocysteine residue into a polypeptide.

All components involved in selenoprotein biosynthesis in *E. coli* have been isolated and characterized, and can therefore serve as a paradigm for related systems. In the following sections we give a detailed description of the purification procedures and the methods to synthesize selenocysteyl-tRNA^{Sec} from tRNA^{Sec} and to analyze binding of selenocysteyl-tRNA^{Sec} to the SelB protein.

Purification of Selenocysteine-Inserting tRNA^{Sec}

SelC can easily be discriminated from canonical tRNAs by its unusual length of 95 bases. It can be purified in large amounts by using the following procedure worked out by Leinfelder *et al.*[15] For small-scale isolation of tRNA^{Sec}, preparative electrophoresis on urea–polyacrylamide gels can be applied as described by Schön *et al.*[33]

Overproduction

Plasmid pCB2013 is a pUC19 derivative with a chromosomal fragment containing *selC* under the control of its own promoter inserted into the multiple cloning site.[16] *Escherichia coli* MC4100[34] carrying pCB2013 is cultivated overnight in 5-liter Erlenmeyer flasks at 37° with maximal aeration in a medium containing 2% (w/v) tryptone, 1% (w/v) yeast extract, 1% (w/v) NaCl, and ampicillin (100 μg/ml). The cells are harvested by centrifugation and washed in buffer A [50 mM Tris-HCl (pH 7.4), 10 mM magnesium acetate].

[28] M. Rother, personal communication (2000).

[29] T. Mizutani, H. Kurata, and K. Yamada, *FEBS Lett.* **289,** 59 (1991).

[30] M. Guimaraes, D. Peterson, A. Vicari, B. Cocks, N. Copeland, D. Gilbert, N. Jenkins, D. Ferrick, R. Kastelein, J. Bazan, and A. Zlotnik, *Proc. Natl. Acad. Sci. U.S.A.* **93,** 15086 (1996).

[31] M. Rother, R. Wilting, S. Commans, and A. Böck, *J. Mol. Biol.* **299,** 351 (2000).

[32] R. M. Tujebajeva, P. R. Copeland, X. Xu, B. A. Carlson, J. W. Harney, D. M. Driscoll, D. L. Hatfield, and M. J. Berry, *EMBO Rep.* **1,** 158 (2000).

[33] A. Schön, A. Böck, G. Ott, M. Sprinzl, and D. Söll, *Nucleic Acids Res.* **17,** 7159 (1989).

[34] M. J. Casadaban and S. N. Cohen, *Proc. Natl. Acad. Sci. U.S.A.* **76,** 4530 (1979).

Column Chromatography

Crude tRNA, isolated from about 70 g (wet weight) of these cells according to the method of von Ehrenstein,[35] is dissolved in buffer B [50 mM sodium acetate (pH 5.0)] containing 400 mM NaCl, and loaded on a BD-cellulose column (2.5 × 24.5 cm; Serva, Heidelberg, Germany) equilibrated with the same buffer. The column is developed with 1200 ml of a linear gradient of 0.4–1.1 M NaCl in buffer B at a flow rate of 0.5 ml/min. SelC is eluted at about 800 mM NaCl. After finishing the gradient, strongly bound material containing the serine-isoacceptor tRNAs is eluted with 50 mM sodium acetate (pH 5.0)–1.5 M NaCl–14% (v/v) ethanol. Fractions containing SelC as monitored by electrophoresis on polyacrylamide–urea gels are pooled and the RNA is precipitated by the addition of 2 volumes of ethanol and incubation at −20°. After collection of the precipitate by centrifugation, the RNA is dissolved in 10 ml of buffer C [10 mM sodium acetate (pH 4.5), 10 mM MgCl$_2$, 1 mM dithiothreitol (DTT), 1 mM EDTA] containing 1.3 M (NH$_4$)$_2$SO$_4$. The solution is then applied to a Sepharose 4B column (2.6 × 15 cm; Amersham Pharmacia Biotech, Piscataway, NJ), which has been equilibrated with the same buffer. Separation of the tRNA species bound to the column is performed with at least 400 ml of a linear gradient decreasing from 1.3 M (NH$_4$)$_2$SO$_4$ to 0.3 M (NH$_4$)$_2$SO$_4$ in buffer C. Fractions containing SelC, which elutes at about 0.55 M (NH$_4$)$_2$SO$_4$, are pooled and dialyzed against two changes (2 liters each) of 10 mM sodium acetate (pH 4.5). The tRNA is then precipitated by the addition of 0.1 volume of 3 M sodium acetate (pH 4.5) and 1 volume of 2-propanol, collected by centrifugation, washed with 80% (v/v) ethanol, and finally dissolved in 10 mM sodium acetate (pH 5.0)–10 mM MgCl$_2$.

The purification protocol described yields about 30 mg of tRNA^{Sec}.

Purification of Selenophosphate Synthetase

SelD is a monomeric protein with a molecular mass of 37 kDa. It can be purified to apparent homogeneity by the following procedure, which is based on the purification protocol described by Ehrenreich *et al.*[20] An alternative procedure, which ensures the absence of trace amounts of adenylate kinase that might be copurified, is described by Veres *et al.*[21]

Overproduction

Escherichia coli BL21(DE3)[36] is transformed with plasmid pMN340,[37] which contains the *selD* open reading frame under the control of the T7 *Φ10* promoter.

[35] G. von Ehrenstein, *Methods Enzymol.* **12A,** 588 (1967).

[36] F. W. Studier and B. A. Moffat, *J. Mol. Biol.* **189,** 113 (1986).

[37] W. Leinfelder, K. Forchhammer, B. Veprek, E. Zehelein, and A. Böck, *Proc. Natl. Acad. Sci. U.S.A.* **87,** 543 (1990).

Transformants are grown under vigorous shaking in 5-liter Erlenmeyer flasks at 37° in Luria–Bertani (LB) medium containing ampicillin (50 μg/ml). At an optical density at 600 nm (OD_{600}) of 1.0, the expression of *selD* is induced by addition of 400 μM isopropyl-β-D-thiogalactopyranoside (IPTG). After shaking the cultures for another 2 hr, the cells are harvested by centrifugation and washed in buffer D [50 mM potassium phosphate (pH 7.2), 2 mM DTT, 1 mM EDTA].

Crude Extract

About 65 g of cells (wet weight) is resuspended in 200 ml of buffer D containing phenylmethylsulfonyl fluoride (PMSF, 20 μg/ml) and DNase I (10 μg/ml). After three passages through a French press cell at a pressure of 118 MPa, cell debris is removed from the suspension by centrifugation (30,000g for 30 min at 4°). Solid KCl is then slowly added to a final concentration of 800 mM to increase the solubility of SelD, while constantly stirring the solution on ice. After another 60 min of stirring, ultracentrifugation (150,000g for 1.5 hr at 4°) is performed. The resulting supernatant is adjusted to 55% ammonium sulfate saturation (at 0°) and subsequently subjected to centrifugation at 20,000g for 20 min at 4°. The precipitate formed is dissolved in 75 ml of buffer E [20 mM potassium phosphate (pH 7.2), 2 mM DTT, 0.5 mM EDTA].

Column Chromatography

After dialysis against two changes (2 liters each) of buffer E, the solution is loaded on a DEAE-Sepharose Fast Flow anion-exchange column (2.6 × 23.5 ml; Amersham Pharmacia Biotech), which is developed with 1875 ml of a linear KCl gradient (0–500 mM in buffer E) at a flow rate of 2.5 ml/min. SelD is eluted at a KCl concentration of about 210 mM. Fractions containing the protein, as monitored by sodium dodecyl sulfate–polyacrylamide gel electrophoresis (SDS–PAGE), are pooled and solid ammonium sulfate is added to 40% saturation (at 0°). The precipitate formed is removed by centrifugation at 20,000g for 20 min at 4°, and the ammonium sulfate concentration is increased to 55% saturation (at 0°). After recovering the precipitate by centrifugation (20,000g for 20 min at 4°), the pellet is dissolved in 40 ml of buffer E and loaded on a BioGel HTP hydroxyapatite column (2.6 × 22 cm; Bio-Rad, Hercules, CA) equilibrated with the same buffer. Separation of the bound proteins is achieved by applying 960 ml of a linear potassium phosphate gradient (20–200 mM in buffer E) at a flow rate of 1.0 ml/min. SelD is eluted at a concentration of about 80 mM. The peak fractions containing SelD are pooled and adjusted to 60% ammonium sulfate saturation (at 0°). The precipitate is then collected by centrifugation (20,000g for 20 min at 4°). The resulting pellet is dissolved in 40 ml of buffer E, dialyzed against 2 liters of the same buffer and subsequently against 2 liters of the same buffer containing 50% (v/v) glycerol, and stored at −20°.

Alternatively, the enzyme can be stored as an ammonium sulfate suspension at 4° for several months without significant decrease in enzymatic activity. Immediately before use, the suspension is centrifuged and the resulting pellet is dissolved in buffer F [20 mM potassium phosphate (pH 7.0), 5 mM magnesium acetate, 0.5 mM DTT, and 0.1 mM EDTA] to a final protein concentration of about 10 mg/ml. After dialysis against the same buffer, the solution can be used directly for the conversion reaction described below.

The purification procedure yields about 550 mg of pure protein.

Biochemical Properties

SelD exhibits a selenide-dependent ATPase activity.[20] However, unlike most AMP-generating ATPases, it catalyzes the hydrolysis of both phosphoanhydride bonds. The β-phosphate group (**P**) is released as orthophosphate, whereas the γ-phosphate (**P**) reacts with selenide to form selenophosphate:

$$\text{AMP-}P\text{-}P + \text{Se}^{2-} + \text{H}_2\text{O} \longrightarrow \text{AMP} + P_i + \text{Se}P \qquad (1)$$

The K_m value determined for ATP is 0.9 mM. ATP cannot be replaced by any other purine or pyrimidine nucleoside triphosphate, and no reaction is observed when pyrophosphate, polyphosphate, or nonhydrolyzable ATP analogs are used instead of ATP. The K_m for selenide is 7.3 μM. However, care must be taken to keep selenium in a fully reduced state and to work under strictly anaerobic conditions. SelD is competitively inhibited by AMP ($K_i = 170$ μM), whereas no inhibition or only a slight decrease in enzymatic activity is observed on addition of pyrophosphate or selenophosphate, respectively.[21]

Purification of Selenocysteine Synthase

SelA has a derived molecular mass of about 51 kDa. It can be purified to homogeneity by the procedure described below, which is based on the protocol established by Forchhammer *et al.*[17] Alternatively, enrichment of the protein from crude extracts can be performed according to the method of Tormay *et al.*[38]

Overproduction

Escherichia coli BL21(DE3)[36] is transformed with plasmid pWL187[17] carrying the *selA* gene under the control of the T7 *Φ10* promoter. Transformants are grown at 37° in LB medium containing ampicillin (100 μg/ml). At an OD$_{600}$ of 1.0, the production of SelA is induced by addition of IPTG to a final concentration of 100 μM.

[38] P. Tormay, R. Wilting, F. Lottspeich, P. K. Mehta, P. Christen, and A. Böck, *Eur. J. Biochem.* **254,** 655 (1998).

After another 2 hr of cultivation, the cells are harvested by centrifugation and washed in buffer G [50 mM potassium phosphate (pH 7.4), 3 mM DTT, 1 mM EDTA].

Crude Extract

Cells (30 g, wet weight) are resuspended in 90 ml of buffer H (buffer G containing 10 μM pyridoxal 5-phosphate) including PMSF (100 μg/ml). After two passages through a French press cell at a pressure of 118 MPa, the suspension is centrifuged at 30,000g for 30 min at 4°. The supernatant is recovered and solid, ground ammonium sulfate is added to 25% saturation (at 0°). The precipitated proteins are collected by centrifugation at 20,000g for 20 min at 4°, resolubilized in 20 ml of buffer H, and dialyzed against 2 liters of buffer I [20 mM potassium phosphate (pH 7.5), 2 mM DTT, 0.5 mM EDTA, 10 μM pyridoxal 5-phosphate].

Column Chromatography

The protein solution is loaded on a DEAE-Sepharose Fast Flow column (2.6 × 17 cm; Amersham Pharmacia Biotech) equilibrated with buffer I lacking pyridoxal 5-phosphate. SelA is then eluted with 1350 ml of a linear KCl gradient (0–220 mM in buffer I) at a flow rate of 2.5 ml/min. The peak fractions at about 180 mM KCl are pooled and immediately applied to a BioGel HTP hydroxyapatite column (1.8 × 11 cm; Bio-Rad) equilibrated with buffer I. SelA adsorbed to the column is eluted with 200 ml of a linear gradient of 20–200 mM potassium phosphate in buffer I at a flow rate of 0.2 ml/min. The protein is eluted between about 50 and 125 mM potassium phosphate. The fractions containing SelA are pooled and adjusted to 40% ammonium sulfate saturation (at 0°). The precipitate is recovered by centrifugation (20,000g for 20 min at 4°), resuspended in 4.5 ml of buffer I, and dialyzed for 16 hr against the same buffer to completely solubilize SelA. After removal of precipitates that might have formed during dialysis by centrifugation at 20,000g for 20 min at 4°, the cleared solution is loaded on a Sephacryl S300 gel-filtration column (2.0 × 70 cm; Amersham Pharmacia Biotech) equilibrated with buffer J [50 mM potassium phosphate (pH 7.5), 2 mM DTT, 0.5 mM EDTA, 10 μM pyridoxal 5-phosphate]. The proteins are separated at a flow rate of 0.5 ml/min. SelA is eluted close to the exclusion volume. Fractions containing the pure protein are pooled, concentrated in a dialysis bag to a final protein concentration of 1 mg/ml by covering with polyethylene glycol 20,000, and dialyzed against 1 liter of buffer J containing 50% (v/v) glycerol. SelA can be stored at −20° for more than 1 year without significant loss in enzymatic activity.

Biochemical Properties

SelA forms a homodecameric complex with an apparent molecular mass of about 600 kDa consisting of two five-membered rings that are stacked on top of each other.[39] It binds one molecule of pyridoxal 5-phosphate (PLP) per monomer[17] and therefore exhibits a distinct absorption maximum at 412 nm, which is characteristic of PLP-dependent enzymes. The reaction catalyzed can be described by Eq. (2):

$$\text{Seryl-tRNA}^{\text{Sec}} + \text{Se-P} \longrightarrow \text{selenocysteyl-tRNA}^{\text{Sec}} + P_i \qquad (2)$$

where Se-P is selenophosphate. SelA specifically interacts with seryl-tRNA^{Sec}. The complex formed is stable in the absence of selenophosphate and can be isolated by gel filtration.[40] Both selenophosphate and thiophosphate are accepted as substrates, whereas selenide and sulfide are active at unphysiologically high concentrations only. The K_m and k_{cat} values determined are 0.3 μM and 1.3 min^{-1} for selenophosphate and 4.0 μM and 0.052 min^{-1} for thiophosphate, respectively.[38]

Preparation of Elongation Factor SelB

SelB is a monomeric protein with a molecular mass of 69 kDa, which tends to form inclusion bodies during overexpression. It is nearly insoluble in low-salt buffers and only poorly soluble at high-salt concentrations. Therefore, care should be taken to avoid protein concentrations higher than 1 mg/ml. The original purification procedure[41] made use of this peculiar feature by precipitating the protein from an enriched fraction by dialysis against low-salt buffer and by resolubilizing it afterward in the presence of high concentrations of salt and its ligand GDP. However, the final yield and purity varied considerably, and GDP might interfere with subsequent experiments. For these reasons, the precipitation step is replaced by ion-exchange chromatography in the following protocol, which ensures better reproducibility combined with high purity of the protein. Because of its comparatively low affinity for guanosine nucleotides, SelB is obtained in a nucleotide-free state.

Overproduction

Escherichia coli strain BL21(DE3)[36] is transformed with plasmid pWL194,[41] which contains the *selB* gene under control of the T7 *Φ10* promoter. Transformants are grown in a 10-liter fermentor (B. Braun, Melsungen, Germany) at 37° and at maximal aeration in a medium containing 3% (w/v) peptone, 1% (w/v) yeast extract, 0.4% (w/v) glucose, 0.2% (v/v) glycerol, 25 mM NaCl, 2 mM MgCl$_2$, 60 mM potassium phosphate (pH 7.1), and ampicillin (100 μg/ml). At an OD$_{600}$ of 3.0, the expression of *selB* is induced by addition of 100 μM IPTG. After lowering

[39] H. Engelhardt, K. Forchhammer, S. Müller, K. N. Goldie, and A. Böck, *Mol. Microbiol.* **6,** 3461 (1992).

[40] K. Forchhammer, K. Boesmiller, and A. Böck, *Biochimie* **73,** 1481 (1991).

[41] K. Forchhammer, K. Rücknagel, and A. Böck, *J. Biol. Chem.* **265,** 9346 (1990).

the temperature of the medium to 30°, cultivation is continued for 75–150 min. The cells are then rapidly chilled to 0°, harvested by centrifugation, and washed in buffer K [50 mM Tris-HCl (pH 7.4) at 4°, 10 mM MgCl$_2$, 50 mM KCl, 2 mM DTT, and 0.5 mM EDTA].

Crude Extract and Fractionate Ammonium Sulfate Precipitation

The cell pellet (about 110 g wet weight) is resuspended in 1 volume of buffer K containing PMSF (100 μg/ml) and DNase I (20 μg/ml). After two passages through a French press cell at a pressure of 118 MPa, the suspension is cleared of cell debris by two consecutive centrifugations at 30,000g for 30 min at 4°. The supernatant is then subjected to centrifugation at 150,000g for 1 hr at 4°. The resulting pellets, containing SelB, are washed twice with buffer K containing 100 mM NH$_4$Cl and finally resuspended in buffer K containing 1 M NH$_4$Cl to solubilize SelB. After a further centrifugation step (150,000g for 60 min at 4°), the supernatant is subjected to fractionate ammonium sulfate precipitation. To this end, the solution is first adjusted to 45% saturation (at 0°) by addition of solid, ground ammonium sulfate. The precipitate formed is removed by centrifugation (20,000g for 20 min at 4°), and the ammonium sulfate concentration of the remaining supernatant is further increased to 63% saturation. The precipitate, containing SelB, is collected by centrifugation (20,000g for 20 min at 4°) and, after thorough removal of the supernatant, dissolved in 50 ml of buffer L [100 mM potassium phosphate (pH 7.0), 2 mM magnesium acetate, 2 mM DTT, 0.5 mM EDTA]. The resulting solution is then dialyzed against two changes (2 liters each) of the same buffer.

Column Chromatography

After removal of insoluble particles and salt crystals that may have formed during dialysis by centrifugation (30,000g for 10 min at 4°), the clear supernatant is loaded on a BioGel HTP hydroxyapatite column (1.6 × 12.5 cm; Bio-Rad) equilibrated with buffer L. The column is then developed with 500 ml of a linear gradient of 100 to 200 mM potassium phosphate (in buffer L) at a flow rate of 0.5 ml/min. SelB is eluted at a potassium phosphate concentration of approximately 165 mM.

The fractions containing SelB, as monitored by SDS–PAGE, are pooled and dialyzed against two changes (2 liters each) of buffer M [100 mM potassium phosphate (pH 7.0), 5 mM MgCl$_2$, 2 mM DTT, 0.5 mM EDTA]. The solution is then applied to a Q-Sepharose anion-exchange column (1.2 × 8 cm; Amersham Pharmacia Biotech), which is developed with 90 ml of a linear gradient of 0–267 mM KCl (in buffer M) at a flow rate of 1.0 ml/min. SelB is eluted at about 125 mM KCl.

The fractions containing SelB are again pooled, dialyzed against 2 liters of buffer M, and loaded on a SP-Sepharose Fast Flow cation-exchange column (1 × 5 cm; Amersham Pharmacia Biotech). The column is developed with 80 ml of a linear gradient of 0–400 mM KCl (in buffer M) at a flow rate of 1.0 ml/min. SelB is eluted at about 230 mM KCl. The fractions containing SelB are pooled and dialyzed against 2 liters of buffer M and finally against 500 ml of buffer M containing 50% (v/v) glycerol.

SelB can be stored at $-20°$ for several months without significant loss of activity. The yield is about 8 mg of highly pure protein.

Biochemical Properties

SelB is a member of the GTPase superfamily of proteins and specifically interacts with guanosine nucleotides. The K_d values for GTP and GDP have been determined to be 0.74 and 13.4 μM, respectively.[42] The protein exhibits a low intrinsic GTPase activity, which is nearly unmeasurable in the absence of other ligands, but can be considerably stimulated by addition of ribosomes and SECIS RNA. The K_m value for GTP is 55 μM in the presence of ribosomes and 58 μM in the presence of ribosomes combined with the *fdhF* SECIS element.[27] The interaction with the SECIS elements is tight as reflected by a K_d of about 1 nM. A value of 1.26 nM has been determined for the interaction with a methylanthraniloyl-labeled *fdhF* minihelix, using the stopped-flow technique.[42] SelB specifically binds seleno-cysteyl-tRNA[Sec] as determined by protection of the tRNA against RNase A or alkaline hydrolysis.[16,22]

Serylation of tRNA[Sec]

Compared with canonical serine isoacceptors, tRNA[Sec] is aminoacylated by seryl-tRNA synthetase with considerably reduced efficiency.[16] Therefore, a huge excess of enzyme is needed to achieve quantitative serylation within a time span short enough to prevent hydrolysis of the ester bonds already formed.

The reaction is performed in a mixture containing 100 mM HEPES (pH 7.0), 10 mM KCl, 10 mM magnesium acetate, 1 mM DTT, 10 mM ATP, 200 μM L-serine, 83 μM tRNA[Sec], bovine serum albumin (0.1 mg/ml), pyrophosphatase (2.5 units/ml), and seryl-tRNA synthetase (0.84 mg/ml).[43] After incubation at $37°$ for 15 min, the solution is acidified by addition of a 1/30 volume of 3 M sodium acetate (pH 4.6) to stabilize the ester bond and extracted with phenol equilibrated with 100 mM sodium acetate (pH 4.6). Seryl-tRNA[Sec] is precipitated from the aqueous phase by addition of a 1/10 volume of 3 M sodium acetate (pH 4.6) and 1 volume of 2-propanol, and incubation at $-20°$. The precipitate formed is collected by centrifugation (16,000g for 30 min at $4°$), washed with 80% (v/v) ethanol ($-20°$), dried, and dissolved in 10 mM sodium acetate (pH 4.6).

To obtain radioactively labeled seryl-tRNA[Sec], 160 μM L-[14C]serine (171.6 mCi/mmol; NEN DuPont, Boston, MA) is included in the reaction instead

[42] M. Thanbichler, A. Böck, and R. S. Goody, *J. Biol. Chem.* **275**, 20458 (2000).
[43] M. Härtlein, D. Madern, and R. Leberman, *Nucleic Acids Res.* **15**, 1005 (1987).

of the unlabeled compound. The amount of aminoacylated tRNA^{Sec} can then be determined by precipitation in 10% (w/v) trichloroacetic acid followed by filtration through a glass fiber filter and measurement of the radioactivity retained by liquid scintillation counting.

Conversion of Seryl-tRNA^{Sec} to Selenocysteyl-tRNA^{Sec}

The synthesis of selenocysteyl-tRNA^{Sec} from seryl-tRNA^{Sec} is performed in a reaction mixture containing 100 mM piperazine-N,N'-bis(2-ethanesulfonic acid) (PIPES, pH 6.7), 10 mM KCl, 10 mM magnesium acetate, 0.5 mM DTT, 5 mM ATP, 250 μM sodium selenite, 17 μM seryl-tRNA^{Sec}, SelA (0.1 mg/ml), and SelD (5.9 mg/ml). After incubation for 35 min at 37°, a 1/20 volume of 3 M sodium acetate (pH 4.6) and a 1/20 volume of 200 mM DTT are added, and the solution is extracted with 1 volume of phenol equilibrated with 100 mM sodium acetate (pH 4.6). Selenocysteyl-tRNA^{Sec} is then precipitated from the aqueous phase by addition of a 1/10 volume of 3 M sodium acetate (pH 4.6) and 1 volume of 2-propanol and incubation on ice for 1 hr. The precipitate is sedimented by centrifugation at 16,000g for 15 min at 4°, washed with 80% (v/v) ethanol containing 5 mM DTT, dried *in vacuo,* and dissolved in 10 mM sodium acetate (pH 4.6) containing 5 mM DTT.

To specifically label selenocysteyl-tRNA^{Sec}, [^{75}Se]selenite is included in the reaction mixture instead of the nonradioactive compound. Quantification can be performed as described for [^{14}C]seryl-tRNA^{Sec}.

As selenophosphate is oxygen labile, all steps of the conversion reaction should be performed in a glove box under strictly anaerobic conditions. Further experiments involving selenocysteyl-tRNA^{Sec} can be made in the presence of oxygen given that at least 2 mM DTT is included in the reaction mixtures. Selenocysteine synthase is prone to form the by-product alanyl-tRNA^{Sec} because of unspecific reduction of the enzyme-bound aminoacrylyl-tRNA^{Sec} intermediate, if strongly reducing conditions are combined with insufficient supply of the selenium donor selenophosphate. On the other hand, enough DTT must be included in the reaction to ensure that sufficient selenite is reduced to selenide to promote the overall reaction. The best results have been obtained with a DTT concentration of 0.5 mM combined with a huge surplus of SelD. When changing the balance of seryl-tRNA^{Sec}, SelD, DTT, and sodium selenite, the concentrations of the individual components should be optimized again to achieve optimal results.

Preparation of Selenocysteine Insertion Sequence Elements

It has been shown that a minimal SECIS element consisting of 17 bases is sufficient to promote binding of SelB *in vitro* and insertion of selenocysteine *in vivo*. Stem–loop RNA serving as an mRNA ligand for SelB can therefore be easily prepared either by chemical synthesis or by *in vitro* transcription using oligonucleotides as templates.[44,45] However, short RNA molecules able to fold into stem–loop structures have the indigenous tendency to form dimers, which often are thermodynamically more stable but biologically inactive. This behavior is strongly favored by the presence of cations during the folding process. Even trace amounts are sufficient to promote at least partial misfolding. As most commercial preparations and RNA transcribed *in vitro* and eluted from urea–polyacrylamide gels contain considerable amounts of residual salt, desalting of the RNA solutions is strongly recommended before further use. To this end, the oligonucleotides are precipitated by addition of a 1/10 volume of 3 M sodium acetate (pH 4.6) and 1 volume of 2-propanol and incubation at $-20°$. The precipitate is collected by centrifugation at $4°$, washed with 80% (v/v) ethanol, dried, and dissolved in 0.1 mM EDTA. To achieve correct folding, 100 μl of solution containing oligonucleotide at a maximal concentration of 200 μM is heated to $80°$ for 90 sec and then rapidly cooled on ice. Under these conditions, all RNA molecules adopt the stem–loop conformation.

Analysis of Interaction of SelB with Its RNA Ligands

The affinity of SelB for guanosine nucleotides is comparably low.[22,42] Therefore, in contrast to EF-Tu, the protein can easily be obtained in a nucleotide-free state and be loaded with either GDP or GTP without taking special measures. To achieve essentially complete saturation of the nucleotide-binding site, SelB is simply mixed with 100 μM GTP or 1 mM GDP. However, the binary complex formed under these conditions is not stable and cannot be purified by gel filtration. GTPase activity is negligible at low temperatures and incubation times up to several hours.[27]

Binding of the SECIS element RNA takes place with high affinity (K_d of $\sim$1 nM) independent of the nucleotide state of SelB.[42] SelB specifically binds selenocysteyl-tRNA^{Sec}, whereas essentially no interaction can be observed with the serylated tRNA.[22] Binding occurs both in the absence and in the presence of GTP. When loaded with selenocysteyl-tRNA^{Sec}, SelB presumably adopts an alternative conformation leading to an increased affinity for the SECIS element.[42] For thermodynamic reasons, this should concomitantly result in an increase of the affinity of SelB for selenocysteyl-tRNA^{Sec}, which renders the quaternary complex the most stable of all SelB–ligand complexes.

Interaction of SelB with its RNA ligands can be visualized by the mobility shift assay. To detect binding of SelB to SECIS elements, increasing amounts of SELB are incubated in the following reaction mixture.

[44] J. R. Wyatt, M. Chastain, and J. D. Puglisi, *BioTechniques* **11,** 764 (1991).
[45] J. F. Milligan, D. R. Groebe, G. W. Witherell, and O. C. Uhlenbeck, *Nucleic Acids Res.* **15,** 8783 (1987).

SECIS element (mixed with traces of ^{32}P-labeled	
stem–loop RNA)	80 nM
Yeast tRNA	3.6 μM
Potassium phosphate (pH 7.0)	100 mM
MgCl$_2$	5 mM
DTT	2 mM
EDTA	0.5 mM
Glycerol	10% (v/v)
SelB	0–1600 nM

The reaction mixtures are incubated for 5 min at 25°, mixed with a 1/10 volume of loading dye [50% (v/v) glycerol, 0.03% (w/v) bromphenol blue], and applied to a 6% (w/v) polyacrylamide gel (cross-link 30 : 0.8) prepared in a buffer containing 100 mM potassium phosphate (pH 7.0), 5 mM MgCl$_2$, 2 mM DTT, and 0.5 mM EDTA. After separation of the unbound SECIS element from the binary complex by electrophoresis in the same buffer for 3 hr at 4° and a constant voltage of 50 V, the gel is dried and exposed on a film or phosphor screen. A 20-fold molar excess of SelB over the SECIS element usually results in quantitative inclusion of the RNA in the slow-migrating binary complex.

To detect the formation of a quaternary complex, the following components are additionally included in the reaction mixture.

| GTP | 100 μM |
| Selenocysteyl-tRNA[Sec] | 1.6 μM |

All steps are performed as described above except that 100 μM GTP is added to the gel and running buffer. The quaternary complex formed can be clearly separated from the binary SelB · SECIS or the ternary SelB · GTP · SECIS complex, because of its faster migration due to its more negative charge. Moreover, binding of the tRNA leads to an increased affinity of SelB for the SECIS element, which results in a quantitative inclusion of the SECIS element in the complex even at low SelB : SECIS ratios.

[2] Selenocysteine Insertion Sequence Element Characterization and Selenoprotein Expression

By Marla J. Berry, Glover W. Martin III, Roza Tujebajeva,
Elisabeth Grundner-Culemann, John B. Mansell,
Nadya Morozova, and John W. Harney

Early studies in the field of eukaryotic selenoprotein synthesis established that a functional selenocysteine insertion sequence (SECIS) element linked downstream of an open reading frame would direct selenocysteine incorporation at any upstream in-frame UGA codon. Characterization of the activities of multiple wild-type SECIS elements and of site-directed mutants resulted in determination of the critical features for optimal SECIS function. This information provides a means of introducing selenocysteine at any desired position in any protein. Subsequent studies have revealed additional conditions and factors that allow for increased efficiency of selenoprotein expression. The goals of this article are to describe the criteria and assays used for identifying SECIS elements and for evaluating relative SECIS function, and the conditions established for optimal expression of selenoproteins in transiently transfected cells.

Criteria for Identifying Functional Selenocysteine Insertion Sequence Elements in Databases

The sequence and structural features of SECIS elements shown to be required for directing selenocysteine incorporation include, primarily, three short stretches of conserved nucleotides located at specific positions in the stem–loop, the length of the stem, and its propensity for base pairing. The conserved nucleotides A/GUGA at the 5′ base of the stem, AA in the hairpin loop, and GA at the 3′ base of the stem are critical for function (Fig. 1).[1–3] In contrast, the sequences of the stems are not constrained, provided the stem length falls in the range of 9–11 base pairs.[1,3] The boundaries of the minimal sequence required for SECIS function correspond precisely with the conserved 5′ A/GUGA and 3′ GA sequences.[3] The loop can assume either an open (form 1) or base-paired (form 2) configuration.[4–6] Finally, a minimal spacing of ~60 nucleotides between UGA codon and SECIS

[1] M. J. Berry, L. Banu, J. W. Harney, and P. R. Larsen, *EMBO J.* **12,** 3315 (1993).
[2] Q. Shen, J. L. Leonard, and P. E. Newburger, *RNA* **1,** 519 (1995).
[3] G. W. Martin III, J. W. Harney, and M. J. Berry, *RNA* **2,** 171 (1996).
[4] S. C. Low and M. J. Berry, *Trends Biochem. Sci.* **21,** 203 (1996).
[5] E. Grundner-Culemann, G. W. Martin III, J. W. Harney, and M. J. Berry, *RNA* **5,** 625 (1999).
[6] D. Fagegaltier, A. Lescure, R. Walczak, P. Carbon, and A. Krol, *Nucleic Acids Res.* **28,** 2679 (2000).

METHODS IN ENZYMOLOGY, VOL. 347   0076-6879/02 $35.00

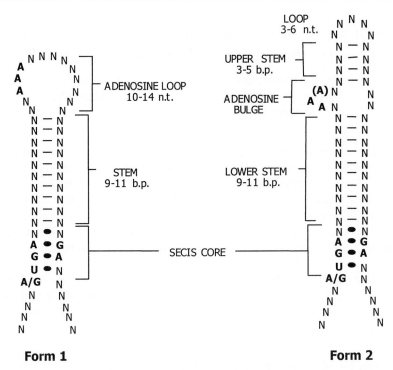

Form 1   **Form 2**

Fig. 1. Consensus SECIS element structures. Conserved sequence and structural features include the SECIS core nucleotides A/GUGA and GA, the stem length, and conserved adenosines in a terminal loop (form 1) or bulge (form 2). Lines indicate Watson–Crick base pairs, dark ovals designate non-Watson–Crick pairing.

element must be present.[1,3] Increasing spacing by insertion of 1.5 kb between the type 1 deiodinase (D1) UGA codon and SECIS element had no effect on SECIS function. A naturally occurring UGA–SECIS spacing of ∼4.5 kb was subsequently identified in the type 2 deiodinase (D2),[7] and as yet, no upper limit has been established. Using the criteria for SECIS function, algorithms have been designed to search sequence databases for these features (see [6] and [8] in this volume[7a,7b] and Refs. 8 and 9). The first step employed in these algorithms involves a SECIS search, using the criteria of the three stretches of conserved nucleotides, the conserved spacing between these regions, and constraints on base pairing.

[7] C. Buettner, J. W. Harney, and P. R. Larsen, *J. Biol. Chem.* **273,** 33374 (1998).

[7a] A. Lescure, D. Gautheret, and A. Krol, *Methods Enzymol.* **347,** [6], 2002 (this volume).

[7b] G. V. Kryukov and V. N. Gladyshev, *Methods Enzymol.* **347,** [8], 2002 (this volume).

[8] G. V. Kryukov, V. M. Kryukov, and V. N. Gladyshev, *J. Biol. Chem.* **274,** 33888 (1999).

[9] A. Lescure, D. Gautheret, P. Carbon, and A. Krol, *J. Biol. Chem.* **274,** 38147 (1999).

Identification of SECIS candidates is followed by a search of the upstream region for an open reading frame containing an in-frame UGA. Next, open reading frames satisfying this requirement are used to search for homologs in other species. Finally, the degree of conservation between species in the region downstream of the UGA codon is examined. If this region is "coding," by virtue of the UGA codon being translated, a significantly higher level of conservation is predicted than if the UGA is decoded as "stop," relegating the downstream region to a 3′ untranslated region (UTR).

Methods for Studying Selenocysteine Insertion Sequence Function and Selenoprotein Synthesis

Once candidate selenoprotein sequences and SECIS elements are identified, how is it verified that a selenoprotein is, in fact, encoded? Several methods have been employed. In both prokaryotes and eukaryotes, constructs have been used that rely on demonstrating incorporation at a UGA codon via expression of a downstream sequence whose expression or bioactivity can be monitored. These include open reading frame–β-galactosidase fusion constructs,[10] and β-galactosidase–UGA–luciferase constructs.[11] A second avenue is the use of selenoenzyme activity assays, where activity is dependent on incorporation of the active site selenocysteine. The D1 assay has been used extensively in this fashion.[3,5,12,13] A complementary approach to that described above is the use of Western blotting to assess relative ratios of selenocysteine incorporation and UGA-termination products. Finally, demonstration of radiolabeled selenium incorporation, dependent on the presence of an SECIS element, is the most convincing proof of a bona fide selenoprotein-encoding gene.

Selenocysteine Insertion Sequence Activity Assays

The D1 activity assay provides a highly quantitative method for comparison of the relative activities of SECIS elements from different selenoproteins, and of the effects of mutations introduced into these elements. This assay has been used extensively for precise definition of the required nucleotides and secondary structural characteristics.[1,3,5,12–14] Constructs are generated containing the D1-coding region linked to heterologous SECIS elements, either subcloned from

[10] F. Zinoni, J. Heider, and A. Bock, *Proc. Natl. Acad. Sci. U.S.A.* **87,** 4660 (1990).

[11] H. Kollmus, L. Flohe, and J. E. McCarthy, *Nucleic Acids Res.* **24,** 1195 (1996).

[12] G. W. Martin III, J. W. Harney, and M. J. Berry, *RNA* **4,** 65 (1998).

[13] G. Bermano, J. R. Arthur, and J. E. Hesketh, *Biochem. J.* **320,** 891 (1996).

[14] M. J. Berry, L. Banu, Y. Y. Chen, S. J. Mandel, J. D. Kieffer, J. W. Harney, and P. R. Larsen, *Nature (London)* **353,** 273 (1991).

other genes, amplified from templates by polymerase chain reaction (PCR), or generated synthetically by amplifying overlapping oligonucleotides. This is followed by transient transfection of the constructs and quantitation of the resulting D1 enzyme activity.

Type 1 Deiodinase Expression Constructs

The basic D1 expression construct consists of the D1-coding region in the mammalian expression vector pUHD10-3,[15] followed by unique restriction sites. This allows subcloning of potential SECIS elements or mutants after amplification by PCR, with the introduction of the unique restriction sites into PCR oligonucleotides. The vector contains a cytomegalovirus (CMV) promoter, which is highly efficient in transfected mammalian cells. In addition, it contains a target site for a tetracycline-regulated *trans*-activator, allowing gene expression to be turned off by the inclusion of tetracycline in media, or induced by simply changing media to remove tetracycline. The tetracycline-regulated *trans*-activator is encoded on a separate plasmid, pUHD15, which must be cotransfected with the pUHD10-based plasmid.

Transient Transfections and Deiodinase Assays

Transient transfections are carried out in human embryonic kidney (HEK-293) cells, using the calcium phosphate method of transfection as described previously.[16] Three days before transfection, HEK-293 cells are plated onto 60-mm culture dishes in Dulbecco's modified Eagle's medium (DMEM) supplemented with 10% (v/v) fetal calf serum. Cells are transfected with 0.5–10 μg of the pUHD10-based expression plasmids and 4 μg of the pUHD15 plasmid. To monitor transfection efficiencies, cells are cotransfected with 3 μg of an expression vector containing the human growth hormone cDNA under control of the herpes simplex virus (HSV) thymidine kinase promoter. Alternatively, a β-galactosidase or other expression plasmid may be used to normalize for transfection efficiency. Medium is changed 1 day after transfection. Two days after transfection, cells are harvested, washed, and resuspended in 0.1 M potassium phosphate (pH 6.9)–1 mM EDTA containing 0.25 M sucrose and 10 mM dithiothreitol (DTT). Cells are sonicated for 5 sec on ice, using a Fisher (Pittsburgh, PA) sonic dismembrator 60. Cell sonicates prepared in this way can be assayed directly, or stored frozen at $-70°$ with no detectable loss in activity following at least one thawing. Cell sonicates are assayed for the ability to 5'-deiodinate ^{125}I-labeled reverse T_3 (3,3',5'-triiodothyronine) as previously described.[17] Reactions typically contain 10–250 μg of protein, 1 μM ^{125}I-labeled

[15] M. Gossen and H. Bujard, *Proc. Natl. Acad. Sci. U.S.A.* **89,** 5547 (1992).

[16] G. A. Brent, J. W. Harney, Y. Chen, R. L. Warne, D. D. Moore, and P. R. Larsen, *Mol. Endocrinol.* **3,** 1996 (1989).

[17] M. J. Berry, J. D. Kieffer, J. W. Harney, and P. R. Larsen, *J. Biol. Chem.* **266,** 14155 (1991).

reverse T$_3$, and 10 mM dithiothreitol in a reaction volume of 300 μl. Reactions are incubated at 37° for 30 min. [125]I release is quantitated as described previously.[17] Deiodinase activities are calculated per microliter of cell sonicate and normalized to the amount of growth hormone secreted into the medium. Constructs are tested in at least three separate transfections and deiodinase assays are performed in duplicate from each transfection. This assay provides highly reproducible data, with standard errors typically in the range of 10% or less. Signal-to-background ratios for D1 assays with constructs containing wild-type SECIS elements versus the absence of a SECIS element typically exceed 50 : 1.

Western Blotting to Assess Selenocysteine Incorporation and UGA-Termination Products

Although enzyme activity assays provide quantitative information about SECIS function, they provide little insight into the amount of UGA termination relative to incorporation. Western blotting of cell sonicates prepared as described above, using antiserum against a peptide in the amino-terminal half of D1, allows determination of these relative ratios and the effects of cotransfected factors. D1 activity and Western assays performed with the same samples show close correlations between the two, a further indication of the reliability of these two methods. For example, cotransfection of a plasmid encoding tRNA[Ser]Sec resulted in dramatic dose-dependent increases in the amount of full-length D1 protein and enzyme activity, and corresponding decreases in the UGA-termination product.[18] Although this approach is useful for D1, it has proved less so for glutathione peroxidase (GPX). In this case, although full-length selenoprotein can be detected, blotting has failed to detect the small ($\sim$5-kDa) UGA-termination product, possibly because of rapid turnover. An alternative scheme employed linking the green fluorescent protein (GFP) sequence to GPX as an amino-terminal fusion, resulting in a large, more stable and easily detectable protein consisting of GFP linked to the glutathione peroxidase UGA-termination or incorporation products.[19]

Another approach to obtaining selenocysteine incorporation and UGA-termination product ratios is the use of a dual-reporter construct in which the two reporters are separated by a UGA codon, with SECIS elements cloned downstream of the second reporter. A construct consisting of the β-galactosidase-coding region followed by an in-frame UGA codon and luciferase-coding region, with downstream restriction sites for introducing candidate SECIS elements, has proved useful in this respect.[11] Luciferase activity is normalized to β-galactosidase activity, and reported relative to activity in the absence of an SECIS element (background). Demonstration of SECIS dependence in this type of assay is crucial, as other

[18] M. J. Berry, J. W. Harney, T. Ohama, and D. L. Hatfield, *Nucleic Acids Res.* **22**, 3753 (1994).
[19] W. Wen, S. L. Weiss, and R. A. Sunde, *J. Biol. Chem.* **273**, 28533 (1998).

rat selenoprotein P

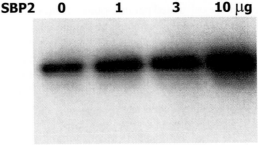

SBP2   0   1   3   10 μg

FIG. 2. Effects of SBP2 on selenoprotein P expression. Cells were transfected with a rat selenoprotein P expression plasmid and the indicated concentrations of an SBP2 expression plasmid (generous gift of P. Copeland and D. Driscoll, Department of Cell Biology, Lerner Research Institute, Cleveland Clinic Foundation, Cleveland, OH). Sodium [75Se]selenite was added the day after transfection and labeling was allowed to proceed for an additional 1 day. Selenoprotein P in the medium was analyzed by SDS–polyacrylamide gel electrophoresis and autoradiography.

mechanisms of UGA suppression can contribute to the level of incorporation product.[20] Activity can also be calculated relative to that obtained with a cysteine codon in place of UGA, to obtain an indication of relative efficiency of translation at this position.

Optimizing Expression in Transiently Transfected Cells

Selenium levels in commercially available serum range from adequate to deficient. As selenium may be particularly limiting during the burst of protein synthesis after transient transfection, supplementation with sodium selenite typically results in significant increases in selenoprotein expression. Increases of up to 7-fold can be obtained with 100 nM sodium selenite added on the day of transfection (see below). In addition to adding selenium to the medium, expression can be increased by cotransfection of plasmids encoding factors involved in selenoprotein synthesis, several of which appear to be limiting in at least some cell lines. Factors shown to increase incorporation include tRNA[Ser]Sec (discussed above), selenophosphate synthetase,[21] and SECIS-binding protein 2 (SBP2).[22] The latter appears to produce the most dramatic effects, consistent with the finding that this factor is of low abundance in most cell lines and tissues (Fig. 2).

[20] J. E. Jung, V. Karoor, M. G. Sandbaken, B. J. Lee, T. Ohama, R. F. Gesteland, J. F. Atkins, G. T. Mullenbach, K. E. Hill, A. J. Wahba, and D. L. Hatfield, *J. Biol. Chem.* **269,** 29739 (1994).

[21] S. C. Low, J. W. Harney, and M. J. Berry, *J. Biol. Chem.* **270,** 21659 (1995).

[22] P. R. Copeland, J. E. Fletcher, B. A. Carlson, D. L. Hatfield, and D. M. Driscoll, *EMBO J.* **19,** 306 (2000).

Incorporation of ^{75}Se into Selenoproteins in Transiently Transfected Cells

[75]Se labeling has the advantage that incorporation of the isotope into the small number of true selenoproteins is highly specific relative to background incorporation via the sulfur metabolic pathways. In addition, the isotope is of high energy. Thus, most of the radiolabel is incorporated into a small number of bands, detectable by autoradiography after relatively short exposure times. The disadvantages are that the isotope is expensive, may not be readily available because of limited production, and requires special handling, shielding, and permits. As ^{75}Se emits strong β and γ radiation, heavy lead shielding or lead bricks should be employed on all sides, in all phases of work—the energy is sufficient to penetrate standard laboratory walls, and Lucite provides an inadequate barrier.

Labeling in Transfected Cells

Incorporation of label can be conducted for short periods of time, with labeling of some proteins detected after 3–6 hr. However, for convenience and timing with the maximum burst of protein synthesis, labeling is typically carried out for 12–18 hr, beginning the day after transfection and continuing to the following day, the day of harvest. Label is added directly to media, the volume of which may be reduced by 50% to economize on label and increase specific activity. Typically, ~6 μCi is added per 60-mm dish. Reducing the amount of unlabeled selenium to increase the specific activity of the label does not produce an increase in labeled

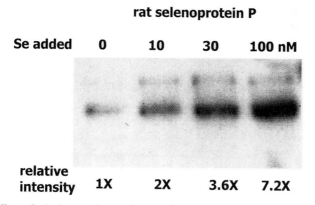

FIG. 3. Effects of selenium supplementation on selenoprotein P expression. Culture medium was supplemented with the indicated concentrations of sodium selenite the day before transfection. Labeling and electrophoresis were carried out as for Fig. 2, followed by densitometric quantitation of the autoradiograph. Relative intensities are indicated below the lanes.

protein, and in fact the opposite has been observed.[23] In the absence of selenium supplementation, low levels of radiolabeled selenoprotein P were produced. With supplementation of 10, 30, or 100 nM selenium, selenoprotein P labeling intensity increased by 2-, 3.6-, and 7.2-fold, despite a decrease in specific activity by a factor of $\sim$100 (Fig. 3).

Endogenous selenoprotein profiles differ significantly in different cell lines, and thus the choice of cell line should be given some consideration. For example, HepG2 and LLCPK1 cells produce readily detectable levels of D1 and seleno-protein P, and Chinese hamster ovary (CHO) cells produce significant levels of selenoprotein P. HEK-293 and COS-7 cells do not produce measurable endoge-nous D1 or selenoprotein P, but readily express these and other selenoproteins after transfection.[23] Most cell lines produce significant levels of endogenous GPX and thioredoxin reductases. In addition, some cell lines such as JEG-3 exhibit a reduced ability to catalyze selenocysteine incorporation, compared with, for ex-ample, COS-7.[24] For secreted selenoproteins, the high level of albumin in media supplemented with serum may present problems in subsequent analysis by gel electrophoresis. This can be avoided by changing to serum-free media during the 18- to 24-hr labeling period, followed by harvesting and direct analysis of seleno-proteins in the medium.

Acknowledgment

Work in the author's laboratory is supported by the National Institutes of Health.

[23] R. M. Tujebajeva, J. W. Harney, and M. J. Berry, *J. Biol. Chem.* **275,** 6288 (2000).
[24] M. J. Berry, A. L. Maia, J. D. Kieffer, J. W. Harney, and P. R. Larsen, *Endocrinology* **131,** 1848 (1992).

[3] Transfer RNAs That Insert Selenocysteine

By BRADLEY A. CARLSON and DOLPH L. HATFIELD

Introduction

Selenium is incorporated into protein as the amino acid selenocysteine (Sec). Sec is indeed the 21st naturally occurring amino acid within the genetic code, as it has its own tRNA and its own code word as well as other specialized transla-tional components for its insertion into protein.[1,2] Its tRNA has the dual function

[1] S. Commans and A. Bock, *FEMS Microbiol. Rev.* **23,** 335 (1999).

of (1) serving as the carrier molecule for the biosynthesis of Sec and (2) donating Sec to protein in response to its codon, UGA. Selenocysteine-inserting tRNA (Sec tRNA) is initially aminoacylated with serine, which serves as the backbone for Sec biosynthesis. This tRNA has been designated tRNA^{Sec} (Commans and Bock[1]) and tRNA$^{[Ser]Sec}$ (Hatfield *et al.*[2]). It has also been called the key molecule[3] and the central component[4] in selenoprotein synthesis as this tRNA has the unique distinction of being responsible for the biosynthesis of an entire class of proteins. Sec tRNA has been characterized in many organisms, most extensively in *Escherichia coli*[1,3] and mammals.[2,4] The focus of this article is on isolation and characterization of these Sec tRNAs.

Primary Structures

The primary structures of *E. coli*[1,3,5] and mammalian Sec tRNAs[2,4] are shown in their cloverleaf models in Fig. 1. The mammalian Sec tRNAs are shown in two cloverleaf forms, a 7/5 (Fig. 1B) and a 9/4 (Fig. 1C) structure (i.e., seven or nine base-paired members in the acceptor stem vs. five or four paired members in the T stem). Evidence of both secondary structures has been presented.[6,7]

Sec tRNAs have several unique characteristics that separate them from all other tRNAs. The most obvious feature is that they are longer than other tRNAs. *Escherichia coli* Sec tRNA contains 95 nucleotides[1,3,5] and vertebrate Sec tRNAs contain 90 nucleotides.[2,4] The longest known Sec tRNA (and thus the longest known tRNA) is that from *Moorella thermoacetica,* which is 100 nucleotides in length.[8] At 90 nucleotides, Sec tRNA from animals is the longest known eukaryotic tRNA. The characteristic length of Sec tRNA is due to the extra nucleotides in the variable arm and to the occurrence of 13 nucleotides in the acceptor-TψC stem helices instead of the 12 normally found in all other tRNAs. In addition, prokaryotic and eukaryotic Sec tRNAs have 6 possible base pairs in the dihydrouracil stem instead of the normal 3 or 4 pairs found in other tRNAs, and they are characterized by relatively few modified nucleotides (3 or 4) compared with as many as 15–17 in other tRNAs. Sec tRNAs from prokaryotes and eukaryotes also have the same anticodon, UCA, that decodes UGA. In higher vertebrates,

[2] D. L. Hatfield, V. N. Gladyshev, S. I. Park, H. S. Chittum, B. A. Carlson, M. E. Moustafa, J. M. Park, J. R. Huh, M. Kim, and B. J. Lee, *in* "Comprehensive Natural Products Chemistry" (J. F. Kelly, ed.), p. 353. Elsevier, New York, 1999.

[3] A. Bock, K. Forchhammer, J. Heider, and C. Baron, *Trends Biochem. Sci.* **16,** 463 (1991).

[4] D. L. Hatfield, I. S. Choi, T. Ohama, J.-E. Jung, and A. M. Diamond, *in* "Selenium in Biology and Human Health" (R. F. Burk, ed.), p. 25. Springer-Verlag, Berlin, 1994.

[5] A. Schon, A. Bock, G. Ott, M. Sprinzl, and D. Soll, *Nucleic Acids Res.* **17,** 7159 (1989).

[6] S. V. Steinberg, A. Ioudovitch, and R. Cedergren, *RNA* **4,** 241 (1998).

[7] N. Hubert, C. Sturchler, E. Westhof, P. Carbon, and A. Krol, *RNA* **4,** 1029 (1998).

[8] P. Tormay, R. Wilting, J. Heider, and A. Bock, *J. Bacteriol.* **176,** 1269 (1994).

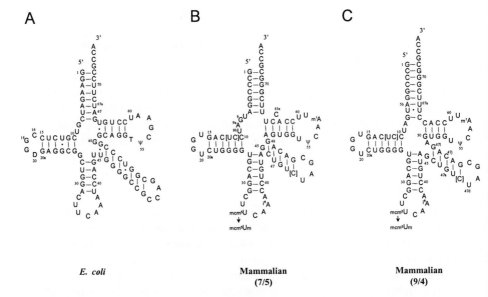

FIG. 1. Cloverleaf models of *E. coli* and mammalian Sec tRNAs. *Escherichia coli* Sec tRNA is shown in (A) and mammalian Sec tRNA is shown in the 7/5 model (B) and the 9/4 model (C). Brackets around bases at positions 11 and 12 and 47c denote variations observed at these positions (all are pyrimidine transitions) in primary structures of Sec tRNA from different mammals. Pseudouridine is designated as ψ at position 55 in *E. coli* and mammalian Sec tRNAs, dihydrouridine as D at position 20 in *E. coli* Sec tRNA, and isopentenyladenosine as i^6A at position 37 in mammalian Sec tRNA. mcm^{5}U and mcm^{5}Um at position 34 and m^{1}A at position 58 in mammalian Sec tRNAs are defined in the text. See text for other details.

the wobble base in the anticodon contains either methylcarboxymethyl-5'-uridine (mcm^{5}U) or methylcarboxymethyl-5'-uridine 2'-*O*-methylribose (mcm^{5}Um) and thus these organisms contain two species of Sec tRNA that differ by a single methyl group on the 2'-*O*-ribosyl moiety at position 34 (see Fig. 1B and C). Zebrafish also contain two species of Sec tRNA,[9] but their primary structures have not been determined. Interestingly, *Drosophila* has two electrophoretically distinguishable Sec tRNA forms, but their primary structures are identical.[10] Thus, the *Drosophila* Sec tRNA population apparently consists of a single species, but in two stable conformations. It lacks the 1-methyladenosine m^{1}A and mcmUm that are found in higher vertebrates.

[9] X. M. Xu, X. Zhou, B. A. Carlson, L. K. Kim, T. L. Huh, B. J. Lee, and D. L. Hatfield, *FEBS Lett.* **454,** 16 (1999).

[10] X. Zhou, S. Park, M. E. Moustafa, B. A. Carlson, P. F. Crain, A. Diamond, D. L. Hatfield, and B. J. Lee, *J. Biol. Chem.* **274,** 18729 (1999).

Selenocysteine-Inserting tRNA Gene, Gene Expression, and Gene Product

Bacteria,[1,3,5] mammals, chickens, frogs, fruit flies, and worms all have one Sec tRNA gene copy,[2,4] whereas zebrafish contains two such copies.[9] Thus, the gene for Sec tRNA occurs in single copy in the genomes of all organisms studied thus far with one exception. The two genes in zebrafish likely arose from gene duplication, a common occurrence with many genes of zebrafish lineage. Pseudogenes of Sec tRNA have been found in a number of mammals[2,4] and may have arisen by reverse transcription of a functional Sec tRNA and its reinsertion into the genome.

The Chinese hamster ovary (CHO) genome has three Sec tRNA pseudogenes and a single functional gene.[11] The sequences of the Sec tRNA gene and gene product in Chinese hamsters differ from those of all other mammalian Sec tRNAs and tRNA genes sequenced to date in that they have two pyrimidine transitions at positions 11 and 12. Because one of the pseudogenes has the CT sequence present in other mammalian Sec tRNA genes, the possibility exists that this pseudogene may have been the primordial gene. A model was proposed whereby the transcript of the primordial gene was edited at positions 11 and 12 before processing, and the altered transcript then reintroduced into the genome.[11] Editing has also been proposed to account for the presence of a minor species of bovine liver Sec tRNA that differed from its gene sequence by pyrimidine transitions at positions 12 and 13.[11,12]

Sec tRNA is transcribed unlike any known tRNA in that transcription begins at the first nucleotide within the coding sequence. All other tRNAs have a leader sequence that must be processed during tRNA maturation. Sec tRNA, therefore, has a 5'-triphosphate guanosine, and the 5'-triphosphate is transferred from the nucleus to the cytoplasm in *Xenopus* oocytes.[13] The possible role of this triphosphate, or even whether it is retained during protein synthesis, is not known. The regulatory elements encoded within the Sec tRNA gene, and the regulatory mechanisms and the transcription factors involved in its expression in higher vertebrates have been examined and reviewed in detail elsewhere.[2] After transcription of mammalian Sec tRNA, its trailer sequence is processed[13] and the modified nucleosides at four positions are synthesized.[14,15] ψ and m^1A at positions 55 and 58, respectively, are synthesized in the nucleus and the other modifications, i^6A at position 37, and

[11] X. M. Xu, B. A. Carlson, L. K. Kim, B. J. Lee, D. L. Hatfield, and A. M. Diamond, *Gene* **239**, 49 (1999).

[12] A. M. Diamond, Y. Montero-Puerner, B. J. Lee, and D. Hatfield, *Nucleic Acids Res.* **18**, 6727 (1990).

[13] B. J. Lee, P. de la Pena, J. A. Tobian, M. Zasloff, and D. L. Hatfield, *Proc. Natl. Acad. Sci. U.S.A.* **84**, 6384 (1987).

[14] I. S. Choi, A. M. Diamond, P. F. Crain, J. Kolker, J. A. McCloskey, and D. Hatfield, *Biochemistry* **33**, 601 (1994).

[15] C. Sturchler, A. Lescure, G. Keith, P. Carbon, and A. Krol, *Nucleic Acids Res.* **2**, 1354 (1994).

mcm^5U and mcm^5Um at position 34, are synthesized in the cytoplasm. Synthesis of mcm^5Um from mcm^5U is the final maturation step.[14] Because the formation of the methyl group at position 34 is responsive to selenium status[16,17] and its presence affects tertiary structure,[16] the effect of the modifications at other positions and tRNA tertiary structure on its synthesis were examined.[18] Efficient synthesis of mcm^5Um from mcm^5U required the prior formation of each modified base and an intact tertiary structure. Synthesis of the bases at other positions, including mcm^5U, however, was not as stringently connected to primary and tertiary structure. These studies, along with those showing enhanced formation of the methyl group with increasing selenium levels and the effect of this group on tertiary structure, suggest that the mcm^5Um and mcm^5U isoforms have different roles in protein synthesis.

The distributions and relative amounts of mcm^5U and mcm^5Um vary in mammalian cells and tissues that, as noted above, are influenced by selenium status.[16,17] Increases as high as 2.5-fold in the total Sec tRNA population were observed in response to selenium and the distribution shifted dramatically toward the mcm^5Um isoform.[16] Selenium-deficient rats that are replenished with selenium manifest restoration of the Sec tRNA population to normal levels and distributions in about 72 hr in several tissues (liver, heart, kidney, and muscle), but the rate of change in distribution varies in the different tissues.[16] The enrichment of the Sec tRNA population in the presence of selenium appears to be due to a reduced turnover rate rather than to enhanced transcription as observed in *Xenopus* oocytes.[14]

Selenocysteine-Inserting tRNA: Carrier Molecule for Biosynthesis of Its Amino Acid

As noted above, Sec tRNA serves as the site of Sec biosynthesis. Prokaryotic and eukaryotic Sec tRNAs are initially aminoacylated with serine, which serves as the backbone for Sec synthesis. Thus, the identity elements in Sec tRNA for its aminoacylation correspond to those in seryl-tRNA synthetase and the identity elements in both *E. coli*[1] and mammalian Sec tRNAs have been examined.[2] Both the long variable arm and the discriminator base are important for the aminoacylation of *E. coli*[1] and mammalian Sec tRNAs[19,20] by seryl-tRNA synthetase.

---

[16] A. M. Diamond, I. S. Choi, P. F. Crain, T. Hashizume, S. C. Pomerantz, R. Cruz, C. J. Steer, K. E. Hill, R. F. Burk, J. A. McCloskey, and D. L. Hatfield, *J. Biol. Chem.* **268,** 14215 (1993).

[17] H. S. Chittum, K. E. Hill, B. A. Carlson, B. J. Lee, R. F. Burk, and D. L. Hatfield, *Biochim. Biophys. Acta* **1359,** 25 (1997).

[18] L. K. Kim, T. Matsufuji, S. Matsufuji, B. A. Carlson, S. S. Kim, D. L. Hatfield, and B. J. Lee, *RNA* **6,** 1306 (2000).

[19] X. G. Wu and H. J. Gross, *Nucleic Acids Res.* **21,** 5589 (1993).

[20] T. Ohama, D. Yang, and D. L. Hatfield, *Arch. Biochem. Biophys.* **315,** 293 (1994).

In addition, the first three 5'-bases that are paired with the last three 3'-bases before the discriminator base in the acceptor stem of *E. coli* Sec tRNA have a role in this process. In mammals, the acceptor stem is also important for aminoacylation, as are the TψC and D stems.[21]

Once serine is attached to Sec tRNA, the serine moiety is then converted to Sec. This biosynthetic pathway has been completely established in *E. coli*[1,3] as described in [1] of this volume.[21a] In mammals, however, the details of this process are lacking. It has been known for sometime that a minor seryl-tRNA, capable of decoding UGA,[22] and was subsequently identified as Sec tRNA,[23] formed phosphoseryl-tRNA.[24] The role of phosphoseryl-tRNA and whether it is an intermediate in the biosynthesis of Sec are not known.

The active form of selenium that is donated to the intermediate in Sec biosynthesis is monoselenophosphate that is synthesized by selenophosphate synthetase from selenide and ATP in bacteria.[25] Mammals likely utilize the same selenium donor, although it has not been characterized.[26-28] Two selenophosphate synthetase genes, Sps1 and Sps2, have been identified in mammals[26-28] and Sps2 is a selenoprotein.[28] Donation of activated selenium to the intermediate completes the biosynthesis of Sec on its tRNA.

Selenocysteine-Inserting tRNA: Donor of Selenocysteine to Protein

The role of Sec tRNA, like the other 20 aminoacyl-tRNAs, is to donate its amino acid to the growing peptide in protein synthesis. The details of this process have been established in *E. coli*[1,3] as described in [1] of this volume.[21a] They have been elucidated more recently in mammals[29-31] and this work is also described

[21] R. Amberg, T. Mizutani, X.-Q. Wu, and H. J. Gross, *J. Mol. Biol.* **263**, 8 (1996).

[21a] M. Thanbichler and A. Böck, *Methods Enzymol.* **347**, [1], 2002 (this volume).

[22] D. L. Hatfield and F. H. Portugal, *Proc. Natl Acad. Sci. U.S.A.* **67**, 1200 (1970).

[23] B. J. Lee, P. J. Worland, J. N. Davis, T. C. Stadtman, and D. L. Hatfield, *J. Biol. Chem.* **264**, 9724 (1989).

[24] D. Hatfield, A. M. Diamond, and B. Dudock, *Proc. Natl. Acad. Sci. U.S.A.* **79**, 6215 (1982).

[25] R. S. Glass, W. P. Singh, W. Jung, Z. Veres, T. D. Scholz, and T. C. Stadtman, *Biochemistry* **264**, 9724 (1993).

[26] S. C. Low, J. W. Harney, and M. J. Berry, *J. Biol. Chem.* **270**, 21659 (1995).

[27] I. Y. Kim and T. C. Stadtman, *Proc. Natl. Acad. Sci. U.S.A.* **92**, 7710 (1995).

[28] M. J. Guimaraes, D. Peterson, A. Vicari, B. G. Cocks, N. G. Copeland, D. J. Gilbert, N. A. Jenkins, D. A. Ferrick, R. A. Kastelein, J. F. Bazan, and A. Zlotnik, *Proc. Natl. Acad. Sci. U.S.A.* **93**, 15086 (1996).

[29] P. R. Copeland, J. E. Fletcher, B. A. Carlson, D. L. Hatfield, and D. Driscoll, *EMBO J.* **19**, 306 (2000).

[30] R. M. Tujebajeva, P. R. Copeland, X. M. Xu, B. A. Carlson, J. W. Harney, D. M. Driscoll, D. L. Hatfield, and M. J. Berry, *EMBO Rep.* **1**, 158 (2000).

[31] D. Fagegaltier, N. Hubert, K. Yamada, T. Mizutani, P. Carbon, and A. Krol, *EMBO J.* **19**, 4796 (2000).

in [2] and [4] of this volume.[31a,b] The incorporation of Sec into protein, therefore, is considered only briefly herein.

The factor that distinguishes UGA as a Sec codon is the presence of a stem–loop structure that occurs downstream of the UGA trinucleotide and is designated as the selenocysteine insertion sequence (SECIS) element[32] (see also [2] in this volume[31a]). In *E. coli*, the SECIS element occurs immediately downstream of the UGA Sec codon, whereas in eukaryotes, it occurs in the 3′-untranslated region. A single protein factor, designated SelB, recognizes both the SECIS element and selenocysteyl-tRNA for Sec insertion into the growing seleno-peptide in response to UGA as described in [1] in this volume.[21a] In eukaryotes,[29–31] and apparently in archaea,[33] two protein factors are utilized for inserting Sec into the nascent selenopeptide. One of these components, designated SECIS-binding protein 2 (SBP2), binds to the SECIS element[29] (see also [4] in this volume[31b]). The other factor, which is the specific elongation factor for selenocysteyl-tRNA and is designated EFsec, recognizes selenocysteyl-tRNA. The complex consisting of SECIS · SBP2 · EFsec · selenocysteyl-tRNA donates Sec to the growing seleno-peptide.

Functional Consequences of Over- and Underexpression of Mammalian Selenocysteine-Inserting tRNA and Expression of an i⁶ A-Deficient Selenocysteine-Inserting tRNA

The consequences of over- and underexpression of the Sec tRNA population on selenoprotein biosynthesis have been studied. Chinese hamster ovary cells were transfected with increasing numbers of mammalian Sec tRNA genes.[34] The highest copy number of transfected gene copies was 10 and the amount of Sec tRNA produced was directly proportional to gene copy number. Transgenic mice carrying as many as 20 extra copies of Sec tRNA genes were generated, but the level of the Sec tRNA population increased only up to about 6-fold, depending on the tissue examined.[35] The increased Sec tRNA population did not result in any detectable change in selenoprotein synthesis in either transfected cells or transgenic mice harboring extra copies of the Sec tRNA gene. These observations suggest that

[31a] M. J. Berry, G. W. Martin III, R. Tujebajeva, E. Grundner-Calemann, J. B. Marwell, N. Morozova, and J. W. Horney, *Methods Enzymol.* **347**, [2], 2002 (this volume).

[31b] P. R. Copeland and D. M. Driscoll, *Methods Enzymol.* **347**, [4], 2002 (this volume).

[32] S. C. Low and M. J. Berry, *Trends Biochem. Sci.* **21**, 203 (1996).

[33] M. Rother, R. Wilting, S. Commans, and A. Bock, *J. Mol. Biol.* **299**, 351 (2000).

[34] M. E. Moustafa, M. A. El-Saadani, K. M. Kandeel, D. B. Mansur, B. J. Lee, D. L. Hatfield, and A. M. Diamond, *RNA* **4**, 1436 (1998).

[35] M. E. Moustafa, B. A. Carlson, M. A. El-Saadani, G. V. Kryukov, Q. A. Sun, J. W. Harney, K. E. Hill, G. F. Combs, L. Feigenbaum, D. B. Mansur, R. F. Burk, M. J. Berry, A. M. Diamond, B. J. Lee, V. N. Gladyshev, and D. L. Hatfield, *Mol. Cell. Biol.* **21**, 3840 (2001).

the Sec tRNA population is not limiting in selenoprotein synthesis. Furthermore, the amount of mcm^5U increased much more dramatically than that of mcm^5Um, suggesting that the methylation step is limiting in tRNA maturation.[34]

The Sec tRNA population was reduced approximately in half in a mouse that was heterozygous for a targeted vector lacking the corresponding gene[36] and in mouse embryonic stem cells that were heterozygous for a similar targeted vector.[37] Glutathione peroxidase levels were measured in mice and in cells lacking one Sec tRNA gene copy and found to be the same as in the corresponding controls. These studies also suggest that the Sec tRNA population is not limiting for selenoprotein synthesis. However, removal of both copies of the Sec tRNA gene from the mouse genome is embryonic lethal, demonstrating an essential requirement for selenoprotein expression in mammals.[36]

In transgenic mice expressing an i^6A-deficient Sec tRNA, the levels of numerous selenoproteins were altered in a protein-specific manner.[35] Glutathione peroxidase and thioredoxin reductase 3 were the most and least affected selenoproteins, respectively. In fact, thioredoxin reductase 3 levels appeared to be elevated in some tissues, while glutathione peroxidase levels were reduced dramatically in all tissues examined. Selenoprotein expression was also affected in a tissue-specific manner, with liver and testes being the most and least affected tissues, respectively. As selenoprotein mRNA levels were largely unaffected in transgenic mice, the defect in selenoprotein expression in these animals occurred at translation. Analysis of the Sec tRNA population in the mutant tRNA mice showed that expression of i^6A^- tRNA resulted in a dramatic change in the distribution of the two major isoforms whereby the synthesis of mcm^5Um was repressed. Interestingly, the level of the i^6A mutant tRNA appeared to increase proportionally with gene copy number, suggesting that the levels of i^6A^- Sec tRNA and wild-type Sec tRNA are regulated independently and that the amount of wild-type tRNA is determined, at least in part, by a feedback mechanism governed by the level of the Sec tRNA population. The variation in expression of selenoproteins in the presence of the i^6A^- Sec tRNA suggests that these transgenic mice may be useful in assessing the biological roles of individual selenoproteins.[35] The effect of the expression of i^6A-deficient Sec tRNA in a transiently transfected cell line has been examined and these studies have shown that the level of deiodinase type 1 selenoprotein expression is inhibited when its gene is cotransfected with that of the mutant i^6A gene.[38]

[36] M. R. Bosl, K. Takadu, M. Oshima, S. Nishimura, and M. M. Taketo, *Proc. Natl. Acad. Sci. U.S.A.* **94,** 5531 (1997).

[37] H. S. Chittum, H. J. Baek, A. M. Diamond, P. Fernandez-Salguero, F. Gonzalez, T. Ohama, D. L. Hatfield, M. Kuehn, and B. J. Lee, *Biochemistry* **36,** 8634 (1997).

[38] G. J. Warner, M. J. Berry, M. E. Moustafa, B. A. Carlson, D. L. Hatfield, and J. R. Faust, *J. Biol. Chem.* **275,** 28110 (2000).

Purification of Selenocysteine-Inserting tRNA

Sec tRNA has been purified from *E. coli* and a variety of animal sources including bovine, rat, and *Xenopus* livers, mouse cells, and *Drosophila*. In mammalian cells and tissues, Sec tRNA represents from ~1 to ~7.5% of the serine tRNA population and the latter represents about 8–10% of the total tRNA population. In *Drosophila* and *E. coli*, the amount of Sec tRNA relative to the total tRNA population is much lower than that in mammals, suggesting that the requirements for Sec tRNA to meet the demands of protein synthesis are much greater in mammals than in these lower life forms.[39]

Sec tRNA can be overproduced in *E. coli* by cloning the tRNA gene into a plasmid, introducing the plasmid into an appropriate *E. coli* strain, and isolating the resulting enriched Sec tRNA by denaturing polyacrylamide gel electrophoresis.[5]

Sec tRNA is isolated from mammalian sources by extracting total tRNA from cells or tissues and resolving the Sec tRNA population by a variety of column chromatographic techniques. The two major Sec tRNA isoacceptors, mcm^{5}U and mcm^{5}Um, are resolved from each other by reversed-phase chromatography.[40] A procedure for the isolation and purification of the Sec tRNA isoforms from bovine liver is detailed below (see Diagram 1).

Reagents and Materials

tRNA Extraction

Calf liver
Liquid nitrogen (Roberts Oxygen, Collingdale, PA)
Phenol (Sigma, St. Louis, MO)
Chloroform (Mallinckrodt, St. Louis, MO)
Commercial blender, 5 liter (Waring)
Centrifuge bottles, 500 ml (Sorvall, Newtown, CT)
Low-speed centrifuge (Sorvall RC5B Plus)
Ethanol (Warner-Graham, Cockeysville, MD)
Extraction buffer: 0.14 *M* sodium acetate (pH 4.5)

tRNA Isolation and Fractionation

DE52 cellulose (Whatman, Clifton, NJ)
DE52 buffer A: 0.1 *M* Tris-HCl, 0.1 *M* NaCl
DE52 buffer B: 0.1 *M* Tris-HCl, 1.0 *M* NaCl
BD-cellulose (Serva, Heidelberg, Germany)

---

[39] D. L. Hatfield, B. J. Lee, and R. M. Pirtle (eds.), "Transfer RNA in Protein Synthesis," Chapters 3–5. CRC Press, Boca Raton, Florida, 1992.

[40] A. D. Kelmers and D. E. Heatherly, *Anal. Biochem.* **44**, 486 (1971).

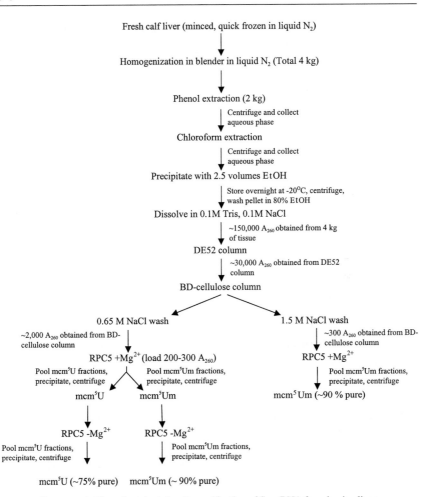

DIAGRAM 1. Flow chart depicting the purification of Sec tRNA from bovine liver.

BD-cellulose buffer A: 0.01 M sodium acetate (pH 4.5), 0.65 M NaCl, 0.01 M magnesium acetate, 0.001 M EDTA

BD-cellulose buffer B: 0.01 M sodium acetate (pH 4.5), 1.5 M NaCl, 0.01 M magnesium acetate, 0.001 M EDTA, 20% (v/v) ethanol

BD-cellulose buffer C: 0.01 M sodium acetate (pH 4.5), 1.5 M NaCl, 0.01 M magnesium acetate, 0.001 M EDTA

Reversed phase chromatographic column 5 (RPC5)[40]

RPC5 starting buffer (plus Mg^{2+}): 0.01 M sodium acetate (pH 4.5), 0.45 M NaCl, 0.01 M magnesium acetate, 0.001 M EDTA

RPC5 starting buffer (without Mg^{2+}): 0.01 M sodium acetate (pH 4.5), 0.5 M NaCl, 0.001 M EDTA

RPC5 buffer A: 0.01 M sodium acetate (pH 4.5), 0.525 M NaCl, 0.01 M magnesium acetate, 0.001 M EDTA

RPC5 buffer B: 0.01 M sodium acetate (pH 4.5), 0.675 M NaCl, 0.01 M magnesium acetate, 0.001 M EDTA

RPC5 buffer C: 0.01 M sodium acetate (pH 4.5), 0.6 M NaCl, 0.001 M EDTA

RPC5 buffer D: 0.01 M sodium acetate (pH 4.5), 0.825 M NaCl, 0.001 M EDTA

Sec tRNA oligonucleotide probe: 5′-CGCCCGAAAGGTGGAATTGA-3′

Human Sec tRNA gene probe (193 bp)[41]

Hybond N+ nylon membrane (Amersham Pharmacia, Piscataway, NJ)

PhosphorImager (Molecular Dynamics, Sunnyvale, CA)

Aminoacylation of tRNA

TAM (10×): 0.5 M Tris-HCl (pH 7.4), 0.06 M ATP (pH 7.0), 0.2 M $MgCl_2$

Unlabeled amino acids, minus serine (for a total of 19, each 0.2 M)

Rabbit reticulocyte synthetases[42]

L-[^{3}H]Serine, 34.0 Ci/mmol, 1 mCi/ml (Amersham)

Trichloroacetic acid, 5% (w/v)

Nitrocellulose filters, 0.45 μm pore size, 25 mm (Millipore, Bedford, MA)

Liquid scintillation counter (Packard, Downers Grove, IL)

Tissue Homogenization

Approximately 4 kg of fresh calf liver is cut into small pieces and blended in a Waring blender in the presence of liquid nitrogen. The frozen powdered liver is then used for extracting tRNA in 2-kg batches and the remainder is stored at $-80°$. Extraction buffer (2800 ml) and 2800 ml of phenol are added to the frozen liver powder and the tissue is homogenized for 3–4 min at high speed in the Waring blender at room temperature. The mixture is then centrifuged at 7000 rpm for 15 min. The aqueous phase is aspirated and extracted with an equal volume of chloroform by vigorous shaking for 5–6 min. The mixture is centrifuged at 7000 rpm for 15 min, the aqueous phase is removed, 2.5 volumes of ethanol is added, and the solution is left overnight at $-20°$. On the following morning, the resulting precipitant has settled to the bottom of the flask and most of the supernatant can be decanted. The precipitate is collected from the remaining solution by centrifugation at 7000 rpm for 15 min. The pellet is washed by resuspending in 80% (v/v) ethanol and recollected by centrifuging. The partially dried pellet

[41] V. A. O'Neill, F. C. Eden, K. Pratt, and D. L. Hatfield, *J. Biol. Chem.* **260**, 2501 (1985).

[42] D. L. Hatfield, C. R. Mathews, and M. Rice, *Biochim. Biophys. Acta* **564**, 414 (1979).

is stored at $-20°$ while the second 2 kg of tissue is processed. The pellets from 4 kg of liver are dissolved in 0.1 M Tris, 0.1 M NaCl and the A_{260} is measured. About 150,000 A_{260} units is typically recovered from 4 kg of liver. The nucleic acid extract is loaded onto a DE52 column (4 × 40 cm) that has been equilibrated in DE52 buffer A. After the sample is loaded, the column is washed in the same buffer until the A_{260} reading is less than 1.0 A_{260}. The tRNA is then eluted with ~1.5 liters of DE52 buffer B, 2.5 volumes of ethanol is added, the solution is left overnight at $-20°$, the resulting precipitate is collected by centrifugation at 7000 rpm for 15 min at $4°$, washed once with 80% (v/v) ethanol, and the pellet is dried and redissolved in H_2O. From 25,000 to 30,000 A_{260} units of bulk tRNA is typically obtained from 4 kg of fresh liver by this procedure.

BD-Cellulose Column Chromatography

Procedure for Obtaining Both Major Selenocysteine-Inserting tRNA Isoforms. About 30,000 A_{260} of bulk tRNA in ~1 liter of BD-cellulose buffer A is loaded onto a BD-cellulose column (30 × 3 cm) that has been equilibrated in the same buffer. The column is washed in this buffer until the A_{260} is reduced to less than 0.1 A_{260}. The tRNA that attaches to the column in 0.65 M NaCl is then eluted with ~1 liter of BD-cellulose buffer B. The eluate is adjusted to less than 1.0 M NaCl with H_2O, 2.0 volumes of ethanol is added, and the solution is left overnight at $-20°$. The resulting precipitate is then collected by centrifugation at 7000 rpm for 15 min at $4°$. About 2000 A_{260} units of tRNA that is highly enriched with both mcm^{5}U and mcm^{5}Um is obtained by this procedure.

Procedure for Obtaining mcm^{5}Um Selenocysteine-Inserting tRNA Isoform. About 30,000 A_{260} of bulk tRNA in ~1 liter of BD-cellulose buffer A is loaded onto a BD-cellulose column (30 × 3 cm) that has been equilibrated in the same buffer. The column is washed in BD-cellulose buffer C until the A_{260} is reduced to less than 0.1 A_{260}. The tRNA that attaches to the column in 1.5 M NaCl is then eluted with ~1 liter of BD-cellulose buffer B. The eluate is adjusted to less than 1.0 M NaCl with H_2O, 2.0 volumes of ethanol is added, and the solution is left overnight at $-20°$. The resulting precipitate is then collected by centrifugation at 7000 rpm for 15 min at $4°$. About 1% of the initial A_{260} units of the bulk tRNA loaded onto the column is recovered by this procedure and the sample is highly enriched for mcm^{5}Um.[43]

RPC5 Chromatography

From 200 to 300 A_{260} units of the high salt–ethanol fractions obtained by BD-cellulose chromatography, precipitated and collected as described above, is maintained in RPC5 buffer A and loaded onto an RPC5 column (30 × 1.5 cm)

---

[43] A. M. Diamond, B. Dudock, and D. L. Hatfield, *Cell* **25**, 497 (1981).

that has been equilibrated in RPC5 starting buffer (plus Mg^{2+}). A 160-ml linear gradient (80 ml of buffer A as the starting buffer and 80 ml of buffer B as the terminal buffer) is run at a flow rate of 2.0 ml/min and 1-ml fractions are collected. The absorbance at A_{260} in the eluted fractions is measured and Sec tRNA is detected either by aminoacylation or hybridization. For detection by aminoacylation, the tRNA must first be precipitated by adding 3 volumes of ethanol, placing the fractions at $-20°$ for 2–3 hr, collecting precipitates on nitrocellulose filters, washing the filters with 80% (v/v) ethanol, drying the filters by air suction, and eluting the tRNA in 500 μl of H_2O. Five microliters of each fraction is then added to a 20-μl solution containing 2.5 μl of 10× TAM, 2.5 μl of [^{3}H]serine, 2.5 μl of 0.2 M unlabeled amino acids (minus serine), and 1.5 μl of rabbit reticulocyte synthetases. The reaction mixture is incubated for 15 min at $37°$ and diluted with 500 μl of H_2O, followed by ~8 ml of 5% (w/v) trichloroacetic acid, and the resulting precipitate is collected on nitrocellulose filters. The filters are dried and then placed in a scintillation cocktail, and the radioactivity is measured in a scintillation counter. The two late-eluting peaks of [^{3}H]serine represent mcm^{5}U (which elutes first) and mcm^{5}Um (which elutes second). Alternatively, a dot blot is prepared by placing 5 μl of each fraction on a nylon membrane and the membrane is hybridized with either a 193-bp probe encoding the human Sec tRNA gene or a 20-nucleotide (oligonucleotide) probe encoding the 3′ end of the mammalian Sec tRNA gene. The two peaks of radioactivity measured in a PhosphorImager represent mcm^{5}U and mcm^{5}Um. Column fractions containing mcm^{5}U or mcm^{5}Um are pooled, the tRNA is precipitated, and the samples are collected either on nitrocellulose filters or by centrifugation as described above. It is recommended that the more highly purified mcm^{5}U and mcm^{5}Um samples be collected by centrifugation, as substantial losses of highly purified samples can occur by collection on filters. Samples containing either mcm^{5}U or mcm^{5}Um are individually loaded onto the RPC5 column in RPC5 starting buffer (without Mg^{2+}). A 160-ml linear gradient (80 ml of buffer C as the initial buffer and 80 ml of buffer D as the terminal buffer) is run as described above. The number of A_{260} units in fractions is measured and the mcm^{5}U and mcm^{5}Um isoforms are detected by aminoacylation or hybridization as described above. Fractions containing mcm^{5}U or mcm^{5}Um are pooled, precipitated with 2.5 volumes of ethanol, collected by centrifugation, dissolved in H_2O, and stored at $-80°$.

Preparation of [^{75}Se]Selenocysteyl-tRNA

Diagram 2 presents a flow chart of the preparation of labeled selenocysteyl-tRNA.

Reagents and Materials

[^{75}Se]Selenious acid (University of Missouri Research Reactor)
Phenol (Sigma)

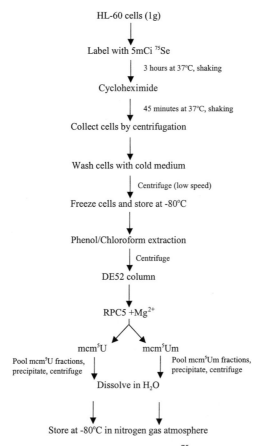

DIAGRAM 2. Flow chart depicting preparation of [^{75}Se]selenocysteyl-tRNA.

Chloroform (Mallinckrodt)

Ethanol (Warner-Graham)

DE52 cellulose (Whatman)

Cycloheximide (Sigma)

Sodium selenite (Sigma)

Extraction buffer: 0.005 M sodium acetate (pH 4.5) 0.225 M NaCl, 0.005 M magnesium acetate, 0.001 M EDTA, 0.2% sodium dodecyl sulfate (SDS), 0.006 M 2-mercaptoethanol

DE52 column equilibration buffer: 0.005 M sodium acetate (pH 4.5), 0.1125 M NaCl, 0.005 M magnesium acetate, 0.001 M EDTA

DE52 elution buffer: 0.01 M sodium acetate (pH 4.5) 1.5 M NaCl, 0.01 M magnesium acetate, 0.001 M EDTA

Reversed-phase chromatographic column 5 (RPC5)

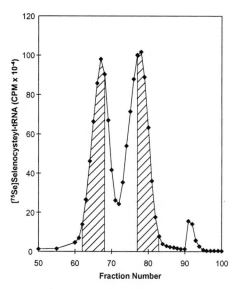

FIG. 2. Fractionation of [^{75}Se]selenocysteyl-tRNA isoforms by RPC5 chromatography. [^{75}Se]Selenocysteyl-tRNA was extracted from HL-60 cells and loaded onto an RPC5 column, and the column was developed as described in text. The initial eluting peak from the column is mcm^{5}U and the latter eluting peak is mcm^{5}Um. The isoforms are pooled as shown by the hatched areas, precipitated from solution with ethanol, and collected as described in text.

RPC5 starting buffer (plus Mg^{2+}): 0.01 M sodium acetate (pH 4.5), 0.5 M NaCl, 0.001 M EDTA

RPC5 buffer A: 0.01 M sodium acetate (pH 4.5), 0.525 M NaCl, 0.01 M magnesium acetate, 0.001 M EDTA

RPC5 buffer B: 0.01 M sodium acetate (pH 4.5), 0.675 M NaCl, 0.01 M magnesium acetate, 0.001 M EDTA

Wizard γ counter (Wallac, Turku, Finland)

^{75}Se Labeling of HL-60 Cells

HL-60 Cells are grown as described[44] in the presence of $5 \times 10^{-7} M$ sodium selenite. One gram of cells is harvested and the cells are washed in fresh growth medium[44] consisting of 1% fetal calf serum (FCS) (without selenium supplementation) and resuspended in 80 ml of the latter medium. Five millicuries of neutralized ^{75}Se is added, the cells are gently shaken for 3 hr at 37°, cycloheximide (1×10^{-4} M) is added, and the cells are shaken for an additional 45 min.

[44] D. L. Hatfield, B. J. Lee, L. Hampton, and A. M. Diamond, *Nucleic Acids Res.* **19,** 939 (1991).

The cells are collected by centrifugation at 800 rpm for 5 min at 4° and washed with cold medium, and the packed cells are stored at −80°.

tRNA Extraction

All steps are carried out at 4° with the exception of the phenol–chloroform extraction and the RPC5 chromatography. A solution containing 3 ml of phenol, 3 ml of chloroform, and 5 ml of extraction buffer is added directly to the frozen cell pellet, and the mixture is vortexed for 5 min and then centrifuged at 4000 rpm for 10 min. The aqueous layer is aspirated, an equal volume of chloroform is added, and the mixture is further vortexed for 2–3 min and centrifuged at 4000 rpm for 10 min. The supernatant is slowly loaded onto a DE52 column (2.5 × 1.8 cm) that has been equilibrated with DE52 column equilibration buffer. The column is washed with several volumes of equilibration buffer and the tRNA is eluted with 10 ml of DE52 elution buffer. The eluate is diluted to a final concentration of 0.45 M NaCl with cold 0.02 M 2-mercaptoethanol and loaded onto an RPC5 column that had been equilibrated with RPC5 starting buffer (plus Mg^{2+}), and the column is washed in 2–3 volumes of starting buffer. A linear 160-ml gradient (80 ml of buffer A, 80 ml of buffer B) is then run as described above. The fractions are analyzed for radioactivity in a γ counter. The mcm^5U and mcm^5Um isoforms are pooled as shown in Fig. 2, 2.5 volumes of ethanol are added, the solution is stored at −20° for 2–3 hr, and the resulting precipitate is collected by centrifugation at 5000 rpm for 30 min at 4°. The samples are washed in 80% (v/v) ethanol, gently dried, and dissolved in 100 μl of H_2O. A_{260} values and counts per minute per microliter are measured and the [^{75}Se]selenocysteyl-tRNA is stored at −80° after gently blowing nitrogen gas over the sample before sealing the tube. From 10 × 10^6 to 20 × 10^6 total cpm of [^{75}Se]selenocysteyl-tRNA is typically recovered by this procedure.

[4] Purification and Analysis of Selenocysteine Insertion Sequence-Binding Protein 2

By PAUL R. COPELAND and DONNA M. DRISCOLL

Introduction

The insertion of selenocysteine (Sec) into selenoproteins is a novel cotranslational event in which a UGA codon is decoded as Sec rather than as a stop codon. The recoding of UGA as Sec in mammalian cells requires well-defined *cis*-acting sequences known as selenocysteine insertion sequence (SECIS) elements found in the 3′ untranslated region (3′ UTR) of selenoprotein mRNAs. However, the *trans*-acting factors involved in directing Sec incorporation from the 3′ UTR remained elusive until more recently.

SECIS-binding protein 2 (SBP2) is a sequence-specific RNA-binding protein that interacts with the conserved GA quartet within the SECIS element. SBP2 was originally detected in UV cross-linking studies using the phospholipid hydroperoxide glutathione peroxidase (PHGPx) 3′ UTR.[1] This protein was subsequently purified from rat testicular extracts[2] and shown to be essential for the recoding of UGA as Sec *in vitro*.[3] SBP2 has been further characterized in terms of domain structure, ribosome-binding activity, and identification of a critical amino acid residue required for RNA binding.[4] The use of testis in the identification of SBP2 was an important advantage as the protein is not highly expressed in any other tissue that we have examined.[3,5] This is likely to be a result of the high level of expression of PHGPx protein in this tissue, thus making testis extracts a potentially rich source for all of the components required for efficient selenoprotein synthesis.

This article describes the methods used to purify native and recombinant SBP2 and the assays that we developed to analyze the functions of SBP2, including RNA binding, Sec insertion, and interaction with ribosomes.

Purification of Native Selenocysteine Insertion Sequence-Binding Protein 2

Preparation of Testicular Extracts

Trimmed rat testicles are shipped on wet ice from Pel Freez (St. Louis, MO). It is important to avoid freezing of the tissue before extraction as this eliminates

---

[1] A. Lesoon, A. Mehta, R. Singh, G. M. Chisolm, and D. M. Driscoll, *Mol. Cell. Biol.* **17,** 1977 (1997).

[2] P. R. Copeland and D. M. Driscoll, *J. Biol. Chem.* **274,** 25447 (1999).

[3] P. R. Copeland, J. E. Fletcher, B. A. Carlson, D. L. Hatfield, and D. M. Driscoll, *EMBO J.* **19,** 306 (2000).

most of the SBP2 binding activity. Initial disruption of the tissue is performed with a fresh razor blade on a Parafilm surface. Testes are obtained and extracted in batches of 50 ($\sim$50 g, total weight). The minced tissue is transferred to buffer A [20 mM KPO$_4$ (pH 7.2), 100 mM KCl, 2 mM dithiothreitol (DTT), 0.1 mM phenylmethylsulfonyl fluoride (PMSF), 0.05% (v/v) Tween 20, 5% (v/v) glycerol] at 2 ml/g of tissue, and homogenized with a BioSpec (Bartlesville, OK) Bio-Homogenizer. The use of PMSF alone is sufficient for the isolation of SBP2, but the inclusion of other protease inhibitors may be desirable for other preparations. Also note that the inclusion of DTT is essential for maintaining SBP2 activity. Crude extract is spun at 10,000g for 20 min at 4$°$ in a Beckman (Palo Alto, CA) JA-17 rotor. Supernatants are transferred to ultracentrifugation bottles and spun at 100,000g for 1 hr at 4$°$ to yield S100 extracts ranging in protein concentration from 10 to 15 mg/ml.

S-Sepharose Chromatography and Ammonium Sulfate Precipitation

All column chromatography is performed on the PE Biosystems (Framingham, MA) BioCad Sprint chromatography system. Columns are jacketed with copper coils for cooling with a circulating water bath. S100 extract (25 ml) is diluted 1 : 1 with 20 mM KPO$_4$ and immediately applied to a 1.6 $\times$ 26 cm S-Sepharose column (bed volume of 32 ml) equilibrated in buffer A. Proteins are eluted with a linear gradient from 100 to 800 mM KCl at 2 ml/min. Fractions (2 ml) are collected and an aliquot of each (10 μl) is assayed for SBP2 by UV cross-linking (see below). Active fractions are pooled, diluted 1 : 1 with ice-cold water, and immediately brought to 35% (w/v) saturation with solid ammonium sulfate. The precipitate is incubated on ice with occasional mixing for 1 hr followed by centrifugation at 18,000g for 10 min at 4$°$. Pellets are resuspended in 1/20th of the original volume and either dialyzed or used directly for analysis or further purification.

RNA Affinity Chromatography

We have taken advantage of the fact that SBP2 binds to a wild-type PHGPx 3$'$ UTR but not to a mutant SECIS element that lacks the conserved AUGA motif (ΔAUGA).[1] Wild-type and ΔAUGA mutant RNAs are synthesized in 200-μl reactions with Ribomax (Promega, Madison, WI) transcription reagents and immobilized on CNBr-activated Sepharose 4B (AP Biotech, Piscataway, NJ) according to the procedure published by Kaminski *et al.*[6] The beads are swollen in 1 mM HCl and washed extensively with water on a vacuum filtration device fitted with a 0.2-μm pore size filter. Washed beads are rinsed in 200 mM

[4] P. R. Copeland, V. A. Stepanik, and D. M. Driscoll, *Mol. Cell. Biol.* **21,** 1491 (2001).

[5] P. R. Copeland, C. A. Gerber, and D. M. Driscoll, unpublished observations (2000).

[6] A. Kaminski, S. L. Hunt, J. G. Patton, and R. J. Jackson, *RNA* **1,** 924 (1995).

2-(N-morpholino) ethanesulfonic acid, brought to pH 6.0 with KOH (MES–KOH), and then scraped from the filter into a 15-ml conical tube. RNA (150–200 μg) is diluted to 2 ml with MES–KOH and added to the swollen beads. This mixture is incubated at 4° overnight on an end-over-end rotator. The RNA-coated beads (150–200 μg of RNA per milliliter of beads) are directly loaded into an AP Biotech 0.5 × 5 cm fast protein liquid chromatography (FPLC) column and equilibrated in buffer A plus 170 mM KCl. Approximately 1 mg of protein from the ammonium sulfate fraction is loaded at 0.2 ml/min, and the column is washed in buffer A plus 170 mM KCl until the A_{280} has returned to baseline. Proteins are eluted (0.5 ml/min) with a linear gradient (0.27–1.7 M KCl) in 10 min, followed by a step to 2 M KCl. For activity analysis, 0.5-ml fractions are collected directly into 0.28 ml of saturated ammonium sulfate. After collection, fractions are incubated on ice for 30 min and spun at 18,000g in a microcentrifuge at 4° for 10 min. The supernatant is discarded and the pellets are resuspended in 50 μl of buffer A and stored at −80°. For analysis by sodium dodecyl sulfate–polyacrylamide gel electrophoresis (SDS–PAGE), 0.5-ml fractions are collected directly into 40 μl of 100% trichloroacetic acid (TCA), incubated on ice for 15–30 min, and centrifuged at 18,000g for 10 min at 4°. Protein pellets are resuspended in 15 μl of 0.2 N NaOH and SDS sample buffer [50 mM Tris-HCl (pH 6.8), 4% (w/v) SDS, 0.2% (w/v) bromphenol blue, 20% (v/v) glycerol, 2 M 2-mercaptoethanol]. Proteins are resolved by 8% (w/v) SDS–PAGE.

Figure 1 illustrates a typical elution profile from the wild-type RNA column as analyzed by SDS–PAGE followed by silver staining. Although the fractions contained a complex mixture of polypeptides, a protein of the expected size for SBP2 (120 kDa) eluted in fractions 17–22 and copurified with SBP2 binding activity. This protein was not detected in the equivalent fractions eluted from the ΔAUGA mutant RNA affinity column. Sufficient protein was obtained for peptide sequencing by tandem electrospray mass spectrometry, and this information was used in molecular cloning experiments to isolate a full-length SBP2 cDNA.[3]

In addition to SBP2, there are several prominent proteins retained by both the wild-type and mutant PHGPx 3′ UTR columns that have been identified as indicated in Fig. 1. The spliceosomal accessory proteins (SAPs) comprise part of the U2 small nuclear ribonucleo protein (snRNP) required for mRNA splicing.[7] DNA-binding protein B (DBPB) is a nonspecific RNA binding protein that has previously been proposed to bind SECIS elements.[8] We have been unable to detect any function or SBP2 binding activity for the SAPs,[9] and the role of DNAB in Sec insertion has not been demonstrated.[8]

[7] B. K. Das, L. Xia, L. Palandjian, O. Gozani, Y. Chyung, and R. Reed, *Mol. Cell. Biol.* **19,** 6796 (1999).
[8] Q. Shen, R. Wu, J. L. Leonard, and P. E. Newburger, *J. Biol. Chem.* **273,** 5443 (1998).
[9] P. R. Copeland and D. M. Driscoll, unpublished observations (2000).

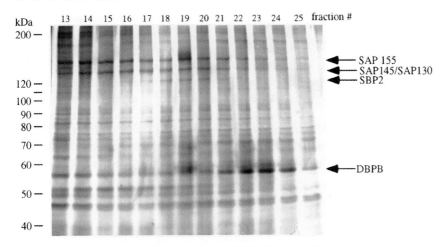

FIG. 1. SDS–PAGE analysis of proteins eluting with SBP2 during RNA affinity chromatography. Proteins were detected by silver stain. Arrows indicate the proteins that have been identified by mass spectrometric peptide sequencing: spliceosome accessory proteins (SAP 155, 145, and 130), SECIS-binding protein 2 (SBP2), and DNA-binding protein B (DBPB). SAP 130 and 145 are contained within the same band. See text for details regarding each protein.

Purification of Recombinant Selenocysteine Insertion Sequence-Binding Protein 2

Both full-length and the fully functional C-terminal half (amino acids 399–846) of SBP2 can be expressed in *Escherichia coli* and purified by the Strep-tag II expression, detection, and purification system (Genosys, The Woodlands, TX). The yield of full-length SBP2 is about an order of magnitude less than that of the C-terminal fragment. The Strep-tag II system uses an eight-amino acid sequence (WSHPQFEK) that forms a biotin-like structure that binds with high affinity to a modified streptavidin termed StrepTactin.[10] Thus, high yields of purified protein can be obtained in a single-step procedure utilizing commercially available StrepTactin agarose (Genosys).

Expression Conditions

The 3' half (1.3 kb) of the SBP2 coding region is subcloned into pASK-IBA7 downstream of the Strep-tag and transformed into BL21 (DE3) (Invitrogen, Carlsbad, CA). Cultures (500 ml) are inoculated with 1 ml of an overnight culture derived from a glycerol stock of transformed bacteria. Cultures are grown at 37° to an OD_{600} of about 1.0 and then induced with 50 μl of 2 mM anhydrotetracycline.

[10] S. Voss and A. Skerra, *Protein Eng.* **10,** 975 (1997).

The incubation temperature is reduced to 30° and growth is continued for 1 hr. Cells are pelleted and resuspended in buffer W [100 mM Tris-HCl (pH 8.0), 1 mM EDTA, 25 mM NaCl] supplemented with a protease inhibitor tablet (Roche, Indianapolis, IN). The samples are frozen at $-80°$ overnight and then sonicated until clarified and no longer viscous. To deplete the endogenous biotin, 75 μg of egg white avidin (Sigma, St. Louis, MO) is added to an amount of sonicated mixture derived from 1 liter of bacterial culture. The avidin–extract mixture is incubated on ice for 15 min and subjected to centrifugation at 17,000g for 15 min at 4°. The supernatant is removed and aliquoted for storage at $-80°$. More than 50% of the recombinant SBP2 is retained in the soluble fraction.

Purification of Strep-Tagged C-Terminal Selenocysteine Insertion
Sequence-Binding Protein 2

StrepTactin agarose is packed into an AP Biotech 0.5 × 5 cm FPLC column and equilibrated with buffer W. Approximately 2 ml of the SBP2-containing crude extract is filtered with a 0.2-μm pore size syringe filter and applied to the column at 0.5 ml/min. The column is washed with buffer W at 2 ml/min until the baseline at A_{280} is recovered. SBP2 is then eluted from the column at 2 ml/min with 2.5 mM desthiobiotin in buffer W. The elution of SBP2 is followed by monitoring the absorbance at 280 nm, and the peak fractions are collected and stored on ice. The desthiobiotin is removed from the fractions on a fast desalting column (AP Biotech) run at 4 ml/min, and SBP2 is concentrated with a Centricon-10 ultrafiltration device (Amicon, Danvers, MA). This procedure usually results in more than 90% pure SBP2. It is likely that further purification can be obtained with any of the procedures used to purify native SBP2 as mentioned above. The protein derived from this procedure is active in both the RNA-binding assays and functional assays described below, but the specific activity is significantly lower than that of *in vitro*-translated SBP2. This suggests that a pool of inactive SBP2 is included in the purified preparation that would likely be excluded during RNA affinity purification.

RNA-Binding Assays for Selenocysteine Insertion
Sequence-Binding Protein 2

SBP2 was originally identified by its ability to UV cross-link to a radiolabeled PHGPx 3′ UTR *in vitro*. This type of analysis was useful for identifying SBP2 in crude extracts and for determining its molecular mass, but the assay is not linear or quantitative. We developed an electrophoretic mobility shift assay (EMSA) for analyzing RNA–protein interactions. This assay works well for *in vitro*-translated SBP2 and purified recombinant protein but the results are less clear with more complex protein mixtures such as testis S100 extract, for which the UV cross-linking assay is preferable.

UV Cross-Linking Assay

Materials

Phosphate-buffered saline (PBS), 1×: 150 mM NaCl, 150 mM NaPO$_4$, pH 7.4; interchangeable with buffer A

Binding mix, 6.7×: 1× PBS with soybean trypsin inhibitor (330 μg/ml; Sigma), 67 mM DTT, yeast tRNA (1.67 μg/ml)

[^{32}P]UTP-labeled *in vitro*-transcribed target RNA (20 fmol/μl) RNase A, 10 mg/ml: Mix with RNase T1 for GC-rich target RNAs

Reaction

Binding mix	3 μl
^{32}P-labeled target RNA	1 μl
Testis S100 extract (10 mg/ml)	2 μl
PBS, 1×	14 μl

Incubate the reaction mixture in a polystyrene round-bottom 96-well assay plate (Corning, Cambridge, MA) covered with sealing tape at 37° for 30 min. Place the uncovered tray on a bed of ice at a level approximately 7 cm from the UV bulbs in a GS Gene Linker (Bio-Rad, Hercules, CA) or equivalent. Expose to UV light for 10 min and then add 2 μl of RNase A (10 mg/ml) and incubate at 37° for 15–30 min. Add SDS sample buffer and resolve proteins by SDS–PAGE. Expose the gel to X-ray film.

Electrophoretic Mobility Shift Assay

The reaction conditions for the EMSA are exactly the same as those for UV cross-linking. The procedures diverge at the UV cross-linking step, at which point the EMSA samples are augmented with 5 μl of EMSA loading buffer [50% (v/v) glycerol, 10 mM Tris-HCl (pH 7.5), 0.1% (w/v) each bromphenol blue and xylene cyanol]. An aliquot of each sample (10 μl) is loaded onto a 4% (w/v) nondenaturing acrylamide gel (29 : 1 acrylamide : bisacrylamide) that has been prerun at 180 V for at least 30 min. The gel is made and run in 1× TGE [50 mM Tris-HCl (pH 8.0), 380 mM glycine, 2 mM EDTA]. The gel is run in the cold (4°) at 20 mA for approximately 2.5 hr or until the bromphenol blue reaches the bottom of the gel. Note that a maximum of 2 μl of an *in vitro* translation reaction can be used in the EMSA as the reticulocyte lysate inhibits binding at higher levels.

In Vitro Translation of Phospholipid Hydroperoxide Glutathione Peroxidase

Much of the advancement we have made in understanding the mechanism of Sec insertion has been greatly accelerated by our development of a reliable *in vitro* assay that is highly responsive to added SBP2. The assay is based on only slight but essential modifications to a standard rabbit reticulocyte lysate *in vitro* translation assay. The key changes include the use of approximately 10-fold less input PHGPx mRNA and the increase in incubation temperature from 30 to 37°.[3] We have also successfully used this assay to examine the translation of luciferase mRNA to which the PHGPx 3′ UTR has been added, and we expect that any selenoprotein for which some method of purification exists (e.g., immunoprecipitation) would be efficiently translated.

Materials

Micrococcal nuclease-treated reticulocyte lysate (Promega)
m[7]GpppG-capped *in vitro*-transcribed PHGPx mRNA
[35S]Methionine, 1000 Ci/mmol (10 μCi/μl; AP Biotech; translation grade)
SBP2 protein, recombinant or *in vitro* translated
RNAGuard (AP Biotech) or equivalent
Amino acid mix minus methionine (Promega), 1 m*M*
PBSD, 1×: 1× PBS with 2 m*M* DTT
Bromosulfophthalein *S*-glutathione agarose (BSP-agarose; Sigma)

Reaction

Reticulocyte lysate	17.5 μl
Amino acid mix minus methionine	0.5 μl
RNAGuard	0.5 μl
[35S]Met	1 μl
PHGPx mRNA (10 ng/μl)	1 μl
SBP2 protein or mRNA	0–4.5 μl
Sterile water to	25-μl total volume

Procedure

1. Incubate at 37° for 50 min.
2. Remove 1 μl for SDS–PAGE analysis of SBP2 levels if desired (PHGPx product will not be detectable).
3. Add 40 μl of 50% BSP-agarose (prewashed with PBSD).
4. Add 250 μl of PBSD.
5. Rotate at 4° for 30 min.
6. Wash the beads twice with 1 ml of PBSD.
7. Add SDS sample buffer, heat the samples, and analyze products by 15% (w/v) SDS–PAGE.
8. Expose the gel to X-ray film.

This assay was originally performed with [75] Se-labeled Sec-tRNA[Sec] [labeled and purified by B. A. Carlson and D. A. Hatfield, the National Institutes of Health

(NIH, Bethesda, MD)[11]] to demonstrate that the translation products were authentic selenoproteins. The results we have obtained with [^{35}S]Met are identical to the original observations with [^{75}Se]Sec-tRNA^{Sec} and are obtained with considerably less expense and labor. With saturating amounts of SBP2, this procedure usually yields about 20 times more PHGPx than without added SBP2. Note that SBP2 can be added in a variety of forms. If saturating amounts are desired, then one can simply cotranslate SBP2 mRNA in the same reaction. If quantitative amounts of highly active SBP2 are needed, then it is best to pretranslate SBP2 *in vitro* and quantitate the amount of product by PhosphorImager analysis, using known amounts of spotted [^{35}S]Met as a standard. Note that the calculation of yield should include the estimated amount of unlabeled methionine in the reticulocyte lysate, which according to the manufacturer is ~5 μM. SBP2 derived from bacterial expression is also functional in this assay; however, it is of much lower specific activity. For high-throughput screening, the reaction can successfully be performed at one-half the stated amounts.

Ribosome-Binding Assay

Sec insertion requires a multitude of components that are likely involved in spatially and temporally regulated complexes. At this point the Sec insertion complex (SIC) is known to consist of the following: the Sec codon (UGA), the SECIS element, the Sec-tRNA^{Sec}, SBP2, eEFSec (the Sec-specific elongation factor), and, of course, the translating ribosome. However, the configuration of a functional complex is for the most part unknown. In terms of interactions, it has already been established that SBP2 and eEFSec interact *in vivo*,[12] but this interaction may be indirect and meditated by the SECIS element. Our attempts to find an SBP2-binding protein have been unsuccessful and have led to our discovery that SBP2 stably associates with ribosomes *in vivo* and *in vitro*.[4] The procedure described below is a departure from a standard analysis of polysomes because the association of SBP2 with the ribosome is sensitive to the artificially high concentration of magnesium necessary to stabilize polysomes during sedimentation. We have thus turned to the use of glycerol gradients made up in PBS.

Materials

Glycerol, 10 and 30% (v/v) in 1× PBSD
Ultracentrifuge tubes (14 × 89 mm; Beckman)
Swing-out ultracentrifuge rotor (e.g., SW41)

[11] A. M. Diamond, I. S. Choi, P. F. Crain, T. Hashizume, S. C. Pomerantz, R. Cruz, C. J. Steer, K. E. Hill, R. F. Burk, J. A. McCloskey, and D. L. Hatfield, *J. Biol. Chem.* **268**, 14215 (1993).
[12] R. M. Tujebajeva, P. R. Copeland, X.-M. Xu, B. A. Carlson, J. W. Harney, D. M. Driscoll, D. L. Hatfield, and M. J. Berry, *EMBO Rep.* **1**, 1 (2000).

Extracts from cells transfected with SBP2
EDTA-free protease inhibitor tablet (Roche)
TCA (10%, w/v)–Tween 20 (0.05%, w/v)

Procedure

Gradients are made by layering 5.5 ml of 10% (v/v) glycerol–PBSD over an equal volume of 30% (v/v) 1× PBSD followed by carefully rotating the tube to a horizontal position. The gradient is left to form for 1–2 hr at room temperature, and the tubes are turned upright and cooled at 4° while the extracts are made.

McArdle 7777 cells, a rat hepatoma cell line, are transiently transfected with Strep-tagged SBP2, using LipofectAMINE (Life Technologies, Bethesda, MD). Forty hours posttransfection, the cells are washed with 1× PBS and harvested by scraping into 1× PBSD containing protease inhibitors. The cells are disrupted by 15 strokes with a high-clearance Dounce homogenizer. The extracts are spun at 14,000g for 10 min at 4°. The supernatants are diluted to 1 mg/ml and 500 μl of extract is loaded onto a 10–30% (v/v) linear glycerol gradient made in PBSD. The gradients are spun at 210,000g in an SW41 rotor for 3.5 hr at 4°. Fractions (0.6 ml) are pulled from the top of each gradient, and 200 μl from each fraction is subjected to TCA precipitation [10% (w/v) TCA, 0.05% (v/v) Tween 20]. Precipitated proteins are resolved by 12% (w/v) SDS–PAGE, blotted to nitrocellulose, blocked with 3% (w/v) bovine serum albumin (BSA) in PBST [1× PBS plus 0.2% (v/v) Tween 20] and, for Strep-tagged proteins, probed with a 1 : 4000 dilution of alkaline phosphatase conjugated with streptavidin (AP Biotech). For untagged proteins, the blots are probed with a 1 : 2000 dilution of anti-SBP2 polyclonal antibody followed by a 1 : 5000 dilution of alkaline phosphatase-conjugated anti-rabbit IgG secondary antibody. Blots are developed in NBT/BCIP (Western Blue; Promega). To identify the positions of the 18S and 28S ribosomal RNAs, fractions are analyzed by agarose gel electrophoresis and ethidium bromide staining.

Summary and Perspectives

Our success in isolating SBP2 was due primarily to the identification of a tissue source that appears to be enriched for these factors and by the development of a high-affinity purification step. The fact that SBP2 exists in mammalian cells as a large macromolecular weight complex suggests that other factors remain to be identified. In future, RNA affinity chromatography may prove fruitful for identifying RNA-binding proteins that bind to the AAA SECIS element that is required for Sec insertion but not for SBP2 binding. Similarly, novel proteins that interact with SBP2 may be isolated by SBP2 affinity chromatography. The development of an *in vitro* system for Sec insertion will facilitate the functional analysis of "new players" as they are isolated.

Acknowledgments

We thank Mike Kinter and Belinda Willard for peptide sequencing; Julia Fletcher for work on the *in vitro* translation assay; and Vince Stepanik for work on SBP2 mutants.

[5] Nonsense-Mediated Decay: Assaying for Effects on Selenoprotein mRNAs

By XIAOLEI SUN and LYNNE E. MAQUAT

Introduction

Selenoprotein mRNAs contain at least one UGA codon that, together with a *cis*-residing selenocysteine insertion sequence (SECIS), cotranslationally direct the incorporation of the unusual amino acid selenocysteine (Sec). Sec incorporation requires *trans*-acting factors that are involved in either the synthesis or function of Sec-charged tRNA, including the newly described SECIS-binding protein (SBP2)[1] and associated elongation factor.[2,3] It makes sense that dietary selenium (Se) regulates the enzymatic activities and levels of selenoproteins given the dependence of selenoprotein synthesis and, often, selenoprotein activity on Sec, the synthesis of which is dependent on selenium. Interestingly, however, dietary selenium also regulates the levels of selenoprotein mRNAs, with some being regulated more than others in ways that cannot be predicted by the degree to which a particular tissue retains selenium. For example, selenium deficiency decreases the abundance of mRNA for classic glutathione peroxidase 1 (GPx1) in rat liver 20-fold[4–8] and selenoprotein P in rat muscle 4-fold.[9] In contrast, selenium deficiency reduces the abundance of mRNA for phospholipid hydroperoxidase glutathione peroxidase

[1] P. R. Copeland, J. E. Fletcher, B. A. Carlson, D. L. Hatfield, and D. M. Driscoll, *EMBO J.* **19,** 306 (2000).

[2] D. Fagegaltier, N. Hubert, K. Yamada, T. Mizutani, P. Carbon, and A. Krol, *EMBO J.* **19,** 4796 (2000).

[3] R. M. Tujebajev, P. R. Copeland, X. M. Xu, B. A. Carlson, J. W. Harney, D. M. Driscoll, D. L. Hatfield, and M. J. Berry, *EMBO Rep.* **1,** 158 (2000).

[4] X. G. Lei, J. K. Evenson, K. M. Thompson, and R. A. Sunde, *J. Nutr.* **125,** 1438 (1995).

[5] M. Saedi, C. G. Smith, J. Frampton, I. Chambers, P. R. Harrison, and R. A. Sunde, *Biochem. Biophys. Res. Commun.* **53,** 855 (1988).

[6] H. Toyoda, S.-I. Himeno, and N. Imura, *Biochim. Biophys. Acta* **1049,** 213 (1990).

[7] S. L. Weiss, J. K. Evenson, K. M. Thompson, and R. A. Sunde, *J. Nutr.* **126,** 2260 (1996).

[8] P. M. Moriarty, C. C. Reddy, and L. E. Maquat, *Mol. Cell. Biol.* **18,** 2932 (1998).

[9] S. C. Vendeland, M. A. Beilstein, J. Y. Yeh, W. Ream, and P. D. Whanger, *Proc. Natl. Acad. Sci. U.S.A.* **92,** 8749 (1995).

(GPx4) by at most 20% in rat liver and testis[4,10–12] and iodothyronine 5′-deiodinase by only 50% in rat liver.[10]

In theory, selenium could regulate selenoprotein mRNA abundance by affecting any step required for the biogenesis or stability of nuclear mRNA, the efficiency of nuclear mRNA export to the cytoplasm, or the stability of cytoplasmic mRNA. In view of the finding that the level of SECIS element-directed incorporation of Sec at a UGA codon is at most ~3% the level of amino acid incorporation at a standard codon,[13] it became evident that selenoprotein mRNAs could be natural targets for the mRNA surveillance pathway called nonsense-mediated decay (NMD).[14–17] For those that are natural targets, selenium deficiency would be expected to increase the extent of NMD even further by increasing the efficiency with which a UGA Sec codon is recognized as nonsense, that is, by increasing the premature termination of selenoprotein mRNA translation. Generally, the premature termination of translation results in mRNA degradation provided that an exon–exon junction resides more the 50–55 nucleotides downstream of the termination site.[18–24] Models accounting for the role of exon–exon junctions in NMD envision the process of splicing leaving one or more proteins near the junctions that, as constituents of messenger ribonucleo protein (mRNP), mediate NMD either directly or by recruiting other proteins.[18–21,23–28] Methods described below have been used to demonstrate that selenium deficiency elicits the NMD of cytoplasmic GPx1 mRNA without affecting other aspects of GPx1 RNA metabolism. These

[10] G. F. Bermano, F. Nicol, J. A. Dyer, R. A. Sunde, G. J. Beckett, J. R. Arthur, and J. E. Hesketh, *Biochem. J.* **311**, 425 (1995).

[11] G. Bermano, J. R. Arthur, and J. E. Hesketh, *FEBS Lett.* **387**, 157 (1996).

[12] X. Sun, X. Li, P. M. Moriarty, T. Henics, J. P. LaDuca, and L. E. Maquat, *Mol. Biol. Cell* **12**, 1009 (2001).

[13] H. Kollmus, L. Flohé, and J. E. G. McCarthy, *Nucleic Acids Res.* **24**, 1195 (1996).

[14] L. E. Maquat, *RNA* **1**, 453 (1995).

[15] L. E. Maquat, *in* "Translational Control of Gene Expression" (N. Sonenberg, J. W. B. Hershey, and M. B. Mathews, eds.), p. 849. Cold Spring Harbor Laboratory Press, Cold Spring Harbor, New York, 2000.

[16] S. Li and M. F. Wilkinson, *Immunity* **8**, 135 (1998).

[17] M. W. Hentze and A. E. Kulozik, *Cell* **96**, 307 (1999).

[18] J. Cheng, P. Belgrader, X. Zhou, and L. E. Maquat, *Mol. Cell. Biol.* **14**, 6317 (1994).

[19] M. S. Carter, S. Li, and M. F. Wilkinson, *EMBO J.* **15**, 5969 (1996).

[20] J. Zhang, X. Sun, Y. Qian, and L. E. Maquat, *RNA* **4**, 801 (1998).

[21] J. Zhang, X. Sun, Y. Qian, J. P. LaDuca, and L. E. Maquat, *Mol. Cell. Biol.* **18**, 5272 (1998).

[22] E. Nagy and L. E. Maquat, *Trends Biochem. Sci.* **23**, 198 (1998).

[23] R. Thermann, G. Neu-Yilik, A. Deters, U. Frede, K. Wehr, C. Hagenmeier, M. W. Hentze, and A. E. Kulozik, *EMBO J.* **12**, 3484 (1998).

[24] X. Sun, P. M. Moriarty, and L. E. Maquat, *EMBO J.* **19**, 4734 (2000).

[25] H. Le Hir, M. J. Moore, and L. E. Maquat, *Genes Dev.* **14**, 1098 (2000).

[26] H. Le Hir, E. Izaurralde, L. E. Maquat, and M. J. Moore, *EMBO J.* **19**, 6860 (2000).

[27] J. Lykke-Anderson, M.-D. Shu, and J. A. Steitz, *Cell* **103**, 1121 (2000).

[28] G. Serin, A. Gersappe, J. D. Black, R. A. Aronoff, and L. E. Maquat, *Mol. Cell. Biol.* **21**, 209 (2000).

methods can be applied to examine the possibility that other selenoprotein mRNAs are subject to NMD. They can also be used to investigate why some selenoprotein mRNAs appear to be immune to NMD.[12]

Rationale for Analyzing Transiently Expressed, *in Vitro*-Modified Glutathione Peroxidase 1 Alleles in Cultured Cells

The transient transfection of cultured cells with plasmids expressing GPx1 alleles provides a means to analyze GPx1 gene expression under a number of conditions that could lend insight into how the gene is normally regulated. For example, the GPx1 gene could be (1) expressed in different cell types to determine whether cell type-specific factors contribute to differences in expression, (2) mutated *in vitro* to test the function of particular nucleotides in GPx1 RNA production and metabolism, (3) driven by an inducible promoter that provides a means to measure GPx1 mRNA half-life without the use of inhibitors of transcription, or (4) coexpressed with putative modulators of the GPx1 gene.[8,24,29] Usually, transient transfections are performed in order to extend the limited analyses that can be undertaken with the endogenous gene of animals or cultured cells. As an example, studies that coupled reverse transcription (RT) and the polymerase chain reaction (PCR) to measure GPx1 RNA metabolism in the livers of rats fed a Se-adequate or Se-deficient diet revealed that selenium deficiency had no effect on the level of nuclear GPx1 pre-mRNA and nuclear GPx1 mRNA but reduced the level of cytoplasmic GPx1 mRNA to ~3% of normal.[8] These findings, together with studies by Bermano and co-workers[11] demonstrating that selenium deficiency reduces the stability of GPx1 mRNA in total liver RNA, indicated that selenium deficiency reduces the half-life of cytoplasmic GPx1 mRNA without affecting the metabolism of nuclear GPx1 RNA. Subsequently, transient transfections involving GPx1 alleles in which the TGA Sec codon had been mutagenized *in vitro* to either a TAA nonsense codon or a TGC cysteine codon were required to determine whether the reduction mediated by selenium deficiency is completely or partially Sec codon-dependent NMD.

Methods Used to Determine the Mechanism by Which Selenium Concentration Affects Glutathione Peroxidase 1 Gene Expression

Generating Plasmid-Based Gene Expression Vector

As described above, one *cis*-acting element that is generally required for NMD is a splicing-generated exon–exon junction. Therefore, studies of NMD necessitate

[29] X. Sun, H. A. Perlick, H. C. Dietz, and L. E. Maquat, *Proc. Natl. Acad. Sci. U.S.A.* **95,** 10009 (1998).

that the mRNA under analysis derive from the corresponding intron-containing gene rather than from cDNA.[21,30] The gene can be driven by its natural promoter, provided the promoter is active in the cells to be utilized, or one of the strong and constitutively active viral promoters, such as the cytomegalovirus (CMV) promoter. When using a heterologous promoter, the transcription start site should be as close to that of the gene under study as possible: for example, if an unnatural 5' untranslated region inhibits translation, it would also inhibit NMD.[31]

Mutagenesis

The single TGA Sec codon of the rat GPx1 gene can readily be converted to either a TAA nonsense codon or a TGC cysteine codon by using one of two methods: overlap-extension PCR,[32] or oligonucleotide-directed mutagenesis of single-stranded phagemids.[33,34] Each method should be designed to generate a mutation-bearing restriction fragment of the GPx1 gene that can be used to substitute for the corresponding fragment of the plasmid used to express the wild-type gene. Notably, the fragment should be sequenced in entirety after mutagenesis in order to ensure the presence of the desired mutation and the absence of undesired mutations.

Choosing Cell Type for Transient Transfection

Ideally, the cell type chosen should readily take up DNA by lipofection or some other easily performed transfection method. The cell type should also enable analysis of the transiently introduced test gene and reference gene (see below) without interference from endogenous cellular genes. In many cases, RNA blot hybridization using a DNA restriction fragment from the 3' untranslated region as probe provides a species-specific assay (i.e., way to differentiate exogenous transcripts from endogenous transcripts). If not, then it will be necessary to quantitate the exogenous transcript by using RT-PCR and PCR primers that anneal to sequences that have diverged between species. The simultaneous quantitation of test and reference transcripts provides a means to control for variations in transfection efficiency and RNA recovery between samples. This is possible only if the reference transcript is unaffected by selenium concentration. Globin (Gl) and major urinary protein (MUP) genes driven by a CMV promoter (see below) usually generate convenient reference transcripts because they are not expressed in most cell types.

[30] P. M. Moriarty, C. C. Reddy, and L. E. Maquat, *RNA* **3**, 1369 (1997).
[31] P. Belgrader, J. Cheng, and L. E. Maquat, *Proc. Natl. Acad. Sci. U.S.A.* **90**, 482 (1993).
[32] S. N. Ho, H. D. Hunt, R. M. Horton, J. K. Pullen, and L. R. Pease, *Gene* **77**, 51 (1989).
[33] D. E. Hill, A. R. Oliphant, and K. Struhl, *Methods Enzymol.* **155**, 558 (1986).
[34] T. A. Kunkel, J. D. Roberts, and R. A. Zakour, *Methods. Enzymol.* **154**, 376 (1987).

Transient transfections

Generally, cells are propagated in minimal essential medium α (MEM-α) containing 10% (v/v) fetal bovine serum (FBS; GIBCO-BRL, Gaithersburg, MD). One day before transfection, 1–1.5 × 10^6 cells per 100-mm dish are cultured in antibiotic-free medium containing 10% (v/v) FBS. After reaching 60–70% confluency, the cells are transfected with, for example, LipofectAMINE PLUS reagent (GIBCO-BRL) according to the manufacturer directions. After transfection, cells are cultured in either selenium-deficient or selenium-supplemented medium. Selenium-deficient medium consists of a 1 : 1 (v/v) mixture of Ham's F12 (GIBCO-BRL) and Dulbecco's modified Eagle's medium (DMEM) plus bovine holotransferrin (25 μg/ml), bovine insulin (10 μg/ml), mouse epidermal growth factor (10 ng/ml), and human high-density lipoprotein (25 μg/ml). Selenium-supplemented medium consists of selenium-deficient medium plus sodium selenite (50 ng/ml). For experiments that do not vary selenium concentration, cells are cultured in MEM-α containing 10% (v/v) FBS.

Isolation of Nuclear and Cytoplasmic RNA

Cells from four to six 100-mm plates per transfection mix are harvested, pooled, pelleted at 2500 rpm and 4° for 10 min in a Sorvall HS-4 rotor, and suspended in 5 ml of ice-cold 25 mM citric acid in 1% (v/v) Triton X-100.[35] After swelling on ice for 5 min, samples are homogenized by 10 strokes of a type B Dounce homogenizer, layered over a 5-ml pad of 0.88 M sucrose in 25 mM citric acid, and centrifuged at 2000 rpm in an SW41 rotor at 4° for 5 min. To prepare cytoplasmic RNA, the solution above the cushion is transferred to a new tube. After adding 5X NETS [50 mM Tris-HCl (pH 7.4), 500 mM NaCl, 2.5% (w/v) sodium dodecyl sulfate (SDS), 25 mM EDTA] to a final concentration of 1×, the solution is extracted sequentially with phenol, phenol–CIA [24 : 1(v/v) chloroform–isoamyl alcohol], and CIA, and the aqueous phase of the last extraction is removed and mixed with 1.1 g of CsCl$_2$ per 2.5 ml of sample. To prepare nuclear RNA, the pellet is suspended in 5 ml of ice-cold 0.25 M sucrose in 25 mM citric acid, layered over a 5-ml pad of 0.88 M sucrose in 25 mM citric acid, centrifuged at 2000 rpm in an SW41 rotor at 4° for 5 min, and suspended in 6 ml of guanidinium isothiocyanate (GITC) by passing it through an 18-gauge needle at least 40 times. Each cytoplasmic and nuclear RNA preparation is layered over 5 ml of 5.8 M CsCl$_2$ in 0.5 M EDTA (pH 7.6) and centrifuged in an SW41 rotor at 30,000 rpm and 23° for 18 hr. RNA pellets are dissolved in water (after carefully pipetting off the DNA-containing gradients that form above the pellets), precipitated in ethanol, dissolved in 100 μl of water, and quantitated by measuring A_{280}/A_{260}.

[35] G. D. Birnie, *in* "Nuclear Structures: Isolation and Characterization" (A. J. Macgillivray and G. D. Birnie, eds.), p. 13. Butterworths, London, 1986.

Semiquantitative Reverse Transcription-Polymerase Chain Reaction

RT-PCR is used when RNA blot hybridization, a simpler and quantitatively more reliable method, cannot be used. In preparation for RT-PCR, RNA (10 μg) is treated with RQ1 DNase I (2 U; Promega, Medison, WI) in 20 μl of reaction buffer [40 mM Tris-HCl (pH 7.9), 10 mM NaCl, 6 mM MgCl$_2$, 10 mM CaCl$_2$] at 37° for 1 hr. The RNA is then extracted sequentially with phenol, phenol–CIA, and CIA, subsequently precipitated with ethanol, and finally dissolved in 10 μl of water. RNA (2.5 μg) in 10 μl of water is boiled and quenched on ice–water.

Reverse Transcriptase Cocktail for cDNA Synthesis

Random hexamers (0.5 μg/μl; Promega)	0.8 μl
First-strand buffer (BRL, Gaithersburg, MD), 5×	4 μl
dNTPs, 5 mM	2 μl
RNasin (40 U/μl; Promega)	0.3 μl
Dithiothreitol 0.1 M	2 μl
SuperScript II RT (200 U/μl; BRL)	1 μl

For cDNA synthesis, RT cocktail (10 μl) is added to 10 μl of each RNA sample, incubated at 37° for 2 hr, and then inactivated at 95° for 5 min.

Polymerase Chain Reaction Cocktail for cDNA Amplification

PCR buffer (Promega), 10×	5 μl
MgCl$_2$, 25 mM	3 μl
Sense and antisense primers to amplify test RNA (100 pmol; 1 μg/μl)	0.4 μl of each
Sense and antisense primers to amplify reference RNA (100 pmol; 1 μg/μl)	0.4 μl of each
dATP, dCTP, dGTP, and dUTP, 5 mM each	0.4 μl of each
[^{32}P]dATP (3000 Ci/mmol; Amersham, Arlington Heights, IL)	1.2 μl
Taq DNA polymerase (Promega; 5 U/μl) 0.4 μl	
Water	to 42 μl

For PCR, PCR cocktail (42 μl) is added to each 6–8 μl of RT reaction.

To ensure a quantitative analysis, at least three dilutions of an RNA sample that will generate RT-PCR products having intensities above and below the intensities of test and reference RNAs should be included. The use of 0.75, 1.25, 2.5, and 5 μg of RNA is often adequate. PCR usually involves 19 cycles of three sequential steps: 94° for 30 sec, 55–60° for 30 sec, and 72° for 1 min. A fraction of each RT-PCR sample (10 μl) is electrophoresed in a 5% (w/v) native gel, and RT-PCR products are quantitated by PhosphorImaging (Molecular Dynamics, Sunnyvale, CA) or some equivalent method (Fig. 1).

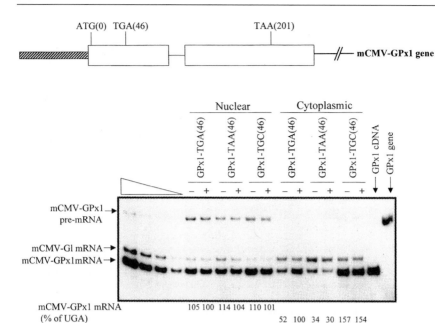

FIG. 1. *Top:* Structure of the mCMV-GPx1 gene. The shaded box represents the mCMV promoter, open boxes represent GPx1 exons, the intervening line represents the single GPx1 intron, and the rightmost line represents 3′-flanking DNA. ATG(0), TGA(46), and TAA(201) specify, respectively, the initiation, Sec, and termination codon. *Bottom:* NIH 3T3 cells were transfected with the specified pmCMV-GPx1 test plasmid and the pmCMV-Gl reference plasmid and, subsequently, propagated in selenium-deficient (−) or selenium-supplemented (+) medium. Nuclear and cytoplasmic RNAs were purified, and RT-PCR was used to quantitate mCMV-GPx1 and mCMV-Gl transcripts. The levels of nuclear mCMV-GPx1 pre-mRNA, nuclear mCMV-GPx1 mRNA, and cytoplasmic mCMV-GPx1 mRNA were normalized to the level of mCMV-Gl mRNA. The ratio of normalized nuclear mCMV-GPx1 mRNA to normalized nuclear mCMV-GPx1 pre-mRNA or the level of normalized cytoplasmic mCMV-GPx1 mRNA in selenium-deficient cells was then calculated as a percentage of the corresponding normalized value for the TGA-containing allele in selenium-supplemented cells, which was defined as 100. The level of nuclear mCMV-GPx1 mRNA was always calculated relative to nuclear mCMV-GPx1 pre-mRNA because the amount of pre-mRNA from each allele varied (for reasons not understood) in a way that was unaffected by selenium and not relevant to the purpose of the studies. The leftmost four lanes contain 2-fold serial dilutions of RNA from selenium-supplemented cells and demonstrate a linear relationship between the amounts of input RNA and RT-PCR products. The two rightmost lanes, which analyze intronless and intron-containing pmCMV-GPx1 DNA by using PCR, provide molecular weight standards for GPx1 pre-mRNA and mRNA, respectively.

hUpf1p (R844C) Assay for NMD

NMD in mammalian cells requires Upf1 protein (p), a group 1 RNA helicase.[29] Expression of human (h) hUpf1p harboring arginine instead of cysteine at position 844 (R844C) has a dominant-negative effect on NMD[29] and provides a general

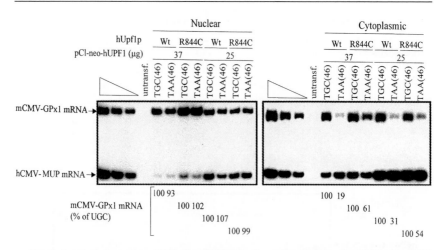

FIG. 2. Transient expression of hUpf1p (R844C) in Cos cells abrogates the NMD of mCMV-GPx1(UAA) mRNA. Cos cells were transiently transfected with the pmCMV-GPx1 test plasmid [either TGC(46) or TAA(46)], the phCMV-MUP reference plasmid (which harbors the mouse major urinary protein gene), and pCI-neo-hUPF1 [either wild type (Wt), which harbors an unmutagenized hUpf1p reading frame, or (R844C), which harbors the arginine-to-cysteine change at amino acid position 844]. The amounts of the three plasmids were, respectively, 10, 3, and 37 μg or 19, 3, and 25 μg. An appropriate amount of pUC13 DNA was added to each mixture to bring the total amount of DNA to 50 μg. RNA was purified from nuclear and cytoplasmic fractions and analyzed by RT-PCR and PhosphorImaging. The level of mCMV-GPx1 mRNA from each allele was normalized to the level of MUP mRNA. Normalized values were then calculated as a percentage of the normalized value of mCMV-GPx1(UGC) mRNA in the presence of either the Wt or hUPF1 (R844C) gene, each of which was defined as 100.

assay for NMD. For example, if a particular selenoprotein gene generates mRNA that is a natural substrate for NMD, then the abundance of that mRNA should be increased when a plasmid expressing hUpf1p R844C is transiently introduced into mammalian cells. If the gene has yet to be isolated but is expressed in cultured cells, then the hUpf1p R884C expression plasmid can be stably introduced into the cell genome and assayed for an effect on mRNA produced by the gene. Notably, the extent to which hUpf1p R844C abrogates NMD is incomplete. For example, relative to the level of cytoplasmic mCMV-GPx1(UGC) mRNA, the level of cytoplasmic mCMV-GPx1(UAA) mRNA is 19% when wild-type hUpf1p (i.e., protein without a dominant-negative effect on NMD) is coexpressed but only 54–61% (rather than 100%) when hUpf1p R844C is expressed (Fig. 2).

Summary

If the abundance of a particular selenoprotein mRNA is reduced during selenium deprivation, then the mRNA is likely to be a natural substrate for NMD. One assay for NMD involves changing the TGA Sec codon(s) to either a TGC cysteine codon or a TAA nonsense codon. If selenium deprivation elicits NMD and has no other effect on selenoprotein gene expression, then, regardless of selenium concentration, the level of UGC-containing mRNA should be most abundant, the level of UGA-containing mRNA should be intermediate in abundance, and the level of UAA-containing mRNA should be least abundant. Furthermore, the level of UGA-containing mRNA should be decreased by a decrease in selenium concentration, while the levels of UGC- and UAA-containing mRNAs should be unaffected by selenium concentration. A different assay for NMD involves coexpression of the particular selenoprotein gene and a vector expressing a dominant-negative version of hUpf1p. This assay is simpler and more versatile than the first assay because it can be used to assay any cellular gene *in situ* provided the cells can be stably transfected with the hUpf1p expression vector.

Acknowledgment

Work in the Maquat laboratory is supported by Public Health Service Research Grants DK 33933 and GM 59614 from the National Institutes of Health.

[6] Novel Selenoproteins Identified from Genomic Sequence Data

By Alain Lescure, Daniel Gautheret, and Alain Krol

Introduction

Selenocysteine is encoded by an in-frame UGA codon. Distinction of UGA stop from UGA selenocysteine codons is mediated by the selenocysteine insertion sequence (SECIS) element, an RNA stem–loop structure residing in the 3′ untranslated region (UTR) of eukaryotic selenoprotein mRNAs.

In the postgenomics era, computer-assisted searching is an invaluable tool for the identification of new proteins. However, selenoproteins constitute a particular case because identification of the relevant open reading frames (ORFs) is obviously more challenging than for other cDNAs where the UGA codon strictly signals the end of the message. Therefore, one can surmise that, in some cases, UGA codons in nucleic acid databases have been misinterpreted, meaning that the coding region could be extended past the supposedly UGA stop. Because the SECIS element is mandatory for selenocysteine incorporation, it constitutes a hallmark of selenoprotein-encoding cDNAs. We thus propose the use of a program capable of identifying selenoprotein-encoding cDNAs by searching for the SECIS secondary structure in databases. We previously applied this program to the identification of novel selenoproteins.[1] In a second step, the selenocysteine-encoding ability of the relevant cDNAs must be confirmed by the experiment.

Computational Detection of Selenocysteine Insertion Sequence Motifs

Problems of RNA Detection

A number of highly efficient tools are available for the identification of protein motifs in sequence databases. Arguably, the most important reason for this success has been the development of sophisticated amino acid substitution models such as those implemented in the PAM (percent accepted mutations) matrix and BLOSUM (blocks substitution matrix). Used in conjunction with appropriate sequence alignment programs, these models allow for accurate detection of protein sequence signatures, even under conditions of poor sequence conservation. The field of RNA detection lies, in comparison, in a rather primitive stage. First, nucleotide bases carry little structural information compared with amino acid residues and, second, the structure (hence the function) of an RNA molecule is in large part defined by distant interactions rather than by the linear sequence. Therefore, neither sophisticated substitution models nor the prevailing sequence alignment procedures can be applied to RNA detection. This is especially true for the SECIS element, which possesses a rather conserved secondary structure, but little sequence conservation.[2,3]

RNA Search Methods

Currently available methods for RNA motif identification can be classified into three categories. In the first category are programs especially developed for the detection of a specific RNA molecule, such as tRNA. Detection efficiency in this approach will strongly depend on the ability of the original author to translate structural knowledge into computer code. Once this conversion is performed, the resulting program is not easily updated to incorporate new knowledge. This explains why the method has been confined to tRNA,[4] a molecule of well-established sequence properties. The next category of program allows biologists to describe virtually any type of RNA motif, using a special descriptor language. Several languages and corresponding search engines have been developed.[5–9] All provide convenient syntactic structures for describing the usual RNA building blocks,

[1] A. Lescure, D. Gautheret, P. Carbon, and A. Krol, *J. Biol. Chem.* **274,** 38147 (1999).

[2] R. Walczak, E. Westhof, P. Carbon, and A. Krol, *RNA* **2,** 367 (1996).

[3] D. Fagegaltier, A. Lescure, R. Walczak, P. Carbon, and A. Krol, *Nucleic Acids Res.* **28,** 2679 (2000).

[4] G. Fichant and C. Burks, *J. Mol. Biol.* **220,** 659 (1991).

[5] D. Gautheret, F. Major, and R. Cedergren, *Comput. Appl. Biosci.* **6,** 325 (1990).

namely helices and single-stranded elements, with optional specification of conserved sequences, base mismatches, and length intervals. These versatile software programs, notably RNAMOT[5,6] and PatScan,[9] are commonly used in RNA motif searches. However, they are limited by their requirement for good prior knowledge of secondary structure and sequence constraints. Insufficient input constraints result in large numbers of false-positive hits. On the other hand, constraints should not be overly tightened if new instances of the molecule, with possible variations, are to be discovered. A correct balance between specificity and sensitivity is better achieved by using a statistical model of the RNA sequences under study.

The third category of programs, based on stochastic context free grammars (SCFGs) is an attempt to derive such a statistical model automatically from sequence data. Context free grammars (CFGs) describe objects in the form of a set of production rules representing the expected structural elements. New instances of the grammar are then identified by a parsing algorithm. Stochastic CFGs[10,11] are an evolution of this model associating probabilities to production rules, which permits evaluation of the significance to any new instance. SCFGs have been successful in helping to identify new small nucleolar RNAs (snoRNAs),[12] but have not been applied to a SECIS search yet.

Mining for Selenocysteine Insertion Sequence Elements

Candidate SECIS elements are sought in the 3′ untranslated regions (UTRs) of RNAs. The most comprehensive source for 3′ UTR sequences is the dbEST database,[13] the division of GenBank containing "single-pass" cDNA sequences, or expressed sequence tags, from a number of organisms. Release 121500 (December 2000) of dbEST contains about 2,800,000 human and 1,900,000 mouse EST sequences. ESTs are highly redundant, with a total number of sequences much higher than the estimation of 30,000 human genes. To avoid needless searches, gene indices, such as the Institute for Genomic Research (TIGR, Rockville, MD) Human Gene Index (HGI) can be used.[14] Release 6.0 of the HGI (June 30, 2000) contains 388,000 tentative human contigs and mRNA sequences, much higher than the upper bound for gene counts due to the presence of partial sequences and

[6] A. Laferrière, D. Gautheret, and R. Cedergren, *Comput. Appl. Biosci.* **10**, 211 (1994).

[7] B. Billoud, M. Kontic, and A. Viari, *Nucleic Acids Res.* **24**, 1395 (1996).

[8] G. Pesole, S. Liuni, and M. D'Souza, *Bioinformatics* **5**, 439 (2000).

[9] R. Overbeek, PatScan web server. (2000): http://www-unix.mcs.anl.gov/compbio/PatScan/HTML/patscan.html.

[10] Y. Sakakibara, M. Brown, R. Hughey, I. S. Mian, K. Sjolander, R. C. Underwood, and D. Haussle, *Nucleic Acids Res.* **22**, 5112 (1994).

[11] S. Eddy and R. Durbin, *Nucleic Acids Res.* **22**, 2079 (1994).

[12] T. M. Lowe and S. Eddy, *Science* **283**, 1168 (1999).

[13] M. S. Boguski, T. M. Lowe, and C. M. Tolstoshev, *Nat. Genet.* **4**, 332 (1993).

[14] J. Quackenbush, F. Liang, I. Holt, G. Pertea, and J. Upton, *Nucleic Acids Res.* **28**, 141 (2000).

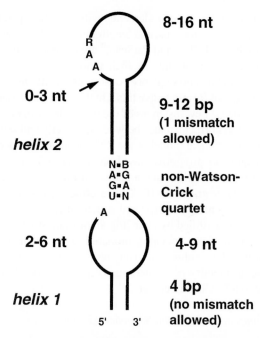

FiG. 1. Consensus secondary structure for the SECIS element, derived from R. Walczak, E. Westhof, P. Carbon, and A. Krol, *RNA* **2**, 367 (1996). The sequence and structure constraints were used to establish the descriptor shown in Fig. 2. R, Purine; B, any nucleotide, except A; N, any nucleotide. Shown are the lengths allotted to single strands (nucleotides, nt) and helices (bp). [Reproduced with permission from A. Lescure, D. Gautheret, P. Carbon, and A. Krol, *J. Biol. Chem.* **274**, 38147 (1999).]

multiple alternate mRNA forms. A more specialized UTR databank (UTRdb[15]) has been compiled from the EMBL databank. Release 12.0 of UTRdb contains 46,881 nonredundant eukaryotic 3′ UTRs (10,207 human). This data set does not incorporate EST-derived sequences and is thus more a control set than a source for gene discovery. In spite of their large size (HGI, 150 Mb; dbEST human, 1,240 Mb) assembled or full cDNA databases can be scanned entirely in reasonable time with current descriptor-based RNA motif search programs.

Describing Selenocysteine Insertion Sequence Signature

Detailed primary and secondary structure constraints on SECIS elements have been obtained from enzymatic/chemical probing, site-directed mutagenesis, and sequence comparisons of a number of SECIS elements.[2,16,17] These constraints are summarized in the consensus structure shown in Fig. 1.[1,2] Salient features include helix 2 with four consecutive non-Watson–Crick base pairs containing a central

[15] G. Pesole, S. Liuni, G. Grillo, F. Licciulli, A. Larizza, W. Makalowski, and C. Saccone, *Nucleic Acids Res.* **28**, 193 (2000).

```
h1 s1 s2 h2 s3 s4 s5 h2 s6 s7 h1

h1 4:4 0
h2 9:12 1
s1 2:6
s2 5:5 AUGAN
s3 0:3
s4 3:3 AAR
s5 8:16
s6 4:4 BGAN
s7 4:9

W 4
M 1
```

FIG. 2. An RNAMOT descriptor for SECIS. The first line describes the organization of secondary structure elements (h, helices; s, single strands). The following lines provide constraints on each element, such as minimal and maximal size, allowed number of mismatches (for helices), and sequence requirements (using the standard IUPAC/IUB code). The last two lines indicate the total number of wobble pairs (W) and mismatched pairs (M) permitted within a complete solution. Other symbols are as in Fig. 1.

G · A tandem, and the shorter helix 1. Loops have variable lengths and sequences, except for the conserved AAR (R, purine) at the 5′ end of the apical hairpin and an unpaired A/G,[3,18] 5′ to the non-Watson–Crick base pair quartet. These features can be easily expressed using most RNA description languages. An RNAMOT[5,6] descriptor for SECIS is shown as an example in Fig. 2. A number of SECIS elements possess an additional helix 3, giving rise to form 2 versus form 1 SECIS.[3,19] However this additional helix was not considered in the descriptor because it is not a crucial discriminant for the computational screen.

Database Searches

In analyzing the collection of SECIS elements available in 1999, we derived the primary/secondary signature shown in Fig. 1 and translated it into the RNAMOT descriptor language, as shown in Fig. 2. RNAMOT[5,6] was then used to find occurrences of the SECIS element in a database of 600,300 3′ EST sequences (222 Mb). We identified 376 candidates (153 mouse, 101 human, 92 *Brugia malayi*, 30 others), most of which were redundant. We then used multiple sequence alignments to aggregate redundant sequences and obtain a reduced list of candidates.

[16] R. Walczak, P. Carbon, and A. Krol, *RNA* **4**, 74 (1998).
[17] G. W. Martin, J. W. Harney, and M. J. Berry, *RNA* **4**, 65 (1998).
[18] C. Buettner, J. W. Harney, and M. J. Berry, *J. Biol. Chem.* **274**, 21598 (1999).
[19] E. Grundner-Culemann, G. W. Martin, J. W. Harney, and M. J. Berry, *RNA* **5**, 625 (1999).

Finally, we discarded low-complexity sequences (i.e., AU or GC rich) prone to secondary structure artifacts, as well as structures with multiple consecutive G · U base pairs. This further reduced the number of sequences to be tested experimentally to 21, as described in Lescure et al.[1] Kryukov et al.[20] used an ad hoc program adapted from PatScan[9] and identified in a first stage 32,652 SECIS candidates in dbEST (1,253,000 sequences at this time). In a second stage, the folding free energies of putative hits were computed on the basis of standard thermodynamic parameters. Solutions were retained only if their stability was better than −7 kcal/mol for helix 1 and internal loop, and −5 kcal/mol for helix 2 and apical loop. This free energy cutoff, derived from measures of known SECIS elements, reduced the number of selected SECIS candidates by more than 6-fold (to 974), increasing the proportion of known selenoprotein genes in the final result from 10 to 70%.

All the laboratories that performed computational screens for SECIS elements concluded that their computer programs and descriptors lacked specificity. On the basis of the analysis of random sequences, we evaluated the number of false-positive hits in our procedure to 3 per 10 Mb, which is significant compared with the number of solutions. Pesole et al.[8] have presented an RNA pattern search algorithm similar to those implemented in RNAMOT or PatScan and applied it to SECIS and other RNA motifs. Interestingly, the SECIS motif is the least selective of all, with 65% of hits in UTRdb not attributed to known selenoprotein UTRs. Likewise, Kryukov et al.[20] acknowledged that their procedure was not efficient in finding new selenoprotein mRNAs exclusively, and that identification of additional cis-acting elements may be necessary for selenoprotein gene sequence recognition. For these reasons, additional screens have been sought before entering the lengthy and expensive experimental phase. Kryukov et al.[20] introduced the criteria that homologous mammalian genes should also have SECIS elements in their 3′ UTR and similar selenocysteine-flanking sequences in their coding regions. Although the number of false positives may suggest that a few selenoprotein genes remain to be discovered, it also prompts for a wider use of statistical tools in RNA identification, such as SCFGs. Only this type of approach will provide the scoring information necessary to properly discriminate candidates for SECIS function.

Functional Screen of Selected Selenocysteine Insertion Sequence Hits

General Principle

The selenocysteine insertion activity of the computer-selected SECIS candidates can be assayed in vivo by using a suitable reporter system. The DNAs of the SECIS candidates are obtained by polymerase chain reaction (PCR) amplification

---

[20] G. V. Kryukov, V. M. Kryukov, and V. N. Gladyshev, J. Biol. Chem. **274**, 33888 (1999).

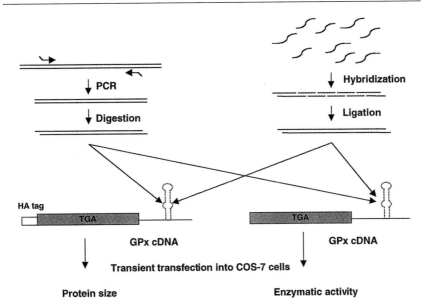

FIG. 3. Cloning strategy to construct reporter cDNAs aiming at testing SECIS activity *in vivo*. The reporter construct is the HA-tagged (or untagged) glutathione peroxidase (GPx) cDNA. The residing SECIS element is replaced, in either type of reporter construct, by the individual SECIS candidate sequences, obtained by PCR of genomic DNA or oligodeoxynucleotide cassette constructions. Western blotting with the anti-HA antibody (protein size), or determination of the GPx enzymatic activity, is used to assay the function of the SECIS candidates (displayed as the stem–loop structure).

of genomic DNA, or constructed with oligonucleotide cassettes. The DNAs are subsequently introduced in place of the residing SECIS element into a reporter cDNA encoding a selenoprotein (Fig. 3). If the introduced SECIS element is functional, then it will promote readthrough of the UGA selenocysteine (Sec) codon and synthesis of a full-length protein. In contrast, insertion of an inactive SECIS sequence will lead to premature termination of translation at the UGA Sec codon read as a stop. The sizes of the full-length and truncated polypeptides can be distinguished by Western blot analysis, taking advantage of a tagged selenoprotein reporter. In addition, because selenocysteine is involved in the catalytic activity of most selenoenzymes, its specific insertion into the protein can be monitored by an enzymatic assay. In our case, we have used the selenoprotein glutathione peroxidase (GPx) as the reporter.[1]

Step 1: Introduction of Selenocysteine Insertion Sequence Candidates into Reporter cDNA

The candidate sequences are amplified by PCR from genomic or cDNA libraries (starting with 1 μg of DNA), using selected primer pairs. PCR conditions should be adjusted to each specific sequence. Alternatively, the SECIS candidates are constructed by hybridization and ligation of overlapping oligonucleotides.[16] The oligonucleotides are designed to add restriction sites at each end of the sequence, which permits oriented insertion into the reporter cDNAs. Engineered by site-directed mutagenesis, the same restriction sites flank the residing SECIS element in the 3′ UTR of the reporter cDNA, allowing easy swapping of the SECIS candidates to be analyzed. Two versions of the reporter constructs are made that give rise to different N-terminal proteins: one carries an N-terminal tag to allow determination of the size of the protein by immunodetection, the untagged version being used to assay the enzymatic activity of the glutathione peroxidase (Fig. 3). Both constructs are cloned into a eukaryotic expression vector.

Step 2: Transient Expression of Reporter cDNA in Mammalian Cells

Expression of the GPx reporter protein is assayed by transfection of the cDNA into COS-7 cells by a modification of the calcium phosphate precipitation method of Jordan *et al.*[21] COS-7 cells are cultured in Dulbecco's modified Eagle's medium (DMEM) supplemented with 10% (v/v) fetal bovine serum, 2 mM L-glutamine, and gentamicin (0.1 mg/ml) at 37° under 5% CO_2 saturation, according to standard cell culture procedures. Transfections are generally performed in triplicate. The day before transfection, 3×10^5 cells are plated onto 100-mm culture dishes in 6 ml of medium. Immediately before transfection, cells are rinsed twice with 6 ml of TBS-CM [25 mM Tris-HCl (pH 7.4), 750 μM Na_2HPO_4, 150 mM NaCl, 5 mM KCl, 1 mM $CaCl_2$, 0.5 mM $MgCl_2$] and fresh medium containing 10 nM sodium selenite is added. Cells are transfected with a DNA mix containing the plasmid to be tested, a selenocysteine tRNA expression vector, and a plasmid encoding β-galactosidase to monitor transfection efficiency. The total amount of DNA is adjusted to 10 μg with an empty DNA plasmid.

    DNA mix

Test DNA	4 μg
tRNA^{Sec} expression vector	2 μg
Cytomegalovirus (CMV)-LacZ, the β-galactosidase expression vector	1 μg
pBluescript, to make up the total amount to 10 μg of DNA	3 μg
TE 1/10 [1 mM Tris-HCl (pH 8), 0.1 mM EDTA]	440 μl
$CaCl_2$, 2.5 M	50 μl

Add, drop by drop, 500 μl of DNA mix into 500 μl of buffer HBS 2×[50 mM HEPES–NaOH (pH 7), 140 mM NaCl, 1.5 mM NaH_2PO_4]. The precipitate forms

[21] M. Jordan, A. Schallhorn, and F. M. Wurm, *Nucleic Acids Res.* **24**, 596 (1996).

within 1 min. Add rapidly the precipitate to the culture plate and homogenize. Sixteen hours posttransfection, the cells are washed twice in TBS-CM, and fresh medium supplemented with 10 nM sodium selenite is added. Twenty-four hours later, cells are washed twice in TBS-CM, harvested by scraping in 1 ml of TBS-CM, centrifuged, and resuspended in 50 μl of 100 mM Tris-HCl, pH 8. Lysis is carried out by four freeze–thaws in liquid nitrogen. The crude cell extract is then centrifuged at 4° for 5 min at 13,000g to remove cell debris. The supernatant is used for subsequent analysis and can be stored at -80°.

The transfection efficiency is normalized to the β-galactosidase activity. For this purpose, 5 μl of cell extract is added to 500 μl of buffer Z [100 mM phosphate buffer (pH 7.4), 10 mM KCl, 1 mM MgSO$_4$, 50 mM 2-mercaptoethanol], 100 μl of o-nitrophenyl-β-D-galactopyranoside [0.4% (w/v) in 100 mM phosphate buffer, pH 7.4] is added, and the mixture is incubated at 37° for 10–20 min until a faint yellow color has developed. Stop the reaction by addition of 500 μl of 1 M Na$_2$CO$_3$. The OD$_{420}$ is measured versus a standard performed without cell extract.

Step 3: Determination of Length of Reporter Protein by Western Blotting

An amount of cell extract corresponding to a given number of β-galactosidase units is fractionated on sodium dodecyl sulfate (SDS)–12% polyacrylamide gels. Proteins are transferred onto an Immobilon-P polyvinylidene difluoride (PVDF) membrane (Millipore, Bedford, MA) and then blocked for 1 hr in TBS 1× [0.5 M Tris-HCl (pH 8), 1.5 M NaCl], 5% (w/v) nonfat dry milk, and 0.075% (v/v) Tween 20. If a hemagglutinin (HA)-tagged GPx is used as the reporter, the membrane is incubated for 1 hr at room temperature with a 1 : 5000 dilution of the anti-HA antibody (12CA5; Roche, Meylan, France) in the same TBS buffer. The membrane is rinsed three times in TBS 1×–0.075% (v/v) Tween 20 for 15 min, and then incubated for another 1 hr with a secondary goat anti-mouse antibody conjugated to horseradish peroxidase, diluted 1 : 10,000 in TBS 1×–5% (w/w) nonfat dry milk–0.05% (v/v) Tween 20. After three additional washes in TBS 1×–0.05% (v/v) Tween 20 for 15 min, the proteins are revealed with an enhanced chemiluminescence (ECL) kit (Amersham, Bucks, England).

Step 4: Glutathione Peroxidase Assay

The GPx enzymatic activity can be measured by monitoring the oxidation of NADPH to NADP$^+$ in a glutathione reduction assay.[22] Gently mix 2 ml of GPx reaction buffer [100 mM Tris-HCl (pH 7.4), 5 mM EDTA, 0.1% (v/v) Triton X-100] with 20 μl of 20 mM NADPH, 20 μl of 300 mM glutathione, 7 μl of glutathione reductase (0.7 U), and 50 μl of cell extract. Divide the mix into two 1-ml volume and incubate at 37° for 2 min. One of the aliquots (the blank) is further

---

[22] A. Roveri, M. Maiorino, and F. Ursini, *Methods Enzymol.* **233,** 202 (1994).

incubated in the absence of *tert*-butyl hydroperoxide and therefore measures the GPx-independent NADPH oxidation in the extract. For the other aliquot, the reaction is initiated by adding 20 μl of 7.3 M *tert*-butyl hydroperoxide. Oxidation of NADPH is monitored at 340 nm in a double-beam spectrophotometer for 5 min at 37°. Data are normalized to the β-galactosidase (β-Gal) activity and subjected to linear regression. The slope of the curve, yielding the number of micromoles of NADPH oxidized per minute, is used as a measure of the GPx activity. Assays are performed from three to six transfection experiments for each test.

The low endogenous GPx activity arising from mock-transfected cell extracts is subtracted from the activity of the transfected constructs.

Alternative Methods

Other methods than using a selenoprotein cDNA can be employed to monitor the activity of the SECIS candidates. For example, use can be made of a construct containing the luciferase cDNA carrying an engineered UGA selenocysteine codon.[23] Another possibility could be the utilization of the construct harboring the β-Gal and luciferase cDNAs fused in frame, separated by a UGA codon.[24] If the SECIS element is active, the UGA codon will be read as selenocysteine, enabling detection of both the β-gal and luciferase activities. These methods are less tedious but less sensitive than measuring actual GPx activities.

Obtaining and Analyzing Selenoprotein cDNA Sequences

If the full-length cDNA carrying the selected SECIS candidate is available in databases or by sequencing, inspection of its sequence will provide clues as to its ability to encode a selenoprotein, because an in-frame UGA Sec codon is the hallmark of a selenoprotein cDNA. Another feature is that the SECIS element must be located within the 3' untranslated region, at least 60 nucleotides downstream of the stop codon. Although no exception to this rule has been observed so far, we cannot exclude the possibility that SECIS elements could be found at other positions than the 3' UTR, especially in viruses with highly compact genomic organization.

When the full cDNA sequence is not directly available, it might be possible to reconstruct it from sequences present in EST databases. Sequences contiguous to the SECIS DNA can be identified by iterative search for overlapping fragments, using the BLASTn program. The sequences obtained are then aligned and assembled into one continuous cDNA sequence, which is used for subsequent sequence analysis.

---

[23] A. Lesoon, A. Mehta, R. Singh, G. M. Chisolm, and D. M. Driscoll, *Mol. Cell. Biol.* **17**, 1977 (1997).
[24] H. Kollmus, L. Flohé, and J. E. G. McCarthy, *Nucleic Acids Res.* **24**, 1195 (1996).

Further Characterization of Selenoproteins

Once the coding sequence linked to the SECIS element has been found, its ability to encode a selenoprotein must be established. Two criteria can be used.

Selenoprotein Synthesis Dependent on Presence of Selenocysteine Insertion Sequence Element

As UGA Sec readthrough is dependent on the presence of the SECIS element, its deletion should lead to a premature stop of translation at the selenocysteine codon, generating a shorter polypeptide. To evaluate the SECIS dependency, a sequence tag is introduced at the N terminus of the protein by site-directed mutagenesis. This construct, including or not including the SECIS element, is cloned into a eukaryotic expression vector. Each construct is transfected into COS-7 cells, as described previously. The amount of DNA to be transformed depends on the expression level of each protein, and should therefore be adapted to each case. After transfection, the cell extract is fractionated on SDS–polyacrylamide gels, and proteins are revealed by Western blotting (Fig. 4A),[1,16] as described above. It should be noticed that transfection conditions usually lead to saturation of the selenocysteine insertion machinery, explaining the appearance of the shorter polypeptide (with a variable intensity), even in presence of the SECIS element. Figure 4A shows that deletion of the SECIS element provokes an intensification of the band containing the shorter polypeptide product. In addition, under certain conditions of transfection, a faint amount of full-size protein can appear despite the absence of the SECIS element. This is due to nonselenocysteine readthrough of the UGA Sec codon.[1,25]

[75]Se Labeling of Selenoproteins

The actual incorporation of selenium into the identified protein can be confirmed *in vivo* by radiolabeling with [75]Se. COS-7 cells are transfected as described above with the expression vector carrying the selenoprotein-encoding cDNA. Sixteen hours later, the cells are washed twice with 6 ml of TBS-CM and incubated in fresh medium containing 5–10 μCi of Na_2[75]SeO_3 (2400–2600 Ci/g; University of Missouri Research Reactor) for 24 to 36 hr. Cells are harvested and lysed as described above. To avoid the presence of contaminating bands due to the labeling of endogenous selenoproteins, and to facilitate the detection of the ^{75}Se-labeled products, the protein of interest can be tagged and thus immunoprecipitated (Fig. 4B).

[25] M. J. Guimaraes, D. Peterson, A. Vicari, B. G. Cocks, N. G. Copeland, D. J. Gilbert, N. A. Jenkins, D. A. Ferrick, R. A. Kastelein, J. F. Bazan, and A. Zlotnik, *Proc. Natl. Acad. Sci. U.S.A.* **93**, 15086 (1996).

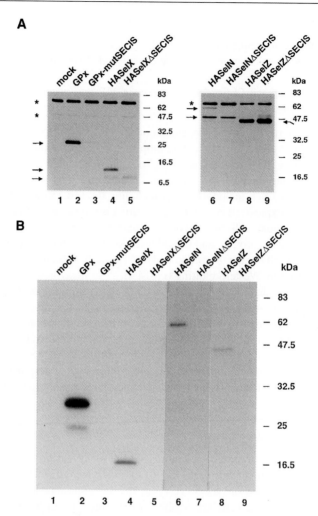

FIG. 4. Two different ways to analyze the coding activity of putative selenoprotein cDNAs. (A) The various HA-tagged proteins are revealed by Western blotting with anti-HA antibody. Mock, untransfected COS-7 cells; SelX, SelN, and SelZ, novel selenoproteins.[1] GPx-mutSECIS is the construct containing a deleterious point mutation in the SECIS element, leading to abolition of GPx synthesis [R. Walczak, P. Carbon, and A. Krol, *RNA* **4,** 74 (1998)]. HASelXΔSECIS, HASelNΔSECIS, and HASelZΔSECIS are constructs lacking the SECIS element. Asterisks indicate unspecific products. The arrow indicating a protein in lane 2 shows GPx, and those in lane 4 point to the full-length and shorter product of SelX. For SelN, the major product in lane 6 has the lowest intensity. Because the UGA Sec codon is the penultimate one in SelZ, both the full-length protein and the product arising from the ΔSECIS construct have the same mobility (lanes 8 and 9). (B) Detection by ^{75}Se labeling of the immunoprecipitated selenoproteins shown in (A). [Reproduced with permission from A. Lescure, D. Gautheret, P. Carbon, and A. Krol, *J. Biol. Chem.* **274,** 38147 (1999).]

Binding of Anti-Hemagglutinin Antibody to protein A–Sepharose. Washed protein A–Sepharose (200 mg; Pharmacia, France) is suspended in 50 mM Tris-HCl, pH 7.4. One milliliter of beads is mixed to 100 μg of the 12CA5 anti-HA antibody (Roche) and incubated for 2 hr at room temperature with gentle agitation. The beads are then rinsed five times with 10 ml of 50 mM Tris-HCl, pH 7.4, and stored at 4°.

Immunoprecipitation. The cell extract is adjusted to 20 mM HEPES–NaOH (pH 7.9), 12.5 mM MgCl$_2$, 150 mM KCl, 0.1 mM EDTA, 10% (v/v) glycerol, and 0.5% (v/v) Tween 20. Thirty microliters of freshly centrifuged cell extract is combined with 175 μl of HEMGT-150 buffer [20 mM HEPES–NaOH (pH 7.9), 12.5 mM MgCl$_2$, 150 mM KCl, 0.1 mM EDTA, 10% (v/v) glycerol, and 0.5% (v/v) Tween 20] and 25 μl of anti-HA bound to protein A–Sepharose beads. Binding is continued for 1 hr at room temperature with gentle agitation. The beads are spun down, washed four times for 15 min each with 200 μl of HEMGT-150 buffer, mixed with 25 μl of loading buffer [100 mM Tris-HCl (pH 6.8), 150 mM dithiothreitol (DTT), 4% (w/v) SDS, 20% (v/v) glycerol, 0.2% (w/v) bromphenol blue], heated in boiling water for 3 min, and centrifuged. The proteins are then fractionated on SDS–polyacrylamide gels, and the ^{75}Se-labeled proteins are revealed by autoradiography with an intensifying screen (Fig. 4B).

Alternative Method. The ^{75}Se labeling establishes the presence of a selenium atom in the polypeptide chain, but not formally the occurrence of a selenocysteine residue. This issue can be more specifically addressed by proteolysis, followed by separation and analysis of the peptides by high-performance liquid chromatography (HPLC)-coupled mass spectrometry (LC-MS) or matrix-assisted laser desorption ionization time-of-flight mass spectrometry (MALDI-TOF), as described by Arnér *et al.*[26]

Conclusion

Identification of novel selenoproteins constitutes an essential step toward a better understanding of the physiological role of selenium. Indeed, the function of the selenoenzymes characterized to date falls short of explaining the pleiotropic effects observed in cases of selenium deficiency.

The application of RNAMOT[5,6] by Lescure *et al.*[1] permitted the discovery of four novel selenoproteins with a single RNA hairpin, the SECIS element. Importantly, and in contrast to the conventional programs devoted to the analysis of nucleic acid sequences, the computational screen we have developed enables discrimination, in translated nucleic acid databases, of proteins actually ending at UGA stop codons from those artifactually ending at UGA Sec codons.

[26] E. S. J. Arnér, H. Sarioglu, F. Lottspeich, A. Holmgren, and A. Böck, *J. Mol. Biol.* **292**, 1003 (1999).

One interesting observation resulting from the identification of novel seleno-proteins is that their number increases in the course of evolution and culminates in mammals (and presumably all vertebrates) with about 20 proteins identified so far. Many mammalian selenoproteins have homologs in lower organisms, but in this case it is a cysteine that is found in place of the selenocysteine residue. One can cite, for example, glutathione peroxidase, SelR/X, or SelT.[1,20] Thioredoxin reductase is of particular interest because it is the only protein in which the seleno-cysteine residue is found in *Caenorhabditis elegans* and vertebrates. Moreover, it is the only selenoprotein in *C. elegans*,[18] implying that the selenocysteine incor-poration machinery has been maintained in this animal for the synthesis of a single selenoprotein.

Acknowledgments

This work was supported by grants from the Centre de Recherche Volvic sur les Oligo-Eléments, the Ligue contre le Cancer, and the Association pour la Recherche contre le Cancer (ARC).

[7] Semisynthesis of Proteins Containing Selenocysteine

By ROBERT J. HONDAL and RONALD T. RAINES

Introduction

Selenocysteine (Sec or U) is often referred to as the 21st amino acid.[1] Yet, unlike other nonstandard amino acid residues (such as hydroxyproline), selenocysteine is not created by posttranslational modification. Instead, selenocysteine shares many features with the 20 common amino acids. Like those, selenocysteine (1) has its own codon, (2) has its own unique tRNA molecule, and (3) is incorporated into proteins cotranslationally.[2]

Currently there are more than 20 known mammalian proteins that contain selenocysteine.[3,4] Detailed structural and functional information about most of these proteins is lacking because of their low natural abundance and the ab-sence of a method to produce selenium-containing proteins by recombinant DNA

---

[1] A. Bock, K. Forchhammer, J. Heider, W. Leinfelder, G. Sawers, B. Veprek, and F. Zinoni, *Mol. Microbiol.* **5**, 515 (1991).

[2] D. Hatfield and A. Diamond, *Trends Genet.* **9**, 69 (1993).

[3] A. Lescure, D. Gautheret, P. Carbon, and A. Krol, *J. Biol. Chem.* **274**, 38147 (1999).

[4] D. Behne and A. Kyriakopoulos, *Annu. Rev. Nutr.* **21**, 453 (2001).

technology. There is one report of formate dehydrogenase being overproduced in *Escherichia coli*[5] and a description of the production of rat thioredoxin reductase (TR) in *E. coli*.[6] The gene for tRNA^{Sec} also had to be overexpressed to achieve formate dehydrogenase production. Moreover, formate dehydrogenase could be produced because it is native to *E. coli* and the elaborate machinery required for selenocysteine insertion is present in its gene. To produce TR, a gene fusion was created that placed a bacterial selenocysteine insertion sequence (SECIS) element immediately downstream of the UGA stop codon. There, the SECIS element allowed decoding of the UGA codon as one for selenocysteine. This strategy is possible only because the UGA codon in TR is proximal to the 3' end of its mRNA. This strategy is not viable for other selenocysteine-containing proteins in which the UGA codon is distal from the 3' end. Selenium incorporation into TR was low, even with the overexpression of several accessory genes.

The limitations to the biosynthesis of selenium-containing proteins are severe. Chemical modification has been used to overcome these limitations in one instance. Specifically, the active-site serine residue of subtilisin was converted to selenocysteine by activation with phenylmethylsulfonyl fluoride (PMSF) and reaction with hydrogen selenide.[7] This approach is not general, however, as it relies on the especially high reactivity of the active-site serine residue in subtilisin.

A semisynthetic approach for incorporating selenocysteine residues into proteins has become evident to us. Our approach makes use of "native chemical ligation."[8,9] Native chemical ligation allows for two peptide chains to be joined chemoselectively through the use of a C-terminal thioester on one peptide and an N-terminal cysteine residue on the other peptide (Fig. 1). Several proteins have been synthesized using this method.[10] A technique related to native chemical ligation is called "expressed protein ligation."[11] This technique makes use of an engineered intein. Inteins are a type of mobile genetic element at the protein level.[12] *In vivo,* inteins catalyze their own excision from a larger precursor. The flanking regions of this precursor are referred to as exteins.[13] (Inteins also have a homing endonuclease activity that allows for the insertion of their DNA into the host.) The protein splicing ability of inteins is a powerful tool for protein engineering. When a target

[5] G. T. Chen, M. J. Axley, J. Hacia, and M. Inouye, *Mol. Microbiol.* **6,** 781 (1992).

[6] E. S. Arner, H. Sarioglu, F. Lottspeich, A. Holmgren, and A. Bock, *J. Mol. Biol.* **292,** 1003 (1999).

[7] Z.-P. Wu and D. Hilvert, *J. Am. Chem. Soc.* **111,** 4513 (1989).

[8] T. Wieland, E. Bokelmann, L. Bauer, H. U. Lang, and H. Lau, *Liebig's Ann. Chem.* **583,** 129 (1953).

[9] P. E. Dawson, T. W. Muir, I. Clark-Lewis, and S. B. Kent, *Science* **266,** 776 (1994).

[10] P. E. Dawson and S. B. H. Kent, *Annu. Rev. Biochem.* **69,** 923 (2000).

[11] T. W. Muir, D. Sondhi, and P. A. Cole, *Proc. Natl. Acad. Sci. U.S.A.* **95,** 6705 (1998).

[12] A. A. Cooper and T. H. Stevens, *Trends Biochem. Sci.* **9,** 351 (1995).

[13] F. B. Perler, E. O. Davis, G. E. Dean, F. S. Gimble, W. E. Jack, N. Neff, C. J. Noren, J. Thorner, and M. Belfort, *Nucleic Acids Res.* **22,** 1125 (1994).

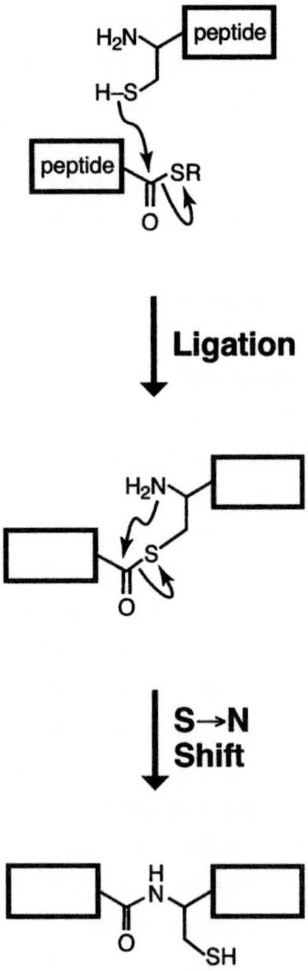

FIG. 1. Scheme for native chemical ligation, which is the chemoselective ligation of peptide segments via a C-terminal thioester and an N-terminal cysteine residue.

protein is fused to an intein, the intein catalyzes the formation of a thioester at the C terminus of the target protein (Fig. 2). This thioester exists in equilibrium with the amide starting material. Addition of exogenous thiol drives the equilibrium toward the thioester, resulting in cleavage of the target protein from the intein. The cleaved target protein now has a thioester moiety at its C terminus. The reactivity of this thioester can be used to ligate an exogenous peptide containing an N-terminal cysteine to the target protein (Fig. 2).

Strategies for Synthesis of Proteins Containing Selenocysteine

The methods we describe herein for the synthesis of selenocysteine-containing proteins are variations of the peptide ligation method first described by Wieland et al.[8] and later developed by Kent and co-workers.[9] This method makes use of

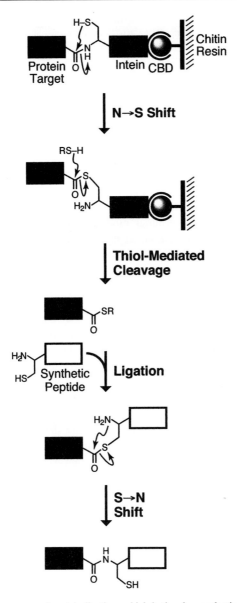

FIG. 2. Scheme for expressed protein ligation, which is the chemoselective ligation of a protein segment with a C-terminal thioester and a peptide segment with an N-terminal cysteine (or, here, selenocysteine) residue. CBD, Chitin-binding domain.

TABLE I

STRATEGIES FOR SEMISYNTHESIS OF PROTEINS CONTAINING SELENIUM BY CHEMICAL
LIGATION OF PEPTIDE/PROTEIN FRAGMENTS

	Fragment		Selenocysteine participates in ligation reaction?	Target selenoprotein
Strategy	N Terminal	C Terminal		
1	H$_2$N · · · Sec · · · C(O)SR	Cys · · · C(O)OH	No	Selenoprotein W
2	H$_2$N · · · C(O)SR	Cys · · · Sec · · · C(O)OH	No	Thioredoxin reductase
3	H$_2$N · · · C(O)SR	Sec · · · C(O)OH	Yes	C110U RNase A
4	H$_2$N · · · C(O)SR	SecOH	Yes	Sec125 RNase A

peptide ligation through a C-terminal thioester and an N-terminal cysteine residue (Fig. 1). Using this general method, we envision four distinct strategies for introducing selenocysteine into a protein, as summarized in Table I.

1. A peptide that contains an embedded selenocysteine residue and C-terminal thioester could be ligated to a peptide that contains an N-terminal cysteine residue. This strategy could be applied to the semisynthesis of selenoprotein W, which has a single selenocysteine residue proximal to its N terminus.[14]

2. A selenocysteine residue could be embedded within a peptide that has an N-terminal cysteine residue such that the selenocysteine does not participate directly in the ligation reaction. This peptide could then be ligated to a thioester fragment. This strategy is viable for TR, which has a single selenocysteine residue proximal to its C terminus.

3. A selenocysteine residue could participate directly in the ligation reaction by using selenocysteine rather than cysteine as the N-terminal residue in the C-terminal fragment. In the example described elsewhere[15] and below, we show that selenocysteine can indeed replace cysteine in the ligation reaction to create a semisynthetic RNase A molecule containing a selenocysteine residue.

4. Selenocysteine itself could be used to cleave a target protein–intein fusion. The result is the target protein with an extra C-terminal selenocysteine residue. This approach is useful for adding a nucleophile with orthologous reactivity to a protein. Below, we show how to add a selenocysteine residue to the C terminus of RNase A.

---

[14] S. C. Vendeland, M. A. Beilstein, C. L. Chen, O. N. Jensen, E. Barofsky, and P. D. Whanger, *J. Biol. Chem.* **268**, 103 (1993).

[15] R. J. Hondal, B. L. Nilsson, and R. T. Raines, *J. Am. Chem. Soc.* **123**, 5140 (2001).

Embedding selenocysteine within a peptide (so as to effect strategies 1 and 2) is not described explicitly herein, but is a straightforward extension of the methods described below.

Preparation of Selenocysteine for Peptide Synthesis

Synthesis of Disodium Diselenide

A 1 M solution of Na_2Se_2 is prepared by the procedure of Klayman and Griffin.[16] Elemental selenium (4.5 g; 56 mmol) is added to water (25 ml) in a stoppered three-necked flask with magnetic stirring. Sodium borohydride (4.5 g; 119 mmol) dissolved in water (25 ml) is added dropwise to the slurry of elemental selenium. The flask may be chilled in an ice bath to prevent boiling. After all of the sodium borohydride is added and the solution has become colorless, additional elemental selenium (4.5 g; 56 mmol) is added to the solution. The solution should be reddish brown, which is characteristic of disodium diselenide. The flask is then stoppered and flushed with Ar(g) or N_2(g).

Synthesis of Selenocystine

The following procedure is essentially the same as that described by Tanaka and Soda.[17] β-Chloro-L-alanine (5.0 g; 31 mmol) is dissolved in water (40 ml), and the pH of the resulting solution is adjusted to pH 9.0. The resulting solution is then added dropwise over 30–60 min to the solution of Na_2Se_2 through one of the septa in the three-necked flask. The mixture is stirred under Ar(g) or N_2(g) at 37° for 12–16 hr. The solution is then acidified with concentrated HCl until vigorous reaction stops, and hydroxylamine (0.33 g; 9.7 mmol) is added to reduce remaining elemental selenium. Additional concentrated HCl is added until there is no more vigorous reaction with the solution. The resulting solution is flushed for at least 1 hr with Ar(g). (The hydrogen selenide exhaust can be trapped with a saturated aqueous solution of lead acetate.) The solution is then filtered, and the pH of the yellow filtrate is adjusted to pH 6–6.5. If the concentration of selenocystine is high, then a yellow precipitate forms immediately. If the concentration is low, then the solution is cooled to 4° overnight and yellow crystals of selenocystine are collected. The yellow crystals of selenocystine may contain some black material. Selenocystine can be recrystallized by dissolving the crystals in the smallest volume possible of 2 N HCl. This dark yellow solution is filtered to remove the black material. The pH of the yellow filtrate is then increased to pH 6–6.5 by the addition of 10 N NaOH. Selenocystine is isolated as a bright yellow crystalline solid in 60% yield (4.1 g from 5.0 g of β-chloro-L-alanine).

---

[16] D. L. Klayman and T. S. Griffin, *J. Am. Chem. Soc.* **95**, 197 (1973).
[17] H. Tanaka and K. Soda, *Methods Enzymol.* **143**, 240 (1987).

Synthesis of Sec(PMB)OH

A *p*-methoxybenzyl (PMB) group is used to protect the selenium, according to the examples of Koide *et al.*[18] and Besse and Moroder.[19] The following procedure was adapted from Koide *et al.*[18] Their procedure uses an excess of *p*-methoxybenzyl chloride in the reaction under highly basic conditions. In our hands, this procedure always results in a dibenzylated product in which the nitrogen as well as the selenium of selenocysteine are alkylated with a PMB group. To produce a product that is alkylated only at selenium, the following protocol is used. Selenocystine (1.8 g; 5.3 mmol) is dissolved in 0.5 *N* NaOH (5 ml). NaBH$_4$ (1.7 g; 43 mmol) dissolved in water (10 ml) is added dropwise and with stirring to the solution of selenocystine in a 100-ml round-bottom flask. The flask can be chilled to prevent boiling. After the vigorous reaction has subsided (the solution turns from yellow to colorless), glacial acetic acid is added dropwise until the pH is near 6.0. *p*-Methoxybenzyl chloride (1.44 ml; 10.64 mmol) is then added dropwise to the solution. The reaction proceeds quickly and is complete in 30 min. White crystals of selenium-benzylated selenocysteine are apparent along with some re-oxidized selenocystine, which appears as yellow crystals. The solution is acidified with concentrated HCl to complete the formation of the white precipitate, which is purified by crystallization from hot water. Sec(PMB)OH–HCl is isolated as a white solid in 72% yield (2.5 g from 1.8 g of selenocystine).

Synthesis of Fmoc-Sec(PMB)OH

The procedure is from Koide *et al.*[18] but is again modified. Sec(PMB)OH–HCl (1.2 g; 3.7 mmol) is dissolved in water (10 ml) to make a slurry. Triethylamine (TEA; 0.27 ml; 3.7 mmol) is added to the slurry in a 100-ml round-bottom flask. Fmoc-*O*-succinimide (where "Fmoc" refers to 9-fluorenylmethoxycarbonyl; 1.25 g; 3.7 mmol), dissolved in acetonitrile (10 ml), is added to the solution and another equivalent of TEA is then added. Additional acetonitrile should be added until all of the solutes are dissolved completely. The resulting solution is stirred at room temperature for 1 hr, and reaction progress is monitored by thin-layer chromatography on silica plates. In dichloromethane, Fmoc-Sec(PMB)OH will not migrate, but impurities from the reaction, especially dibenzofulvene, will have high mobility. After 1 hr, the reaction mixture is acidified with 1 *N* HCl (5 ml) and then extracted with ethyl acetate. The organic layer is washed (three times) with 1 *N* HCl. The resulting, combined aqueous phases are then extracted (three times) with ethyl acetate. The ethyl acetate extracts are combined and dried over

---

[18] T. Koide, H. Itoh, A. Otaka, H. Yasui, M. Kuroda, N. Esaki, K. Soda, and N. Fujii, *Chem. Pharm. Bull.* **41**, 502 (1993).

[19] D. Besse and L. Moroder, *J. Peptide Sci.* **3**, 442 (1997).

$$
\begin{array}{cc}
 & 2 \\
1 & 0
\end{array}
$$

K E T A A A K F E R Q H M D S S T S A A
S S S N Y C N Q M M K S R N L T K D R C
K P V N T F V H E S L A D V Q A V C S Q
K N V A C K N G Q T N C Y Q S Y S T M S
I T D C R E T G S S K Y P N C A Y K T T
Q A N K H I I V A **C E G N P Y V P V H F**
D A S V

$$
\begin{array}{c}
1 \\
2 \\
4
\end{array}
$$

FIG. 3. Primary sequence of RNase A. To produce C110U RNase A, residues 1–109 are synthesized as a fusion protein to the *Mxe* GyrA intein, as shown in Fig. 2. RNase A(1–109) is cleaved from the fused intein and then ligated to a synthetic peptide (underlined) with selenocysteine replacing cysteine (boldface).

$MgSO_4(s)$. After filtering, the organic layer is concentrated under vacuum to produce a yellow oil. This oil is dissolved in dichloromethane (10–20 ml) and purified by chromatography on a column of silica gel (20 cm × 20 cm^2). The impurities in the reaction, primarily dibenzofulvene, are eluted when the column is washed extensively with CH_2Cl_2, and Fmoc-Sec(PMB)OH is eluted with methanol–CH_2Cl_2 (1 : 4, v/v). The solvent is removed under vacuum. Fmoc-Sec(PMB)OH is isolated as a slightly yellow crystalline solid in 53% yield [1.0 g of Fmoc-Sec(PMB)OH from 1.2 g of Sec(PMB)OH–HCl].

Solid-Phase Synthesis of Peptide Containing Selenocysteine

A methylbenzhydrylamine polystryrene resin functionalized with a 4-hydroxymethylphenoxy acid-labile linker that had been loaded with the C-terminal amino acid is used for all syntheses. Cycles of *O*-benzotriazol-1-yl-*N*,*N*,*N′*,*N′*-tetramethylunonium hexafluorophosphate/diisopropylamine (HBTU/DIEA) activation of the carboxylic acid group, followed by piperidine deprotection of the Fmoc group, are used to couple monomers. Syntheses are done on a 25-μmol scale with a 3-fold excess of each amino acid monomer, using an Applied Biosystems (Foster City, CA) model 432A synthesizer.

As a model protein to demonstrate the efficacy of our methods, we chose ribonuclease A (RNase A; EC 3.1.27.5), which has been the object of much seminal work in protein chemistry and enzymology.[20] RNase A has 124 residues, including 8 cysteine residues that form 4 disulfide bonds in the native enzyme (Fig. 3). Of these eight cysteine residues, Cys-110 is closest to the C terminus. A semisynthetic

[20] R. T. Raines, *Chem. Rev.* **98,** 1045 (1998).

RNase A has already been constructed by expressed protein ligation to form the peptide bond between Ala-109 and Cys-110.[21] The sequence of the wild-type RNase A peptide used in our ligation reactions is CEGNPYVPVHFDASV (which corresponds to residues 110–124) and the sequence of the selenocysteine variant is UEGNPYVPVHFDASV, where "U" refers to selenocysteine (Fig. 3).

Deprotection of Peptides Containing Selenocysteine

Deprotection of the wild-type peptide (which does not contain selenocysteine) is achieved by using a cleavage cocktail containing trifluoroacetic acid–ethanedithiol–H_2O (95 : 2.5 : 2.5, v/v/v) for 3 hr at room temperature. When this cocktail is used for the selenocysteine-containing peptide, only partial removal of the PMB group is achieved. The PMB group is removed successfully, however, using conditions reported by Koide et al.[18] The cleavage cocktail for the selenium-containing peptide contains m-cresol–thioanisole–trifluoroacetic acid–trimethylsilyl trifluoromethane sulfonate (50 : 120 : 690 : 194, v/v/v/v). After purification by high-performance liquid chromatography (HPLC), the intact peptide is observed, along with some dehydrated peptide (Fig. 4). Dehydration most likely occurs during deprotection.

Semisynthesis of a Protein Containing Selenocysteine

One way to incorporate selenocysteine into a protein is to ligate a protein fragment with a C-terminal thioester to a peptide that contains an N-terminal selenocysteine residue (strategy 3 in Table I).[15] To incorporate a selenocysteine residue into RNase A, a protein is produced in which residues 1–109 of RNase A are fused to the *Mxe* GyrA intein and a chitin-binding domain.[21] Plasmid pTXB1-RNase (a kind gift from New England BioLabs, Beverly, MA) is transformed into ER2566 *E. coli* cells. Luria–Bertani (LB) medium (0.10 liter) containing ampicillin (0.10 mg/ml) is inoculated with a single colony and grown for 8 hr at 37°. Six 2-liter flasks that each contain 1 liter of LB medium containing ampicillin (0.10 mg/ml) are then inoculated with 10 ml of the 8-hr culture, and grown at 37° until the A_{600} equals 0.6. The flasks are then cooled on ice for 10 min, and expression is induced by the addition of isopropyl-β-D-thiogalactopyranoside (IPTG, to 0.5 mM). The flasks are shaken at room temperature for 5 hr. Cells are harvested by centrifugation, and the wet cell pellet is frozen for storage. Frozen cells are thawed and homogenized in 50 mM Tris-HCl buffer, pH 8.5, containing NaCl (0.50 M), and then lysed by sonication. Lysed cells are subjected to centrifugation at 6000g for 30 min at 4°. The supernatant is passed over a 30-ml column of chitin resin

---

[21] T. C. Evans, Jr., J. Benner, and M.-Q. Xu, *Protein Sci.* **7**, 2256 (1998).

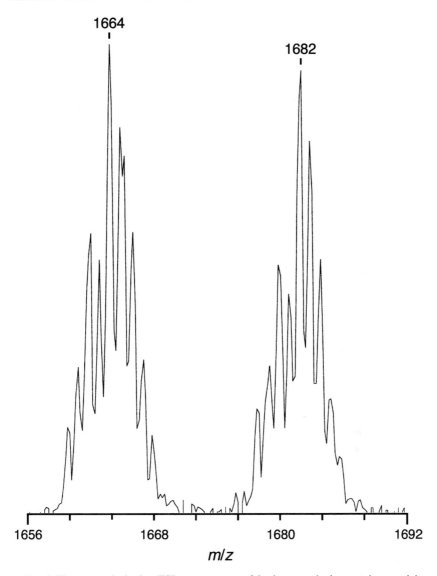

FIG. 4. Electrospray-ionization (ESI) mass spectrum of the deprotected selenocysteine-containing peptide corresponding to residues 110–124 of RNase A (m/z = 1682). The dehydrated peptide is also present (m/z = 1664).

(New England BioLabs) that had been equilibrated with the same buffer. After loading, the column is washed with 5 column volumes of buffer to elute non-specific proteins. Cleavage of RNase A(1–109) is initiated by equilibrating the column in cleavage buffer, which is 50 mM Tris-HCl buffer, pH 7.4, containing NaCl (0.10 M) and N-methylmercaptoacetamide (NMA; 50 mM) as the cleavage reagent. We have found NMA to be better than 2-mercaptoethanol, dithiothreitol (DTT), or 2-mercaptoethanesulfonic acid for the cleavage of target protein–intein fusions (data not shown). The chitin resin is incubated with the cleavage buffer for 2 hr at room temperature. Fractions containing the thioester-tagged RNase A(1–109) are then collected. The protein solution is concentrated with a Centriplus 3000 apparatus (Amicon, Danvers, MA) to a volume of 5 ml. For ligation reactions with the peptide UEGNPYVPVHFDASV, 5 ml of 20 mM Tris-HCl buffer, pH 7.0, containing guanidine hydrochloride (7 M) is added to the protein solution containing RNase A(1–109). The solution is then added to the lyophilized peptide (15 mg; 10 μmol) and the resulting solution is incubated overnight at room temperature. The protein is unfolded by adding 10 ml of 20 mM Tris-HCl buffer, pH 8.0, containing guanidine hydrochloride (7 M). Dithiothreitol is added (to 0.10 M) to reduce any disulfide or selenosulfide bonds, and the resulting solution is incubated for 1 hr at room temperature. This solution is then dialyzed overnight at 4° against folding buffer, which is 50 mM Tris-HCl buffer, pH 7.8, containing NaCl (0.10 M), reduced glutathione (GSH; 1 mM), and oxidized glutathione (GSSG; 0.2 mM). The protein solution is dialyzed for an additional 4 hr against fresh refolding buffer and then overnight against 50 mM sodium acetate buffer, pH 5.0.

Intact RNase A has a much higher affinity for guanidine 3′-diphosphate (GDP) than does RNase A(1–109). Hence, affinity chromatography using GDP–Sepharose resin (Sigma, St. Louis, MO) is used to purify the intact protein. A 5-ml column of GDP–Sepharose equilibrated with 10 mM sodium acetate buffer, pH 5.0, is loaded with the protein solution and then washed with 50 ml of the same buffer. The bound protein is then eluted with buffer containing NaCl (1 M). Fractions containing active RNase A are dialyzed against 50 mM sodium acetate buffer, pH 5.0, overnight at 4°. Approximately 1 mg of pure C110U RNase A can be prepared in this manner from 6 liters of culture.

pH of Ligation Reactions

At low pH, the nucleophilic attack of selenocysteine on a thioester is much faster than that of cysteine. Ligation reactions with cysteine-containing peptides are typically performed at pH 8.0. We recommend using a lower pH for ligation reactions with selenocysteine-containing peptides. For example, we prepare C110U RNase A at pH 7.0. This lower pH suppresses β elimination of the selenol group from selenocysteine but is high enough to allow for the intramolecular attack of nitrogen on the intermediate selenoester to form an amide (Fig. 1).

Characterization of Proteins Containing Selenocysteine

The atomic mass of selenium is 47 amu greater than that of sulfur. Hence, mass spectrometry is a useful tool for demonstrating the replacement of sulfur by selenium. Matrix-assisted laser desorption ionization time-of-flight (MALDI-TOF) mass spectral analysis (Fig. 5) shows that the mass of folded wild-type RNase A is 13,820 Da (13,812 Da predicted) and that of folded C110U RNase A is 13,865 Da (13,820 Da predicted). Not only are the experimental values in excellent agreement with the predicted values, but also the mass difference of 45 amu is close to that expected for the replacement of sulfur by selenium.

A critical measure of the successful ligation and folding of a selenium-containing protein is the demonstration of function. For C110U RNase A, that demonstration involves measuring ribonucleolytic activity. The k_{cat}/K_m for the cleavage of 6-FAM~dArU(dA)$_2$~6-TAMRA[22] is $(1.13 \pm 0.06) \times 10^7 \ M^{-1} \ sec^{-1}$ for wild-type RNase A and $(1.1 \pm 0.1) \times 10^7 \ M^{-1} \ sec^{-1}$ for C110U RNase A. The similarity of these values indicates that selenium can replace sulfur with minimal perturbation to function and (presumably) structure.

An important problem to consider when using selenocysteine in ligation reactions is the redox state of the selenium in the folded protein. RNase A contains eight cysteine residues that are oxidized to form four disulfide bonds in the native enzyme. When replacing cysteine with selenocysteine in RNase A, misfolding could result in the irreversible formation of an improper selenosulfide bond. Selenosulfide bonds are more difficult to reduce than disulfide bonds,[23] and thus could limit the yield of properly folded protein. Indeed, the overall yield of C110U RNase A is low. Yet, the main difficulty encountered is the recovery of sufficient amounts of thioester-tagged RNase A(1–109). As noted by Evans et al.[21] the cleavage of RNase A with an alanine residue at the junction with the intein results in poor yields in recovery of thioester-tagged proteins. The low yield of C110U RNase A may not be attributable to selenocysteine because the yields of other variants of RNase A produced by expressed protein ligation are similar (R. J. Hondal and R. T. Raines, unpublished results, 2000).

Use of Selenocysteine as Cleavage Reagent

Another way to incorporate a selenocysteine residue into a protein is to use selenocysteine itself to cleave the thioester produced by intein fusion (strategy 4 in Table I). A 30-ml solution of 0.10 M selenocysteine is prepared as follows.

[22] B. R. Kelemen, T. A. Klink, M. A. Behlke, S. R. Eubanks, P. A. Leland, and R. T. Raines, *Nucleic Acids Res.* **18,** 3696 (1999).

[23] D. Besse, F. Siedler, T. Diercks, H. Kessler, and L. Moroder, *Angew. Chem. Int. Ed. Engl.* **36,** 883 (1997).

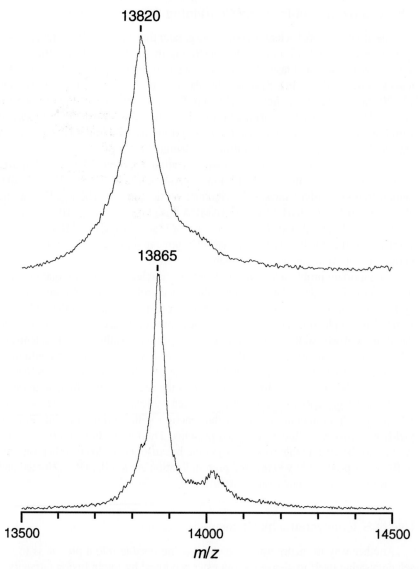

FIG. 5. MALDI-TOF mass spectra of semisynthetic wild-type RNase A (*top; m/z* = 13,820) and C110U RNase A (*bottom; m/z* = 13,865).

A 25-ml solution of 0.2 N NaOH is flushed with Ar(g) for 1 hr. Selenocystine (507 mg) is dissolved in this solution. Water (5 ml) is flushed with Ar(g) for 1 hr, and tris(2-carboxyethyl)phosphine hydrochloride (TCEP; 430 mg; 1.5 mmol) is dissolved in the water. The two aqueous solutions are mixed, and the pH of the resulting solution is adjusted to pH 6.0 by the addition of 10 N NaOH. This 0.10 M solution of selenocysteine is then applied to a target protein–intein fusion that has been purified by affinity chromatography with a chitin resin. To do so, the buffer from the chitin resin is drained until the resin is nearly dry. The resin is transferred to the solution of selenocysteine and incubated overnight at room temperature. The slurry of resin is filtered, and the protein in the filtrate has one selenocysteine residue at its C terminus. (*Note:* The selenocysteine residue may be in the form of a mixed diselenide with free selenocysteine.) Using this method, we have prepared RNase A containing an additional residue, Sec-125, at its C terminus (R. J. Hondal and R. T. Raines, unpublished results, 2001).

Summary

The methods described herein can be used to incorporate one or more selenocysteine residues into a protein. These methods enable detailed structure–function analyses of proteins containing selenocysteine. In addition, the methods provide a means to incorporate a selenol into a protein.

Acknowledgments

We thank Dr. U. Arnold, B. L. Nilsson, and C. L. Jenkins for advice and assistance. R.J.H. was supported by Postdoctoral Fellowship GM20180 (NIH). This work was supported by Grant GM44783 (NIH).

[8] Mammalian Selenoprotein Gene Signature: Identification and Functional Analysis of Selenoprotein Genes Using Bioinformatics Methods

By GREGORY V. KRYUKOV and VADIM N. GLADYSHEV

Introduction

The major form of selenium in selenium-containing proteins is selenocysteine (Sec). Sec, the 21st amino acid in proteins, is cotranslationally incorporated into polypeptide chains in ribosome-based protein synthesis. It is encoded by a UGA codon, and thus UGA has a dual function of dictating either Sec or the termination of translation.[1] For UGA to dictate Sec incorporation, a special stem–loop structure, designated the selenocysteine insertion sequence (SECIS) element (Fig. 1), must be present within the corresponding mRNA[1,2] (Fig. 2). In eukaryotes, this stem–loop structure is tightly associated with SECIS-binding protein 2,[3] which in turn recruits the Sec-specific elongation factor[4,5] and Sec tRNA for insertion of this amino acid at in-frame UGA codons. In bacteria, one multidomain protein serves SECIS-binding and Sec elongation functions. Bacterial SECIS elements are located immediately downstream of Sec-encoding UGA codons,[1] whereas in eukaryotes, SECIS elements are present in the 3′ untranslated regions (3′ UTRs) of selenoprotein mRNAs.[2]

The eukaryotic SECIS element possesses both primary sequence and secondary structure features essential to its biological function. However, the primary consensus sequence of the SECIS element is limited to six or seven nucleotides distributed in three locations within the stem–loop structure[2,6,7] (Fig. 1). This makes identification of SECIS elements on the basis of primary sequence consensus virtually impossible, and their identification by a combination of sequence and structural criteria challenging.

All known eukaryotic selenoproteins contain a single Sec residue except one. The exception is selenoprotein P, which contains 10–17 Sec residues depending

[1] A. Bock, K. Forchhammer, J. Heider, and C. Baron, *Trends Biochem. Sci.* **16**, 463 (1991).

[2] S. C. Low and M. J. Berry, *Trends Biochem. Sci.* **6**, 203 (1996).

[3] P. R. Copeland, J. E. Fletcher, B. A. Carlson, D. L. Hatfield, and D. M. Driscoll, *EMBO J.* **19**, 306 (2000).

[4] R. M. Tujebajeva, P. R. Copeland, X. M. Xu, B. A. Carlson, J. W. Harney, D. M. Driscoll, D. L. Hatfield, and M. J. Berry, *EMBO Rep.* **1**, 158 (2000).

[5] D. Fagegaltier, N. Hubert, K. Yamada, T. Mizutani, P. Carbon, and A. Krol, *EMBO J.* **19**, 4796 (2000).

[6] R. Walczak, E. Westhof, P. Carbon, and A. Krol, *RNA* **2**, 367 (1996).

[7] D. Fagegaltier, A. Lescure, R. Walczak, P. Carbon, and A. Krol, *Nucleic Acids Res.* **28**, 2679 (2000).

METHODS IN ENZYMOLOGY, VOL. 347

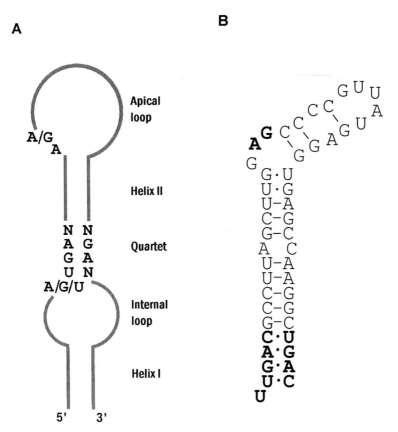

FIG. 1. Eukaryotic SECIS elements. (A) Consensus sequence and structure of the eukaryotic SECIS element. The location of the non-Watson core portion of the element, Quartet, Apical and Internal loops, and Helices I and II, are indicated. (B) Predicted SECIS element in human thioredoxin reductase 2.

on the organism, and whose mRNA has two functional SECIS elements.[8,9] Selenoproteins have diverse functions and lack amino acid sequence or structural features specific for this class of proteins. Therefore, selenoproteins cannot be identified by simply searching sequence databases for homologies or amino acid sequence signatures or patterns. However, in each of the functionally characterized selenoproteins, Sec is present in the enzyme active center and is essential for the catalyzed reactions. Thus a common feature of the Sec residue is that it is directly involved in protein function.

[8] R. M. Tujebajeva, D. G. Ransom, J. W. Harney, and M. J. Berry, *Genes Cells* **5,** 897 (2000).
[9] G. V. Kryukov and V. N. Gladyshev, *Genes Cells* **5,** 1049 (2000).

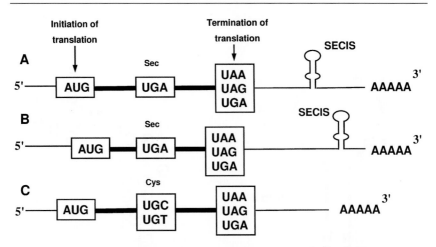

FIG. 2. Structure of mammalian selenoprotein mRNAs. (A) Selenoprotein mRNA. An in-frame UGA codon encodes Sec and Sec insertion sequence (SECIS) element is present in the 3′ UTR of selenoprotein mRNA. In the absence of the SECIS element, the in-frame UGA would signal termination of translation. (B) mRNA that encodes a selenoprotein that is homologous to the selenoprotein shown in (A). SECIS elements and Sec locations are conserved between two proteins. (C) mRNA that encodes a homologous protein that contains Cys in place of Sec. SECIS element is absent in this mRNA.

An in-frame Sec-encoding UGA codon, which apparently occurs rather infrequently, complicates correct recognition and annotation of genes. Examples of misinterpretation of TGA codons as both Sec[10] and stop signals[11,12] are known and no algorithms are currently available that would correctly assign in-frame TGA codons. However, bioinformatics analyses of TGA codons, together with analyses of SECIS elements and homologies in Sec-flanking sequences, help to determine the functional identity of these codons and to identify and characterize selenoprotein genes. Here we describe a set of criteria and bioinformatics procedures that can be used for these purposes.

Eukaryotic Selenoproteins

For many years, only a few selenoproteins, primarily of a bacterial origin, were known, and the discovery of each additional selenoprotein was an extraordinary

[10] L. Cataldo, K. Baig, R. Oko, M. A. Mastrangelo, and K. C. Kleene, *Mol. Reprod. Dev.* **45,** 320 (1996).

[11] V. N. Gladyshev, K. T. Jeang, and T. C. Stadtman, *Proc. Natl. Acad. Sci. U.S.A.* **93,** 6146 (1996).

[12] M. J. Guimaraes, D. Peterson, A. Vicari, B. G. Cocks, N. G. Copeland, D. J. Gilbert, N. A. Jenkins, D. A. Ferrick, R. A. Kastelein, J. F. Bazan, and A. Zlotnik, *Proc. Natl. Acad. Sci. U.S.A.* **93,** 15086 (1996).

event. However, in more recent years, the list of known selenoproteins has rapidly grown and the majority of new selenoproteins were found in vertebrates. Currently, 22 eukaryotic selenoproteins are known (Table I), excluding highly similar seleno-proteins formed by genome duplication in zebrafish.[9] These proteins belong to 12 families of homologous proteins [glutathione peroxidase (GPx),[13,14] thiore-doxin reductase (TR),[15] deiodinase (DI),[16,17] selenoprotein N (SelN),[18] SelP,[19] SelR,[18,20] SelT,[20] SelW,[21] selenophosphate synthetase (SPS),[22] 15-kDa seleno-protein (Sep15),[23] BthD,[24] and G-rich[24]], each having 1–4 members.

The majority of known eukaryotic selenoproteins can be placed in two groups according to the location of Sec in these proteins (Table I). One such group includes selenoproteins that contain Sec within N-terminal sequences of relatively small $\alpha\beta$ proteins or domains. The proteins of this group contain a redox-active Sec residue that is often separated from a conserved cysteine (Cys) by two other amino acids. This group includes glutathione peroxidases, BthD, and selenoproteins W, W2, T, T2, P, and Pb.

In the second selenoprotein group, Sec is located near the C terminus. Such proteins include thioredoxin reductases and *Drosophila* G-rich, in which Sec is the C-terminal penultimate residue. In the remaining selenoproteins, a prominent feature is the separation of Sec from Cys by one amino acid. However, additional studies are required to further classify the remaining seven selenoproteins shown in Table I. Comparison of Sec-flanking sequences for eukaryotic selenoproteins reveals the abundance of glycines (Gly) and Cys (Table I). In fact, the majority of selenoproteins contain Gly adjacent to Sec. The chemistry associated with this flanking occurrence is not well understood.

[13] R. A. Sunde, *in* "Selenium in Biology and Human Health" (R. F. Burk, ed.), p. 146. Springer-Verlag, New York, 1994.

[14] F. Ursini, M. Maiorino, and A. Roveri, *Biomed. Environ. Sci.* **10**, 327 (1997).

[15] Q.-A. Sun, Y. Wu, F. Zappacosta, K.-T. Jeang, B. J. Lee, D. L. Hatfield, and V. N. Gladyshev, *J. Biol. Chem.* **274**, 24522 (1999).

[16] P. R. Larsen, *Biochem. Soc. Trans.* **25**, 588 (1997).

[17] D. L. St. Germain and V. A. Galton, *Thyroid* **7**, 655 (1997).

[18] A. Lescure, D. Gautheret, P. Carbon, and A. Krol, *J. Biol. Chem.* **274**, 38147 (1999).

[19] R. F. Burk and K. E. Hill, *Bioessays* **21**, 231 (1999).

[20] G. V. Kryukov, V. M. Kryukov, and V. N. Gladyshev, *J. Biol. Chem.* **274**, 33888 (1999).

[21] S. C. Vendeland, M. A. Beilstein, C. L. Chen, O. N. Jensen, E. Barofsky, and P. D. Whanger, *J. Biol. Chem.* **268**, 17103 (1993).

[22] M. J. Guimaraes, D. Peterson, A. Vicari, B. G. Cocks, N. G. Copeland, D. J. Gilbert, N. A. Jenkins, D. A. Ferrick, R. A. Kastelein, J. F. Bazan, and A. Zlotnik, *Proc. Natl. Acad. Sci. U.S.A.* **93**, 15086 (1996).

[23] E. Kumaraswamy, A. Malykh, K. V. Korotkov, S. Kozyavkin, Y. Hu, S. Y. Kwon, M. E. Moustafa, B. A. Carlson, M. J. Berry, B. J. Lee, D. L. Hatfield, A. M. Diamond, and V. N. Gladyshev, *J. Biol. Chem.* **275**, 35540 (2000).

[24] F. J. Martin-Romero, G. V. Kryukov, A. V. Lobanov, B. J. Lee, V. N. Gladyshev, and D. L. Hatfield, *J. Biol. Chem.* **276**, 29798 (2001).

TABLE I

Characteristics of Eukaryotic Selenocysteine-Containing Proteins[a]

Protein	Protein length	Sec position	Location of sec
Cytosolic glutathione peroxidase	201	47	...ASL U GTT...
Gastrointestinal glutathione peroxidase	190	40	...ASL U GTT...
Plasma glutathione peroxidase	226	73	...ASY U GLT...
Phospholipid hydroperoxide glutathione peroxidase	197	73	...ASQ U GKT...
Cytosolic thioredoxin reductase (TR1)	499	498	...AGC U G
Thioredoxin reductase expressed in testis (TR2)	615	614	...KGC U G
Mitochondrial thioredoxin reductase (TR3)	524	523	...TGC U G
Thyroid hormone deiodinase 1	250	126	...SCT U PSF...
Thyroid hormone deiodinase 2	266	133	...SAT U PPF...
Thyroid hormone deiodinase 3	279	144	...SCT U PPF...
Plasma selenoprotein P	381	59	...QAS U YLC...
Selenoprotein Pb[b]	265	64	...KAS U HFC...
Selenoprotein W	87	13	...CGA U GYK...
Selenoprotein W2[b]	94	13	...CGA U GYE...
Selenoprotein T	163	17	...CVS U GYR...
Selenoprotein T2[b]	165	19	...CVS U GYK...
Selenoprotein R	116	95	...SRF U IFS...
Selenoprotein N	556	428	...QSC U GSG...
Selenoprotein G-rich[c]	109	110	...GGG U G
Selenoprotein BthD[c]	249	37	...CRS U RVF...
15-kDa Selenoprotein	162	93	...VCG U KLG...
Selenophosphate synthetase	448	60	...MKG U GCK...

[a] All selenoproteins are human unless noted otherwise. Filled boxes illustrate length of selenoproteins. Sec (U) and Sec-flanking regions illustrate the location of Sec in a protein sequence. For several smaller selenoproteins, small triangles below sequences indicate the exact location of Sec.

[b] Zebrafish selenoproteins.

[c] Drosophila melanogaster selenoproteins.

Selenocysteine/Cysteine Pair in Homologous Sequences

Thus far, the largest number of known selenoproteins has been found in zebrafish. The reason why zebrafish (and probably other fish) accumulated many selenoproteins is not clear. It is possible that this phenomenon is related to a relatively uniform distribution of selenium in seas and oceans. Interestingly, zebrafish accumulated multiple homologs of many mammalian selenoproteins (GPx1, GPx4, SelW, SelP, and SelT) through gene and genome duplications. In addition, in at least one of these proteins, it replaced Cys residues with Sec. This protein is zebrafish selenoprotein Pa, which contains 17 Sec residues, the largest number of Sec in any known selenoprotein. A homolog of this protein in mammals contains 10–12 Sec residues, and the additional Sec residues in the zebrafish protein correspond to Cys residues in mammalian selenoprotein P.

The observation of a larger number of selenoproteins encoded in vertebrate genomes compared with that encoded in the genomes of lower eukaryotes suggests that there has been a tendency in evolution toward an expansion in selenoprotein number in vertebrates.[25] Currently, vertebrate selenoproteins account for all known eukaryotic selenoproteins (i.e., those found in lower eukaryotes also occur in vertebrates, but not vice versa), and several selenoproteins, such as selenoproteins P and N, have been found exclusively in vertebrates. It should be noted that the examples of the loss of Sec during evolution are also known. For example, *Saccharomyces cerevisiae* and *Arabidopsis thaliana* genomes do not encode selenoproteins, but selenoproteins were likely present in a primordial organism that gave rise to bacteria, archaea, and eukaryotes.

Independent of the mechanism for Sec evolution, the location of Sec in selenoproteins generally indicates an important catalytic or regulatory site. As such, this information helps to functionally characterize these proteins. An additional feature for the majority of known eukaryotic selenoproteins is the occurrence of homologs that contain Cys in place of Sec (Fig. 2). This phenomenon is clearly seen when vertebrate selenoproteins are compared with their homologs in lower eukaryotes. Sec and Cys exhibit similar chemical properties but also have important differences. While the majority of Cys residues are protonated at the physiological pH, Sec residues are ionized. In addition, Sec is usually a better nucleophile and reductant than Cys.[26] Thus, selenoproteins utilize the unique chemistry of Sec, and this is consistent with the observation that the replacement of Sec with Cys in natural selenoproteins functionally penalize these proteins,[26] whereas the replacement of Cys with Sec in natural Cys-containing homologs of selenoproteins may result in superior catalysts.[27]

[25] V. N. Gladyshev and G. V. Kryukov, *Biofactors* **14,** 87 (2001).
[26] T. C. Stadtman, *Annu. Rev. Biochem.* **65,** 83 (1996).
[27] S. Hazebrouck, L. Camoin, Z. Faltin, A. D. Strosberg, and Y. Eshdat, *J. Biol. Chem.* **275,** 28715 (2000).

The presence of Sec in selenoproteins in positions that are occupied by Cys in homologous lower eukaryotic sequences may be viewed as an evolutionary advancement of protein function that is manifested by Sec insertion in place of Cys. This process may be contrasted with the evolutionary pressure of eliminating Sec utilization as this amino acid is rarely used in protein synthesis, has numerous unique factors dedicated solely to its biosynthesis and incorporation into protein, and the availability of selenium to support Sec biosynthesis and insertion is extremely low. This view of evolutionary advancement (to achieve superior function) versus pressure (to remove an infrequently, noncanonically used amino acid when it can be substituted by one of the standard 20 amino acids) is consistent with the presence of Sec/Cys pairs in homologous sequences of lower and higher organisms and possibly explains the fact that selenoprotein genes are present in all domains of life, but are also absent in certain representatives of each of these domains.

There are notable exceptions with respect to the presence of the Sec/Cys pair in homologous sequences. For example, selenoprotein N is specific for vertebrates and does not have known homologs, including Cys-containing homologs.[18] In fact, only a human cDNA sequence is currently available for this protein, which makes it difficult to fully analyze selenoprotein N with respect to common features found in selenoprotein genes and their homologs. Another vertebrate selenoprotein that lacks Cys homologs is thyroid hormone deiodinase. However, three isoenzyme genes are known for this protein that contain conserved Sec-encoding UGA codons and SECIS elements. Thus, the use of the Sec/Sec pair instead of the Sec/Cys pair helps to define deiodinases as selenoproteins.

Mammalian Selenocysteine Insertion Sequence Element

Although Sec-encoding UGA codons and Cys/Sec pairs help to identify and characterize selenoprotein genes, the only functional element that is present in all known selenoprotein genes and has features that allow direct computational searches is the SECIS element. Comparison of SECIS elements in known mammalian selenoprotein genes and experimental probing of these structures resulted in the identification of a mammalian SECIS element consensus sequence and structure[2,6,7,20,28] (Fig. 1A).

The mammalian SECIS element is composed of Helices I and II, Internal and Apical loops, and a non-Watson–Crick base-paired SECIS core, the Quartet. Conserved primary sequences of the stem–loop structure include TGA in the 5' segment and GA in the 3' segment of the Quartet, and an unpaired AR (R is purine) in the Apical loop. The most characteristic secondary structure feature found in mammalian SECIS elements is the length of Helix II, which separates the

---

[28] E. Grundner-Culemann, G. W. Martin III, J. W. Harney, and M. J. Berry, *RNA* **5**, 625 (1999).

Quartet and AR in the Apical loop by 11 or 12 nucleotides. Although the initial SECIS element consensus included additional conserved sequences, these features can no longer be viewed as strictly conserved. These formerly conserved sequences include an A directly preceding the Quartet and an AAA motif in the Apical loop. Figure 1B shows the predicted SECIS element in the human thioredoxin reductase 2 gene, which is the example of the SECIS element that lacks these features.

SECIS elements may be divided into two related but distinct subfamilies, type I and type II.[28] These subfamilies are different in the area of the Apical loop; that is, type II SECIS elements have an additional ministem that places AR in the bulge. These two types of SECIS elements are interconvertible by mutations that remove or create the ministem, suggesting a similar structure and function of both SECIS element types.

SECIS element-binding protein(s) bind to SECIS elements[3] and the resulting complex is most likely present as a relatively stable structure during every step in protein synthesis. This property allows utilization of criteria for the free energy of this stem–loop structure in addition to criteria for primary and secondary consensus structures (Fig. 1A). When the free energies predicted from a computer analysis of mRNA sequences were analyzed for known SECIS elements, all stem–loop structures exhibited similar free energy parameters.[20] No differences were observed in the free energy values for type I and type II SECIS elements, suggesting that the ministem serves to stabilize the stem–loop structure. The ministem seems to be formed when an Apical loop is large enough to destabilize the structure, but it may not be required for the SECIS elements with smaller Apical loops. In addition, the ministem may serve to maintain thermodynamic stability or to nucleate SECIS element folding.[28] The commonality in the energetic criteria for SECIS elements allowed us to maintain a single model, shown in Fig. 1A, for designing an algorithm that searches for novel SECIS elements.[20]

Identification of Selenocysteine Insertion Sequence Elements

Candidate SECIS elements can be found in nucleotide sequences by using the SECISearch program specifically designated for this purpose.[20] The algorithm currently utilized by SECISearch involves three stages (Fig. 3).

1. Initially, a collection of nucleotide sequences is searched for the SECIS element primary sequence consensus: DTGA/11–12 nucleotides/AR/18–27 nucleotides/GA. This consensus is weak and does not allow identification of SECIS elements per se, but it significantly reduces the number of sequences for further analyses and anchors candidate sequences for subsequent secondary structure consensus analyses.

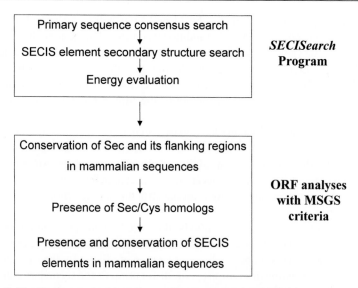

FIG. 3. Identification of selenoprotein genes with SECISearch and MSGS. Shown is a strategy that is a used to search for new selenoprotein genes in large nucleotide sequence databases. A collection of sequences is analyzed with the SECISearch program, which searches for primary consensus sequence, secondary consensus structure, and the free energy parameters characteristic of SECIS elements in known selenoprotein genes. Sequences selected by the program are manually analyzed for the presence of open reading frames that satisfy mammalian selenoprotein gene signature (MSGS) criteria.

2. The second step verifies whether candidate sequences selected in the first step fit to the SECIS element secondary structure consensus that includes Helices I and II and Internal and Apical loops (Fig. 1A).

3. In the last step, the free energies of candidate structures are evaluated and if they fit the assigned range, candidate sequences are reported as putative SECIS elements.

The first module of the SECISearch program that is responsible for recognition of SECIS element primary sequence and secondary structure consensuses is based on the PatScan program (*http://www-unix.mcs.anl.gov/compbio/PatScan/HTML/ patscan.html*). The core of the free energy evaluating module is the Vienna RNA package RNAfold program[29] (*http://www.tbi.univie.ac.at/~ivo/RNA*). It separately estimates the free energies for Helix I plus Internal loop and Helix II plus Apical loop regions of the putative SECIS element (Fig. 1). The free energy cutoff parameters are based on free energy minimization calculations of SECIS elements

[29] I. L. Hofacker, W. Fontana, P. F. Stadler, S. Bonhoeffer, M. Tacker, and P. Schuster, *Monat. Chem.* **125,** 167 (1994).

in previously characterized eukaryotic selenoprotein mRNAs and are typically set as -7 kcal/mol for Helix I and Internal loop and -5 kcal/mol for Helix II and Apical loop. Changing the SECISearch parameters modulates the specificity and sensitivity of the program, which in turn allows tuning of the program either for SECIS element search in a candidate selenoprotein mRNA or gene (that demands increased sensitivity) or for genome-wide screening for potential selenoprotein genes (that requires increased specificity).

Mammalian Selenoprotein Gene Signature

Three features emerge from the above-described analyses of structural, mechanistic, and evolutionary properties of selenoprotein genes: the Sec-encoding UGA codon, the SECIS element (Fig. 1), and the Sec/Cys pair (Fig. 2) in homologous sequences. Analysis of these features allows definition of criteria that formally describe selenoprotein genes (Fig. 2). As these criteria identify a signature for mammalian selenoprotein genes, we have designated them as the mammalian selenoprotein gene signature (MSGS). According to the MSGS, a protein-encoding gene contains a Sec codon if the following criteria are met.

1. Sec is conserved in orthologs and is flanked by homologous sequences, that is, conservation of the Sec-encoding TGA and of sufficiently long Sec-flanking amino acid sequences within the gene for this protein among mammals. This requirement is consistent with the fact that Sec is essential for biological activity of selenoproteins and is located in functionally important regions of polypeptide chains.

2. There are distinct Cys- and/or Sec-containing homologs, that is, occurrence of genes (usually in lower eukaryotes) that contain a Cys codon in place of TGA in the gene for this protein (or the occurrence of distinct homologous genes that conserve TGA) and meet an additional criterion of encoding sufficiently long homologous sequences on both sides of the Cys/Sec (or Sec/Sec) pair. The presence of Cys-containing homologs is an important criterion for selenoprotein identification.

3. The SECIS element is conserved, that is, conservation of the SECIS element sequence and structure in the 3' UTRs of mammalian genes for this protein. The requirement for the SECIS element to achieve Sec incorporation imposes strong selective constraints. At the same time it should be noted that the position of the SECIS element in the 3' UTR does not need to be conserved.

According to MSGS, a typical new mammalian selenoprotein will have not only a conserved SECIS element in the 3' UTR of mRNA and conserved Sec and Sec-flanking regions in the protein sequence in other mammals, but it will also have eukaryotic homologs that contain a Cys residue in place of Sec (or distinct homologs that conserve Sec).

Applicability of Mammalian Selenoprotein Gene Signature Criteria

The applicability of MSGS was tested by searching the entire database of 3' UTRs of human genes with SECISearch. Sequences selected by the program were further analyzed with MSGS criteria through homology analyses of putative SECIS elements and Sec-flanking areas. All known human selenoprotein genes were found by these procedures. However, we did not detect other sequences in this nonredundant database that satisfied all MSGS criteria but could be classified as nonselenoproteins. Thus, MSGS is highly specific for previously characterized human selenoprotein genes.

In addition to mammalian and other vertebrate selenoproteins, MSGS analysis correctly identified selenoprotein genes in a number of lower eukaryotic organisms, such as *Dictyostelium discoideum* (the slime mold), *Drosophila melanogaster,* *Caenorhabditis elegans, Schistosoma mansoni* (trematode), and *Leishmania major* (parazitic protozoa). These data suggest that the SECIS element structure (and possibly the Sec incorporation system) are conserved in all eukaryotes, and that MSGS may be applied for identification of selenoprotein genes in all eukaryotes. However, here we designate selenoprotein gene criteria only for mammalian systems, because Sec incorporation has not been studied in detail in nonmammalian eukaryotes, and it is not known whether alterations from MSGS occur in certain eukaryotic genomes. It should also be noted that SECIS element primary sequences do not exhibit significant homology for distantly related eukaryotic species.

Other Approaches for Identification of Selenoprotein Genes

The previous method of choice for identification of new selenoproteins has been the isolation of ^{75}Se-labeled proteins followed by chromatographic identification of Sec in protein hydrolysates. This approach is labor intensive, includes the use of radioactive materials, and requires relatively high expression levels of selenoproteins. The use of bioinformatics approaches appears to be advantageous in the current situation, when all most abundant selenoproteins are already known.[30]

In addition to the development of SECISearch by our group, a group led by A. Krol employed a similar strategy of searching for selenoproteins by identification of SECIS elements.[18] These efforts resulted in the identification of two new selenoproteins, selenoproteins N and X (the latter protein is identical with selenoprotein R that was identified by SECISearch), and two alternative splicing forms of a mitochondrial thioredoxin reductase. In this study, experimental screens were used to test whether candidate stem–loop structures were functional SECIS elements.

In addition to SECIS element searches followed by ORF analyses using MSGS criteria, there are possible alternative strategies to identify selenoprotein genes in

---

[30] V. N. Gladyshev and D. L. Hatfield, *in* "Current Protocols in Protein Science," p. 3.8.1 (2000).

TABLE II
EXAMPLES OF IDENTIFICATION OF SELENOPROTEIN GENES IN NUCLEOTIDE
SEQUENCE DATABASES

	Database	
Parameter	Human dbEST[a]	*Drosophila melanogaster* genome[b]
Size of database	~610 Mb	~137 Mb
Candidate SECIS elements selected by SECISearch and analyzed by MSGS	974	179
Number of known selenoprotein genes that satisfied MSGS criteria	12	0
New selenoproteins identified	2	3
Number of false positives (nonselenoprotein genes that satisfied MSGS criteria)	0	0

[a] Human dbEST was analyzed by the initial version of SECISearch when 14 human selenoproteins were known (April 1999) [G. V. Kryukov, V. M. Kryukov, and V. N. Gladyshev, *J. Biol. Chem.* **274**, 33888 (1999)]. The new selenoproteins are SelR and SelT.

[b] *Drosophila* genome was analyzed by the advanced version of SECISearch when the number of known selenoproteins was 17 (May 2000) [F. J. Martin-Romero, G. V. Kryukov, A. V. Lobanov, B. J. Lee, V. N. Gladyshev, and D. L. Hatfield, *J. Biol. Chem.* **276**, 29798 (2001)]. The new selenoproteins are G-rich, BthD, and SPS2.

genomic databases. One such strategy is the use of homology analyses of Sec-flanking sequences (identification of homologous sequences that contain Cys in place of Sec in candidate selenoprotein sequences). This approach has not yet been used in database searches, but the presence of Sec/Cys pairs in homologous sequences is used as one of the MSGS criteria.

Strategy to Search for Selenoproteins in Nucleotide Sequence Databases

The above-described approaches may be used for identification of new seleno-proteins in large nucleotide sequence databases and the general strategy for such searches is outlined in Fig. 3. Below, we show two examples of such analyses (Table II).[20,24]

Human dbEST Analysis Using SECISearch/MSGS Approach

The human expressed sequence tag database (dbEST) was searched with SE-CISearch. Nucleotide sequences that contain putative SECIS elements were further extended by generating dbEST consensus sequences. Sequences corresponding to the known selenoprotein genes were excluded and the remaining extended sequences were analyzed for the presence of open reading frames (ORFs) allowing in-frame TGA codons to be interpreted as either Sec residues or stop signals. ORFs containing in-frame TGA codons were identified and analyzed by applying MSGS. This procedure resulted in selective enrichment of known selenoprotein genes and, in addition, led to identification of two new selenoproteins, SelT and SelR.[20]

Drosophila melanogaster Genome Analysis Using SECISearch/MSGS Approach

The entire *Drosophila melanogaster* genomic sequence was analyzed for the presence of SECIS elements[24] using an advanced version of SECISearch (G. V. Kryukov, A. V. Lobanov and V. N. Gladyshev, unpublished) (Table II). The location of each putative SECIS element in the genome was determined with the annotated version of the *D. melanogaster* genome. The sequences that were present in 5′ UTRs, introns, or coding regions of previously predicted proteins on either DNA strand were discarded. MSGS criteria were applied to analyze the upstream ORFs closest to the predicted candidate SECIS elements. Such *D. melanogaster* genome analysis resulted in the identification of three selenoprotein genes: selenophosphate synthetase 2 homolog and two selenoproteins, 12-kDa G-rich and 28-kDa BthD, with no homology to known proteins.[24]

Strategy to Test Whether a Newly Isolated Gene Encodes a Selenoprotein

Besides the analysis of nucleotide sequence databases for new selenoprotein genes, MSGS criteria may be used to determine the functional identity of in-frame UGA codons in newly isolated genes (Fig. 4).[30] Such a test may result in the identification of a new selenoprotein, or if MSGS criteria are not satisfied, in assigning a terminator function to the UGA. When a potential selenoprotein gene is found, several questions may be asked that relate to the conserved features observed in known selenoprotein genes. Does this gene have a putative SECIS element in the 3′ UTR? If the gene has an in-frame TGA codon and a putative SECIS-like structure, are these features conserved in orthologs in other mammals? Are there distinct homologous proteins that have a Cys codon, TGC or TGT, in place of the Sec codon? If answers to these questions are negative, it is unlikely that the new gene encodes a selenoprotein. It should be noted, however, that insufficient information about the new gene or its homologs might prevent definitive answers to some of these questions. For example, SECIS elements are often located in the very 3′ ends of mRNAs. If this portion of cDNA is lacking, a SECIS element cannot be easily identified. In addition, variations from conserved features found in known SECIS elements may be anticipated to occur in at least some new selenoprotein genes. Additional difficulties in answering questions in Fig. 4 may arise if a new

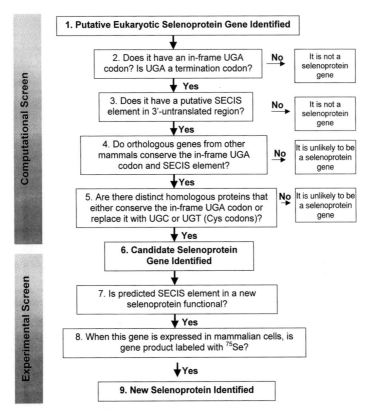

FIG. 4. Strategy to test whether a new gene encodes a selenoprotein. See text for details.

gene does not have known homologous genes in other species. Therefore, further experimental verification of a new selenoprotein gene may be required.

The experimental design in identifying new Sec-containing protein genes may include determining whether the gene product incorporates selenium when it is expressed in mammalian cells. One technique to address this question is to label cells with ^{75}Se that are transfected with a construct that expresses the new gene. However, the molecular mass for a new selenoprotein should not correspond to the known masses for major natural selenoproteins expressed in host cells, such as thioredoxin reductase (57 kDa), glutathione peroxidase 1 (25 kDa), glutathione peroxidase 4 (20 kDa), and Sep15 (15 kDa). This can be achieved by expressing the new gene as a fusion with a tag protein, such as the green fluorescent protein.[20] If all four questions in the computational screen are answered positively, and if the overexpressed protein is labeled with ^{75}Se, the data may be considered as proof for identification of a new selenoprotein gene.

Functional Analyses of New Selenoproteins

The use of bioinformatics methods is not limited to identification of new selenoprotein genes. Often, important information about functional properties of a new protein or about the metabolic process in which it is involved may be gained from further bioinformatics analyses. Such methods go beyond sequence analyses of new proteins, which are homologous to proteins of known function. We further illustrate several methods, including gene clustering,[31] domain fusion,[32,33] and phylogenetic profiles,[34] using selenoprotein R (SelR) as an example (G. V. Kryukov, V. N. Gladyshev, unpublished). SelR was identified by a human dbEST search using the SECISearch/MSGS approach, and it does not have homology to any identified protein of known function.[20] Homologs of mammalian SelR in bacteria, archaea, and lower eukaryotes contain Cys in place of Sec.

Phylogenetic Profiles: Correlated Evolution

If two or more proteins are involved in the same branch or pathway of cellular metabolism, they tend to be either all present or all absent in a genome, and such gene conservation is called correlated evolution. Similarity in patterns of distribution of genes in completely sequenced genomes, phylogenetic profiles, allow metabolic linkages of genes. Comparison of phylogenetic profiles for SelR homologs and peptide methionine sulfoxide reductase (MsrA) in completely sequenced genomes reveals that they are identical (Fig. 5A). Such phylogenetic pattern is not seen for any other genes in encoded in these genomes. This level of correlation implies a strong linkage between functions for these two proteins.

Gene Clusters

In bacteria, genes of proteins that are involved in a common metabolic pathway are often clustered to allow synchronous regulation. If two genes are adjacent to each other in several analyzed genomes, especially in distantly related organisms, it is suggested that proteins encoded by these genes may participate in the same or related metabolic pathways. If the function of one of these genes is known, it helps to determine the function of the second protein. Analysis of bacterial SelR homologs reveals that genes encoding SelR homologs are often located immediately upstream or downstream of the MsrA gene (Fig. 5B).

---

[31] R. Overbeek, M. Fonstein, M. D'Souza, G. D. Pusch, and N. Maltsev, *Proc. Natl. Acad. Sci. U.S.A.* **96,** 2896 (1999).

[32] E. M. Marcotte, M. Pellegrini, H. L. Ng, D. W. Rice, T. O. Yeates, and D. Eisenberg, *Science* **285,** 751 (1999).

[33] A. J. Enright, I. Iliopoulos, N. C. Kyrpides, and C. A. Ouzounis, *Nature (London)* **402,** 86 (1999).

[34] M. Pellegrini, E. M. Marcotte, M. J. Thompson, D. Eisenberg, and T. O. Yeates, *Proc. Natl. Acad. Sci. U.S.A.* **96,** 4285 (1999).

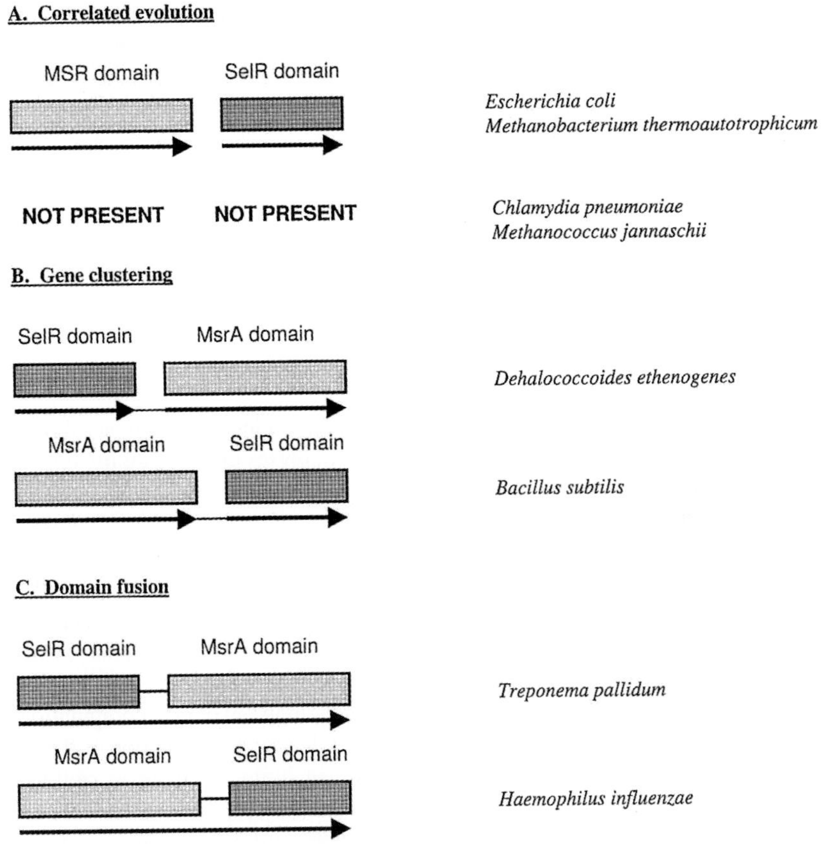

A. Correlated evolution

MSR domain   SelR domain

Escherichia coli
Methanobacterium thermoautotrophicum

NOT PRESENT   NOT PRESENT

Chlamydia pneumoniae
Methanococcus jannaschii

B. Gene clustering

SelR domain   MsrA domain

Dehalococcoides ethenogenes

MsrA domain   SelR domain

Bacillus subtilis

C. Domain fusion

SelR domain   MsrA domain

Treponema pallidum

MsrA domain   SelR domain

Haemophilus influenzae

FIG. 5. Functional analyses of SelR that functionally link this protein to peptide methionine sulfoxide reductase (MsrA). (A) Analyses of correlated evolution (phylogenetic patterns). (B) Analyses of gene clustering (phylogenetic profiles). (C) Domain fusion analyses. Open boxes indicate MsrA, and filled boxes indicate SelR domains in organisms shown on the right. Arrows indicate the $5' \rightarrow 3'$ direction of the genes.

Domain Fusion

Many proteins that physically interact or are involved in consecutive or closely interrelated metabolic steps may be fused into a single polypeptide in certain organisms. Thus, the occurrence of a fusion protein, which in other species is encoded by two genes, allows linking the two genes in the same or related pathway or biological process. Domain fusion analyses may be performed using Protein Families (Pfam), Clusters of Orthologous Groups (COG), or other bioinformatics tools and databases. Analysis of SelR reveals that in many organisms SelR homologs are fused either upstream or downstream of the MsrA domain (Fig. 5C).

In a rare case, when the potential partner protein is predicted by the three independent methods, the reliability of prediction is high, but if only one of the methods predicts interaction or function, additional computational or experimental analyses may be required. Additional computational techniques for assessing protein function may include analyses of correlated expression profiles (e.g., correlation of changes in mRNA expression for two or more genes under different cellular conditions assessed by cDNA microarrays or microchips) and analyses of genome-wide protein–protein interactions. The function of SelR is not yet known; however, it is clearly linked to that of peptide methionine sulfoxide reductase by the three described bioinformatics approaches.

Concluding Remarks

MSGS is a set of criteria to recognize mammalian selenoprotein genes through identification of SECIS elements and homology analyses of Sec-flanking regions. MSGS is also useful for distinguishing between TGA terminator signals and TGA Sec codons within ORFs. MSGS is specific for mammals, and likely other eukaryotic selenoprotein genes. MSGS may predict, if sufficient information is available for new cDNAs or genomic DNAs, whether a given sequence encodes a Sec-containing protein. Thus, MSGS should be useful for analyses of nucleotide sequences generated by genome and EST sequencing projects and for characterization of new eukaryotic cDNA sequences. The use of the described bioinformatics tools (SECISearch, MSGS criteria) and further development of new computational approaches that assess protein function will likely lead to identification of the majority of selenoproteins in mammalian genomes, including humans, as well as to elucidating the function of many of these proteins. This functional information about human selenoproteins should provide new scientific directions that will attempt to link selenoprotein functions to the biological and biomedical effects of selenium.

Acknowledgment

This work was supported by NIH GM61603 (to V.N.G.).

[9] Estimation of Individual Types of Glutathione Peroxidases

By REGINA BRIGELIUS-FLOHÉ, KIRSTEN WINGLER, and CORDULA MÜLLER

Introduction

For more than a decade the classic glutathione peroxidase (cGPx),[1] the gene product of *gpx1*, remained not only the only type of glutathione peroxidase but also the only selenoprotein identified in mammals. At present 19 mammalian selenoproteins have been described,[2] among them 4 distinct glutathione peroxidases: cGPx, the plasma GPx (pGPx), the phospholipid hydroperoxide GPx (PHGPx), and the gastrointestinal GPx (GI-GPx). This redundancy, at first glance, has led to the impression that nature considers removal of hydroperoxides so important that it has created multiple lines of defense to make sure that every single hydroperoxide molecule is destroyed immediately. Several aspects, however, argue against this assumption.

1. The individual types of glutathione peroxidases are not evenly distributed in the organism. cGPx is the most ubiquitous GPx. pGPx is an extracellular enzyme found in blood plasma; it is released from the kidney into the plasma. Other sites of synthesis are placenta, large intestine, or lung (reviewed by Brigelius-Flohé[3]). It thus might be responsible for extracellular hydroperoxide removal, especially at interfaces protecting tissues from environmental influences. This interpretation is, however, problematic because of the low extracellular concentration of glutathione, which cannot guarantee a sustained reduction of hydroperoxides. PHGPx is preferentially expressed in testis. It has, therefore, been postulated that this particular enzyme is responsible for the protection of sperm from oxidative damage. High PHGPx activity is, however, found only in immature spermatids, not in mature sperm cells.[4] Sperm cells contain a high concentration of PHGPx protein, which obviously is inactive. A putative role for GI-GPx was deduced from its preferential expression in the gastrointestinal epithelium[5]: prevention of hydroperoxide absorption. The identification of GI-GPx in tumor cell lines not derived from the intestine,[6] the increase in its mRNA[7] and protein[8] in colon cancer cells, its

---

[1] G. C. Mills, *J. Biol. Chem.* **229**, 189 (1957).

[2] L. Flohé, J. R. Andreesen, R. Brigelius-Flohé, M. Maiorino, and F. Ursini, **49**, 411 (2000).

[3] R. Brigelius-Flohé, *Free Radic. Biol. Med.* **27**, 951 (1999).

[4] M. Maiorino, J. B. Wissing, R. Brigelius-Flohè, F. Calabrese, A. Roveri, P. Steinert, F. Ursini, and L. Flohè, *FASEB J.* **12**, 1359 (1998).

[5] F. F. Chu, J. H. Doroshow, and R. S. Esworthy, *J. Biol. Chem.* **268**, 2571 (1993).

[6] R. S. Esworthy, M. A. Baker, and F. F. Chu, *Cancer Res.* **55**, 957 (1995).

METHODS IN ENZYMOLOGY, VOL. 347   0076-6879/02 $35.00

localization in Paneth cells in human ileum,[8] and its specific intracellular localization in cells in the apical part of colon crypts,[8,9] however, hardly support the barrier hypothesis.

2. The individual glutathione peroxidases do not respond equally to selenium supply. This phenomenon has been addressed in the "hierarchy of selenoproteins." Selenoproteins ranking low in the hierarchy respond to selenium deprivation with a rapid loss of activity, protein, and even mRNA levels. Those ranking high in the hierarchy respond slowly in terms of activity and protein, and the mRNA levels remain stable. The hierarchy of glutathione peroxidases is as follows[3]: GI-GPx > PHGPx > pGPx = cGPx.

3. Reverse genetics ruled out the notion that cGPx is a vital enzyme.[10] cGPx$^{-/-}$ mice developed and grew normally. The hypothesis that cGPx represents a major antioxidant device, however, could be corroborated. cGPx$^{-/-}$ mice were more sensitive to poisoning with the redox cycling herbicides paraquat[11,12] and diquat.[13] cGPx$^{-/-}$ mice died within hours irrespective of selenium supplementation as did selenium-deficient wild-type mice, whereas selenium-adequate wild-type mice survived. This indicates that none of the other selenoproteins, notably none of the other glutathione peroxidases, can protect mice against oxidative stress as efficiently as cGPx. Thus, only cGPx functions as an antioxidative enzyme; the other glutathione peroxidases must have different roles.

This has been convincingly demonstrated for PHGPx. Apart from modulating intracellular signaling, reviewed by Brigelius-Flohé et al.,[14] it has been discovered as an enzymatically inactive structural protein in spermatozoa.[15] Whether the other types of glutathione peroxidases, pGPx and GI-GPx, similarly exert surprising

[7] H. Mörk, O. H. al-Taie, K. Bähr, A. Zierer, C. Beck, M. Scheurlen, F. Jakob, and J. Köhrle, Nutr. Cancer **37,** 108 (2000).

[8] S. Florian, K. Wingler, K. Schmehl, G. Jacobasch, O. J. Kreuzer, W. Meyerhof, and R. Brigelius-Flohé, Free Rad., in press (2001).

[9] R. Brigelius-Flohé, C. Müller, J. Menard, S. Florian, K. Schmehl, and K. Wingler, Biofactors **14,** 101 (2001).

[10] Y. S. Ho, J. L. Magnenat, R. T. Bronson, J. Cao, M. Gargano, M. Sugawara, and C. D. Funk, J. Biol. Chem. **272,** 16644 (1997).

[11] J. B. de Haan, C. Bladier, P. Griffiths, M. Kelner, R. D. O'Shea, N. S. Cheung, R. T. Bronson, M. J. Silvestro, S. Wild, S. S. Zheng, P. M. Beart, P. J. Hertzog, and I. Kola, J. Biol. Chem. **273,** 22528 (1998).

[12] W. H. Cheng, Y. S. Ho, B. A. Valentine, D. A. Ross, G. F. Combs, Jr., and X. G. Lei, J. Nutr. **128,** 1070 (1998).

[13] Y. Fu, W. H. Cheng, J. M. Porres, D. A. Ross, and X. G. Lei, Free Radic. Biol. Med. **27,** 605 (1999).

[14] R. Brigelius-Flohé, F. Ursini, M. Maiorino, and L. Flohé, in "Handbook of Antioxidants: Biochemical, Nutritional, and Clinical Aspects" (E. Cadenas and L. Packer, eds.). Marcel Dekker, New York, 2001.

[15] F. Ursini, S. Heim, M. Kiess, M. Maiorino, A. Roveri, J. Wissing, and L. Flohé, Science **285,** 1393 (1999).

functions remains to be elucidated. One of the prerequisites to identifying the roles of these enzymes is the availability of assays to differentiate the isotypes, by either activity, RNA, or protein level.

Determination by Activity

Assay System

Estimation of glutathione peroxidase activity is not simple. Several test systems have been described and discussed[16–18] and are not to be repeated here in detail. Instead, some important points are noted that must be considered in order to obtain a reliable determination. Activity is usually measured in a test coupled to the glutathione reductase-catalyzed reduction of oxidized glutathione (GSSG) at the expense of NADPH

$$ROOH + 2GSH \xrightarrow{GPx} ROH + GSSG + H_2O \qquad (1)$$

$$GSSG + NADPH + H^+ \xrightarrow{GR} 2GSH + NADP^+ \qquad (2)$$

where GR is glutathione reductase, and GSH is reduced glutathione. Reaction (1) is a two-substrate reaction characterized as a ping–pong mechanism.[19] ROOH is reduced first, and then the resulting alcohol is released. The oxidized enzyme (containing a selenenic acid in the active center) is reduced back by a stepwise reaction with two molecules of GSH.[19] For bovine cGPx infinite limiting maximum velocities and Michaelis constants for both substrates, hydroperoxide and GSH, have been determined.[16] In consequence, conditions for estimation of international units requiring "saturating" concentrations of all substrates are not applicable. This is a general phenomenon characteristic of all glutathione peroxidases tested in this respect so far.[20,21] The implication of the peculiar kinetics of glutathione peroxidases must be considered in the choice of proper substrate concentrations for the assay and in the definition of the unit of activity.

Reaction (1) is dependent not only on the concentration of the hydroperoxide but also on the concentration of GSH. The apparent maximum velocities and Michaelis constants for one substrate are linear functions of the co-substrate concentration. Under such conditions, a reasonable compromise

[16] L. Flohé, G. Loschen, W. A. Günzler, and E. Eichele, *Hoppe Seylers Z. Physiol. Chem.* **353**, 987 (1972).

[17] L. Flohé and W. A. Günzler, *Methods. Enzymol.* **105**, 114 (1984).

[18] R. Brigelius-Flohé, K. Lötzer, S. Maurer, M. Schultz, and M. Leist, *Biofactors* **5**, 125 (1995).

[19] L. Flohé, in "Glutathione: Chemical, Biochemical and Medical Aspects" (D. Dolphin, R. Poulson, and O. Avramovic, eds.), Part A, p. 643. John Wiley & Sons, New York, 1989.

[20] F. Ursini, M. Maiorino, and C. Gregolin, *Biochim. Biophys. Acta* **839**, 62 (1985).

[21] R. S. Esworthy, F.-F. Chu, A. W. Girotti, and J. H. Doroshow, *Arch. Biochem. Biophys.* **307**, 29 (1993).

is to measure GPx activities at a fixed regenerated GSH concentration and at a hydroperoxide concentration that is high enough not to substantially influence the initial reaction rate.

Reaction (1) strongly depends on the pH. The pH optimum is pH 8.7, with a steep increase in the pH dependence of the reaction rate starting from pH 7.[16] Slight variations in the pH of the assay system thus result in major variations in activities.

The temperature optimum is near the stability limit at 43°,[22] with a steep increase between 30 and 43°. This means that a minor deviation from a given temperature results in major differences in activities.

Hydroperoxides react with GSH spontaneously. The rate of the spontaneous reaction is different for different hydroperoxides and also increases with pH. It must be estimated separately and subtracted from the overall reaction rate obtained in the presence of enzyme.

Assay conditions (final concentrations) providing reproducible results both for tissue extracts and cultured cells, for example, are as follows: 100 mM Tris-HCl, pH 7.6; 5 mM EDTA; 1 mM sodium azide; 3 mM GSH; 0.1 mM NADPH; 0.1% (v/v) peroxide-free Triton X-100; and 600 mU of glutathione reductase (GR), for example, from Sigma (St. Louis, MO) type III baker's yeast (117 U/ml). GSH, NADPH, and GR must be freshly prepared daily. The reaction is started by the addition of 10 μl of 5 mM hydroperoxide (final concentration, 50 μM). These assay conditions result in a pseudo zero-order rate because GSH remains constant due to regeneration and the hydroperoxide concentration is close to the apparent V_{max} (see comments to Table I). The turnover of hydroperoxides is calculated from the decrease in NADPH measured at 340 nm, using Beer's law.

Comments. GPx activities are comparable only if measured under identical conditions. As the reaction rate depends on the steady state level of GSH, the assay coupled with the regeneration of GSH by glutathione reductase, reaction (2), is most suitable for reproducible results. To avoid misinterpretations, enzymatic units expressed as ΔNADPH per minute must be given together with clearly defined conditions applied in the test system. As hydroperoxide is a substrate for all glutathione peroxidases, H_2O_2 or *tert*-butyl hydroperoxide (*t*-BOOH) can be used. *t*-BOOH has a lower spontaneous reaction rate with GSH than H_2O_2 and therefore is the preferred substrate to measure low GPx activities.

Rate constants for the reaction with both substrates differ between GPx isotypes (Table I). The implications of the kinetic constants compiled in Table I with respect to GPx determinations are as follows.

1. If equivalent concentrations of the oxidizing and the reducing substrate are chosen, the enzyme is present close to 100% as the oxidized form. In consequence,

---

[22] F. Schneider and L. Flohé, *Hoppe Seylers Z. Physiol. Chem.* **348,** 540 (1967).

TABLE I
RATE CONSTANTS FOR REACTIONS OF GLUTATHIONE PEROXIDASES[a]

Substrate	k_{+1}				k_{+2}	Ref.
	H_2O_2	t-BOOH	Cumene-OOH	PCOOH[b]	(GSH)	
cGPx	59	7.5	12.8	—	0.45	16[c]
	50	11.7	17	—	0.80	27[d]
	86			—	0.25	21[e]
PHGPx	3.0	1.2	1.8	12.1	0.006–0.15	20[f]
pGPx	33			ND	0.075	21[e]

[a] Apparent second-order rate constant k_{+1} defined by $v = k_{+1}[ROOH][GPx_{red}]$ for the reaction of glutathione peroxidases with different hydroperoxides, and k'_{+2} defined as $k_{+2} + k_{+3}$ in the rate equation $v = k_{+2}[GPx_{ox}][GSH] + k_{+3}[GPx_{ox} - GSH][GSH]$, which describes the two successive reactions of glutathione peroxidases with GSH. The rate constants are given in μM^{-1} sec^{-1}.

[b] —, No reaction; ND, Not determined.

[c] cGPx from bovine blood measured at pH 6.7.

[d] cGPx from hamster liver measured at pH 7.6.

[e] pGPx from human plasma and cGPx from human erythrocytes (Sigma) measured at pH 7.6.

[f] PHGPx from pig heart measured at pH 7.6.

the activity measurement is determined by k'_{+2}; more precisely the activity is then defined as $v = k'_{+2}[GSH][GPx_{ox}]$. For practical purposes equimolar substrate concentrations are not feasible, for example, because of solubility limits of most organic hydroperoxides. In most cases, constant substrate consumption is nevertheless obtained if the hydroperoxide concentration is not lower than 2% of the GSH concentration and the reaction is made pseudo first order by keeping the GSH concentration constant through regeneration. At lower relative concentrations of ROOH, the activity becomes codetermined by k_{+1}, which means the slopes deviate from linearity.

2. Appropriate test conditions are difficult to recommend without reliable knowledge of the kinetic constants, as, for example, for GI-GPx. Differences of rate constants between species also cannot be excluded. A suitable choice of substrate concentrations according to the rules outlined above must be validated in each case.

3. The differences in the rate constants between the isotypes of GPx preclude an estimation of enzyme molarities in mixtures of unknown relative compositions.

4. Because activity measurements ideally reflect the k'_{+2} values, identical "units" represent more moles of pGPx or PHGPx than cGPx.

Differentiation by Choice of Substrates

Individual glutathione peroxidases exert different levels of specificity for hydroperoxide substrates. All glutathione peroxidases, however, use H_2O_2 and

t-BOOH, which are the most commonly used substrates for the estimation of GPx activity. The respective kinetic constants (k_{+1} and k'_{+2}) for different types of glutathione peroxidases differ (see Table I), but are not pronounced enough to allow differential analysis. With some precautions, PHGPx can be measured individually by using complex hydroperoxides such as phosphatidylcholine hydroperoxide (PCOOH).[18,23,24] pGPx has also been shown to react with PCOOH.[25] As an extracellular enzyme it is present only when GPx activities are to be measured in crude tissue homogenates. An unequivocal differentiation can only be made in samples solely containing cGPx and pGPx. Other glutathione peroxidases cannot be differentiated by activity thus far. There are, however, hints of a certain specificity of GI-GPx for 13-hydroperoxyoctadecadienoic acid (13-HPODE).[26] In GI-GPx expressing cells (HepG2) a GPx activity with 13-HPODE could be determined in selenium deficiency. Selenium-deficient HepG2 cells express reasonable amounts of GI-GPx, whereas cGPx is completely lost.[26] In ECV cells, which do not express GI-GPx, little activity with 13-HPODE was detected. PHGPx activity was comparable in both cell types. This might indicate that 13-HPODE is a substrate preferentially used by GI-GPx. As long as purified GI-GPx is not available this can, however, not be stated with certainty.

Comments. PCOOH is synthesized as described in Maiorino *et al.*,[23] a procedure that requires oxygenation of phosphatidylcholine by soybean lipoxygenase in the presence of bile salts, for example, 3 mM deoxycholate (DOC). At the end of the reaction DOC must be removed, because it is a strong inhibitor of PHGPx. Whereas cGPx is not inhibited by deoxycholate up to a concentration of 10 mM, PHGPx is inhibited by 50% at 1 mM DOC.[23] Thus, care must be taken that the deoxycholate concentration is kept below 0.1 mM in the test system. Every freshly prepared batch of PCOOH should be checked to determine that the DOC concentration is low enough. This can be done by comparing the initial velocities of the enzymatic reaction obtained with different dilutions of the respective PCOOH batch. The concentration of PCOOH is easily determined by letting the coupled test run to completion in the presence of purified PHGPx or, alternatively, a testis extract. DOC concentrations are low enough if the initial velocities of the reaction at different dilutions of PCOOH are the same or similar (±10% might be acceptable). Only if the initial slopes are comparable, is the DOC concentration low enough not to interfere with PHGPx activity.

[23] M. Maiorino, C. Gregolin, and F. Ursini, *Methods Enzymol.* **186,** 448 (1990).

[24] S. Maurer, C. Friedrich, M. Leist, M. Maiorino, and R. Brigelius-Flohé, *Z. Ernährungswiss.* **37,** 110 (1998).

[25] Y. Yamamoto and K. Takahashi, *Arch. Biochem. Biophys.* **305,** 541 (1993).

[26] K. Wingler, C. Müller, K. Schmehl, S. Florian, and R. Brigelius-Flohé, *Gastroenterology* **119,** 420 (2000).

[27] J. Chaudière and A. L. Tappel, *Arch. Biochem. Biophys.* **226,** 448 (1983).

Determination by RNA Analysis

Northern Blotting

Northern blots have been widely used for qualitative as well as quantitative determination of GPx expression. Northern blots are less sensitive than, for example, polymerase chain reaction (PCR)-based methods, and the specificity depends on the probe used for hybridization. In our hands, DNA probes spanning almost the complete coding sequence plus the 3′-untranslated region provided the best results[28]:

cGPx (human): nt 76–629 (accession no. M21304), 554 bp
PHGPx (pig): nt 42–773 (accession no. X76009), 732 bp
GI-GPx (human): nt 29–925 (accession no. X68314), 897 bp

The probe for pig PHGPx also hybridizes with human and rat PHGPx and can be used to detect PHGPx expression in species different from pig. Probes can be obtained by reverse transcriptase (RT)-PCR and subsequent cloning of the PCR products.[28]

Comments. The low sensitivity of Northern analysis implies that low-level transcription may be overlooked. For example, GI-GPx mRNA was not detected in rodent liver by Northern blots.[5] RT-PCR, however, revealed that GI-GPx mRNA is indeed expressed in rat liver (Fig. 1), as was also confirmed by Western blotting (see below).

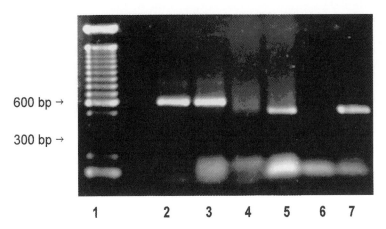

FIG. 1. GI-GPx expression in human cell lines and various rat tissues as detected by RT-PCR. Total RNA was isolated by using the SV total RNA isolation system (Promega, Madison, WI) according to the manufacturer instructions. PCR products from human cell types HepG2 (lane 2), CaCo-2 (lane 3), and ECV (lane 4) were obtained with the human GI-GPx primer pair, resulting in the 602-bp fragment. PCR products from rat tissues, colon (lane 5), spleen (lane 6), and liver (lane 7), were obtained by means of the mouse GI-GPx primer pair, resulting in the 570-bp fragment. Other conditions were as described in Procedures under Determination by RNA Analysis.

Reverse Transcriptase-Polymerase Chain Reaction

The structural relatedness between the various mammalian GPx isotypes is so low that they can easily be differentiated by PCR by choosing suitable primers. Appropriate PCR primers are as follows.

cGPx (human) (M21304)
Forward primer: 5'-AGTCGGTGTATGCCTTCTCG-3', nt 76–95
Reverse primer: 5'-TTGAGACAGCAGGGCTTCGAT-3', nt 629–609, 554 bp
GI-GPx (human) (X68314)
Forward primer: 5'-TCACTCTGCGCTTCACCATG-3', nt 18–37
Reverse primer: 5'-AGCAGTTCACATCTATATGGC-3', nt 619–599, 602 bp
GI-GPx (mouse) (X91864)
Forward primer: 5'-GGCTTACATTGCCAAGTCGTTC-3', nt 42–63
Reverse primer: 5'-CTAGATGGCAACTTTGAGGAGCCGT-3', nt 611–587, 570 bp
PHGPx (human) (X71973)
Forward primer: 5'-ATGAGCCTCGGCCGCCTTTG-3', nt 81–100
Reverse primer: 5'-AGCTAGAAATAGTGGGGCAGG-3', nt 676–656, 596 bp

Procedures

Any method for the isolation of total RNA can be used. Care should be taken that the RNA does not contain DNA, that is, use methods that contain a DNA digestion step.

Appropriate conditions for RT-PCR (e.g., Access RT-PCR system from Promega, Madison, WI) are as follows: 10 μl of 5× reaction buffer, dNTPs (0.2 mM each), 3 mM MgSO$_4$, 2% (v/v) dimethyl sulfoxide (DMSO), 50 pmol of each primer, 5 U of avian mycloblastosis virus (AMV) reverse transcriptase, and 5 U of *Thermus flavus* DNA polymerase, in a final volume of 50 μl. Overlay the samples with mineral oil (PerkinElmer, Weiterstadt, Germany). Program a thermal cycler [e.g., a Biometra (Göttingen, Germany) thermal cycler T3] to 45 min at 48°, 2 min at 95°, 45 sec at 94°, 1 min at 63°, and 1 min at 68°, (40 cycles) and a final elongation step of 7 min at 68°. An aliquot of the RT-PCR sample can then be run on a 1% (w/v) agarose gel and the fragments stained with ethidium bromide.

The human and mouse primers (which work also for rat GI-GPx) listed above were used to analyze GI-GPx mRNA in human cell lines and various rat tissues according to the described procedure (Fig. 1). GI-GPx mRNA was clearly detectable in HepG2 (Fig. 1, lane 2) and CaCo-2 (Fig. 1, lane 3) cells but not in ECV

---

[28] K. Wingler, M. Böcher, L. Flohé, H. Kollmus, and R. Brigelius-Flohé, *Eur. J. Biochem.* **259**, 149 (1999).

cells (Fig. 1, lane 4), as expected. In rats, GI-GPx mRNA was present in colon (Fig. 1, lane 5) but not in spleen (Fig. 1, lane 6), as known from previous studies.[5] Surprisingly, GI-GPx mRNA was also detectable in rat liver (Fig. 1, lane 7). This is in contrast to the results from Chu *et al.*,[5] who did not find GI-GPx mRNA in rat but only in human liver by Northern blotting. This demonstrates the preference for RT-PCR over Northern analyses if the expression levels are low.

Comments. Because of the different ranking of glutathione peroxidases in the hierarchy of selenoproteins, the amount of mRNA does not necessarily reflect the amount of protein being translated therefrom. Protein and mRNA levels can be correlated only when the selenium status of the tissues investigated is adequate. RT-PCR can, however, be used to study the actual amount of RNA in tissues and cells at different selenium states, if an appropriate method to quantify the PCR products is available.

Determination by Immunochemical Methods

Western Blotting

Western blotting allows both qualitative and quantitative detection of glutathione peroxidase proteins. It indicates the real amount of a certain GPx that is translated into the protein from the respective mRNA. RNA and protein content do not necessarily correlate, especially when measured in cells or tissues grown in a limited supply selenium (see above).

A major problem in the detection by immunochemical methods is that specific antibodies are not commercially available for all types of GPx. Only for cGPx is a small number of antibodies or antisera available with, at least in our hands, insufficient specificity. Antisera against pig PHGPx have been raised in rabbits[29] and against rat liver PHGPx.[30] We have had the opportunity to test the former and found a high specificity. Because of the small amount of PHGPx in most cell lines and tissues, it must be concentrated before the assays. This can be done by running the samples through a small column filled with bromosulfophthalein Sepharose and eluting PHGPx with a high-salt buffer [e.g., three times through a small volume of 25 mM Tris-HCl, (pH 8.0), 0.5 M KCl, 10% (v/v) glycerol]. An exception is testis, which has the highest PHGPx levels of all mammalian tissues. For GI-GPx only two antisera have been described so far.[31,32] The first has been raised against a recombinant GI-GPx protein in which the selenocysteine was replaced by cysteine.[31] The second was raised against the 17 C-terminal amino acids of GI-GPx.[32] This peptide is identical in humans, rats, and mice. The antiserum can therefore be used for the detection of GI-GPx in a variety of species.

---

[29] A. Roveri, M. Maiorino, and F. Ursini, *Methods Enzymol.* **233,** 202 (1994).

[30] M. Nakashima, S. Komura, N. Ohishi, and K. Yagi, *Biochem. Mol. Biol. Int.* **29,** 1139 (1993).

[31] R. S. Esworthy, K. M. Swiderek, Y.-S. Ho, and F.-F. Chu, *Biochim. Biophys. Acta* **381,** 213 (1998).

[32] M. Böcher, T. Böldike, M. Kiess, and U. Bilitewski, *J. Immunol. Methods* **208,** 191 (1997).

Procedures

SAMPLING. (1) Cultured cells are homogenized and extracts prepared as described.[18] Wash harvested cells in phosphate-buffered saline (PBS) and centrifuge at 150g for 8 min at 4°. For enzymatic determinations homogenize cell pellets containing 2–4 × 10^7 cells in 0.7–1.5 ml of Tris buffer (100 mM Tris-HCl, 300 mM KCl, pH 7.6) containing 0.1% (v/v) peroxide-free Triton X-100 by sonification with 10 strokes at 70% energy and 30% duty cycle. (2) Tissues are taken from anesthesized animals and immediately frozen in liquid nitrogen. Transfer small pieces of frozen tissue samples into ice-cold buffer [100 mM Tris (pH 7.6), 300 mM KCl, 0.1% (v/v) peroxide-free Triton X-100; 1 ml/500 mg tissue] and homogenize with an Ultra-Turrax (e.g., an Ultra Turrax T25 from IKA, Staufen, Germany). Clear homogenates by centrifugation (10,000g, 10 min, 4°). Separate 100–150 μg of protein on a 12.5% (w/v) polyacrylamide gel according to standard electrophoresis protocols.

BLOTTING. Estimation of cGPx and GI-GPx requires different procedures. For cGPx determination, the nitrocellulose membrane and six blotting papers (Schleicher & Schuell, Dassel, Germany) are mounted between the cathode and anode of a semidry blotting system as follows: cathode, three blotting papers equilibrated in buffer C (see below), polyacrylamide gel, membrane wetted with buffer B, one blotting paper equilibrated in buffer B, and two blotting papers equilibrated in buffer A. For GI-GPx determination the arrangement is the same but only buffer D is needed. Protein transfer is performed at 0.8 mA/cm². The blotting time for cGPx is 8 hr; the blotting time for GI-GPx is 2 hr. Other glutathione peroxidases can be blotted like GI-GPx.

Buffer A: 300 mM Tris-HCl, 20% (v/v) methanol, pH 10.4
Buffer B: 20 mM Tris-HCl, 20% (v/v) methanol, pH 10.4
Buffer C: 25 mM Tris-HCl, 20% (v/v) methanol, 40 mM aminocaproic acid, pH 9.4
Buffer D: 25 mM Tris-HCl, 190 mM glycine, 15% (v/v) methanol, pH 8.5

IMMUNOCHEMICAL DETECTION

1. Blocking of nonspecific binding sites on the membrane. The nitrocellulose membrane is kept overnight in 2% (cGPx) or 5% (GI-GPx) low-fat milk powder in TTBS [50 mM Tris-HCl, 150 mM NaCl, 0.1% (v/v) Tween 20, pH 7.5] at 4° with gentle shaking. Excess milk powder is removed by washing with TTBS.

2. An appropriate GPx antibody must be selected and used as the first antibody. The antibody or serum, respectively, is diluted in TTBS. TTBS for dilution may contain 3% (w/v) bovine serum albumin. The reaction is performed for 3 hr at room temperature with gentle shaking. Thereafter, excess antiserum is removed by washing (four times, 10 min each) with TTBS.

3. As second antibody, a horseradish peroxidase-coupled sheep anti-rabbit IgG fraction (Dako, Hamburg, Germany) is used. This is diluted 1 : 3000 with TTBS and the blot is incubated with the dilution for 1 hr at room temperature. Thereafter, the blot is washed with TTBS (four times, 10 min each).

4. The chemiluminescence reaction is initiated by treating the blot with enhanced chemiluminescence (ECL) solution (Amersham, Braunschweig, Germany) for 2 min. Chemiluminescence is made visible either by autoradiography (X-Omat, Kodak, Stuttgart, Germany) for various times or by densitometry (in a Fuji LAS-1000-CCD-camera system) (Raytest, Straubenhardt, Germany).

According to the procedures described, various tissues from rat and three cultured cell lines were investigated for cGPx and GI-GPx expression. Two cell lines (HepG2 and CaCo-2) are known to express GI-GPx, and one (ECV, a human bladder carcinoma epithelial cell line) is known not to express GI-GPx. As first antibody for GI-GPx the antiserum described by Böcher et al.[32] was used, and for cGPx an antiserum kindly provided by Q. Shen (University of Massachusetts, Worcester, MA) was used. Anti-cGPx antiserum was diluted 1 : 1000 and anti-GI-GPx antiserum 1 : 5000, both in TTBS. As shown in Fig. 2 (*middle*) all cells

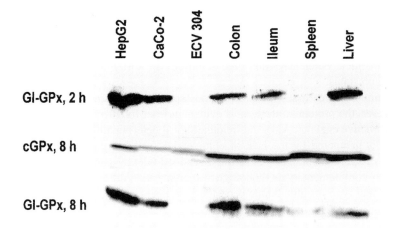

FIG. 2. cGPx and GI-GPx in various cell lines and rat tissues as detected by antisera against the respective enzymes. Tissues were sampled from male Wistar rats housed in individual conventional cages under controlled conditions of temperature and humidity. Animals were given free access to water and standard pelleted rat diet (C 1000; Altromin, Lage, Germany). Rats were anesthesized and perfused *in situ* with ice-cold sterile 0.15 *M* KCl. The tissues were then quickly removed and the intestinal lumen was flushed with ice-cold sterile phosphate-buffered saline. The tissues were cut into small pieces of ~50–100 mg, immediately frozen in liquid nitrogen, and stored at −80°. Samples were prepared and analyzed as described in Procedures under Determination by Immunochemical Methods. *Top:* Blotting under conditions required for GI-GPx and detection by anti-GI-GPx serum. *Middle:* Blotting under conditions required for cGPx and detection by anti-cGPx serum. *Bottom:* Blotting under conditions required for cGPx and detection by anti-GI-GPx serum.

and tissues express cGPx although in different amounts. GI-GPx was found in HepG2 and CaCo-2 cells, but not in ECV cells, as expected (Fig. 2, *top*). It was also found in rat colon and ileum and, surprisingly, in rat liver. This shows that GI-GPx is indeed expressed in liver from a rodent species that has previously been excluded.[5] Cross-reactivity of GI-GPx antiserum with cGPx or PHGPx was checked earlier in blots prepared for GI-GPx determination (2 hr) and excluded.[26] To rule out misinterpretations due to the different blotting times required for cGPx and GI-GPx, specificity of the antisera used was reevaluated. Figure 2 (*middle*) shows the blot analyzed with anti-cGPx antiserum (blotting time, 8 hr) and indicates cGPx is present in all cells and tissues, whereas Fig. 2, (*bottom*) demonstrates that anti-GI-GPx antiserum detects GI-GPx only in HepG2 and CaCo-2 cells and in rat colon, ileum, and liver. ECV cells and rat spleen did not react with GI-GPx antiserum, indicating that it does not cross-react with cGPx.

Conclusions

Differential determination of individual glutathione peroxidases by activity measurements is complicated by overlapping specificities. Only PHGPx can be differentiated in this way from other intracellular glutathione peroxidases by means of the specific substrate PCOOH.

The mRNAs can, of course, be estimated specifically by Northern blotting or, more sensitively, by PCR. Unfortunately, the mRNA levels do not correlate well with pertinent enzyme levels, because the different mRNAs are not transcribed with identical efficiencies and their stabilities are differentially affected by the selenium status.

Immunological assays promise the most specific measurement of GPx isotypes. However, the protein content, as is evident at least for PHGPx, does not always reflect active enzyme. Also, the immunological determination of individual GPx types is hampered by the lack of commercial availability of most of the antibodies required. The analysis of the diversified GPx system must still therefore employ complementary methodological approaches. A differentiation between the individual types of glutathione peroxidases appears mandatory, because they are evidently engaged in distinct biological roles.

Acknowledgments

This work was supported by the Deutsche Forschungsgemeinschaft (INK 26/B1-1 and SPP 1087, Br 778/5-1).

[10] High-Throughput 96-Well Microplate Assays for Determining Specific Activities of Glutathione Peroxidase and Thioredoxin Reductase

By ALLEN D. SMITH and ORVILLE A. LEVANDER

Introduction

Conventional cuvette-based assays for determining protein concentrations and enzyme activities are laborious, time consuming, and can significantly impede laboratory productivity when large numbers of samples need to be analyzed. Modern microplate readers can perform both end-point and kinetic measurements, thus enabling end-point and kinetic measurements of multiple samples at considerably reduced cost compared with clinical analyzers. One drawback of using the 96-well format is that the path length is less than 1 cm, thus precluding direct comparison with values obtained with a standard 1-cm cuvette. We describe here three assays designed to work in a 96-well format that allows for the rapid determination of both protein concentration and enzyme activity that is corrected for the difference in path length between the microplate and a standard 1-cm cuvette. The first assay is a modified microplate version of the Micro bicinchoninic acid (BCA) protein assay (Pierce, Rockford, IL)[1] with improved sensitivity and range. Enzyme assays for glutathione peroxidase (GPX; glutathione: H_2O_2 oxidoreductase, EC 1.11.1.9)[2] and thioredoxin reductase (TR; NADPH:oxidized thioredoxin oxidoreductase, EC 1.6.4.5; adapted from Tamura and Stadtman[3] as originally described in Holmgren[4]) have been modified to work in a 96-well format with a microplate reader that employs a unique path length determination protocol as part of its software. This allows data obtained using the microplate format to be expressed in standard units that are based on a 1-cm path length. Although spectrophotometrically derived path lengths are the most accurate, calculated path lengths based on assay volume and well size can be used with only a small amount of error introduced.

GPX catalyzes the breakdown of peroxides with reduced glutathione (GSH) as its source of reducing equivalents as shown in reaction (1).

$$2GSH + ROOH \xrightarrow{\text{GPX}} GSSG + ROH + H_2O \tag{1}$$

[1] P. K. Smith, R. I. Krohn, G. T. Hermanson, A. K. Mallia, F. H. Gartner, M. D. Provenzano, E. K. Fujimoto, N. M. Goeke, B. J. Olson, and D. C. Klenk, *Anal. Biochem.* **150,** 76 (1985).
[2] P. A. McAdam, V. C. Morris, and O. A. Levander, *Fed. Proc.* **43,** 867 (1984) (abstract).
[3] T. Tamura and T. C. Stadtman, *Proc. Natl. Acad. Sci. U.S.A.* **93,** 1006 (1996).
[4] A. Holmgren, *J.Biol.Chem.* **252,** 4600 (1977).

The oxidized glutathione produced is reduced to GSH by the action of glutathione reductase [GR, reaction (2)], utilizing NADPH as its source of reducing equivalents.

$$\text{GSSG} \xrightarrow{\quad\text{GR}\quad} \text{2GSH} \tag{2}$$

$$\text{NADPH} + \text{H}^+ \qquad\qquad \text{NADP}^+$$

By coupling reactions (1) and (2) together GPX activity can be measured *in vitro* by monitoring the oxidation of NADPH spectrophotometrically.

Thioredoxin reductase (TR) catalyzes the reduction of oxidized thioredoxin (Trx-S$_2$) to its reduced form [reaction (3)].

$$\text{Trx-S}_2 \xrightarrow{\qquad\qquad} \text{Trx-(SH)}_2 \tag{3}$$

$$\text{NADPH} + \text{H}^+ \qquad\qquad \text{NADP}^+$$

Mammalian TR can use several other substrates in addition to Trx including 5,5'-dithiobis(2-nitrobenzoic acid) (DTNB).[4] The use of DTNB as a substrate provides an inexpensive and easy method to measure TR activity spectrophotometrically *in vitro* by monitoring the conversion of DTNB to TNB as shown in reaction (4).

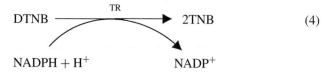

$$\text{DTNB} \xrightarrow{\quad\text{TR}\quad} \text{2TNB} \tag{4}$$

$$\text{NADPH} + \text{H}^+ \qquad\qquad \text{NADP}^+$$

Methods

Sample Preparation

Virtually any tissue, serum, plasma, or isolated cells can serve as a source material for the assays. Typically, for assaying GPX1 or TR activities, frozen tissues are thawed, weighed, and homogenized in 4 or 9 volumes of KCl (0.154 mol/liter). Isolated cells are thawed, resuspended in 0.5–1.0 ml of hypotonic potassium phosphate buffer [5 mM (pH 7.4), 0.5 mM EDTA], and disrupted by sonication. After a low-speed spin to remove cellular debris, the samples are ultracentrifuged at 105,000g for 2 hr at 4° to generate postmicrosomal supernatants and then frozen at −80° until assayed.

[5] K. E. Hill, G. W. McCollum, and R. F. Burk, *Anal. Biochem.* **253,** 123 (1997).

Micro Bicinchoninic Acid Protein Assay with Enhanced Sensitivity and Range

Materials

Micro BCA protein assay (Pierce)[*]
Bovine serum albumin (BSA) stock solution (provided by Pierce), 2 mg/ml
Plate sealers (Costar, Cambridge, MA)
Incubator capable of reaching 60°
96-Well flat-bottom plates [Greiner Labortechnik (Frickenhausen, Germany) or equivalent]
96-Well microplate reader (SPECTRAmax PLUS; Molecular Devices, Sunnyvale, CA)[*]
Racked microtubes (Bio-Rad, Hercules, CA), titerblocks (U.S.A. Scientific, Ocala, FL), or equivalent
Multichannel pipettor
Sample buffer

Assay

The standard microplate version of the Pierce Micro BCA protein assay, performed as described by the manufacturer (37° for 2 hr, 1 : 1 sample-to-reagent ratio), has a reported range of 1–20 μg/ml. Enhanced sensitivity and an increased working range can be achieved by changing the ratio of sample to reagent from 1 : 1 to 2 : 1 and increasing the incubation temperature from 37 to 60°. A working BSA stock solution is prepared by diluting the stock 2-mg/ml BSA solution to 0.4 mg/ml with sample buffer. BSA standards are prepared by diluting the 0.4-mg/ml BSA solution with sample buffer to generate standards in the appropriate range from 0 to 160 μg/ml. Add 200 μl of standard or sample to the wells of a 96-well microplate. The efficiency of this step can be increased by using racked microtubes or titerblocks set up in a 96-well format for making dilutions and transferring samples with a multichannel pipettor. Standards should be run at least in duplicate. Serial dilutions of unknowns can be prepared to ensure that at least two dilutions fall within the standard curve. Using a multichannel pipettor, add 100 μl of the BCA assay reagent to each well. Mix by shaking for 30 sec, and then cover with a plate sealer and incubate for 1 hr or, for maximum sensitivity, 2 hr at 60°. Most modern plate readers allow templates to be created that designate specific wells as blanks, standards, or unknowns. Create a template and then measure the absorbance at or near 562 nm. Subtract the average absorbance of the blanks from all standards and unknowns. Plot the concentration of the standards versus absorbance at 562 nm and extrapolate the protein concentrations of the unknowns from the standard curve.

Modern microplate readers should provide software that automatically performs these functions based on the template well designations and allows transfer of data to computer spreadsheets for compilation with the kinetic data.

Glutathione Peroxidase and Thioredoxin Reductase Assays

Materials

General
    96-Well flat-bottom plates (Greiner Labortechnik, or equivalent)
    96-Well microplate reader (SPECTRAmax PLUS; Molecular Devices)
    Multichannel pipettor

GPX1 Assay: Prepare all reagents with 18.2-MΩ deionized water:
    Potassium phosphate (250 mM), 25 mM EDTA, pH 7.4
    Reduced glutathione (40 mM; Roche Molecular Biologicals, Indianapolis, In), dissolved in water
    Glutathione reductase (Sigma, St Louis, MO): Concentration varies from lot to lot
    NADPH (40 mM; Sigma), dissolved in 0.1% (w/v) sodium bicarbonate
    H_2O_2 (25 mM), diluted into water
    Sodium azide (40 mM), dissolved in water
    Sample dilution buffer: 10 mM potassium phosphate, 154 mM potassium chloride, BSA (1 mg/ml), pH 7.0

TR Assay: Prepare all reagents with 18.2-MΩ deionized water:
    Potassium phosphate (1 M), 100 mM potassium chloride, 20 mM EDTA, pH 7.0
    NADPH (24 mM; Sigma) dissolved in 0.1% (w/v) sodium bicarbonate
    5,5'-Dithiobis(2-nitrobenzoic acid) (DTNB, 50 mM; Calbiochem, La Jolla, CA) dissolved in absolute ethanol; prepare fresh and protect from light
    Bovine serum albumin (20 mg/ml; Roche Molecular Biologicals) dissolved in water
    Aurothioglucose (10 mM; Sigma), dissolved in water
    Sample dilution buffer: 10 mM potassium phosphate, 154 mM potassium chloride, BSA (1 mg/ml), pH 7.0

Assays

Prepare a 1.25× assay mix without the final components added as stated below and preheat to 37°. Place 50 μl of appropriately diluted samples into a 96-well microplate. Add the final components to the assay mix and immediately start the reactions by adding 200 μl of the completed 1.25× assay mix to wells containing samples. Measure the change in optical density over time at 340 nm for GPX or at 412 nm for TR, using the kinetics program of a microplate spectrophotometer with the reading chamber temperature set at 37°. Collect data for a minimum of 3 min with plate readings occurring every 15 sec. To serve as a blank, add 50 μl of buffer to wells in place of sample and subtract the average blank values from all samples. This often can be achieved by designating specific wells as blanks that are averaged and subtracted from all samples by the plate reader software.

For the TR assays, data collection starts 1 min after the initiation of the reaction to allow nonenzymatic reduction of DTNB to go to completion. Contributions of nonthioredoxin reductase activities are determined by assaying each sample in the presence of 20 μM aurothioglucose (ATG), a potent inhibitor of TR activity.[5] Samples are preincubated in the presence of ATG (100 μM) for 10 min before addition of assay mix (final ATG concentration, 20 μM). ATG, in the absence of sample, also catalyzes the slow reduction of DTNB. Therefore, data from wells containing buffer plus ATG must also be collected and this rate subtracted from the background rate obtained when both ATG and sample are present to give the corrected background rate due to nonthioredoxin reductase activity. This background rate is then subtracted from the rate obtained with sample in the absence of ATG to give the final corrected TR rate used in calculating the specific activities of samples.

The final concentrations of the reagents in the GPX1 assay after addition to the samples in the microplate are as follows.

Potassium phosphate, 50 mM (pH 7.4), 5 mM EDTA
Reduced glutathione, 2 mM
Glutathione reductase, 2.0 U/ml
NADPH, 0.4 mM
Hydrogen peroxide, 0.25 mM
Sodium azide, 1 mM

NADPH, glutathione reductase, and hydrogen peroxide are added just before dispensing completed 1.25 × assay mix into the microplate.

The final reagent concentrations for the TR assay after addition to the samples in the microplate are as follows.

Potassium phosphate, 500 mM (pH 7.0), 50 mM potassium chloride, 10 mM EDTA
Bovine serum albumin, 0.2 mg/ml
NADPH, 0.24 mM
DTNB, 2.5 mM
ATG, 20 μM: Add to sample for determining nonspecific contributions to TR assay

NADPH and DTNB are added just before dispensing completed 1.25× assay mix into the microplate.

To calculate specific activities in standard units, which are based on a 1-cm path length, path lengths for each well of the microtiter plate can be determined at the end of the kinetic run, using the microplate reader and software (SPECTRAmax PLUS and Softmax Pro; Molecular Devices) according to the manufacturer instructions. In the absence of a plate reader with path length determination capabilities, estimated average path lengths can be calculated by using the total assay volume, the well radius, and the formula

$$\text{Path length (cm)} = \frac{\text{volume (ml)}}{\pi r(\text{cm})^2}$$

The enzyme activity in standard units can then be calculated by inserting the path length value (d) into the formula

$$U = \frac{(\Delta A / \min_s - \Delta A / \min_{bg}) \times \text{TV}}{\varepsilon d}$$

where $\Delta A / \min_s$ is the change in absorbance per minute for the sample; $\Delta A / \min_{bg}$ is the change in absorbance per minute for the blank; TV is the total volume of the assay (ml); ε is the extinction coefficient (for NADPH, 6.22 ml μmol^{-1}; for TNB, 13.6 ml μmol^{-1}. Because 2 mol of TNB is generated for every mole of DTNB consumed, the effective extinction coefficient is 27.2); and d is the light path length (cm).

One enzyme unit of GPX1 activity is defined as 1 μmol of NADPH oxidized per minute at 37° and 1 unit of TR activity is defined as 1 μmol of 5-thio-2-nitrobenzoic acid formed per minute at 37°. Specific activities are expressed as units per milligram protein.

The simple modifications described above to the manufacturer protocol for the Micro BCA protein assay result in enhanced sensitivity and increased assay range while maintaining the high-throughput capabilities of the microplate format (Fig. 1). Maximum sensitivity (0.25 μg/ml; $A_{562} = 0.014 \pm 0.002$, $n = 5$) is achieved with a 2-hr incubation at 60° but a 1-hr incubation at 60° achieves a sensitivity of 0.5 μg/ml (0.017 ± 0.005, $n = 6$), suitable for most applications. Because of the shorter path length associated with use of the microplates, the 1-hr modified microplate (reagent : sample ratio of 1 : 2) version of the Micro BCA protein assay does not achieve absorbance values above 0.7 for protein concentrations of 20 μg/ml and the standard curve can be extended to 160 μg/ml. Thus, it is possible to achieve a high degree of sensitivity and a wide protein working range with a single assay system while maintaining the convenience and high throughput of the microplate format.

Shown in Figs. 2 and 3 are standard curves for the GPX1 and TR assays generated with mouse liver postmicrosomal supernatants. The microtiter plate version of the GPX1 assay is linear ($R^2 = 0.996$) up to about 0.22 ΔOD/min over the 3-min data collection period (Fig. 2). Rates higher than 0.22 ΔOD/min

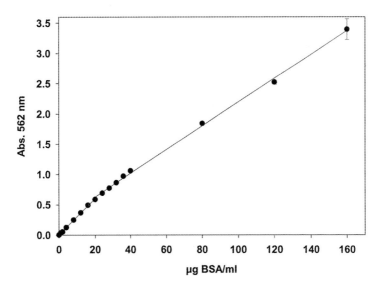

FIG. 1. Modified Micro BCA protein assay. Extended range standard curve was generated with a sample-to-reagent ratio of 2 : 1 and a 1-hr incubation at 60°. Each value represents the mean ± SD of three replicate wells. The solid line is the linear regression fit of the data for two ranges, 0–20 and 20–160 μg/ml.

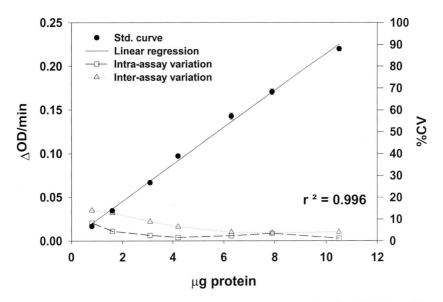

FIG. 2. Standard curve, and inter- and intraassay variation for the microplate GPX1 assay. The standard curve is plotted as ΔOD/min (left y axis) versus the amount of mouse liver postmicrosomal supernatant protein added to each assay. Each value represents the mean ± SD of eight replicate wells. The solid line is the linear regression fit of the data. The intra- and interassay variation, expressed as the average %CV ($n = 3$–4), is plotted on the right y axis.

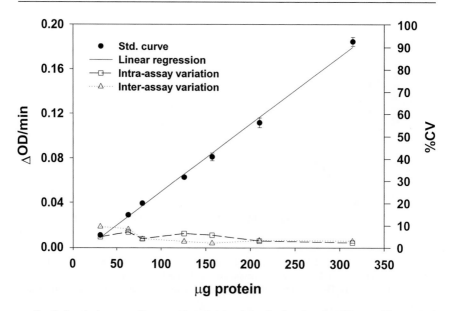

FIG. 3. Standard curve and inter- and intraassay variation for the microplate TR assay. The standard curve is plotted as ΔOD/min (left y axis) versus the amount of mouse liver postmicrosomal supernatant protein added to each assay. Each value represents the mean $\pm$ SD of eight replicate wells. The solid line is the linear regression fit of the data. The intra- and interassay variation, expressed as the average %CV ($n = 3$), is plotted on the right y axis.

result in depletion of substrates and loss of linearity after 1–2 min. The microtiter plate version of the TR assay is linear ($R^2 = 0.996$) up to at least 0.18 ΔOD/min over the 3-min data collection period (Fig. 3). Linearity up to 0.25 ΔOD/min has been routinely observed (data not shown). The average intra- and interassay %CV (Coefficient of variation, $n = 3$–4) for each point in the standard curves are also shown. The GPX1 intra- and interassay variation is highest (8 and 14%) at the lowest enzyme rates but drops off with increasing enzymatic rates. TR intraassay variation is between 2 and 7% across the entire standard curve. Interassay variation is highest (9%) at the lowest enzyme rates but drops off with increasing enzymatic rates.

The specific activities of samples measured by the microplate version of the GPX and TR assays coupled with path length determinations gave specific activities equivalent to those measured using the standard cuvette-based (1-cm path length) assays.[6] Thus the two assay systems are interchangeable.

Plate readers with path length correction capabilities are not always available. Therefore, we determined the accuracy of the values obtained with a calculated

[6] A. D. Smith, V. C. Morris, and O. A. Levander, *Int. J. Vitam. Nutr. Res.* **71**, 87 (2001).

path length of 0.65 cm for a 250-μl assay using 96-well plates with 7-mm-diameter wells. The mean ($\pm$ SD) path length as determined by the plate reader with software correction and a 250-μl assay volume is 0.642 ± 0.029 cm ($n = 288$) with a range of 0.519–0.794 cm. Thus, an average error of approximately 1% would be introduced by using a calculated path length as opposed to using an experimentally determined one. However, if the path length varies by 2 standard deviations from the mean path length, the error increases to approximately 10% and much larger errors would occur with sample path lengths at each end of the observed range. The difference between the calculated and empirical path lengths is most likely due to pipetting errors. Nevertheless, a calculated path length can be determined on the basis of the assay volume and well diameter that will give a close approximation to the average real path length. Thus, even microplate readers that do not have path length determination capabilities can be used to generate specific activities in standard units.

The software that accompanies these microplate readers usually includes the ability to designate specific wells as containing standards or unknowns, to generate standard curves, to calculate the values of unknowns from the standard curves, and to determine the ΔOD/min for kinetic measurements. Furthermore, the data generated by these software programs can be readily interfaced with other software spreadsheets. Data from protein assays and kinetic assays can be combined in software spreadsheets for the calculation of specific activities without having to manually enter data or perform repetitive calculations, thus increasing efficiency and decreasing the chances of data errors. As a result, high sample throughput and rapid data analysis can be achieved without investing in expensive equipment such as a clinical analyzer.

[11] Selenoprotein P

By GAVIN E. ARTEEL, LARS-OLIVER KLOTZ, DARIUS P. BUCHCZYK, and HELMUT SIES

Introduction

The presence of a serum protein distinct from plasma glutathione peroxidase that incorporates ^{75}Se was first described in the 1970s.[1,2] This protein was relatively abundant and was calculated to contain more than 50% of total plasma

[1] K. R. Millar, *N. Z. J. Agric. Res.* **15,** 547 (1972).
[2] R. F. Burk, *Proc. Soc. Exp. Biol. Med.* **143,** 719 (1973).

selenium.[3] The initial attempts to purify this new selenium-containing protein by conventional chromatography met with little success.[4,5] Subsequent chromatographic studies with immobilized antibodies yielded the purified protein[6–8]; this isolation technique was then standard for a number of years. More recent chromatographic methods involving heparin affinity, ion exchange, and then metal affinity, have been described to attain pure protein from human plasma[9,10]; a modification of this technique has also been described to separate and estimate selenium-containing proteins in plasma by high-performance liquid chromatography (HPLC).[11]

Structure

Selenoprotein P is unique among the selenoproteins in that it contains multiple selenol amino acids per polypeptide,[12,13] whereas all other characterized selenoproteins contain only one residue. Full-length human selenoprotein P contains 10 selenocysteine residues as predicted by the mRNA sequence.[12] The gene for human selenoprotein P is located on chromosome 5q31.[14] Although produced largely by the liver, expression of selenoprotein P in mammalian species appears to occur in many tissues, including testes, brain, and gut (e.g., see Refs. 15–17); receptors for selenoprotein P were also found in numerous tissues in the rat.[18] Human selenoprotein P is highly glycosylated. Although human selenoprotein P cDNA predicts a protein size of 41 kDa, purified human selenoprotein P has a molecular mass of ~60 kDa when run on reducing sodium dodecyl sulfate (SDS)-polyacrylamide gels[8]; however, when N-deglycosylated, the molecular mass of 40 kDa is as expected.[9] Selenoprotein P also contains two polyhistidine domains,

[3] J. T. Deagen, J. A. Butler, B. A. Zachara, and P. D. Whanger, *Anal. Biochem.* **208**, 176 (1993).

[4] J. L. Herrman, *Biochim. Biophys. Acta* **500**, 61 (1977).

[5] M. A. Motsenbocker and A. L. Tappel, *Biochim. Biophys. Acta* **719**, 147 (1982).

[6] J. G. Yang, J. Morrison-Plummer, and R. F. Burk, *J. Biol. Chem.* **262**, 13372 (1987).

[7] B. Akesson and B. Martensson, *Int. J. Vitam. Nutr. Res.* **61**, 72 (1991).

[8] B. Akesson, T. Bellew, and R. F. Burk, *Biochim. Biophys. Acta* **1204**, 243 (1994).

[9] V. Mostert, I. Lombeck, and J. Abel, *Arch. Biochem. Biophys.* **357**, 326 (1998).

[10] Y. Saito, T. Hayashi, A. Tanaka, Y. Watanabe, M. Suzuki, E. Saito, and K. Takahashi, *J. Biol. Chem.* **274**, 2866 (1999).

[11] H. Koyama, K. Omura, A. Ejima, Y. Kasanuma, C. Watanabe, and H. Satoh, *Anal. Biochem.* **267**, 84 (1999).

[12] K. E. Hill, R. S. Lloyd, J. G. Yang, R. Read, and R. F. Burk, *J. Biol. Chem.* **266**, 10050 (1991).

[13] P. Steinert, M. Ahrens, G. Gross, and L. Flohé, *Biofactors* **6**, 311 (1997).

[14] K. E. Hill, M. Dasouki, J. A. Phillips, and R. F. Burk, *Genomics* **36**, 550 (1996).

[15] J. W. Kasik and E. J. Rice, *Placenta* **16**, 67 (1995).

[16] M. Koga, H. Tanaka, K. Yomogida, J. Tsuchida, K. Uchida, M. Kitamura, S. Sakoda, K. Matsumiya, A. Okuyama, and Y. Nishimune, *Biol. Reprod.* **58**, 261 (1998).

[17] P. Steinert, D. Bachner, and L. Flohé, *Biol. Chem.* **379**, 683 (1998).

[18] B. Gomez and A. L. Tappel, *Biochim. Biophys. Acta* **979**, 20 (1989).

as well as a number of other histidine and lysine residues, and has a predominance of basic compared with acidic amino acids (see Burk and Hill[19]). Work has shown that commercial antibodies against histidine residues react with relative specificity with selenoprotein P in cell culture.[20]

Properties

Matrix-bound heparin retains selenoprotein P on chromatography columns.[7,8] These characteristics are common for proteins that bind proteoglycans (e.g., Borza and Morgan[21]). Studies with surface plasmon resonance identified two binding sites for heparin, one high-affinity, low-capacity site, and one lower affinity, higher capacity site.[22] Binding at both sites is sensitive to pH and ionic strength, and the high-affinity site is abolished by histidine carbethoxylation with diethyl pyrocarbonate.[22] The dependence on pH and ionic strength supports the hypothesis that the binding of selenoprotein P is electrostatic. Binding of selenoprotein P to glycosaminoglycans may be the mechanism of the accumulation of selenoprotein P on the surface of cells (e.g., endothelial cells[23,24]). Selenoprotein P was identified as a survival-promoting factor in cultured neurons with a half-maximal effective concentration at 0.2 nM,[25] and cultured astrocytes produce selenoprotein P.[26] It is possible that this neurotrophic effect is mediated by selenoprotein P serving as a ligand to glycosaminoglycans, because the effective concentrations were similar to the determined binding constant for selenoprotein P and heparin.[22] Selenoprotein P can form complexes with transition metals (e.g., Cu, Hg, and Cd)[27–29]; it has been proposed that this action may be a mechanism of binding and thereby detoxifying these metals *in vivo*.

Studies with both monoclonal antibodies against the protein and polyclonal antibodies raised against the N terminus of the protein have identified at least two isoforms of selenoprotein P when isolated from human plasma (e.g., Akesson *et al.*[8] and Mostert *et al.*[9]). Rat plasma also contains multiple isoforms of selenoprotein P.[30]

[19] R. F. Burk and K. E. Hill, *J. Nutr.* **124**, 1891 (1994).
[20] R. M. Tujebajeva, J. W. Harney, and M. J. Berry, *J. Biol. Chem.* **275**, 6288 (2000).
[21] D. B. Borza and W. T. Morgan, *J. Biol. Chem.* **273**, 5493 (1998).
[22] G. E. Arteel, S. Franken, J. Kappler, and H. Sies, *Biol. Chem.* **381**, 265 (2000).
[23] D. S. Wilson and A. L. Tappel, *J. Inorg. Biochem.* **51**, 707 (1993).
[24] R. F. Burk, K. E. Hill, M. E. Boeglin, F. F. Ebner, and H. S. Chittum, *Histochem. Cell. Biol.* **108**, 11 (1997).
[25] J. Yan and J. N. Barrett, *J. Neurosci.* **18**, 8682 (1998).
[26] X. Yang, K. E. Hill, M. J. Maguire, and R. F. Burk, *Biochim. Biophys. Acta* **1474**, 390 (2000).
[27] S. Yoneda and K. T. Suzuki, *Biochem. Biophys. Res. Commun.* **231**, 7 (1997).
[28] C. Sasakura and K. T. Suzuki, *J. Inorg. Biochem.* **71**, 159 (1998).
[29] U. Sidenius, O. Farver, O. Jons, and B. Gammelgaard, *J. Chromatogr. B. Biomed. Sci. Appl.* **735**, 85 (1999).
[30] H. S. Chittum, S. Himeno, K. E. Hill, and R. F. Burk, *Arch. Biochem. Biophys.* **325**, 124 (1996).

It should be noted that by using an isolation technique similar to that of Mostert *et al.*,[9] Saito *et al.*[10] detected only one isoform of purified selenoprotein P from human plasma. The reasons for the differences in number of isoforms found in human plasma are unclear. However, isolation of selenoprotein P in the presence of diisopropyl fluorophosphate as described[10] still yielded two isoforms of the protein, and these isoforms were both detectable with the monoclonal antibody described in Saito *et al.*[10] (and G. E. Arteel and H. Sies, unpublished observations, 2000). The interesting possibility that there exists genetic heterogeneity in the number of isoforms found in human plasma has not yet been specifically addressed.

Proposed Functions

Because of the high selenium content of selenoprotein P and the fact that it is one of the early selenoproteins to increase during selenium supplementation, it was proposed initially that its function was that of a selenium-carrying protein; however, the energy required to form the protein and subsequently release it on delivery makes this process highly inefficient (see Burk and Hill[31] for a more detailed discussion). Selenoprotein P was also proposed to be an antioxidant protein; this was based on the observation of a correlation between hepatic damage due the redox-cycler diquat and selenoprotein P plasma levels in rats *in vivo.*[32] Selenoprotein P concentrations have been shown to be decreased in patients with alcoholic liver injury,[33] a condition in which low plasma selenium concentrations[34,35] and oxidative stress[36] occur. Selenoprotein P in human plasma protects against peroxynitrite *in vitro,*[37] suggesting that it may serve as a protectant against peroxynitrite in human blood. This protein was also shown to have phospholipid hydroperoxide glutathione peroxidase activity, using thiol-containing small molecules [e.g., glutathione (GSH)] as reducing equivalents.[10] Although the thiol concentration in plasma is low, the coating of endothelial cells by selenoprotein P may serve as a protective barrier against oxidants and enable transfer of reducing equivalents from the plasma membrane to maintain a catalytic cycle of antioxidant defense. The observation that selenoprotein P expression can be regulated by

---

[31] R. F. Burk and K. E. Hill, *Bioessays* **21,** 231 (1999).

[32] R. F. Burk, K. E. Hill, J. A. Awad, J. D. Morrow, T. Kato, K. A. Cockell, and P. R. Lyons, *Hepatology* **21,** 561 (1995).

[33] R. F. Burk, D. S. Early, K. E. Hill, I. S. Palmer, and M. E. Boeglin, *Hepatology* **27,** 794 (1998).

[34] B. M. Dworkin, W. S. Rosenthal, G. G. Gordon, and R. H. Jankowski, *Alcohol. Clin. Exp. Res.* **8,** 535 (1984).

[35] R. F. Burk, D. S. Early, K. E. Hill, I. S. Palmer, and M. E. Boeglin, *Hepatology* **27,** 794 (1998).

[36] R. G. Thurman, B. U. Bradford, Y. Iimuro, K. T. Knecht, H. D. Connor, Y. Adachi, C. Wall, G. E. Arteel, J. A. Raleigh, D. T. Forman, and R. P. Mason, *J. Nutr.* **127,** 903S (1997).

[37] G. E. Arteel, V. Mostert, H. Oubrahim, K. Briviba, J. Abel, and H. Sies, *Biol. Chem.* **379,** 1201 (1998).

stress-response genes supports the hypothesis that selenoprotein P plays a role in protecting normal tissue,[38–41] possibly by preventing oxidative stress. The expression of recombinant rat selenoprotein P in a transiently transfected human epithelial kidney cell line allows for high levels of the purified protein to be studied to better clarify the functions of the protein.[20]

Acknowledgments

H.S. is a Fellow of the National Foundation for Cancer Research (Bethesda, MD). G.A. is a Fellow of the National Institute of Alcohol Abuse and Alcoholism (NIAAA).

[38] I. Dreher, T. C. Jakobs, and J. Köhrle, *J. Biol. Chem.* **272,** 29364 (1997).
[39] V. Mostert, I. Dreher, J. Köhrle, and J. Abel, *FEBS Lett.* **460,** 23 (1999).
[40] K. Hesse-Bahr, I. Dreher, and J. Köhrle, *Biofactors* **11,** 83 (2000).
[41] T. Tanaka, S. Kondo, Y. Iwasa, H. Hiai, and S. Toyokuni, *Am. J. Pathol.* **156,** 2149 (2000).

[12] Iodothyronine Deiodinases

By JOSEF KÖHRLE

Introduction

Thyroid hormones are essential for normal development, growth, and metabolic functions in higher vertebrates. The thyroid gland produces and secretes mainly the prohormone L-thyroxine (T_4), which is further metabolized by three deiodinase enzymes (5'DI, 5'DII, and 5DIII) to yield thyromimetically active hormones and inactive metabolites or derivatives with regulatory properties in thyroid hormone economy. The essential trace element iodine is required for thyroid hormone synthesis,[1] and iodine contributes 65% of the mass of L-thyroxine. A further essential trace element, selenium, is absolutely required for thyroid hormone synthesis and metabolism of thyroid hormones.[2] Probably a coevolution occurred during terrestrial life of higher vertebrates, involving both trace elements, which are deficient in wide areas of most continents due to postglacial erosion. The deiodinase enzyme family has been characterized in more detail, cDNA and genomic clones have been analyzed, and mutants have been constructed to identify essential structural and functional elements in their genes, mRNAs, and proteins that are required for expression and for enzymatic thyroid hormone deiodination. Most of

[1] J. E. Dobson, *Geogr. Rev.* **88,** 1 (1998).
[2] J. Köhrle, *Biochimie* **81,** 527 (1999).

the effects of thyroid hormones are exerted by 3,3',5-triiodo-L-thyronine (T_3), the 5'-deiodination product of T_4. T_3 is also secreted to a minor extent by the thyroid in iodine deficiency, where the relative T_3-to-T_4 ratio secreted by the thyroid increases. T_3 binds to three forms of T_3 receptors, members of ligand-dependent transcription factors of the c-ErbA family.[3] Functional T_3 receptors have also been identified in mitochondria in addition to the well-established nuclear receptors.[4] Effects of iodothyronines, especially T_4 and reverse-T_3 (rT_3), the 5-deiodination product of T_4, have been observed at the plasma membrane level and are linked to reorganization of the actin filament cytoskeleton.[5] So far, no human gene defects for any of the three deiodinase enzymes have been identified,[6,7] but in a mouse strain a variant of the 5'DI gene with functional consequences in the form of decreased expression levels has been found.[8] In vertebrates and especially in humans, only severe selenium deficiency leads to alterations in expression and function of deiodinases. Activity of 5'DI in liver, kidney, and several other organs decreases whereas no major effects are observed in the thyroid, the central nervous system, and several other endocrine organs. Mild selenium deficiency provokes no major, direct alterations of expression and function of 5'DI and 5DIII. However, expression and function of the deiodinase isoenzymes are sensitive to thyroid hormone status, various cytokines and growth factors, severe illness, reactive oxygen species, a variety of hormones and signaling compounds, circadian rhythm, and pharmacological agents, and therefore might be useful in sensor function of physiology and pathophysiology.

Historical Reminiscence

T_3, the iodothyronine with the highest thyromimetic potency, was discovered in 1953,[9,10] almost 50 years after T_4 had been chemically characterized.[11] Many attempts were made to understand the formation of T_3, which might originate also from incomplete iodination of iodothyronines in the thyroid hormone synthesis and storage protein thyroglobulin, which shows preferential T_3 hormonogenic sites among other T_4 hormonogenic domains. Photochemical random

[3] J. Zhang and M. A. Lazar, *Annu. Rev. Physiol.* **62,** 439 (2000).

[4] F. Casas, P. Rochard, A. Rodier, I. Cassar-Malek, S. Marchal-Victorion, R. J. Wiesner, G. Cabello, and C. Wrutniak, *Mol. Cell. Biol.* **19,** 7913 (1999).

[5] J. L. Leonard and A. P. Farwell, *Thyroid* **7,** 147 (1997).

[6] J. Köhrle, T. C. Jakobs, and C. Schmutzler, *Thyroid* **7,** 687 (1997) (Abstract).

[7] N. Toyoda, N. Kleinhaus, and P. R. Larsen, *J. Clin. Endocrinol. Metab.* **81,** 2121 (1996).

[8] A. L. Maia, M. J. Berry, R. Sabbag, J. W. Harney, and P. R. Larsen, *Mol. Endocrinol.* **9,** 969 (1995).

[9] J. Gross and R. Pitt-Rivers, *Lancet* **I,** 439 (1952).

[10] J. Gross, *Thyroid* **3,** 161 (1993).

[11] C. R. Harington, *Biochem. J.* **20,** 293 (1926).

deiodination, enhanced by exposure to oxygen, was demonstrated soon after the discovery of T_3, and several lower iodinated iodothyronines, the diiodothyronines and monoiodothyronines, were discovered thereafter. For many years, random deiodination occurring in a light- and oxygen-dependent radicalreaction has been assumed to generate most of the iodothyronine metabolites. Unequivocal enzymatic generation of T_3 was shown in 1970 by Braverman *et al.*, who identified T_3 in sera of athyrotic patients treated with synthetic thyroxine.[12,13] This observation supported nonrandom sequential deiodination reactions of thyroxine to T_3, T_2, T_1, and ultimately T_0 (Fig. 1).[14,15] Using *in vitro* enzyme assays, specific radioimmunoassays for iodothyronines, as well as liver perfusions, incubations of intact hepatocytes or their subcellular fractions, enzymatic formation of T_3 and later on reverse T_3, $3,3'-T_2$, $3,5-T_2$, $3',5'-T_2$ and some iodothyronines could unequivocally be shown. This series of experiments clearly proved the concept of sequential monodeiodination and established that at least three enzymes with different biochemical characteristics are involved in deiodinative T_4 and iodothyronine metabolism. Characterization of these enzyme reactions, development of specific affinity labeling protocols, and purification and enrichment of deiodinases supported this concept. However, attempts to isolate, purify, and crystallize deiodinase proteins failed. A major breakthrough occurred with the identification of type I deiodinase as a selenoprotein in 1990.[16,17] Shortly thereafter, cloning of the deiodinases of several species established the concept of deiodinases as a second important family of selenoenzymes[18] after the discovery of the glutathione peroxidase family and, more recently, the family of mammalian thioredoxin reductases.[19]

Type I 5′-Deiodinase

Subcellular Location and Protein Characteristics of Type I 5′-Deiodinase

Type I 5′-deiodinase (5′DI) is a selenoenzyme encoded by a 2.1-kb mRNA.[20] The functional enzyme is composed as a homodimer of two 27-kDa polypeptides

[12] L. E. Braverman, S. H. Ingbar, and K. Sterling, *J. Clin. Invest.* **49**, 855 (1970).

[13] L. E. Braverman, *Exp. Clin. Endocrinol.* **102**, 355 (1994).

[14] T. Sakurada, M. Rudolph, S. L. Fang, A. G. Vagenakis, L. E. Braverman, and S. H. Ingbar, *J. Clin. Endocrinol. Metab.* **46**, 916 (1978).

[15] T. J. Visser, D. Fekkes, R. Docter, and G. Hennemann, *Biochem. J.* **174**, 221 (1978).

[16] D. Behne, A. Kyriakopoulos, H. Meinhold, and J. Köhrle, *Biochem. Biophys. Res. Commun.* **173**, 1143 (1990).

[17] J. R. Arthur, F. Nicol, and G. J. Beckett, *Biochem. J.* **272**, 537 (1990).

[18] J. Köhrle, *Cell. Mol. Life Sci.* **57**, 1853 (2000).

[19] J. Köhrle, R. Brigelius-Flohe, A. Bock, R. Gartner, O. Meyer, and L. Flohe, *Biol. Chem.* **381**, 849 (2000).

[20] M. J. Berry, L. Banu, and P. R. Larsen, *Nature (London)* **349**, 438 (1991).

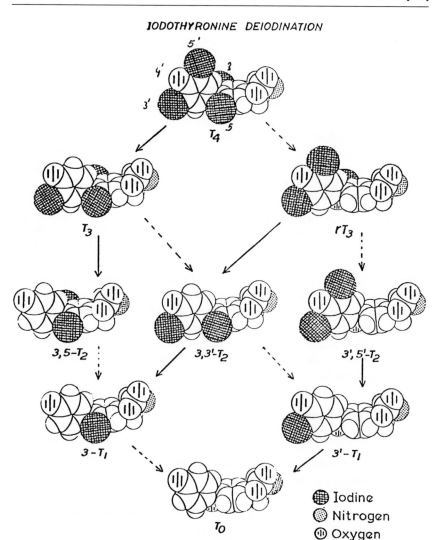

FIG. 1. Schematic presentation of the sequential deiodination cascade of L-thyroxine (T_4) and its iodothyronine metabolites to L-thyronine (T_0). Dark spheres illustrate iodine atoms which are removed in 5'-position (or the chemically equivalent 3'-position) from the phenolic ring of iodothyronines by two 5'-deiodinase enzymes (solid arrows), type I 5'-deiodinase (5'DI) and type II 5'-deiodinase (5'DII). Deiodination at the 5-position of the tyrosyl-ring (broken arrow) is catalyzed by the third deiodinase enzyme, type III 5-deiodinase (5DIII). All three deiodinases are integral membrane enzymes, which cleave C-I bonds in a reductive reaction, supported in vitro by strong reducing agents such as the vicinal dithiol DTT. The molecular details of the deiodinase reactions are not fully understood, but involve active site essential selenocysteine-residues at least in the 5'DI and 5DIII and probably also in the 5'DII enzyme reaction.

with a sedimentation coefficient ($s_{20,w}$) of 3.5–3.7.[21–23] The orientation of the two p27 subunits is head to head and tail to tail, and evidence has been presented that the putative membrane anchor in the N-terminal region of the p27 subunit of 5′DI is not essential for functional assembly of the dimeric enzyme, and membrane association is not essential for maintenance of functional activity. These conclusions are compatible with previous reports that limited tryptic digestion of 5′DI from rat liver microsomal membrane liberates a functional 5′DI preparation if low detergent concentrations are present to maintain functional enzyme activity.[24,25] The 5′DI is located at the cytosolic surface of the smooth and rough endoplasmic reticulum (ER) in liver [26–29]; in kidney epithelial cells a subcellular location at the inner leaflet of the basolateral plasma membrane has unequivocally been demonstrated.[23,30] No convincing data on subcellular location are available for thyroid, the anterior pituitary or other tissues with high expression of 5′DI. For bovine anterior pituitary, a location in the endoplasmic reticulum has been suggested,[31] and in the thyroid, some evidence has been put forward for location of 5′DI at the basolateral intracellular surface of the plasma membrane.[32,33] Reports on a putative cytosolic, nuclear, mitochondrial, or even lysosomal localization of 5′DI or the other deiodinases were not supported by appropriate analysis of corresponding marker enzymes, adequate functional 5′DI tests, or relevant techniques in cell biology.

Attempts of many groups to prepare solubilized functionally active 5′DI were rarely successful, because of the extremely low abundance of this integral membrane enzyme. Solubilization has been achieved with specific detergents, for

[21] J. L. Leonard and I. N. Rosenberg, *Biochim. Biophys. Acta* **659**, 205 (1981).

[22] M. Safran and J. L. Leonard, *J. Biol. Chem.* **266**, 3233 (1991).

[23] J. L. Leonard, T. J. Visser, and D. M. Leonard, *J. Biol. Chem.* **276**, 2600 (2001).

[24] J. Köhrle, "Schilddruesenhormonstoffwechsel 1983—Untersuchungen zur enzymatischen Dejodierung von L-Thyroxin und anderen Jodthyroninen durch Lebergewebe der Ratte," 1st Ed. Ferdinand Enke Verlag, Stuttgart, 1983.

[25] J. Köhrle, R. Ködding, and R. D. Hesch, *in* "Thyroid Research VIII" (J. R. Stockigt and S. Nagataki, eds.), p. 371. Australian Academy of Science, Canberra, 1980.

[26] J. Köhrle and R. D. Hesch, *Ann. Endocrinol. (Paris)* **40**, 22A (1979) (Abstract).

[27] D. Auf dem Brinke, R. D. Hesch, and J. Köhrle, *Biochem. J.* **180**, 273 (1979).

[28] D. Fekkes, E. van Overmeeren-Kaptein, R. Docter, G. Hennemann, and T. J. Visser, *Biochim. Biophys. Acta* **587**, 12 (1979).

[29] D. Fekkes, E. van Overmeeren, G. Hennemann, and T. J. Visser, *Biochim. Biophys. Acta* **613**, 41 (1980).

[30] J. L. Leonard, D. M. Ekenbarger, S. J. Frank, A. P. Farwell, and J. Köhrle, *J. Biol. Chem.* **266**, 11262 (1991).

[31] F. Courtin, G. Pelletier, and P. Walker, *Endocrinology* **117**, 2527 (1985).

[32] F. Santini, L. Chiovato, P. Lapi, M. Lupetti, A. Dolfi, F. Bianchi, N. Bernardini, G. Bendinelli, C. Mammoli, P. Vitti, I. J. Chopra, and A. Pinchera, *Mol. Cell. Endocrinol.* **110**, 195 (1995).

[33] D. Prabakaran, R. S. Ahima, J. W. Harney, M. L. Berry, P. R. Larsen, and P. Arvan, *J. Cell Sci.* **112**, 1247 (1999).

example, deoxycholate, taurodeoxycholate, Brij 56, and a few others.[21,24,25,29,34,35] Treatment of microsomal membranes with detergent concentrations liberating absorbed hydrophobic proteins [0.05% (w/v) deoxycholate] or treatment of membranes with alkaline pH did not liberate soluble 5'DI but increased specific deiodinase activity of the remaining membrane preparation, indicating an integral membrane nature of 5'DI.[24,25]

Structure and Function of Type I 5'-Deiodinase Genes

The human 5'DI gene is located on chromosome 1p32-p33, comprises approximately. 18 kb, and is composed of four exons.[36] The essential selenocysteine residue is encoded by exon 2, and the selenocysteine insertion sequence (SECIS) element is in exon 4. The promoter of the human 5'DI gene lacks both a TATA box and a CAAT box but several other putative regulatory elements have been identified by sequence analysis.[37,38] So far, complex hormone response elements have been identified and characterized in more detail, among them a DR4+2 thyroid response element/retinoic acid response element (TRE/RARE) and a DR10 TRE/RARE.[37,39,40] Further putative recognition sites for transcription factors have not been characterized in detail. The mouse 5'DI gene has been mapped to a region of chromosome 4 syntenic to the human 1p32 range. In the C3H mouse strain, the 5'DI promoter contains a 21-bp insert including five CTG repeats. Apparently, this insertion leads to markedly decreased promoter activity and impaired 5'DI function in this strain, which has markedly decreased circulating T_3 and elevated T_4 levels.[8,41,42]

Mechanism of Reaction of Type I 5'-Deiodinase

5'DI catalyzes deiodination of T_4 and its metabolites by a two-substrate ping-pong mechanism of reaction[43-47] (Table I). In the first half-reaction, T_4 or other iodothyronines interact with a selenolate anion, to form an intermediate E–Se–I

[34] D. Fekkes, G. Hennemann, and T. J. Visser, *Biochim. Biophys. Acta* **742**, 324 (1983).

[35] J. A. Mol, T. P. van den Berg, and T. J. Visser, *Mol. Cell. Endocrinol.* **55**, 149 (1988).

[36] T. C. Jakobs, M. R. Koehler, C. Schmutzler, F. Glaser, M. Schmid, and J. Köhrle, *Genomics* **42**, 361 (1997).

[37] T. Jakobs, C. Schmutzler, J. Meissner, and J. Köhrle, *Eur. J. Biochem.* **247**, 288 (1997).

[38] I. Dreher, T. Jakobs, and J. Köhrle, *J. Endocrinol. Invest.* **20**, 112 (1997) (Abstract).

[39] N. Toyoda, A. M. Zavacki, A. L. Maia, J. W. Harney, and P. R. Larsen, *Mol. Cell. Biol.* **15**, 5100 (1995).

[40] C.-Y. Zhang, S. Kim, J. W. Harney, and P. R. Larsen, *Endocrinology* **139**, 1156 (1998).

[41] C. H. Schoenmakers, I. G. Pigmans, A. Poland, and T. J. Visser, *Endocrinology* **132**, 357 (1993).

[42] M. J. Berry, D. Grieco, B. A. Taylor, A. L. Maia, J. D. Kieffer, W. Beamer, E. Glover, A. Poland, and P. R. Larsen, *J. Clin. Invest.* **92**, 1517 (1993).

[43] T. J. Visser, *Biochim. Biophys. Acta* **569**, 302 (1979).

[44] T. J. Visser, D. Fekkes, R. Docter, and G. Hennemann, *Biochem. J.* **179**, 489 (1979).

TABLE I

Properties of Deiodinase Enzymes

Enzyme characteristic	Type I 5'-deiodinase	Type II 5'-deiodinase	Type III 5-deiodinase
Function	Systemic > local T_3 production, degradation of rT_3 and sulfated iodothyronines	Local > systemic T_3 production	Inactivation of T_4 and T_3 and their sulfates
Expression	Liver, kidney, thyroid, pituitary; heart	(Hypothyroid) pituitary, brain, brown adipose tissue, skin, placenta; thymus, pineal and harderian gland; glial cells and tanycytes	Placenta, brain; many tissues, except pituitary, thyroid, kidney, adult healthy liver
Cosubstrate	DTT or DTE *in vitro* (K_m, 2–5 mM); not glutathione or thioredoxin *in vivo*	DTT or DTE *in vitro* (K_m, 5–10 mM); higher concentrations than for 5'DI	DTT or DTE *in vitro* (K_m, 10–20 mM); higher concentrations than for 5'DI
Subcellular location	Endoplasmic reticulum in liver, inner plasma membrane in kidney and thyroid	Inner plasma membrane; p29 subunit associated with F-actin in respect to perinuclear vesicles	Endoplasmic reticulum
Cloned in species	Human, rat, mouse, dog, chicken; not expressed in *Rana catesbeiana, Oreochromis niloticus* (tilapia), rainbow trout	Human, rat, mouse, chicken, *Rana catesbeiana, Fundulus heteroclitus* (teleost), rainbow trout	Human, rat, mouse, chicken, *Rana catesbeiana, Xenopus laevis*
Essential amino acid residues	Histidine, selenocysteine, cysteine, phenylalanine	Selenocysteine	Selenocysteine
Enzyme induction	T_3, retinoids; TSH and cAMP in thyroid only; testosterone in liver	cAMP; FGF; phorbol esters via PKC; ANP and CNP via cGMP in glial cells	T_3, FGF, EGF
Stimulation	Selenium, carbohydrate	β-Adrenergic agonists, nicotine	Selenium?
Repression	Ca^{2+}–PI pathway in thyroid; dexamethasone	T_3	
Inhibition	PTU, iodoacetate, aurothioglucose, iopanoate	T_4, rT_3, iopanoate	Iopanoate

Abbreviations: T_4, thyroxine; T_3, 3,3',5-triiodo-L-thyronine; rT_3, reverse T_3; DTT, dithiotreitol; DTE, dithio-erythreitol; TSH, thyroid-stimulating hormone; FGF, fibroblast growth factor; EGF, epidermal growth factor; PKC, protein kinase C; ANP, atrial natriuretic protein; CNP, C-type natriuretic protein; PTU, 6-n-propyl-2-thiouracil.

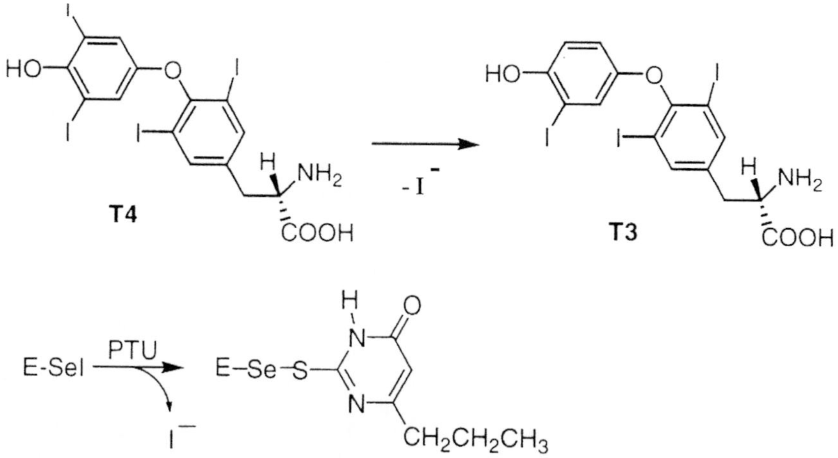

FIG. 2. 5'-deiodination of T_4 yields the thyromimetic hormone T_3. This reaction is catalyzed by both the 5'DI and the 5'DII enzyme, albeit with different mechanisms. The 5'DI enzyme reacts in a ping-pong two substrate mechanism of reaction leading to an enzyme-selenenyl-iodide (E-SeI). This intermediate reacts with the antithyroid drug 6-n-propyl-2-thiouracil (PTU) to form a stable selenenyl-sulfide suicide complex. The 5'DII enzyme proceeds in a sequential two substrate mechanism without formation of an enzyme-iodide intermediate and is also not sensitive to PTU-inhibition.

with an oxidized selenium atom.[48] *In vitro,* this intermediate is reduced by vicinal dithiols such as dithiothreitol (DTT) or dithioerythritol (DTE); monothiols such as mercaptoethanol, or other reducing agents including glutathione, thioredoxin, and related compounds, are poor cosubstrates for the second half-reaction. Initially this selenolate had been characterized as a highly reactive essential SH residue of the active site.[49,50] The E–Se–I intermediate also reacts with thiouracils such as 6-*n*-propyl-2-thiouracil (PTU) to form mixed E–Se–S–PTU compounds (Fig. 2), which irreversibly inactivate 5'DI.[45,51,52] Regeneration of this inactive complex is possible to some extent *in vitro* by cleavage of the Se–S bond at high pH in the presence of strong reducing agents such as DTT. Inhibition of 5'DI by PTU (Fig. 3) is a

[45] J. L. Leonard and I. N. Rosenberg, *Endocrinology* **107,** 1376 (1980).

[46] J. L. Leonard and T. J. Visser, *in* "Thyroid Hormone Metabolism" (G. Hennemann, ed.), p. 189. Marcel Dekker, NY, 1986.

[47] J. L. Leonard, *in* "Thyroid Hormone Metabolism—Regulation and Clinical Implications" (S. Y. Wu, ed.), p. 1. Blackwell Scientific, Cambridge, Massachusetts, 1991.

[48] M. J. Berry, J. D. Kieffer, J. W. Harney, and P. R. Larsen, *J. Biol. Chem.* **266,** 14155 (1991).

[49] J. L. Leonard and I. N. Rosenberg, *Endocrinology* **106,** 444 (1980).

[50] J. L. Leonard, *in* "The Thyroid Gland" (M. A. Greer, ed.), p. 285. Raven, New York, 1991.

[51] J. L. Leonard and I. N. Rosenberg, *Endocrinology* **103,** 2137 (1978).

[52] T. J. Visser, S. Frank, and J. L. Leonard, *Mol. Cell. Endocrinol.* **33,** 321 (1983).

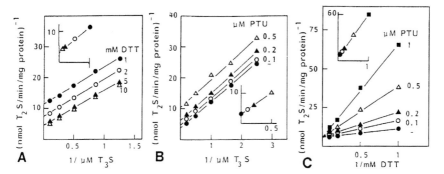

FIG. 3. Lineweaver-Burk initial rate kinetic analysis of the 5'DI reaction in human liver microsomes using T_3-sulfate (T_3S) as substrate. Parallel lines typical for the ping-pong mechanism of reaction are obtained for variation of the iodothyronine substrate T_3S at different cosubstrate DTT concentrations. The 5'DI-specific inhibitor PTU acts competitively versus the cosubstrate DTT and uncompetitively with T_3S. The inset illustrates a replot of the intercepts revealing limiting K_M and K_I-values [reprinted from T. J. Visser, E. Kaptein, O. T. Terpstra, and E. P. Krenning, *J. Clin. Endocrinol. Metab.* (1988)].

key property to distinguish 5'DI from type II 5' deiodinase (5'DII), which is insensitive to PTU and reacts by a sequential mechanism (see below). So far, the putative physiological cosubstrate regenerating 5'DI after deiodination has not been identified. SH-regenerating systems such as glutathione and glutathione redox reductase or thioredoxin and thioredoxin reductase or NADPH-dependent regeneration of GSH entertains the 5'DI reaction *in vitro* using semipurified microsomal membrane preparations, albeit at low efficiency and turnover numbers.[51,53–55] Divergent results have been published concerning the effect of intracellular glutathione depletion on cellular 5'-deiodination[56–59] (for review see Leonard and Visser[46]). Some evidence in intact cell deiodinase assays suggests that depletion of intracellular glutathione (GSH) in respect to formation of glutathione mixed sulfides impairs deiodinase activity of the corresponding cells. Whether these observations are direct proof for contribution of GSH-dependent intracellular cofactor systems to enzyme activity, or merely reflect association of altered intracellular redox control to altered deiodinase activity, remains to be established.[59]

[53] K. Sawada, B. C. W. Hummel, and P. G. Walfish, *Endocrinology* **117,** 1259 (1985).

[54] A. Balsam and S. H. Ingbar, *J. Clin. Invest.* **63,** 1145 (1979).

[55] Y. Imai, K. Yamauchi, and M. Nishikima, *Endocrinol. Jpn.* **27,** 201 (1980).

[56] K. Sato and J. Robbins, *Endocrinology* **109,** 844 (1981).

[57] T. J. Visser, *in* "Glutathione: Metabolism and Physiological Functions" (J. Vina, ed.), p. 317. CRC Press, Boca Raton, Florida, 1994.

[58] L. A. Gavin, F. A. McMahon, and M. Moeller, *J. Clin. Invest.* **65,** 943 (1980).

[59] S. Melmed, M. Nelson, N. Kaplowitz, T. Yamada, and J. M. Hershman, *Endocrinology* **108,** 970 (1981).

There is ample evidence to suggest that a productive 5′DI reaction might lead to suicide inactivation of 5′DI without regeneration of active enzyme *in vivo*. Although in low abundance in the major tissues expressing 5′DI, that is, liver, kidney, thyroid, and pituitary, they contain many more 5′DI catalytic units than needed to metabolize T$_4$ and all of its derivatives with lower iodine content including conjugates.[18,50,60–64] As the biological half-life of 5′DI enzyme is in the range of 20 to 30 hr, and that of its mRNA is between 10 and 20 hr, neosynthesis of functional 5′DI might fully account for maintenance of sufficient 5′DI enzyme activity levels. This consideration might also be supported by the observation that *in vivo* administration of PTU exerts its antithyroid action mainly via inhibition of thyroid hormones synthesis[65] and only to a minor extent by inhibition of deiodination, both in thyroid and extrathyroidal tissues.[49]

*Inhibitors of Type I 5′-Deiodinase Directed Toward Selenocysteine
Residue of Active Site*

5′DI is also effectively inhibited by alkylating agents such as iodoacetate or iodoacetamide,[45,51,66] known inhibitors of selenoenzymes. Similarly, alkylating agents such as *N*-bromoacetyl-T$_4$ (BrAcT$_4$), -T$_3$, or -rT$_3$, used as highly specific affinity ligands to identify especially the p27 subunit of 5′DI of several species, and also the ligand-binding subunits of 5′DII and 5DIII, bind to and inactivate 5′DI.[63,64,67–69] However, when using BrAcT$_4$ or related affinity labels care must be paid to the specificity of reaction; a detailed analysis of the kinetics of inhibition of the deiodinase reaction is needed, as is protection from affinity labeling by substrates or competitive inhibitors. As BrAcT$_4$ also reacts rapidly with activated SH groups, such as are found in the active site of creatine kinase,[70] and also with

---

[60] J. Köhrle, *Exp. Clin. Endocrinol.* **102,** 63 (1994).

[61] J. Köhrle, *Mol. Cell. Endocrinol.* **151,** 103 (1999).

[62] T. J. Visser, E. Kaptein, O. T. Terpstra, and E. P. Krenning, *J. Clin. Endocrinol. Metab.* **67,** 17 (1988).

[63] J. Köhrle, U. B. Rasmussen, H. Rokos, J. L. Leonard, and R. D. Hesch, *J. Biol. Chem.* **265,** 6146 (1990).

[64] J. Köhrle, U. B. Rasmussen, D. M. Ekenbarger, S. Alex, H. Rokos, R. D. Hesch, and J. L. Leonard, *J. Biol. Chem.* **265,** 6155 (1990).

[65] A. Taurog and M. L. Dorris, *Endocrinology* **124,** 3038 (1989).

[66] J. L. Leonard and T. J. Visser, *Biochim. Biophys. Acta* **787,** 122 (1984).

[67] J. A. Mol, R. Docter, E. Kaptein, G. Jansen, G. Hennemann, and T. J. Visser, *Biochem. Biophys. Res. Commun.* **124,** 475 (1984).

[68] C. H. H. Schoenmakers, I. G. A. J. Pigmans, and T. J. Visser, *Mol. Cell. Endocrinol.* **107,** 173 (1995).

[69] C. H. H. Schoenmakers, I. G. A. J. Pigmans, and T. J. Visser, *Biochim. Biophys. Acta* **1121,** 160 (1992).

[70] M. Wyss, T. Walliman, and J. Köhrle, *Biochem. J.* **292,** 463 (1993).

protein disulfide isomerase (PDI),[71,72] a suggestion that PDI is the 5'-deiodinase has been put forward[73] but subsequently disproven.[74] BrAcT$_4$ or its congeners react with other thyroid hormone-binding sites in T$_3$ receptors or serum and cellular binding proteins.[75-78] Photoaffinity labeling of the deiodinase active site, using p-nitrophenyl-2-diazo-3,3,3-trifluoropropionate coupled to T$_3$, also targets the p27 subunit of rat 5'DI, but selectivity and specificity of labeling are limited compared with that of BrAcT$_4$-protocols.[79]

The specific inhibitor of selenolates, aurothioglucose (ATG), is also a competitive inhibitor of 5'DI.[48] At low concentration, aurothioglucose is also used, similar to PTU, to distinguish between 5'DI and 5'DII, the latter of which is less sensitive also to aurothioglucose inhibition. Cysteine mutants of 5'DI are markedly less inhibited by aurothioglucose[48]; however, the inhibition of 5'DI by PTU and aurothioglucose does not necessarily imply that inhibition is selective for the active site selenocysteine residue. The observation that the selenocysteine-containing 5'DI enzyme from the teleost fish tilapia is insensitive to PTU and ATG[80] indicates that additional residues in the context of the active site environment also determine the specificity and selectivity of inhibitors, irrespective of the presence of selenocysteine or cysteine at the active site.

Inhibition of Type I 5'-Deiodinase by Thiouracils

Type I 5'-deiodinase activity is affected by many compounds acting either as competitive or noncompetitive inhibitors. The selenoprotein nature of 5'DI already predicts inhibition of this enzyme by aurothioglucose as well as by thiouracil derivatives. In addition, several phenolic and aromatic compounds similar to the phenolic or tyrosyl ring system of iodothyronines affect the enzyme reaction.

In addition, potent SH reagents such as mono- or divalent cations, as well as iodoacetate and other alkylating compounds, affect 5'DI activity. Of special importance is the inhibition of 5'DI by thiouracils. Structure–activity relationships

[71] S.-Y. Cheng, Q.-H. Gong, C. Parkison, E. A. Robinson, E. Appella, G. T. Merlino, and I. Pastan, *J. Biol. Chem.* **262**, 11221 (1987).

[72] K. Yamauchi, T. Yamamoto, H. Hayashi, S. Koya, H. Takikawa, K. Toyoshima, and R. Horiuchi, *Biochem. Biophys. Res. Commun.* **146**, 1485 (1987).

[73] R. J. Boado, D. A. Campbell, and I. J. Chopra, *Biochem. Biophys. Res. Commun.* **155**, 1297 (1988).

[74] C. H. H. Schoenmakers, I. G. A. J. Pigmans, H. C. Hawkins, R. B. Freedman, and T. J. Visser, *Biochem. Biophys. Res. Commun.* **162**, 857 (1989).

[75] S.-Y. Cheng, M. Wilchek, H. J. Cahnmann, and J. Robbins, *J. Biol. Chem.* **252**, 6076 (1977).

[76] V. M. Nikodem, S.-Y. Cheng, and J. E. Rall, *Proc. Natl. Acad. Sci. U.S.A.* **77**, 7064 (1980).

[77] J. S. Siegel, L. Korcek, and M. Tabachnik, *Endocrinology* **113**, 2173 (1983).

[78] U. B. Rasmussen, J. Köhrle, H. Rokos, and R.-D. Hesch, *FEBS Lett.* **255**, 385 (1989).

[79] J. D. Kieffer and P. R. Larsen, *Endocrinology* **129**, 1042 (1991).

[80] J. P. Sanders, S. Van Der Geyten, E. Kaptein, V. M. Darras, E. R. Kühn, J. L. Leonard, and T. J. Visser, *Endocrinology* **138**, 5153 (1997).

revealed that for inhibition their free thiol group is essential, aromatic structures of thiouracils are important, and polar hydrogen atoms proximal to the thiol group are mandatory.[81,82] In contrast to thiouracils, methimazole and carbimazole are poor inhibitors of type I 5'-deiodinase, although both thiouracils and methimazole derivatives are potent blockers of thyroid hormone synthesis catalyzed by thyroperoxidase (TPO). In spite of identification of 5'DI as a selenoprotein, the exact mechanism of thiouracil derivatives has not yet been explained, and no convincing evidence for selective inhibition of 5'DI by thiouracils compared with the low sensitivity of 5'DII and 5DIII to this group of inhibitors has been presented. 6-Anilino-2-thiouracils are more potent in inhibiting placental 5DIII than PTU.[83] However, unlike PTU, no regeneration of active enzyme has been achieved for placental 5DIII by an excess of reducing agent. *In vivo* administration of 6-anilino-2-thiouracil to rats also inhibited hepatic 5'-deiodinase, but no effect on thyroid has been observed, in contrast to PTU. Not only thiouracils but also selenouracil derivatives inhibit deiodinases with high potency, but no selective inhibition or major advantage over thiouracils has been reported yet.[84,85]

Evidence has been put forward,[86] for the first time verifying the proposed mechanism of inhibition of 5'DI by thiouracils utilizing two selenenyl iodides stabilized by steric protection and internal chelation. These model studies revealed that PTU reacts only with a covalently bound E–Se–I intermediate, but not with native enzyme, and imply that basic amino acid residues such as histidine near the active center may kinetically activate the selenium–iodine bond or act as a general base for abstraction of HI during inhibition. These model observations support experimental evidence from kinetic studies utilizing thyroid hormone analogs and derivatives and pH-dependent analysis of the mechanism of competition of T_4 deiodination with respect to the formation of analog products, using rat liver microsomal membranes.[87] These studies revealed that observed structure–activity relationships, substrate : product ratios, pH-dependent reaction characteristics, and shifts from phenolic to tyrosyl ring deiodination might be explained by assumption of an imidazolium–selenolate ion pair in the active site of 5'DI. Initially, an imidazolium–thiolate ion pair was postulated[87] before 5'DI had been identified as a selenocysteine-containing protein. This complex might explain the peculiar reaction characteristics of 5'DI, which is not completely specific for phenolic ring deiodination, but by some degenerate promiscuous mechanism is also able to

[81] T. J. Visser, E. Overmeeren, D. Fekkes, R. Docter, and G. Hennemann, *FEBS Lett.* **103,** 314 (1979).

[82] R. Harbottle and S. J. Richardson, *Biochem. J.* **217,** 485 (1984).

[83] T. Nogimori, L. E. Braverman, A. Taurog, S. L. Fang, G. Wright, and C. H. Emerson, *Endocrinology* **118,** 1598 (1986).

[84] K. Ohkawa, T. Hatano, N. Takizawa, K. Shinmoto, K. Yamada, M. Matsuda, and K. Takada, *In Vitro Cell Dev. Biol.* **28A,** 449 (1992).

[85] A. Taurog, M. L. Dorris, W. X. Hu, and F. S. J. Guziec, *Biochem. Pharmacol.* **49,** 701 (1995).

[86] W.-W. du Mont, G. Mugesh, C. Wismach, and P. G. Jones, *Angew. Chem. Int. Ed.* **40,** 2486 (2001).

[87] J. Köhrle and R. D. Hesch, *Horm. Metab. Res. Suppl.* **14,** 42 (1984).

perform tyrosyl ring deiodination under specific substrate and reaction conditions, that is, alkaline pH.

Essential Amino Acid Residues of Type I 5′-Deiodinase

Studies employing site-directed mutagenesis support the essential role of a histidine residue, probably at position 174, in the 5′-deiodinase reaction, compatible with formation of the selenolate–imidazolium ion pair and the *in vitro* model studies using thiouracil-based inhibitors. However, PTU sensitivity is not an obligatory characteristic of type I 5′-deiodinase, as demonstrated for the tilapia enzyme,[80] whereas mammalian type I 5′-deiodinase appears to be potently inhibited by PTU. Cloning and functional characterization including site-directed mutagenesis revealed that 5′DI of the teleost fish tilapia, although acting as a 5′-deiodinase containing a selenocysteine residue in the active site and two putative SECIS elements in the 3′ untranslated region (UTR), shows low sensitivity to PTU, iodoacetate, and aurothioglucose, characteristics typical of mammalian 5′DI.[80] Substrate specificity is comparable to that of mammalian 5′DI, with rT_3 markedly preferred over T_4, T_3 sulfate, and T_3 as substrate. These observations underscore the fact that sensitivity of deiodinases, to PTU, iodoacetate, and aurothioglucose cannot be used simply to classify deiodinase isoenzymes without further mechanistic and kinetic characterization. In addition, sensitivity to these inhibitors also cannot be taken as an indicator of the selenocysteine nature of the protein. The latter statement is of importance with respect to characterization of cysteine mutants of selenoproteins, which normally are of markedly lower catalytic efficiency compared with wild-type selenoproteins not only in the deiodinase family, but also in glutathione peroxidase and thioredoxin reductase enzymes. Site-directed mutagenesis in the tilapia selenoprotein type I 5′-deiodinase (insensitive to PTU) also revealed that a conserved proline residue at position 128 is important in 5′DI activity because mutation of Pro-128 to serine markedly decreased 5′-deiodinase activity without affecting 5-deiodinase activity and PTU insensitivity.

Apart from selenocysteine-126 in human 5′DI, three other amino acid residues exert crucial roles in 5′DI function. Biochemical analysis of the pH-dependent kinetic properties of 5′DI had already indicated the contribution of an essential histidine residue in the active site; on the basis of these studies, the formation of an essential imidazolium–thiolate ion pair was postulated.[87] Later on, the identification of deiodinases as selenoenzymes would suggest that not thiolate but selenolate contributes to these specific active site arrangements, which can explain the peculiar reaction characteristics of 5′DI in the context of different substrates (see below). Using histidine-directed reagents such as diethyl pyrocarbonate (DEPC) and rose bengal[88] as well as *in vitro* mutagenesis,[89] the contribution of histidine residues to the active site of rat 5′DI has been clearly shown. In particular, histidine residue 174 is critical to rat 5′DI function and hormone binding. A further histidine residue at position 158 appears to alter protein confirmation or catalysis. Two further histidine residues at positions 185 and 253 in rat 5'DI are inert to mutation with respect to enzyme activity. A cysteine residue at position 124 is also crucial for catalysis in the rat 5'DI enzyme; this vicinal cysteine residue probably forms an intermediate mixed selenosulfide or facilitates regeneration of the oxidized selenium of selenocysteine-126 after the deiodination step.[90] Mutations of selenocysteine-126 to cysteine reduce catalytic activity by more than two orders of magnitude, similar to loss of function in glutathione peroxidases with mutated essential selenocysteine residues. Whether a further cysteine residue at position 194 in 5'DI, which is also conserved in the other deiodinases, is crucial for the enzyme reaction remains to be established. Replacement of Cys-194 by alanine had only minor effects on the kinetic properties of this mutant,[91] whereas deiodination by double mutants was impaired. However, the results of these two reports are not in full agreement, as no impairment of deiodination by the cysteine mutants was observed in intact cells transfected with these constructs. Apparently, low 5'-deiodination efficiencies of these *in vitro*-expressed 5'DI mutants in comparison with natural enzymes in rat or human tissues render interpretation of deiodinase assays difficult.

A fourth amino acid residue, Phe-65, appears to be essential for 5'-deiodination of the substrates rT_3 and $3,3'-T_2$ sulfate, whereas deiodination of other ligands is not affected if Phe-65 is mutated.[92] This observation is supported by the fact that in dog 5'DI, the corresponding residue to Phe-65 is leucine, and dog 5'DI is unusual in its 30-fold higher K_m for rT_3 compared with human or rat 5'DI. These findings suggest that the active site adapts to its various substrates and that Leu-65, in contrast to Phe-65, lacks potential $\pi-\pi$ interactions between the aromatic rings. Furthermore, the more flexible reverse T_3 loses its high-affinity substrate property compared with the more rigid T_4 if this putative $\pi-\pi$ interaction is missing in the dog 5'DI or rat and human mutants, where Phe-65 is replaced.

Substrate Preference and Kinetic Parameters of Type I 5'-Deiodinase Reaction

Table II[23,46,106,169,185,193,197] summarizes the kinetic parameters of the 5'DI reaction. The substrate preference is $rT_3 > T_4 > T_3$ sulfate $> T_3$ with limiting K_m values in the submicromolar (rT_3, 0.2–1 μM) to micromolar range (T_4, 2 μM; T_3, 8–10 μM). For the *in vitro* cosubstrate DTT, limiting K_m values were determined in the millimolar range (2–10 mM). The pH optimum for 5'DI is at pH 6.8 in the acid range and also *in vitro* perfusions of rat liver indicate slightly acid pH optimum

[88] J. A. Mol, R. Docter, G. Hennemann, and T. J. Visser, *Biochem. Biophys. Res. Commun.* **120,** 28 (1984).

[89] J. P. Chanoine, M. Safran, A. P. Farwell, P. Tranter, S. M. Ekenbarger, S. Dubord, and S. Alex, *Endocrinology* **130,** 479 (1992).

[90] B. C. Sun, J. W. Harney, J. B. Marla, and P. R. Larsen, *Endocrinology* **138,** 5452 (1997).

[91] W. Croteau, J. E. Bodwell, J. M. Richardson, and D. L. St. Germain, *J. Biol. Chem.* **273,** 25230 (1998).

[92] N. Toyoda, J. W. Harney, M. J. Berry, and P. R. Larsen, *J. Biol. Chem.* **269,** 20329 (1994).

TABLE II
KINETIC PARAMETERS OF DEIODINASE REACTIONS[a]

Enzyme	Substrate	Limiting K_m (nM)	V_{max} (pmol/mg/min)	Substrate preference
5'DI	T_4	2000–4000	5–30	$rT_3 > 3',5'$-$T_2 > T_4 \geq T_3$ sulfate $> T_3$
	rT_3	60–1250	96–1780	
	$3',5'$-T_2	500	270	
5'DII	T_4	1.3–2.2	0.035–0.127	$T_4 > rT_3$
	rT_3	1.2	11–64	
5DIII	T_4	1900	18	T_4 sulfate $> T_3$ sulfate $> T_3 > T_4 >> rT_3$
	T_3	3.6–6200	0.016–4550	
	T_4 sulfate	300	527	
	T_3 sulfate	4700	1050–1400	

[a] In various models. Data are from several authors and reviews: See J. L. Leonard, T. J. Visser, and D. M. Leonard, *J. Biol. Chem.* **276**, 2600 (2001); J. L. Leonard and T. J. Visser, *in* "Thyroid Hormone Metabolism" (G. Hennemann, ed.), p. 189, Marcel Dekker, NY, 1986; J. L. Leonard and J. Köhrle, *in* "Werner and Ingbar's The Thyroid—A Fundamental and Clinical Text" (L. E. Braverman and R. D. Utiger, eds.), 8th Ed., p. 136, Lippincott Williams & Wilkins, Philadelphia, 2000; C. Valverde-R, C. Aceves, and E. Reyes, *Endocrinology* **132**, 867 (1993); J. P. Sanders, S. van der Geyten, E. Kaptein, V. M. Darras, E. R. Kuhn, J. L. Leonard, and T. J. Visser, *Endocrinology* **140**, 3666 (1999); N. Toyoda, E. Kaptein, M. J. Berry, J. W. Harney, P. R. Larsen, and T. J. Visser, *Endocrinology* **138**, 213 (1997); and K. A. Mol, S. van der Geyten, V. M. Darras, T. J. Visser, and E. R. Kuhn, *Endocrinology* **138**, 1787 (1997).

and pH-dependent regulation of 5'DI in the intact liver.[93,94] Remarkably, 5'DI is not completely specific with respect to both substrate preference and position of deiodination. *In vitro,* incubation of T_4 with 5'DI-containing enzyme preparations at alkaline pH in the presence of DTT also leads to 5-deiodination at the tyrosyl ring, yielding rT_3. This indicates that the active site shows some "wobble" toward the iodothyronine substrate, in contrast to the 5'DII enzyme (see below). Therefore, the use of both the optimal substrate, rT_3, which is almost exclusively deiodinated at the 5'-position, and of an appropriate pH of incubation, pH 6.8–7.0, warrants reasonable readouts of 5'DI activity if, in addition, the correct ratios of excess substrate versus enzyme are employed to allow for observation of initial rate kinetics and substrate depletion and/or product inhibition are avoided. Unfortunately, these principal rules of enzyme theory frequently have been neglected and led to the postulation of so-called low-K_m 5'-deiodinase in tissues such as adult liver, kidney, and human thyroid, which may be devoid of this artifactual activity.[95–99] The reason

[93] J. Köhrle, M. J. Müller, R. Ködding, H. J. Seitz, and R. D. Hesch, *Biochem. J.* **202**, 669 (1982).
[94] B. Höffken, R. Ködding, and R. D. Hesch, *Clin. Chim. Acta* **78**, 261 (1977).
[95] A. Goswami and I. N. Rosenberg, *J. Clin. Invest.* **74**, 2097 (1984).
[96] A. Goswami and I. N. Rosenberg, *Endocrinology* **123**, 2774 (1988).
[97] R. J. Boado and I. J. Chopra, *Endocrinology* **124**, 2245 (1989).
[98] G. B. Bhat, K. Iwase, B. C. W. Hummel, and P. G. Walfish, *Biochem. J.* **258**, 785 (1989).
[99] J. Sharifi and D. L. St. Germain, *J. Biol. Chem.* **267**, 12539 (1992).

for these observations might be that in the presence of high 5'DI activity in these tissues an initial rapid release of iodide from substrate employed is misinterpreted as evidence of a second 5'-deiodinase, which is insensitive to inhibition by PTU or ATG, and has a low K_m for T_4 or rT_3. Actually, this activity most probably represents the first half-reaction of the 5'DI activity[45,50] according to the ping–pong mechanism of reaction generating the oxidized enzyme selenenyl iodide, which then might undergo *in vitro* regeneration by reduced vicinal dithiols such as DTT.

A further unsettled issue is the limited information about available free substrate concentrations at the active site of the 5'DI. As iodothyronines are hydrophobic molecules with high octanol : water coefficient, their interaction with phospholipid bilayers in membranes usually employed for deiodinase assays and their tendency to stick to plastic and glass materials need careful attention, especially if assays are performed at low substrate concentrations or in a mode using "tracer only," as in many initial studies in deiodinase research. In an attempt to account for free iodothyronine substrates, Heinen *et al.* determined the K_m of the rat liver microsomal enzyme at 9.7 nM free T_4 and two binding sites of T_4, one with a K_d of 7.5 nM and a second, lower affinity site with a K_d of 1.7 μM. rT_3 competed for the higher affinity site with a K_d of 45 nM and T_3, with a K_d of 850 nM, competed for the lower affinity site.[100] The higher affinity site has been assumed to represent the 5'DI. Analysis of 5'DI enzyme kinetics by progress curves in intact cells such as hepatocytes or their homogenates also reveals kinetic parameters different from those reported from conventional double-reciprocal Lineweaver–Burk or Eadie–Hofstee plots, also indicating that available iodothyronine substrate concentrations are in the nM range (J. Köhrle, unpublished results, 2001).

Extensive analysis of structure–activity relationships of the active site of rat liver microsomal 5'DI, using iodothyronine substrates, their analogs, and various classes of naturally occurring and synthetic inhibitors, provided a detailed picture of the structural and functional demands of the 5'DI active site for productive substrates as well as for substrate- and/or cosubstrate-competitive inhibitors. In short, iodothyronines or isosteric analogs with a negative net charge of the alanine-derived side chain are preferred ligands; substituents in the 3' and 5' positions as well as the 3 and 5 positions with a volume similar to iodine (Br, NO_2) as well as a negative net charge at the 4' position favor interaction with 5'DI. Also, molecules different from thyronines, such as flavonoids and various (poly)cyclic aromatic or phenolic complex rings systems, are potent ligands provided they can adopt the skewed three-dimensional conformation of T_4 or the antiskewed structure of rT_3.[101–106] Surprisingly, these active site requirements closely resemble the binding

[100] E. Heinen, M. Basler, J. Herrmann, D. Hafner, and H. L. Krüskemper, *Endocrinology* **107**, 1198 (1980).

[101] J. Köhrle, M. Auf'mkolk, H. Rokos, R. D. Hesch, and V. Cody, *J. Biol. Chem.* **261**, 11613 (1986).

[102] M. Auf'mkolk, J. Köhrle, R. D. Hesch, and V. Cody, *J. Biol. Chem.* **261**, 11623 (1986).

site of transthyretin for T_4, which therefore can be used to model potent synthetic 5'DI inhibitors such as the flavonoid F21388.[104,107] F21388 also proved to be a valuable tool with which to study thyroid hormone metabolism under *in vivo* conditions and especially maternal–fetal interrelationships in thyroid hormone economy in rat models.[108–110]

The stereospecificity of the 5'-deiodinase reaction has also been analyzed. Here, data indicate that 5'DI preferentially accepts L-T_4 as substrate, whereas D-T_4 preferentially undergoes tyrosyl ring deiodination at the 5 position to yield D-rT_3 in rat liver *in vitro*.[111,112] Whether this stereospecific pattern of deiodination is of relevance to physiology or pharmacology needs to be established. *In vivo* studies of stereospecificity are controversial, as in humans no preferential 5-deiodination of D-T_4 could be found.[113] However, compared with L-T_4 the deiodination of D-T_4 to both D-T_3 and D-rT_3 was more rapid. No evidence exists on *in vivo* racemization of D-T_4 to L-T_4 or vice versa. These data already indicate that administration of D-T_4 or D-T_3 preparations, although used previously to treat hypercholesterinemia, should be avoided as both compounds are not biologically inert and are deiodinated to yield high amounts of iodide, and thus initial assumptions concerning the use of D-T_4 or D-T_3 preparations with low or absent thyromimetic potency are not justified, and such treatment should be avoided.

Type II 5'-Deiodinase Activity

Evidence of the existence of a second thiol-dependent 5'-deiodinase activity has been clearly presented, on the basis of enzyme kinetic distinction of this 5'-deiodinase activity (Fig. 4)[117] from the previously known type I 5'-deiodinase

[103] M. Auf'mkolk, J. Köhrle, R. D. Hesch, S. H. Ingbar, and V. Cody, *Biochem. Pharmacol.* **35**, 2221 (1986).

[104] J. Köhrle, M. Spanka, K. Irmscher, and R. D. Hesch, *in* "Plant Flavonoids in Biology and Medicine II: Biochemical, Cellular and Medicinal Properties" (V. Cody, E. Middleton, J. B. Harborne, and A. Beretz, eds.), 1st Ed., p. 323. Alan R. Liss, New York, 1988.

[105] M. Spanka, R.-D. Hesch, K. Irmscher, and J. Köhrle, *Endocrinology* **126**, 1660 (1990).

[106] J. L. Leonard and J. Köhrle, *in* "Werner and Ingbar's The Thyroid—A Fundamental and Clinical Text" (L. E. Braverman and R. D. Utiger, eds.), 8th Ed., p. 136. Lippincott Williams & Wilkins, Philadelphia, 2000.

[107] C. M. Mendel, R. R. Cavalieri, and J. Köhrle, *Endocrinology* **130**, 1525 (1992).

[108] J. P. Schröder-Van der Elst, D. Van der Heide, H. Rokos, G. Morreale de Escobar, and J. Köhrle, *Am. J. Physiol.* **274**, E253 (1998).

[109] J. Schröder-van der Elst, D. Van der Heide, and J. Köhrle, *Am. J. Physiol.* **261**, E227 (1991).

[110] J. P. Schröder van der Elst, D. Van der Heide, H. Rokos, J. Köhrle, and G. Morreale de Escobar, *Endocrinology* **138**, 79 (1997).

[111] M. Grussendorf and M. Ntokalou, *Acta Endocrinol. Suppl.* **240**, 96, 16 (1981).

[112] M. Hüfner and M. Grußendorf, *Akt. Endokr. Stoffw.* **3**, 108 (1982).

[113] I. D. Hay, C. A. Gorman, K. D. Burman, and N.-S. Jiang, *Metabolism* **34**, 266 (1985).

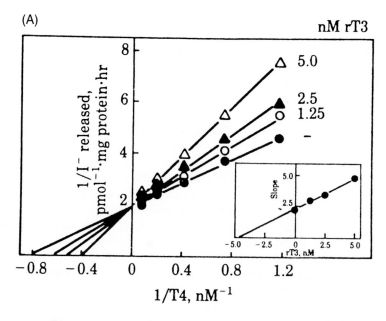

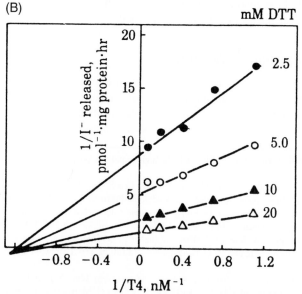

FIG. 4. Lineweaver-Burk initial rate kinetic analysis of the 5′DII reaction in rat cerebral cortex microsomes using T_4 as substrate, DTT as cosubstrate and rT_3 as competitive inhibitor of T_4 in the 5′DII reaction (A). Lines intersecting at the abscissa are typical for the sequential mechanism of reaction of 5′DII when the concentration of T_4 is varied at different DTT cosubstrate concentrations (B). rT_3 competitively inhibits T_4 deiodination by 5′DII as indicated by the intersection of lines on the ordinate (A). The inset illustrates a replot of the slopes revealing limiting K_I-values [reprinted with permission from T. L. Visser, J. L. Leonard, M. M. Kaplan, and P. R. Larsen, *Proc. Natl. Acad. Sci. U.S.A.* **79**, 5080 (1982)].

in liver, kidney, and thyroid.[18,50,106,114–116] A differently regulated enzyme had already been indirectly described (Table I) on the basis of increased deiodination under hypothyroid conditions in the pituitary and brain and the rapid change of deiodinase activity in the brain compared with peripheral tissues in response to alteration of thyroid hormone administration or depletion.[117,118] In addition, the brain and pituitary 5′ activity is insensitive to PTU inhibition, and also nanomolar concentrations of T_4 inhibit enzyme activity if rT_3 is used as substrate. The clear kinetic distinction between 5′DI and 5′DII is based on its nanomolar K_m values for T_4 and rT_3 (Fig. 4 and Table II), the insensitivity to PTU and iodoacetate inhibition, and its sequential two-substrate mechanism of reaction, in contrast to the ping–pong kinetics of 5′DI. The tissue-specific expression of 5′DII is also distinct from that of 5′DI, and only a few tissues express both enzymes. However, location of 5′DI and 5′DII in different cell types has not yet been excluded in this group of tissues. Studies of a few cell lines expressing both 5′DI and 5′DII activity, however, suggest coexistence of both enzymes in some instances.

Biological and physiological regulation of 5′DII is distinct from that of 5′DI, as illustrated in Table I. After the identification of 5′DI as a selenocysteine-containing enzyme, a series of initial observations did not support the concept that 5′DII also contains selenocysteine in its active site.[119–121] Arguments supporting a nonseleno-protein nature of 5′DII included increased activities in selenium-deficient animals, sluggish or no response to selenium-depletion, lack of [^{75}Se]selenite labeling of specific bands associated with alteration of 5′DII activity in several cell lines, lack of correlation of bromoacetyl[^{125}I]T_4 affinity labeling of the p29 subunit of 5′DII in cAMP-stimulated rat glial cells,[119] and low sensitivity to inhibition by aurothio-glucose in several rat tissues and cell lines, a property resembling that of the active site cysteine mutant of 5′DI.[120]

Evidence for Type II 5′-Deiodinase Being a Selenocysteine-Containing Protein

Reverse transcription-polymerase chain reaction (RT-PCR) cloning using homologous primers to 5′DI and cDNA from *Rana catesbeiana* tissues revealed a transcript with homology to rat type I 5′-deiodinase encoding an in-frame

[114] P. R. Larsen, J. E. Silva, and M. M. Kaplan, *Endocr. Rev.* **2**, 87 (1981).

[115] D. L. St. Germain and V. A. Galton, *Thyroid* **7**, 655 (1997).

[116] T. J. Visser, *Acta Med. Austriaca* **23**, 10 (1996).

[117] T. J. Visser, J. L. Leonard, M. M. Kaplan, and P. R. Larsen, *Proc. Natl. Acad. Sci. U.S.A.* **79**, 5080 (1982).

[118] J. L. Leonard, J. E. Silva, M. M. Kaplan, S. A. Mellen, T. J. Visser, and P. R. Larsen, *Endocrinology* **114**, 998 (1984).

[119] M. Safran, A. P. Farwell, and J. L. Leonard, *J. Biol. Chem.* **266**, 13477 (1991).

[120] M. J. Berry, J. D. Kieffer, and P. R. Larsen, *Endocrinology* **129**, 550 (1991).

[121] H. Meinhold, A. Campos-Barros, B. Walzog, R. Köhler, F. Müller, and D. Behne, *Exp. Clin. Endocrinol.* **100**, 87 (1993).

selenocysteine residue, as well as two conserved histidines and a further cysteine residue in addition to an N-regional hydrophobic domain.[122] This structure also had similarity to *Xenopus laevis* type III deiodinase cDNA.[123] Injection of capped RNA transcripts synthesized *in vitro* from the *R. catesbeiana* cDNA into *X. laevis* oocytes increased iodide release from rT_3. This activity could not be inhibited by PTU or ATG, compatible with properties of type II 5'-deiodinase so far reported. Subsequent work identified sequences similar to the *R. catesbeiana* 5'DII gene in several other vertebrate species including humans, rat, mouse, chicken, fish, and amphibia, providing evidence that a family of genes exists resembling the frog 5'DII gene.[122,124–128] Northern and RT-PCR analysis of the transcript levels of this 5'DII gene family in different tissues under different biological and pharmacological conditions partially agrees with the alterations of 5'DII enzyme activity observed previously or in these studies. However, the transcript size of the 5'DII mRNA in mammalian cells and tissues is much longer (4–7 kb) than that of the frog mRNA, as well as the mRNAs encoding 5'DI or 5DIII in other species. Furthermore, several hybridizing bands are detected in mammalian cells and tissues, and some tissues that do not exhibit 5'DII activity express transcripts. A further problem was encountered in that initially no SECIS element could be found in the native 5'DII transcript. *In vitro* expression of functional 5'DII activity was initially possible only by constructing chimeras between the coding region of the 5'DII transcript and fusion to SECIS structures of other deiodinases or selenocysteine-containing proteins. Using these artificial fusion constructs, both in *Xenopus* oocyte injection as well as for transfection, functional 5'DII activity with most of the characteristics described for the native enzyme could be demonstrated, for example, low K_m values for T_4 and rT_3, preference for T_4 over rT_3 as substrate, requirement for high dithiol concentrations for optimal enzyme activity, and low sensitivity to PTU, iodoacetate, ATG, and other reagents highly reactive with selenocysteine residues or activated thiol groups. Also, the mechanism of reaction was sequential, as expected. However, in many of these studies low expression levels have been achieved, and iodide release from substrates used was low. Therefore it could not be excluded in some of these reports that the

[122] J. C. Davey, K. B. Becker, M. J. Schneider, D. L. St. Germain, and V. A. Galton, *J. Biol. Chem.* **270**, 26786 (1995).

[123] D. L. St. Germain, R. A. Schwartzman, W. Croteau, A. Kanamori, Z. Wang, D. D. Brown, and V. A. Galton, *Proc. Natl. Acad. Sci. U.S.A.* **91**, 7767 (1994).

[124] W. Croteau, J. C. Davey, V. A. Galton, and D. L. St. Germain, *J. Clin. Invest.* **98**, 405 (1996).

[125] B. Gereben, T. Bartha, H. M. Tu, J. W. Harney, P. Rudas, and P. R. Larsen, *J. Biol. Chem.* **274**, 13768 (1999).

[126] C. Buettner, J. W. Harney, and P. R. Larsen, *J. Biol. Chem.* **273**, 33374 (1998).

[127] J. C. Davey, M. J. Schneider, K. B. Becker, and V. A. Galton, *Endocrinology* **140**, 1022 (1999).

[128] C. Valverde, W. Croteau, G. J. Lafleur, Jr., A. Orozco, and D. L. Germain, *Endocrinology* **138**, 642 (1997).

iodide release observed might be due to the first half-reaction occurring in the deiodination of low but significant endogenous 5'DI activity in cells and models exploited.

Subsequent follow-up studies then identified a SECIS-like structure far downstream in the cDNA of several mammalian species, and functional analysis of this SECIS element, again in chimeric fusion constructs, revealed its biological role.[126] However, expression levels of the native 5'DII by the long mRNA were low and suggest inefficient translation in this organization of the SECIS element several kilobases downstream of the UGA codon of 5'DII. A further observation was that the cDNA of the chicken 5'DII contains short open reading frames in the 5' untranslated region, and some of the 5'DII cDNAs have two UGA codons in frame. Subsequent analysis, however, demonstrated that mutation of the second C-regional selenocysteine residue 266 does not affect human 5'DII enzyme activity. Mutation of the first selenocysteine residue conserved with respect to 5'DI and 5DIII markedly increases the K_m value for T_4 by three orders of magnitude and leads to a 10-fold lower relative turnover number of the enzyme, again indicating the essential nature of the core selenocysteine residue, similar to other deiodinases or selenoproteins.[129] A series of articles analyzing distribution of transcripts and their regulation by various factors known to alter the expression and function of 5'DII, exploiting known biological and pharmacological models of thyroid hormone metabolism in several species and experimental models, support the idea that the 30- to 32-kDa 5'DII subunit encoded by this cDNA, which contains one or two in-frame UGA codons and in mammals and chicken is expressed as a long transcript, produces an essential subunit of 5'DII.

Unsettled Issues Concerning Type II 5'-Deiodinase Structure and Function

On the other hand, some unexplained observations still exist. First of all, some tissues express high levels of transcripts hybridizing to 5'DII cDNA or probes and lack expression of functional 5'DII enzyme. In addition, hybridization with 5'DII probes does not detect transcripts in all tissues expressing 5'DII activity. A major discrepancy has been that a previously well-characterized p29 subunit of the rat 5'DII enzyme best studied in cAMP-stimulated glial cells, but also other tissues, which has been identified by bromoacetyl-T_4 affinity labeling leading to inactivation of 5'DII as well as several cell biological experiments linking affinity labeling, inactivation, and intracellular translocation of this p29 subunit to functional states of 5'DII or its inactivation, does not correspond to the p30 to 32 protein encoded by the selenocysteine cDNA, which is not affinity labeled by bromoacetyl-T_4.[130-134]

[129] C. Buettner, J. W. Harney, and P. R. Larsen, *Endocrinology* **141**, 4606 (2000).
[130] J. L. Leonard, D. M. Leonard, M. Safran, R. Wu, M. L. Zapp, and A. P. Farwell, *Endocrinology* **140**, 2206 (1999).
[131] A. P. Farwell and J. L. Leonard, *J. Biol. Chem.* **264**, 20561 (1989).

Further evidence suggested that 5'DII is an oligomeric enzyme composed of an additional 60-kDa cAMP-induced subunit required to assemble catalytically reactive enzyme[132] and to direct 5'DII to the inner leaflet of the plasma membrane. Structural analysis and cloning of the cAMP-induced p29 subunit of 5'DII revealed a 3.5-kb mRNA encoding a 277-amino acid hydrophobic protein lacking selenocysteine.[134] This gene product has some homology to the transcription factor Dickkopf 3 (*dkk-3*) but is not identical to this protein involved in Wnt signaling. A series of papers analyzing property, expression, and function of the p29 protein is compatible with a function of this protein in producing enzymatic activity of 5'DII in cAMP-stimulated astrocytes. Injection of adenovirus p29-GFP (green fluorescent protein) constructs into the cerebral cortex of neonatal rats increases brain type II deiodinase activity, and cell culture studies revealed kinetic properties of the enzyme compatible with those previously described for the native 5'DII.[134] Attempts to express the mammalian homolog of frog type II selenodeiodinase, combined with immunological methods such as immunoprecipitation and related techniques, were not successful in demonstrating that the native 7.5-kb type II selenodeiodinase mRNA encodes functional enzyme in cAMP-stimulated astrocytes, and antisera raised against the C terminus of this protein failed to immunoprecipitate deiodinase activity from cAMP-stimulated astrocytes or other tissues expressing functional 5'DII activity.[130] These authors suggest that this long mammalian 5'DII cDNA with the SECIS far downstream has low efficiency in expressing functional enzyme under the conditions tested, and the observation that truncated protein terminated at the first in-frame UGA codon is synthesized abundantly would support the low efficiency of this translation. However, it is possible that some mechanism *in vivo* might increase the efficiency of translation of the selenocysteine-containing p30 to p32 subunit. Further studies are required both to identify the holoenzyme structure of 5'DI and to clarify the possible interaction between the thus far described p29, p30 to p32, and p60 subunits of 5'DII.

A transgene model has been established, exploiting a chimeric fusion construct driven by the α-MHC promoter, the coding region of the human 5'DII p32 subunit hooked to the SECIS element of rat selenoprotein P, and a human growth hormone (hGH)–poly(A) tail.[135] Mice expressing this artificial gene had high 5'DII activity, but no significant alterations of myocardial or plasma T_3 or T_4 growth rate or heart weight. However, heart rate was increased in isolated perfused heart and some gene

[132] M. Safran, A. P. Farwell, and J. L. Leonard, *J. Biol. Chem.* **271**, 16363 (1996).

[133] M. Safran, A. P. Farwell, H. Rokos, and J. L. Leonard, *J. Biol. Chem.* **268**, 14224 (1993).

[134] D. M. Leonard, S. J. Stachelek, M. Safran, A. P. Farwell, T. F. Kowalik, and J. L. Leonard, *J. Biol. Chem.* **275**, 25194 (2000).

[135] J. Pachucki, J. Hopkins, R. Peeters, H. Tu, S. D. Carvalho, H. Kaulbach, E. D. Abel, F. E. Wondisford, J. S. Ingwall, and P. R. Larsen, *Endocrinology* **142**, 13 (2001).

products and functional parameters were altered, indicating that small deviations in synthetic 5′DII activity in the heart or other factors changed in this transgene model lead to biological effects. Whether this transgene model can be interpreted as proof of the role of the p30 to p32 selenoprotein subunit in type II 5′-deiodinase function remains to be analyzed.

Rapid Inactivation of Type II 5′-Deiodinase

One of the most peculiar and interesting features of 5′DII is its rapid inactivation and intracellular translocation after administration of its substrates T_4 and rT_3 as demonstrated initially in rat brain and later on especially in cAMP-stimulated rat astrocytes or glial cells.[118,132,136,137] The contribution of the p29 subunit to this ligand-induced inactivation and intracellular transport has been monitored by bromoacetyl-T_4 labeling, which could be blocked by specific enzyme inhibitors, for example, the flavonoid EMD 21388 and the substrates T_4 and rT_3, whereas iopanoic acid was not a potent inhibitor of bromoacetyl-T_4 labeling in spite of its inhibition of 5′DII activity. Inactivation of 5′DII was shown to promote actin polymerization in thyroid hormone-depleted cells,[5] and inactivation of 5′DII was most efficient by thyroid hormone analogs blocked at their carboxylic group of the site chain.[133] Less potent were analogs with a blocked amine function. A net negative charge of the side chain had a weak effect on both inactivation of 5′DII and associated actin polymerization. Association between p29 and the inner leaflet of the plasma membrane depends on cAMP stimulation of a 60-kDa protein, which leads to translocation of p29 residing in the perinuclear space in unstimulated plasma membranes to the plasma membrane. This translocation coincides with the appearance of stimulated 5′DII activity.[132] The substrate (T_4 and rT_3)-induced inactivation and translocation of 5′DII is not achieved by T_3, which is less than 100-fold less potent. Studies have revealed that thyroxine-dependent 5′DII inactivation is exerted by binding of primary endosomes to actin microfilaments, and that the myosin V motor mediates the actin-based endocytosis of type II deiodinase.[138] Also, proteasome activity appears to be involved in degradation of 5′DII, as shown for rat pituitary tumor cells using FLAG-tagged cysteine-5′DII expression constructs. Ubiquitination of 5′DII has been suggested as contributing to substrate-activated inactivation in this model with overexpression of 5′DII protein.[139,140]

[136] J. L. Leonard, M. M. Kaplan, T. J. Visser, J. E. Silva, and P. R. Larsen, *Science* **214**, 571 (1981).
[137] L. A. Burmeister, J. Pachucki, and D. L. S. Germain, *Endocrinology* **138**, 5231 (1997).
[138] S. J. Stachelek, T. F. Kowalik, A. P. Farwell, and J. L. Leonard, *J. Biol. Chem.* **275**, 31701 (2000).
[139] J. Steinsapir, J. Harney, and P. R. Larsen, *J. Clin. Invest.* **102**, 1895 (1998).
[140] B. Gereben, C. Goncalves, J. W. Harney, P. R. Larsen, and A. C. Bianco, *Mol. Endocrinol.* **14**, 1697 (2000).

5'DII is encoded by two exons, separated by a large, 8.5-kb intron. The gene is located on human chromosome 14q24.3 as a single-copy gene.[141–143] First characterization of the 5' upstream and untranslated regions revealed alternative use of start sites, and cAMP-responsive elements have been functionally characterized in the promoter, which also contains a TATA box.[143–145] The identification of novel splicing variants of the human type II selenocysteine iodothyronine deiodinase mRNA might be one explanation for the observed pattern of several hybridizing signals in Northern blot experiments. In addition to thyroid-stimulating hormone (TSH) stimulation of 5'DII in brown adipocytes of rats,[146] stimulation of human but not rat 5'DII expression has also been demonstrated by thyroid transcription factor 1 (TTF-1).[147] No response to the thyroid-relevant transcription factor Pax-8 has been found. Whether the observed TTF-1 stimulation of human but not rat 5'DI gene expression solely accounts for the observed expression of 5'DII activity, especially in hyperthyroid and Graves' disease human thyroids, remains to be studied in more detail. Because of technical differences in measuring reliable 5'DII activities in the presence of high levels of 5'DI as observed in thyroid, especially under hyperthyroid conditions, so far no clear location of 5'DII in the human thyroid has been reported by the use of specific antibodies, in comparison with 5'DI, in immunohistochemical procedures.

Regulation of Type II 5'-Deiodinase Expression and Activity

Several factors are known to modulate 5'DII activity and expression, and the rapid modulation of 5'DII functional activity by increased T_4 and rT_3 complicates analysis of regulation. Whenever T_4 and/or rT_3 levels are altered this direct posttranslational effect must be excluded. This is of special importance in brain tissues, where T_4, T_3, and rT_3 levels are similar,[148] in contrast to serum or peripheral tissues. Therefore several drugs, stress, hormones, and neurotransmitters as

[141] F. S. Celi, G. Canettieri, D. P. Yarnall, D. K. Burns, M. Andreoli, A. R. Shuldiner, and M. Centanni, *Mol. Cell. Endocrinol.* **14,** 49 (1998).

[142] F. S. Celi, G. Canettieri, D. Mentuccia, L. Proietti-Pannunzi, A. Fumarola, R. Sibilla, V. Predazzi, M. Ferraro, M. Andreoli, and M. Centanni, *Eur. J., Endocrinol.* **143,** 267 (2000).

[143] S. Song, K. Adachi, M. Katsuyama, K. Sorimachi, and T. Oka, *Mol. Cell Endocrinol.* **165,** 189 (2000).

[144] K. Ohba, T. Yoshioka, and T. Muraki, *Mol. Cell Endocrinol.* **172,** 169 (2001).

[145] G. Canettieri, F. S. Celi, G. Baccheschi, L. Salvatori, M. Andreoli, and M. Centanni, *Endocrinology* **141,** 1804 (2000).

[146] M. Murakami, Y. Kamiya, T. Morimura, O. Araki, T. Imamura, T. Ogiwara, H. Mizuma, and M. Mori, *Endocrinology* **142,** 1195 (2001).

[147] B. Gereben, D. Salvatore, J. W. Harney, H. M. Tu, and P. R. Larsen, *Mol. Endocrinol.* **15,** 112 (2001).

[148] A. Campos-Barros, T. Hoell, A. Musa, S. Sampaolo, G. Stoltenburg, G. Pinna, M. Eravci, H. Meinhold, and A. Baumgartner, *J. Clin. Endocrinol. Metab.* **81,** 2179 (1996).

well as circadian rhythms are known to modulate 5'DII activity in the brain,[149–152] but direct transcriptional effects have not been analyzed in most instances. The most prominent stimulation of 5'DII expression is exerted by cAMP and several signaling pathways increasing cAMP concentrations such as TSH and β-adrenergic agents in several cell types.[132,146,149,153–156] Several other agents such as cGMP, nicotine, and several growth factors also increase 5'DII activity and expression.[106,157–159] T_3 might exert a small inhibitory effect on expression of 5'DII[137,155] and substrate-induced degradation of 5'DII by a ubiquitination-dependent proteasome pathway has been discussed as occurring in pituitary cells or cells overexpressing 5'DII or its mutants.[140,160] Fasting does not directly affect 5'DII expression and alterations of its activity might be indirect via altered T_4 and/or rT_3 levels.[161]

Type III Iodothyronine Deiodinase

5DIII catalyzes the inactivation of thyroid hormones by reductively removing iodide from the chemically equivalent 5 or 3 positions of the tyrosyl ring. Highest activities of mammalian 5DIII enzymes are located in the brain, placenta, skin, and some fetal tissues such as liver and intestine (Table I). Like 5'DI and 5'DII, 5DIII is also an integral membrane enzyme.[68] During development of vertebrate organisms, 5DIII plays an essential role in preventing accumulation of potential thyromimetic ligands and also prohormone T_4. Developmental profiles of 5DIII have been described for many tissues in human, rat, fish, chicken, amphibia, and

[149] Y. Kamiya, M. Murakami, O. Araki, Y. Hosoi, T. Ogiwara, H. Mizuma, and M. Mori, *Endocrinology* **140**, 1272 (1999).

[150] M. Eravci, G. Pinna, H. Meinhold, and A. Baumgartner, *Endocrinology* **141**, 1027 (2000).

[151] A. Baumgartner, L. Hiedra, G. Pinna, M. Eravci, H. Prengel, and H. Meinhold, *J. Neurochem.* **71**, 817 (1998).

[152] A. Campos-Barros, A. Musa, A. Flechner, C. Hessenius, U. Gaio, H. Meinold, and A. Baumgartner, *J. Neurochem.* **68**, 795 (1997).

[153] J. L. Leonard, *Biochem. Biophys. Res. Commun.* **151**, 1164 (1988).

[154] V. Pruvost, S. Valentin, I. Cheynel, E. Vigouroux, and M.-F. Bézine, *Horm. Metab. Res.* **31**, 591 (1999).

[155] S. W. Kim, J. W. Harney, and P. R. Larsen, *Endocrinology* **139**, 4895 (1998).

[156] A. Hernandez and M. J. Obregon, *Am. J. Physiol.* **271**, E15 (1996).

[157] A. Gondou, N. Toyoda, M. Nishikawa, S. Tabata, T. Yonemoto, Y. Ogawa, T. Tokoro, N. Sakaguchi, F. Wang, and M. Inada, *Thyroid* **8**, 615 (1998).

[158] A. Gondou, N. Toyoda, M. Nishikawa, T. Yonemoto, N. Sakaguchi, T. Tokoro, and M. Inada, *Endocr. J.* **46**, 107 (1999).

[159] A. M. Lennon, A. Esfandiari, J. M. Gavaret, F. Courtin, and M. Pierre, *J. Neurochem.* **62**, 2116 (1994).

[160] J. Steinsapir, A. C. Bianco, C. Buettner, J. Harney, and P. R. Larsen, *Endocrinology* **141**, 1127 (2000).

[161] S. Diano, F. Naftolin, F. Goglia, and T. L. Horvath, *Endocrinology* **139**, 2879 (1998).

several other species.[162–171] The pertinent regulatory role of 5DIII in the control of locally available thyromimetic activity has been elegantly demonstrated by overexpression of 5DIII in metamorphosing *Xenopus* tadpoles. Here, overexpression of 5DIII leads to incomplete metamorphosis of the tadpole, preventing resorption of gills and tail and reorganization of brain and bone structure, as well as transition from gill breathing to lung development.[172,173]

Subcellular Location and Subunit Composition

5DIII activity has been found in microsomal or endoplasmic reticulum fractions of brain regions,[162,174,175] placenta,[176] and fetal intestine.[177] 5DIII is a phospholipid-requiring enzyme[178] as demonstrated by partial purification and solubilization with detergents such as 3-[(3-cholamidopropyl)-dimethyl-ammonio]-1-propane sulfonate (CHAPS) or *n*-octyl-ß-D-glucopyranoside. Phosphatidylserine, and at lower efficiency also phosphatidylcholine or ethanolamine, reactivate solubilized enzyme partially purified by a gel and ion-exchange chromatography. The relative size of 5DIII of human placenta has been estimated in the range of M_r 40,000. However, no detailed studies confirming these preliminary data have been presented.[178] No detailed information is available for the subunit structure of 5DIII, and affinity labeling experiments using alkylating *N*-bromoacetyl-iodothyronine affinity labels such as BrAc[^{125}I]rT$_3$ or related compounds initially provided the impression that

---

[162] R. Ködding, H. Fuhrmann, and A. Von zur Mühlen, *Endocrinology* **118**, 1347 (1986).

[163] J. G. Eales, J. A. Holmes, J. M. McLeese, and J. H. Youson, *Gen. Comp. Endocrinol.* **106**, 202 (1997).

[164] S. Van Der Geyten, J. P. Sanders, E. Kaptein, V. M. Darras, E. R. Kühn, J. L. Leonard, and T. J. Visser, *Endocrinology* **138**, 5144 (1997).

[165] J. M. Bates, D. L. St. Germain, and V. A. Galton, *Endocrinology* **140**, 844 (1999).

[166] K. B. Becker, K. C. Stephens, J. C. Davey, M. J. Schneider, and Y. A. Galton, *Endocrinology* **138**, 2989 (1997).

[167] V. A. Galton, E. Martinez, A. Hernandez, E. A. St. Germain, J. M. Bates, and D. L. St. Germain, *J. Clin. Invest.* **103**, 979 (1999).

[168] M. J. Escamez, A. Guadano-Ferraz, A. Cuadrado, and J. Bernal, *Endocrinology* **140**, 5443 (1999).

[169] C. Valverde-R, C. Aceves, and E. Reyes, *Endocrinology* **132**, 867 (1993).

[170] A. Kawahara, Y. Gohda, and A. Hikosaka, *Dev. Growth Differ.* **41**, 365 (1999).

[171] E. Krysin, E. Brzezinska-Slebodinska, and A. B. Slebodzinski, *J. Endocrinol.* **155**, 295 (1997).

[172] H. Huang, N. Marsh-Armstrong, and D. D. Brown, *Proc. Natl. Acad. Sci. U.S.A.* **96**, 962 (1999).

[173] D. L. Berry, C. S. Rose, B. F. Remo, and D. D. Brown, *Dev. Biol.* **203**, 24 (1998).

[174] B. Höffken, R. Ködding, A. Von zur Mühlen, T. Hehrmann, H. Jüppner, and R. D. Hesch, *Biochim. Biophys. Acta* **539**, 114 (1978).

[175] F. Courtin, P. Liva, J. M. Gavaret, D. Toru-Delbauffe, and M. Pierre, *J. Neurochem.* **56**, 1107 (1991).

[176] E. Roti, S. L. Fang, K. Green, C. H. Emerson, and L. E. Braverman, *J. Clin. Endocrinol. Metab.* **53**, 498 (1981).

[177] V. A. Galton, P. T. McCarthy, and D. L. St. Germain, *Endocrinology* **128**, 1717 (1991).

[178] F. Santini, I. J. Chopra, D. H. Solomon, and G. N. C. Teco, *J. Clin. Endocrinol. Metab.* **74**, 1366 (1992).

the active site and catalytic subunit of 5DIII might be a 30-kDa protein. Affinity labeling experiments similar to those described for 5′DI and 5′DII in human placenta, rat, fetal and adult brain, as well as other tissues initially revealed a good correlation between affinity labeling of this p32 protein and inactivation of 5DIII by unlabeled $BrAcT_3$.[68,179] However, a p32 affinity-labeled protein has also been detected in other tissues with low or nondetectable 5DIII activity, for example, spleen and fetal liver. Furthermore, amounts of affinity-labeled p32 protein did not correlate with enzyme activity or content, respectively, in these tissues. Furthermore, lack of p32 labeling by $BrAc[^{125}I]T_3$ in embryonic chicken liver is in sharp contrast to the high 5DIII activity in this tissue.[179] p32 also has been labeled by $BrAc[^{125}I]rT_3$, which is not a substrate of 5DIII. In addition, affinity labeling of p32 by $BrAc[^{125}I]T_3$ was not prevented by substrates and inhibitors of 5DIII to the extent expected on the basis of structure–activity analysis of 5DIII activity in the tissues tested. p32 content in the rat brain has been estimated as 12 pmol/mg microsomal protein, whereas the V_{max} of 5DIII deiodination of T_3 in rat brain microsomes is as low as 0.13 pmol/min per mg protein,[180] indicating either that p32 does not represent a true active site catalytic subunit of 5DIII or that the low-abundance 5DIII protein comigrates in one-dimensional sodium dodecyl sulfate–polyacrylamide gel electrophoresis (SDS–PAGE), together with an abundant p32 protein not related to 5DIII; an alternative explanation might also be that the turnover number for T_3 deiodination by 5DIII is extremely low compared with the highly abundant protein p32 in rat brain, as suggested by Schoenmakers et al.[68] Attempts to distinguish between different p32 forms by taking advantage of partial purification permeabilization or solubilization of microsomal membranes confirmed the integral membrane nature of 5DIII as well as of p32. Treatment of microsomal membranes with 0.05% (w/v) deoxycholate (DOC), which removes lumenal proteins,[181] as well as detachment of microsomal proteins lacking transmembrane anchors by treatment of microsomes with carbonate buffer at pH 11.5,[182] removed neither 5DIII nor p32, indicating that both are stable transmembrane proteins. Similar to observations for 5′DI, treatment of placental microsomes with 0.05% (w/v) DOC doubled specific 5DIII activity, indicating latency of part of the 5DIII enzyme in microsomal membranes.[68]

Whereas affinity labeling of 5DIII revealed ambiguous results with respect to correlation of enzyme expression and activity versus affinity labeling of a p32 protein, in vitro studies of bromoacetyl[$^{125}I]T_3$ affinity labeling in cells transiently transfected with the 5DIII cDNA (see below) revealed a 32-kDa protein, and

[179] C. H. Schoenmakers, I. G. Pigmans, E. Kaptein, V. M. Darras, and T. J. Visser, FEBS Lett. 335, 104 (1993).
[180] M. M. Kaplan, T. J. Visser, K. Yaskoski, and J. L. Leonard, Endocrinology 112, 35 (1983).
[181] G. Kreibich, P. Debey, and D. D. Sabatini, J. Cell Biol. 58, 436 (1973).
[182] Y. Fujiki, A. L. Hubbard, S. Fowler, and P. B. Lazarow, J. Cell Biol. 93, 103 (1982).

affinity labeling was blocked by known 5DIII substrates and to minor extent also by aurothioglucose. Again, whereas micromolar aurothioglucose concentrations inhibit 5DIII activity, much higher (millimolar) concentrations are required for efficient competition of bromoacetyl $[^{125}I]T_3$ incorporation. In contrast to 5'DI, so far no clear^{75}Se labeling of the p32 protein associated with 5DIII expression has been achieved.[183]

In spite of these initially disappointing findings, molecular cloning of 5DIII of several species, using a homology cloning approach based on the cDNA sequence of 5'DI, revealed that 5DIII contains a 30- to 32-kDa protein with an active site selenocysteine and a functional SECIS element in the 3' UTR of the corresponding mRNA.[123] Functional analysis initially of *X. laevis* and later on rat, chicken, human, and other species by *in vitro* expression studies, kinetic experiments, and complementary analysis of expression and its regulation at the transcript level revealed that 5DIII is also a selenocysteine-containing enzyme, contains a 30- to 32-kDa catalytic subunit, and is expressed in tissues previously characterized as 5DIII activity-containing organs or cells.[123,184–186] It has been unclear whether there is a major discrepancy between protein expression and enzyme activity, as would be indicated by the discrepant levels of bromoacetyl-labeled p32 and enzyme expression, as the holoenzyme structure of 5DIII is still unknown. It might well be that a second heterologous subunit confirms functional activity or even that 5DIII might be an oligomeric protein complex, as suggested for 5'DII. A still open question is also the sluggish response of 5DIII expression activity to selenium depletion and repletion in spite of its selenocysteine-containing protein nature, which initially led to the assumption that, like 5'DII, 5DIII might not be a selenocysteine-containing enzyme.[121,187–189] However, comparative cloning and characterization of 5DIII from various species, for example, frog, amphibia, fish, birds, and mammals, revealed high sequence identity at the amino acid and DNA levels.[18,115] Expression of recombinant 5DIII in heterologous cells (i.e., COS-1, HEK 293, and other cell types) revealed biochemical and kinetic characteristics similar to previously described 5DIII enzymes (strongest substrate preference is for T_3, K_m in the range of 20 nM), whereas T_4 and rT_3 have lower affinity. The

[183] D. Salvatore, S. C. Low, M. Berry, A. L. Maia, J. W. Harney, W. Croteau, D. L. St. Germain, and P. R. Larsen, *J. Clin. Invest.* **96,** 2421 (1995).

[184] W. Croteau, S. K. Whittemore, M. J. Schneider, and D. L. St. Germain, *J. Biol. Chem.* **270,** 16569 (1995).

[185] J. P. Sanders, S. Van Der Geyten, E. Kaptein, V. M. Darras, E. R. Kuhn, J. L. Leonard, and T. J. Visser, *Endocrinology* **140,** 3666 (1999).

[186] A. Hernández, G. J. Lyon, M. J. Schneider, and D. L. St. Germain, *Endocrinology* **140,** 124 (1999).

[187] M. R. Stulp, J. J. M. de Vijlder, and C. Ris-Stalpers, *Mol. Cell. Endocrinol.* **142,** 67 (1998).

[188] J.-P. Chanoine, S. Alex, S. Stone, S. L. Fang, I. Veronikis, J. L. Leonard, and L. E. Braverman, *Pediatr. Res.* **34,** 288 (1993).

[189] M. Ramauge, S. Pallud, A. Esfandari, J.-M. Gavaret, A.-M. Lennon, M. Pierre, and F. Courtin, *Endocrinology* **137,** 3021 (1996).

inhibitory pattern also is comparable to native 5DIII activity in various tissues, indicating that aurothioglucose is a preferred inhibitor over iodoacetate, and only a low sensitivity of enzyme activity is observed for PTU inhibition.[123,184–186,189] The mouse and human genes encoding 5DIII have been cloned and encode 5DIII in a single exon.[186,190] Two putative transcriptional start sites downstream of a consensus TATA box have been suggested. Upstream elements resembling CAAT and GC boxes have also been mapped. A mRNA encoding 5DIII is the smallest among the deiodinases, 1.6 to 1.8 kb in size. However, larger DIII transcripts (3.3 and 3.6 kb) appeared in rat brain after induction of hyperthyroidism by T_3 administration. Whether these faint signals truly represent 5DIII transcripts or immature splice forms remains to be analyzed. The human 5DIII gene has been localized to chromosome 14q32 and to mouse chromosome 12 F1,[190] next to the region where human 5'DII is located (14q24.2-q24.3).[191]

Determination of Type III Iodothyronine Deiodinase Activity

Typically, 5DIII activity is analyzed with $[3,5-^{125}I]T_3$ or $[3,5-^{125}I]T_3$ sulfate purified as described for other iodothyronines by paper, electrophoresis, or adsorption chromatography. Membrane fractions or protein preparations are incubated in triplicate for approximately 60 min at 37° with 10 nM ^{125}I-labeled substrate (approximately 50,000–100,000 cpm) in a 0.1-ml incubation mixture containing 0.1 M sodium phosphate or 100 mM HEPES buffer at pH 8.0 or higher (alkaline pH optimum), 1 mM EDTA, and 10 mM DTT. Reactions are stopped and $[^{125}I]$iodothyronines are precipitated by successive addition of 0.1 ml of 5% (w/v) BSA and 0.5 ml of 10% (v/v) trichloroacetic acid. After mixing and centrifugation for 10 min at 10,000g, radioiodide contained in the supernatant is separated from residual ^{125}I-labeled substrate and product, using Dowex WX-50 or Sephadex LH-20 minicolumns equilibrated in 0.1 M acetic acid. Alternatively, deiodination products can be analyzed by reversed-phase chromatography, thin-layer chromatography, or paper chromatography. For high-performance liquid chromatography (HPLC) analysis, the reaction is stopped by addition of 0.2 ml of ice-cold methanol, proteins are precipitated by centrifugation, and supernatant is mixed in an equal volume with 0.2 ml of 0.02 M ammonium acetate (pH 4), and an aliquot of this mixture is applied to a reversed-phase column and developed isocratically with a mixture of acetonitrile and 0.02 M ammonium acetate (33 : 66, v/v) at a flow rate of 1 to 1.2 ml/min. Eluates are monitored by radiodetection monitors. Substrates used for 5-deiodine assays are $[3-^{125}I]T_3$ (approximately 2000 Ci/mmol)

[190] A. Hernández, J. P. Park, G. J. Lyon, T. K. Mohandas, and D. L. St. Germain, *Genomics* **53,** 119 (1998).
[191] O. Araki, M. Murakami, T. Morimura, Y. Kamiya, Y. Hosoi, Y. Kato, and M. Mori, *Cytogenet. Cell Genet.* **84,** 73 (1999).

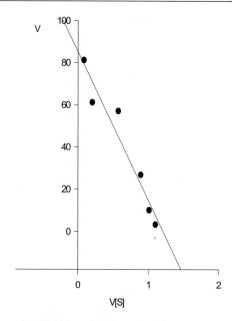

FIG. 5. Eadie Hofstee plot of initial rate kinetic analysis of the 5DIII reaction in rat fetal placental homogenates using 3,3′,5-[^{125}I]-T$_3$ as substrate (3 nM to 3 μM) and 20 mM DTT as cosubstrate. 87 μg of protein was incubated for 90 min at 37° in 100μl of 100 mM Tris buffer, pH 8.0, containing 3 mM EDTA. K$_M$ was determined as 70 nM T$_3$ and V$_{Max}$ as 85 fmole iodide release per mg protein per minute (Unpublished data: A. Baur, Würzburg, 2001). V: fmole iodide $\times$ mg^{-1} $\times$ min^{-1}. [S] : nM T3.

or [3,5-^{125}I]T$_3$ (approximately 3500 Ci/mmol). Radioiodinated iodothyronine sulfates are prepared as described.[192,193] HPLC elution can also be performed with a mixture of methanol and 20 mM ammonium acetate (pH 4.0) in a 45 : 55 (v/v) mixture.

An important control required in these analyses of 5DIII activity is the exclusion of the contribution of 5′DI activity to the 5DIII reaction, a known property of 5′DI. This can be achieved by parallel incubations of the enzyme preparation to be analyzed with 1 μM T$_3$, which saturates the 5DIII because of its nanomolar K_m value, whereas the high-K_m 5′DI activity is not blocked (Table II). Any residual iodide release and T$_3$ degradation in the presence of 1 μM T$_3$ is probably due to contamination or the presence of 5′DI activity.

Figure 5 gives an example of a kinetic analysis of 5DIII activity in rat maternal placenta homogenate. Incubation of 77 μg of placental protein was performed for

[192] M. Moreno, M. J. Berry, C. Horst, R. Thoma, F. Goglia, J. W. Harney, P. R. Larsen, and T. J. Visser, *FEBS Lett.* **344,** 143 (1994).

[193] N. Toyoda, E. Kaptein, M. J. Berry, J. W. Harney, P. R. Larsen, and T. J. Visser, *Endocrinology* **138,** 213 (1997).

90 min at 37° in 0.1 mM phosphate buffer, pH 8.0, in the presence of 1 mM EDTA, 20 mM DTT, and 3 nM to 3 μM T$_3$ substrate concentrations. Kinetic analysis reveals a K_m value of 74 nM and a V_{max} value of 260 fmol/mg per min. A parallel analysis of the fetal placental part revealed a similar K_m value (70 nM) and a lower V_{max} value of 85 fmol/mg per min (A. Baur, unpublished results, 2000).

Regulation and Function of Type III Iodothyronine Deiodinase

Several factors are involved in the regulation of expression of 5DIII in cell culture systems as well as *in vivo*. So far, only one study clearly shows a selenium-dependent expression of 5DIII in an *in vitro* cell culture model of primary cultures of rat astrocytes.[189] In this model, which might still contain contamination by other cell types, selenium depletion leads to a decrease in glutathione peroxidase activity, whereas no decrease in 5DIII activity is observed after culturing astrocytes for several days in selenium-depleted medium. Selenium-dependent alterations of 5DIII activity were, however, observed if astrocytes were stimulated with tetradecanoyl phorbol acetate (TPA), which is known to induce 5DIII activity. Here, selenium depletion reduces TPA-stimulated 5DIII activity, and selenium repletion increases 5DIII activity associated with effects of these agents on cell proliferation and growth. Further stimulators of DIII activity are cAMP, acidic fibroblast growth factor (FGF), and, to a minor extent, T$_4$, T$_3$, and all-*trans*-retinoic acid.[189] As TPA and FGF stimulation of 5DIII expression occurs more rapidly than by T$_3$ and retinoic acid, further analyses were performed that revealed that the extracellular signal-regulated kinase (ERK) signaling cascade and protein kinase C activation are involved in 5DIII regulation. Induction of 5DIII by TPA was blocked by a protein kinase C inhibitor, and effects both of TPA and FGF were partially prevented by a specific inhibitor of the [MAPK (mitogen-activated protein kinase)/ERK kinase] (MEK) and ERK signaling cascade, PD98059.[194] The observation that several growth factors and especially the substrate of 5DIII, T$_3$, stimulates expression and function of 5DIII not only in astrocytes but also in cultured rat brown adipocytes[195] indicates that this enzyme might be actively regulated during development and differentiation of various tissues and cells.[196]

Concordant with these observations are detailed analyses of developmental profiles of 5DIII expression in several brain regions, the developing amphibia, tadpole, rat, chicken, and fish embryos and neonates.[163–166,168,169,196–199] 5DIII

---

[194] S. Pallud, M. Ramaugé, J.-M. Gavaret, A.-M. Lennon, N. Munsch, D. L. St. Germain, M. Pierre, and F. Courtin, *Endocrinology* **140,** 2917 (1999).

[195] A. Hernández, D. L. St. Germain, and M. J. Obregon, *Endocrinology* **139,** 634 (1998).

[196] J. M. Bates, V. L. Spate, J. S. Morris, D. L. St. Germain, and V. A. Galton, *Endocrinology* **141,** 2490 (2000).

[197] K. A. Mol, S. Van Der Geyten, V. M. Darras, T. J. Visser, and E. R. Kühn, *Endocrinology* **138,** 1787 (1997).

[198] S. Van Der Geyten, N. Buys, J. P. Sanders, E. Decuypere, T. J. Visser, E. R. Kühn, and V. M. Darras, *Mol. Cell Endocrinol.* **147,** 49 (1999).

expression has also been detected in neonatal rat brain by *in situ* hybridization, where transcripts are selectively expressed in those areas involved in sexual differentiation of the brain, for example, the bed nucleus of the stria terminalis and preoptic nuclei.[168] These expression patterns are transient and compatible with the assumption that DIII activity is involved in local control of thyromimetic activity, guaranteeing appropriate availability of the prohormone T_4 and its active product T_3 at the right site, the right time, and the correct concentration.

In agreement with these observations are the complex developmental expression patterns of 5DIII transcript protein and activity during amphibian metamorphosis, where 5DIII gene expression is highly regulated in those organs during metamorphosis.[170,172,173,199] Especially high levels are observed before metamorphic changes are initiated. A close correlation of 5DIII expression is observed to the temporal regulation of TRß receptors in various tadpole organs, indicating systemic and local control of available T_3 ligand for the nuclear T_3 receptors. The elegant proof for this hypothesis of function of 5DIII in metamorphosis has been demonstrated in transgenic *X. laevis* tadpoles overexpressing 5DIII.[172] These tadpoles expressing high deiodinase activity are resistant to induction of premature metamorphosis by exogenous T_3. The tadpoles develop normally throughout embryogenesis and premetamorphic stages, but then arrest or retard development at time points when thyroid hormones are reaching highest levels in the appropriate metamorphic tissues. For example, gill and tail resorption are delayed, and intestinal and hind leg development proceeds. Similar patterns of development-dependent expression of 5DIII and also 5'DII were demonstrated in the rat uterus after implantation or induction of pseudopregnancy.[167] Whereas 5DIII expression in brain and other cells and tissue rapidly responds to T_3, the T_3-dependent regulation of placental 5DIII is still controversial. Apparently, more detailed analysis of species differences in placental structures as well as dissection of maternal to fetal placental membranes compartments are required to settle this issue.[200-202] So far, *in vitro* expression and regulation of 5DIII function do not correlate with *in vivo* observations. For example, no 5DIII activity has yet been demonstrated in the brown adipose tissue of rats. However, a series of reports is available on correlation of expression of 5DIII activity and thyroid hormone content in the rat and also the human brain, indicating that high T_3 activity in specific brain regions correlates with low T_3 content. For example, in postmortem autoptic specimens of various human brain areas, high levels of 5DIII activity were observed in the

[199] J. G. Eales, J. M. McLeese, J. A. Holmes, and J. H. Youson, *J. Exp. Zool.* **286,** 305 (2000).
[200] E. Roti, L. E. Braverman, S. L. Fang, S. Alex, and C. H. Emerson, *Endocrinology* **111,** 959 (1982).
[201] K. Mori, K. Yoshida, H. Fukazawa, Y. Kiso, N. Sayama, K. Kikuchi, Y. Aizawa, and K. Abe, *Endocr. J.* **42,** 753 (1995).
[202] C. H. Emerson, G. Bambini, S. Alex, M. I. Castro, E. Roti, and L. E. Braverman, *Endocrinology* **122,** 809 (1988).

hippocampus and temporal cortex, which had the lowest T_3 contents. Also, treatment of rats with various neurological drugs led to alterations in T_3 expression and concomitant changes in thyroid hormone contents.[148,174] Of note also are marked species differences in regulation and function of deiodinases, especially 5DIII. A close correlation between 5DIII expression and maturation of specific brain regions has been observed in the chicken embryo,[169] and in this species growth hormone and dexamethasone decrease hepatic 5DIII activity at the pretranslational level, with the result of increasing circulating T_3 serum levels. Whether GH and glucocorticoids have similar effects in rodents and mammals including humans remains to be analyzed. So far, data on GH and glucocorticoid regulation of deiodinase enzymes are discrepant between different models and systems used for analysis. An important clinical observation is the high (neo)expression of 5DIII in pathological adult human liver tissue, as well as in hemangioma in children.[203] This high expression in hemangioma leads to exaggerated degradation of circulating T_3, which cannot be compensated by administration of high amounts of T_4. The mechanism of overexpression in hemangioma and its consequences are not clear. Whether reexpression of 5DIII in adult human liver, similar to its expression in the fetal liver in pathological conditions, contributes to the manifestation of the so-called low T_3 syndrome characterized by adequate T_4 and low circulating T_3 in addition to decreased hepatic 5'DI activity, remains to be established. Developmental analysis of deiodinase expression in human perinatal liver revealed expression of 5DIII activity not only during the fetal phase but also after birth.[204]

Methods to Determine Deiodinase Activity

Chromatographic Methods: Paper, Thin-Layer, and High-Performance Liquid Chromatography

Initially, deiodination reactions were analyzed by time-consuming chromatographic methods such as descending paper chromatography or, later on, thin-layer chromatography or HPLC methods as established by pioneers such as J. Gross, R. Pitt-Rivers, E. V. Flock, S. Lissitzky, E. C. Albright, and many others in this field of deiodination.[10] Chromatographically pure iodine-131 or ^{125}I labeled T_4 or other iodothyronines were incubated under light (and oxygen) protection with tissues slices, homogenates or other samples containing deiodinase enzymes. The reaction was stopped, tissues were homogenized if slices or organs were incubated, and products were extracted with butanol–NH_4OH or other comparable solvents. Butanol extracts were applied to chromatography paper and developed

[203] S. A. Huang, H. M. Tu, J. W. Harney, M. Venihaki, A. J. Butte, H. P. Kozakewich, S. J. Fishman, and P. R. Larsen, *N. Engl. J. Med.* **343**, 185 (2000).

[204] K. Richard, R. Hume, E. Kaptein, J. P. Sanders, H. Van Toor, W. W. De Herder, J. C. den Hollander, E. Krenning, and T. J. Visser, *J. Clin. Endocrinol. Metab.* **83**, 2868 (1998).

with various solvent systems. Radioactivity was then registered either directly with Geiger–Müller counters, radiometers, or autoradiography or counted after cutting the chromatography paper strips. Quantification and identification of substrates and products required preparation of appropriately labeled standards. A major advantage of these techniques was that several products of the deiodination reaction, including iodide, as well as remaining substrates of the deiodination reaction could be analyzed simultaneously if appropriate separation conditions were applied.[205] These techniques supported the concept of sequential monodeiodination of iodothyronines, in which iodide release from the covalent carbon–iodide bond of various iodothyronines correlated stoichiometrically to the formation of iodothyronine product and to substrate consumption. These studies also revealed that lower iodinated compounds, that is T_3 and T_2 compounds, were not end products of the reaction, but underwent further deiodination along the cascade to yield the iodine-free thyronine T_0. However, the disadvantage of these tedious chromatographic procedures has been interference by light and oxygen leading to nonenzymatic deiodination of sensitive iodothyronine substrates and products. These artifacts as well as the disappearance of iodide caused by oxidation and formation of volatile iodine later on led to the development of more efficient and less time-consuming test systems. Furthermore, sensitivity problems impaired quantification of low enzyme activities in several tissues. Therefore underestimation of deiodination rates was frequently encountered.

Product stability and/or deiodination can be monitored by double-isotope techniques. Frequently the substrate is labeled with ^{125}I and the product tracer can be ^{131}I labeled and added to the reaction mixture to monitor its stability during the test procedure. The short half-life of ^{131}I compared with ^{125}I allows initial monitoring of product stability during incubation and sample work-up and, subsequent to the decay of ^{131}I, the formation of product from the ^{125}I-labeled substrate can be analyzed in the same sample.

Later, HPLC methods for sample analysis were also adopted for determination of deiodinase activities, taking advantage of the possibility to directly quantify substrates and products online by radiometric detector systems or subsequently quantify eluted fractions containing substrate and products, using γ counters. Preferable systems employed here for sample handling involved stopping the incubation reaction by acids (partially after addition of serum or protein to bind labeled iodothyronines) and chromatography of iodothyronines, using reversed-phase technology and various appropriate solvent systems such as acetonitrile and trifluoroacetic acid.[206] The HPLC method also provides the advantage to detect both free and conjugated iodothyronines such as sulfates.[207–213] However, the

[205] E. C. Albright, F. C. Larson, and R. H. Tust, *Proc. Soc. Exp. Biol. Med.* **86,** 137 (1954).
[206] J. M. Koopdonk-Kool, M. C. van Lopik-Peterse, G. J. M. Veenboer, T. J. Visser, C. H. H. Schoenmakers, and J. J. M. de Vijlder, *Anal. Biochem.* **214,** 329 (1993).

disadvantages of this method for routine analysis are the time-consuming HPLC procedure and the need for complex reversed-phase solvent systems as well as for expensive HPLC equipment and HPLC columns.

Thin-layer chromatography (TLC) or HPLC methods are of special importance for analysis of 5-deiodination of the tyrosyl ring if no tracer of sufficiently high specific activity, carrying the label at the tyrosyl ring 5 position, is available. Here one- or two-dimensional TLC, or paper chromatography, or HPLC provide a means to sensitively and quantitatively monitor substrate consumption and product formation without the possibility of analyzing released radioactively labeled iodide.[214–216] Iodothyronine standards are needed to calibrate the model system.[206,217] A new procedure to synthesize iodothyronines with reasonably high specific activity and labeled at the 5 position of the tyrosyl ring provides useful tracers for specific analysis of 5-deiodinase reactions and sensitive quantification of 3,5-T_2.[192,218]

Purification of [131]I- or ^{125}I-Radiolabeled Substrates Employed in Deiodinase Reactions

Purification of iodothyronines used as substrates in deiodinase assays is required before incubation with enzyme preparations. Commercially available [^{131}I]- or [^{125}I]iodothyronines (specific activities, ~1100–1300 μCi/μg or 40–49 MBq/μg) and tracers synthesized by chemical iodination contain free iodide due to oxidative, light- and radiation-induced degradation. To increase sensitivity and specificity of the deiodinase reaction and to allow determination of equimolar formation of iodothyronine products it is necessary to reduce background levels of iodide as much as possible. This is best achieved by batch purification of the radiolabeled substrate used in the reaction either by high-voltage paper electrophoresis[45] or by adsorption to small (1- to 2-ml) Sephadex G-25 or G-50 columns or to LH-20 columns equilibrated in neutral phosphate buffer solutions. Contaminating iodide

[207] T. J. Visser, *Chem. Biol. Interact.* **92**, 293 (1994).
[208] M. H. Otten, J. A. Mol, and T. J. Visser, *Science* **221**, 81 (1983).
[209] D. L. St. Germain, R. Adler, and V. A. Galton, *Endocrinology* **117**, 55 (1985).
[210] M. Rutgers, F. A. Heusdens, and T. J. Visser, *Endocrinology* **125**, 424 (1989).
[211] D. C. Ferguson and A. S. Jennings, *Am. J. Physiol.* **245**, E220 (1983).
[212] S. J. Eelkman Rooda, M. A. C. Van Loon, and T. J. Visser, *J. Clin. Invest.* **79**, 1740 (1987).
[213] R. C. Smallridge and N. E. Whorton, *Metabolism* **33**, 1034 (1984).
[214] J. van Doorn, F. Roelfsema, and D. Van der Heide, *Acta Endocrinol. (Copenh.)* **101**, 386 (1982).
[215] J. van Doorn, D. Van der Heide, and F. Roelfsema, *Endocrinology* **115**, 174 (1984).
[216] J. van Doorn, F. Roelfsema, and D. Van der Heide, *Endocrinology* **117**, 1201 (1985).
[217] K. Sorimachi and J. Robbins, *J. Biol. Chem.* **252**, 4458 (1977).
[218] G. Pinna, H. Meinhold, L. Hierda, R. Thoma, T. Hoell, K.-J. Gräf, G. Stoltenburg-Didinger, M. Eravci, H. Prengel, O. Brödel, R. Finke, and A. Baumgartner, *J. Clin. Endocrinol. Metab.* **82**, 1535 (1997).

is eluted by washing the columns with buffer and distilled water and purified substrate is eluted with an ethanolic solution at alkaline pH or with a methanol–2 N NH$_4$OH (99 : 1, v/v) solution. The eluted radiolabeled substrate, containing less then 0.5% free radioiodide, can be concentrated at room temperature, protected from light, under a stream of nitrogen gas in a hood and then used for incubation in appropriate assay buffers. Care must also be applied to use substrates ^{125}I labeled only in one specific molecule position. Otherwise, no clear allocation and quantification are possible for specific removal of [^{125}I]iodide from chemically and sterically different substitution positions in iodothyronines, for example, the 3' or 5' position of the phenolic ring or the 3 or 5 position of the tyrosyl ring. Therefore, substrates too high in specific activity and radiolabeled in more than one position are of limited value in quantifying specific steps of the monodeiodinase sequence. On the other hand, a sufficiently high specific activity of substrate is needed to enable sensitive detection of low enzyme activities and to avoid "dilution" of unlabeled substrate with high amounts of radiolabeled tracer in kinetic analysis. In general, "carrier-free" tracers that have been specifically synthesized from appropriate precursors such as T$_3$ (for labeled T$_4$), 3,3'-T$_2$ (for labeled rT$_3$), or 3,5-T$_2$ (for labeled T$_3$), which are produced by radioiodination with chloramine T under conditions that avoid iodide exchange reactions, are preferred. Specific tracers of sufficiently high specific activity have also been produced for analysis of 5-deiodination at the tyrosyl ring of T$_3$ or 3,5-T$_2$.[192,218] *In vivo* biosynthesis of radiolabeled iodothyronines after injection of carrier-free Na[131]I or Na[125]I into rats has also been used to generate tracers and standards of reasonably high specific activity, but these substrates need to be extensively purified and contain the isotope at several positions in the iodothyronine.[14]

Radioimmunoassays

A major breakthrough in the analysis of sequential monodeiodination was the introduction by the groups of Hesch, Visser, and Chopra[219–221] of radioimmunoassays for specific analysis of products and substrates of sequential monodeiodination reactions. The use of radioimmunoassays allowed the highly specific, highly sensitive, and quantitative analysis of both substrate consumption and product formation along the monodeiodination cascade, with the exception of simultaneous quantification of iodide formation. However, the radioimmunoassay-based methods were time consuming and tedious, as (ethanolic) extracts of incubation reactions had to be prepared and in a few cases needed partial purification or fractionation in order to allow specific product analysis in radioimmunoassays adapted to application

[219] R. D. Hesch, G. Brunner, and H. D. Soeling, *Clin. Chim. Acta* **59**, 209 (1975).
[220] T. J. Visser, I. van der Does-Tobe, R. Docter, and G. Hennemann, *Biochem. J.* **150**, 489 (1975).
[221] I. J. Chopra, D. H. Solomon, U. Chopra, S. Y. Wu, D. A. Fisher, and Y. Nakamura, *Recent Prog. Horm. Res.* **34**, 521 (1978).

for *in vitro* enzyme reactions. Problems encountered here were similar to those of the chromatographic approaches: the low solubility of thyroid hormones, their tendency to stick to plastic tips and reaction vessels, as well as their light and oxygen sensitivity. In addition, some sensitivity of the deiodination reaction by chromatographic methods was lost because the use of substrates and products labeled with iodide isotopes had to be avoided in order not to interfere with radioimmunometric quantification of the reaction mixture compounds. However, the major advantage has been the possibility of highly specific detection of several metabolites and also consumption of substrates from the same extracts of the incubation mixture. This method unequivocally established the principle of the sequential monodeiodination of iodothyronines as well as their conjugates, such as the sulfates, by three different deiodinase enzymes.

Radiometric Assay: Release of Radioiodide

The most rapid, sensitive, and specific deiodinase assay currently used was introduced by Leonard and Rosenberg in 1980,[45] who took advantage of the observation that reverse T_3 is an excellent substrate for type I 5'-deiodinase and a competitive inhibitor of T_4 5'-deiodination to T_3 as demonstrated by several groups in the early studies characterizing 5'DI. In tissues expressing both 5'- and 5-deiodinase activity, rT_3 is superior as substrate because it yields equimolar concentrations of 3,3'-diiodothyronine and iodide in contrast to T_4, which may yield deiodination products undergoing further deiodination. Therefore iodide formed in stoichiometric amounts with 3,3'-T_2 can easily be quantified and directly reflects deiodination activity. The test procedure is as follows.

Extracts or tissues expressing 5'-deiodinase activity are incubated in HEPES, phosphate, or Tris buffers at the pH optimum of type I 5'-deiodinase (pH 6.8 to 7.0) in the presence of 10 to 20 mM DTT, in the absence or presence of deiodinase inhibitors, for a certain period of time (10 min to 1 hr) at 37°. The reaction is stopped by addition of bovine serum albumin (BSA) or serum, which binds thyroid hormones, and subsequent addition of 10% (v/v) trichloroacetic acid (TCA), which precipitates protein and BSA bound to the majority of thyroid hormones and stops enzymatic reaction. After centrifugation of the protein pellet, the supernatant containing liberated iodide and residual substrate not bound and precipitated by protein and BSA is applied to acid-equilibrated Dowex 50W cation-exchange resins binding nonmetabolized substrate and product iodothyronines, whereas liberated iodide is released and eluted by diluted acidic acid. The released iodide is then quantified in a γ counter. Deiodinase activity can be calculated by taking into account that statistically, one radioactively labeled [^{125}I]iodide reflects only half of the phenolic ring 5' deiodination reactions if monolabeled [^{125}I] rT_3 is employed. To obtain highly sensitive assay conditions, the substrate [^{125}I] rT_3 used in the reaction needs to be prepurified from contaminants and liberated radioiodide, either by high-voltage paper electrophoresis or adsorption/desorption chromatography on Sephadex LH-20 or Dowex 50W as described above.

Although widely used and successfully applied, this radioiodide release assay, which is also applicable to $[^{125}I]T_4$ or other iodothyronine metabolites including sulfates, needs special preconditions and requires observation of basic rules of enzyme kinetics. First, the method relies on equimolar production of $[^{125}I]$iodide and deiodinated iodothyronine product, which, especially if new sources of enzyme activity are evaluated, needs to be monitored and confirmed in order to avoid the possibility that nonspecific liberation of radioiodide due to oxidative dehalogenation, light, or chemically induced destruction of substrates is interpreted as enzyme activity. Furthermore, basic rules of enzymology must be applied with respect to analysis of initial velocities of enzyme reactions, which means sufficient substrate : enzyme ratios (>1000 : 1) allowing determination of true activities, where enzyme is acting as catalyst and not as stoichiometric reactant.[46,50,106] The latter point is frequently neglected and has led to a series of publications announcing so-called low K_m deiodinase activities in tissues definitely devoid of, for example, 5'DII activity, such as kidney, liver, and, more recently controversially discussed, human thyroid.

An observation not yet explained in the type I deiodinase reaction is a rapid initial release of iodide, which so far cannot be accounted for by simultaneous production of iodothyronine metabolites, especially in tissues with high 5'-deiodinase activity, such as liver, kidney, and thyroid.[45,47,50] It might well be that the complex mechanism of reaction of 5'DI, proceeding by a two-substrate ping–pong mechanism, includes rapid release of iodide from substrate by active enzyme, which does not follow linear progress curves of enzyme activities. If this initial iodide release is taken as evidence of low-K_m type II deiodinase activity in corresponding tissues because it is not inhibited by PTU or ATG, this probably reflects a misunderstanding of the principle of enzyme determination by the radioiodide release assay; several publications in the deiodinase field did not observe this complication.

A third problem is encountered if tissues contain both 5'-deiodinase type I and type II activity and if type I activity markedly exceeds that of 5'DII. Under these conditions, the rapid initial iodide release in the first half-reaction of 5'DI cannot be distinguished from low 5'DII activity, and thus the latter activity might be considerably overestimated. This difficulty is encountered in the human thyroid gland, where in addition to high 5'DI expression also functional activity of 5'DII has been reported, in agreement with detection of transcripts for the selenocysteine-p32 5'DII subunit.[222] In contrast, no evidence of 5'DII expression could be found in several human thyroid carcinoma cell lines and human thyroid tumors,[223–225] and also adult rat thyroid does not express functional 5'DII activity or corresponding transcripts.

[222] D. Salvatore, H. Tu, J. W. Harney, and P. R. Larsen, *J. Clin. Invest.* **98**, 962 (1996).

In summary, careful analysis of adequate reaction conditions, enzyme kinetics, product : substrate ratios, as well as equimolarity of iodide release and formation of iodothyronine product are required if new sources of deiodinase enzyme activities are characterized. This especially also holds true for attempts to characterize deiodinase mutants by site-directed mutagenesis, domain swaps, or transient or stable overexpression of wild-type and/or mutant enzymes in cells devoid of activity or already expressing high endogenous deiodinase activities. Unfortunately, many of these precautions frequently are neglected, and several publications have appeared that do not meet the basic criteria of exact determination of deiodinase activity.

Immunosequestration

The technique of immunosequestration of both products and substrates of deiodinase reactions has been extremely useful in the initial setup and detailed analysis of complex sequential deiodination pathways, *in vivo* and *in vitro* kinetic studies, compartmental analysis of deiodinative metabolism in both human and animal models, as well as in perfusion and cell culture studies. The technique takes advantage of the high specificity of antibodies for various iodothyronines in situations where they mutually interfere with each other's metabolism (e.g., T_4 and rT_3), or where iodothyronines comigrate in chromatographic systems, or where enrichment of a minor product is required to overcome an excess of substrate or alternative product. Here antibodies, usually in high concentrations, are added directly to the reaction mixtures for capturing originating product or are applied to the reaction mixture after termination of the reaction to enable immunoextraction. Of advantage here is coupling of antibodies to solid-phase matrices in order to facilitate separation of antibody–ligand complexes either by centrifugation; binding to second antibodies, membranes, or chromatographic columns; or, more recently, magnetic beads. This technique has been helpful to analyze iodothyronine metabolism in the developmental and neonatal phase of several organisms, where limited material is available as enzyme source,[226] in clinical conditions for analysis of rare metabolites in body fluids,[227,228] or for analysis of the regulatory role of rT_3 in the 5′DI reaction and iodothyronine metabolism.[229] An elegant approach to analyze

[223] M. Oertel, R. D. Hesch, and J. Köhrle, *Exp. Clin. Endocrinol.* **97**, 182 (1991).

[224] J. Köhrle, M. Oertel, C. Hoang-Vu, F. Schnieders, and G. Brabant, *Exp. Clin. Endocrinol.* **101** (Suppl. 3), 60 (1993).

[225] R. Schreck, F. Schnieders, C. Schmutzler, and J. Koehrle, *J. Clin. Endocrinol. Metab.* **79**, 791 (1994).

[226] M. Borges, Z. Eisenstein, A. G. Burger, and S. H. Ingbar, *Endocrinology* **108**, 1665 (1981).

[227] J. S. LoPresti, D. W. Warren, E. M. Kaptein, M. S. Croxson, and J. T. Nicoloff, *J. Clin. Endocrinol. Metab.* **55**, 666 (1982).

[228] R. Rajatanavin, D. A. Fisher, L. Chailurkit, W. S. Huang, S. Srisupandit, and S. Y. Wu, *Thyroid* **9**, 989 (1999).

[229] T. Kaminski, J. Köhrle, R. Ködding, and R. D. Hesch, *Acta Endocrinol.* **98**, 240 (1981).

in more detail the relative contribution and intermediate production of iodothyronines in the complex deiodination cascade from T_4 to T_0 has been the technique of immunosequestration, employing analysis of the deiodinase reaction in the presence of specific antibodies or products formed intermediately in the deiodination cascade. Here, especially the simultaneous analysis of relative production of T_3 and rT_3 has been studied, because rT_3 formed from T_4 by 5-monodeiodination is an excellent substrate for 5′ deiodination. Therefore if tissues under investigation contain both 5′- and 5-deiodinase activity, rT_3 is instable and, if formed by 5-deiodinase, undergoes 5′-deiodination. These studies could confirm that during development, there is a shift from 5- to 5′-deiodination, for example, in immature chicken embryo liver.[226] Furthermore, a pH-dependent intermediate formation of rT_3 could be confirmed, and the hypothesis that rT_3 inhibits T_4 5′-deiodination and T_3 formation could also be supported by an immunosequestration approach.[229] Unfortunately, these techniques need high amounts of antibodies or immunoglobulins and therefore have not been systematically evaluated for all steps of the monodeiodination cascade.

Determination of Enzyme Activity in Intact Cells and Organs

Most of the determinations of deiodinase activities have been performed in broken cell preparations, homogenates, or partially purified membranes because so far no sufficiently easy and reproducible method to isolate purified enzymes has been developed. Most reliable results have been obtained by using purified membrane preparations, especially microsomal or plasma membrane preparations devoid of cytosolic contaminations and adsorbed proteins. Regularly increased specific activity can be obtained if peripheral membrane proteins binding to microsomal membranes are removed by low detergent concentration or salt treatment at pH 8.0 or higher. These procedures, of course, also remove putative cofactors of the enzyme contained in cytosolic preparations. However, no convincing evidence has yet been presented that cytosolic cofactor systems based on NADPH-regenerating redox systems, glutathione-dependent redox systems, or even glutaredoxin-, thioredoxin-, or dihydrolipoamide-based cofactor systems efficiently support deiodinase reactions. Reasonable stimulation of *in vitro* enzyme activity by these cofactor systems is seen only with long incubation times and low iodothyronine substrate concentrations, reaching stoichiometric ratios close to those of enzymes employed, conditions not adequate for initial rate determinations.

So far, reliable enzyme activities are determined in the presence of at least 10 mM concentrations of strongly reducing dithiol systems such as DTT or DTE. These observations question whether *in vivo* glutathione- or thioredoxin-based cofactor systems exist at all and regenerate active enzyme. It might well be assumed that deiodinase reaction is irreversible and regeneration of active enzyme does not occur *in vivo* in contrast to the *in vitro* experimental systems for deiodinase activity determination. Support for this observation is given by determination of enzyme activities in intact cell systems, which in general confirm characteristics observed for broken cell preparations. Use of cell lines derived from tissues expressing type I 5'-deiodinase activity, for example, liver, kidney, or thyroid cells, enable determination of enzyme activity also in intact cell preparations, either as suspended isolated cells, primary cultures, or immortalized or tumor cell lines derived from these tissues.[56,59,91,105,208,212,217,219,230]

When using cell lines for determination of enzyme activity, and especially if progress curves are analyzed, a discrepancy is observed between kinetic parameters in intact cell lines versus broken cell preparations. In most instances, apparent K_m values differ between isolated microsomal preparations and intact cells, because of binding of iodothyronines to extracellular and intracellular proteins and membranes. If attempts are made to determine available free hormone concentrations, appropriate kinetic interpretations are feasible; however, they demand major effort and thus have not been routinely employed. In addition, some discrepancies have been observed between microsomal preparations and intact cells with respect to cofactor stimulation of enzyme activity. Whereas in isolated purified microsomal membranes no significant deiodinase activity can be determined in the absence of sufficient amounts of reducing dithiols, deiodination occurs in intact cells in the absence of addition of exogenous dithiols in significant amounts and with characteristics comparable to *in vitro* analysis of microsomal membranes. This observation suggests that endogenous activators of 5'-deiodinase activity are present in sufficient amounts or that at available free substrate concentrations, the endogenous enzyme regeneration does not need an excessively high concentration of strongly reducing dithiols. An advantage of using intact cells for analysis of deiodinase reactions is the possibility to examine the efficacy of relevant inhibitors under *in vivo* conditions[105] or to monitor regulatory effects exerted by cellular metabolism and to maintain cellular integrity and organization of enzymes, thus avoiding artifacts of preparation of microsomal membranes, subfractionation of cell compartments, and possibly the need to reactivate deiodinases by high dithiol concentrations. A disadvantage, of course, is that detailed mechanistic and kinetic studies are impaired because of restricted accessibility of substrates and release of products from active sites of the intracellular enzymes through cellular membranes.

Deiodinase activities have also been determined in isolated organ perfusion models. Here, liver, kidney, and thyroid have been employed, and many of the observations made previously in isolated membrane or cellular systems could

[230] J. T. Hidal and M. M. Kaplan, *J. Clin. Invest.* **76,** 947 (1985).

be confirmed at the level of an intact isolated perfused organ.[93,211,231–234] This especially holds true for substrate specificity, distribution of deiodinase enzymes, pH-dependent regulation in the liver,[93] as well as confirmation of many properties of thyroidal deiodinase activities in various nonprimate and rodent models.[233,235]

Determination of Deiodinase Activity in Intact Organisms

Several models and methods have been developed to analyze deiodinase activity in intact vertebrates, both laboratory animals and humans. As mentioned above, the first clear evidence of the physiological role of T_4 deiodination to T_3 resulted from in vivo studies of thyroidectomized patients, clearly demonstrating enzymatic T_3 production.[12] Later, this principle was expanded by using both radiolabeled tracers and specific chromatographic or radioimmunoassay-based procedures including immunosequestration[59,236,237] in serum, body fluids, or urine to estimate both deiodination rates and ratios in healthy and diseased individuals or after administration of pharmaceuticals. These in vivo studies confirmed in vitro results with respect to identification of at least three deiodinase enzymes, relative and absolute contribution of extrathyroidal metabolism to thyroid hormone economy, and tissue contribution to production and degradation of various iodothyronines and T_4. Of special importance is the clear demonstration that T_3 production occurs not only in liver but also in kidney, thyroid, and several other tissues. rT_3 deiodination is catalyzed mainly by the liver, and formation of rT_3 occurs in extrahepatic tissues and also not in the thyroid.

The quantitative analysis of tissue-specific deiodination and determination of relative contribution of the three deiodinase enzymes and different tissues to circulating hormone levels and local production of hormones require complex models and assumptions.[214–216,238–243] Analysis can be performed under steady state

---

[231] M. J. Müller, J. Köhrle, R. D. Hesch, and H. J. Seitz, Biochem. Int. 5, 495 (1982).

[232] A. S. Jennings, D. C. Ferguson, and R. D. Utiger, J. Clin. Invest. 64, 1614 (1979).

[233] P. Laurberg, Endocrinology 117, 1639 (1985).

[234] K. Inoue, Y. Grimm, and M. A. Greer, Endocrinology 81, 946 (1967).

[235] P. Laurberg, Acta Endocrinol. (Copenh.) Suppl. 236, 1 (1980).

[236] S. M. C. Lum, J. T. Nicoloff, C. A. Spencer, and E. M. Kaptein, J. Clin. Invest. 73, 570 (1984).

[237] A. Pilo, G. Iervasi, F. Vitek, M. Ferdeghini, F. Cazzuola, and R. Bianchi, Am. J. Physiol. 258, E715 (1990).

[238] H. F. Escobar-Morreale, F. Escobar del Rey, M. J. Obregón, and G. Morreale de Escobar, Endocrinology 137, 2490 (1996).

[239] J. J. DiStefano III, M. Jang, T. K. Malone, and M. Broutman, Endocrinology 110, 198 (1982).

[240] J. J. DiStefano III, in "Thyroid Hormone Metabolism" (S.-Y. Wu, ed.), p. 65. Blackwell Scientific, Cambridge, Massachusetts, 1991.

[241] T. T. Nguyen, F. Chapa, and J. J. DiStefano III, Endocrinology 139, 4626 (1998).

condition or conditions of isotopic equilibrium in order to avoid interference of transport and hormone distribution artifacts. This can be achieved by constant infusion of hormone, using implanted minipumps, implanted slow-release pellets, or silastic tubing in animals. For these studies double-isotope techniques are essential to account for substrate and product turnover. Several models have also been employed to perform two-, three-, or multicompartmental analysis of *in vivo* kinetics of thyroid hormone metabolism.[236,237,240,243,244] A major problem encountered during these studies both in animal models and in humans is the possible existence of "hidden pools" of iodothyronines, such as rT_3.[245,246] This term implies that some T_4 metabolites such as rT_3 but also T_3 and the diiodothyronines are formed in intracellular compartments, never reach the circulation, and do not mix with circulation hormone pools. Thus, kinetic parameters determined from tracer or substrate injection studies might be incorrect because these hidden intracellular pools cannot be incorporated into the data sets.[241,247]

Acknowledgments

The expert secretarial help of Mrs. M. Mager is greatfully acknowledged. Dr. A. Baur (Clinical Research Group, Med. Poliklinik, Würzburg) kindly provided results of the kinetic analysis presented in Fig. 5. This work is supported by grants from the Deutsche Forschungsgemeinschaft, Bonn, Kö 922/7-1 and Kö 922/8-1.

[242] R. A. McGuire and M. T. Hays, *J. Clin. Endocrinol. Metab.* **53,** 852 (1981).

[243] A. A. Zaninovich, E. el Tamer, S. el Tamer, M. I. Noli, and M. T. Hays, *Thyroid* **4,** 285 (1994).

[244] P. M. Versloot, D. van der Heide, J. P. Schröder-van der Elst, and L. Boogerd, *Eur. J. Endocrinol.* **138,** 113 (1998).

[245] J. S. LoPresti, K. P. Anderson, and J. T. Nicoloff, *J. Clin. Endocrinol. Metab.* **70,** 1479 (1990).

[246] M. J. Obregon, F. Roelfsema, G. Morreale de Escobar, F. Escobar del Rey, and A. Querido, *Clin. Endocrinol. (Oxf.)* **10,** 305 (1979).

[247] W. J. Carter, K. M. Shakir, S. Hodges, F. H. Faas, and J. O. Wynn, *Metabolism* **24,** 1177 (1975).

[13] Expression and Regulation of Thioredoxin Reductases and Other Selenoproteins in Bone

By Franz Jakob, Katja Becker, Ellen Paar,
Regina Ebert-Duemig, and Norbert Schütze

Bone Physiology

Bone is a complex tissue that is build up in a concerted action of mainly two different cell types of different origin, osteoblasts and osteoclasts.

Osteoblasts

Osteoblasts are mesenchyme-derived cells that develop from mesenchymal stem cells present in the bone microenvironment. Mesenchymal stem cells are the principal precursors for osteoblasts, chondrocytes, adipocytes, myotubes, and fibroblasts. Either pathway of differentiation is dependent on the expression of specific transcription factors and/or the action of growth factors. The essential basic transcription factor core binding factor a1 (Cbfa1) as well as $1,25(OH)_2$-vitamin D_3 and bone morphogenetic proteins such as BMP2 and their cognate receptors promote osteoblast development and differentiation.[1] These processes are as well influenced by systemic factors such as growth hormone, insulin, the insulin-like growth factors and their binding proteins, and steroid hormones. Mature osteoblasts secrete extracellular matrix proteins (mainly type I collagen) and several less abundant matrix-associated signaling molecules.[2] Two of the latter, osteocalcin (bone gla protein, BGP) and matrix gla protein (MGP), undergo vitamin K-dependent γ-carboxylation and are inhibitors of matrix calcification.[3] Others are integrin-mediated signaling molecules such as members of the angiogenic and proliferation-promoting CCN (connective tissue growth factor/cysteine-rich 61/nephroblastoma overexpressed) family of proteins, cysteine-rich protein 61 (hCYR61), and connective tissue growth factor-like protein (CTGFl).[4] The final process of bone remodeling is mineralization of the extracellular matrix. Some osteoblast-derived cells are buried in inorganic bone, where they become osteocytes. These cells develop processes that contain parts of the cellular skeleton and probably are the main mediators of transmission of physical forces into biochemical signals.

[1] P. Ducy, *Dev. Dynam.* **219,** 461 (2000).
[2] P. Ducy, T. Schinke, and G. Karsenty, *Science* **289,** 1501 (2000).
[3] M. J. Shearer, *Curr. Opin. Clin. Nutr. Metabol. Care* **3,** 433 (2000).
[4] D. R. Brigstock, *Endocr. Rev.* **20,** 189 (1999).

Osteoclasts

Osteoclasts (OCs) are derived from myeloic precursor cells under the influence of a series of factors including macrophage colony-stimulating factor (M-CSF) and its receptor (cfms) and the essential system of osteoprotegerin (OPG), receptor activator of nuclear factor κB (RANK), and their common ligand named either OPGL or RANKL. OCs represent one of several stages of terminal macrophage differentiation.[5,6] OC differentiation processes comprise a final fusion of precursor cells leading to giant multinucleated cells. These mature OCs adsorb to the bone surface through a sealing zone, thus creating an "extracellular lysosome" where they secrete protons and proteases, thereby resorbing bone. This process is essential to provide calcium for the maintenance of the serum calcium level and as well for constant bone remodeling. Knockout of different components of the above-discussed signal transduction pathways and consecutive lack of osteoclasts produces osteopetrosis. Being monocyte-macrophage-derived cells, OCs are capable of producing major oxidative bursts and in addition produce NO via inducible nitric oxide (iNO) synthase. Thus the process of bone resorption is accompanied by the production of reactive oxygen species (ROS), mainly the hydroxyl radical and probably peroxynitrite.[7,8] The activity and recruitment of both anabolic osteoblasts and bone-resorbing osteoclasts are coupled through multiple local systems, but they are not principally dependent on each other.

Expression of Selenoproteins in Cells of Bone Microenvironment

The pattern of expression of selenoproteins in various cells and tissues can be analyzed by exploiting radioactive selenium-75 incorporation into proteins. In extracts of human fetal osteoblast (hFOB) cells we obtained a characteristic pattern of selenium-labeled proteins. We detected six markedly labeled bands and at least three additional bands of lower intensity (Fig. 1).[9] A band, $\sim$15 kDa in molecular mass, might be assigned to a small 15-kDa selenoprotein described in T cells.[10] Bands of approximately 22 and 24 kDa can be assigned to cytosolic glutathione peroxidase (cGPx) and plasma glutathione peroxidase (pGPx). An intense band of $\sim$55 kDa may represent thioredoxin reductase α (TrxRα), which we identified as a vitamin D$_3$-responsive gene in hFOB cells (see below). Thioredoxin reductase

---

[5] S. L. Teitelbaum, *Science* **289**, 1504 (2000).

[6] T. Suda, N. Takahashi, N. Udagawa, E. Jimi, M. T. Gillespie, and T. J. Martin, *Endocr. Rev.* **20**, 345 (1999).

[7] D. M. Evans and S. H. Ralston, *J. Bone Min. Res.* **11**, 300 (1996).

[8] J. H. Fraser, M. H. Helfrich, H. M. Wallace, and S. H. Ralston, *Bone* **19**, 223 (1996).

[9] I. Dreher, N. Schütze, A. Baur, K. Hesse, D. Schneider, J. Köhrle, and F. Jakob, *Biochem. Biophys. Res. Commun.* **245**, 101 (1998).

[10] V. N. Gladyshev, K. T. Jeang, J. C. Wootton, and D. L. Hatfield, *J. Biol. Chem.* **273**, 8910 (1998).

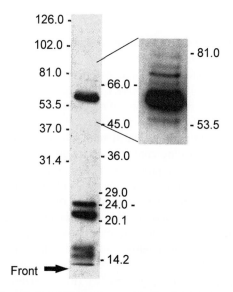

FIG. 1. Selenium-75 metabolic labeling and separation of radiolabeled proteins [reproduced from I. Dreher, N. Schütze, A. Baur, K. Hesse, D. Schneider, J. Köhrle, and F. Jakob, *Biochem. Biophys. Res. Commun.* **245,** 101 (1998) with permission]. hFOB cells were cultured under selenium-deficient conditions (serum-free culture for 3 days). Selenium-75 (specific activity, 1.9 Ci/μg; Research Reactor Facility, University of Missouri, Columbia, MO) was added at 10 n*M* for 24 hr. Labeled proteins from cellular extracts of hFOB cells were separated by denaturing gel electrophoresis. Positions of molecular weight markers are shown on the left and right. After gel electrophoresis the dried gel was exposed to Kodak (Rochester, NY) X-Omat X-ray film for 24 hr.

β (TrxRβ) should as well be present in minor amounts. We found low levels of mRNA expression in Northern blots from hFOB cells (our unpublished results, 2001). Because we detected mRNA for selenoprotein P by reverse transcriptase-polymerase chain reaction (RT-PCR) one ore more bands and a smear in the 60-kDa region might represent selenoprotein P (SeP) isoforms (differentially glycosylated or truncated) as well as some unidentified selenoproteins (Fig. 1, inset). There were no bands in expected regions for the deiodinases at 27–33 kDa, consistent with the fact that 5′-deiodinase activity was undetectable in hFOB extracts. Thus there are at least four unidentified bands of approximately 13, 16, 70, and 80 kDa.

Mesenchymal cells and hematopoietic cells are in close vicinity in the bone microenvironment. Osteoclasts are recruited from myeloic cells and can be differentiated from monocytic cells *in vitro*. We and others have previously shown that monocytic cells express glutathione peroxidases and thioredoxin reductase α.[9,11–14]

[11] N. Schütze, J. Fritsche, R. Ebert-Dümig, D. Schneider, J. Köhrle, R. Andreesen, M. Kreutz, and F. Jakob, *Biofactors* **10,** 329 (1999).
[12] R. Ebert-Dümig, N. Schütze, and F. Jakob, *Biofactors* **10,** 227 (1999).

On labeling with ^{75}Se we obtained at least nine bands in monocytic THP-1 cells, which can be only partially assigned to known selenoproteins (our unpublished results, 2001). At least three of them are secreted into the supernatants. Secreted selenoproteins such as pGPx and SeP and their substrates such as thioredoxins may represent scavenging systems for extracellular ROS in the bone microenvironment. There is presently no information about the expression of selenoproteins in osteoclasts themselves.

Expression and Regulation of Glutathione Peroxidases in Osteoblasts

The family of selenium-dependent glutathione peroxidases (GPx) comprises the selenium-dependent classic cytosolic cGPx, the gastrointestinal cytosolic isozyme giGPx, a secretable plasma isozyme pGPx, and the membrane-associated phospholipid hydroperoxide glutathione peroxidase PHGPx. The former three are involved in the neutralization of H_2O_2 via regeneration of their main substrate glutathione. PHGPx neutralizes oxidized membrane lipids. As we have previously shown, the mRNAs for both cGPx and pGPx are expressed in human fetal osteoblast (hFOB) cells.[15] Selenium-dependent GPx activity (using an assay that does not discriminate between both isoforms) was easily measurable albeit at low levels (Fig. 2).[9] Serum-depleted cell cultures contained $\sim$5 nM selenium, whereas 100 nM selenium was used for selenium supplementation.

Expression and Regulation of Thioredoxin Reductases in Osteoblasts

Response to 1,25(OH)$_2$-Vitamin D$_3$

We have identified thioredoxin reductase α as a vitamin D-responsive gene by exploiting differential display PCR.[16] Using an immortalized fetal osteoblast cell line (hFOB; generous gift from T. Spelsberg, Mayo Clinic, Rochester, MN) we observed an immediate early response at the mRNA level with a maximum at 1–2 hr after treatment with 10 nM 1,25(OH)$_2$-vitamin D$_3$. The mRNA induction was short and transient. AUUUA sequences in the 3' untranslated region may be responsible for this phenomenon, which is well known from other immediate-early genes.[17] By that time we could not find a stimulatory effect of 1,25(OH)$_2$-vitamin D$_3$ treatment on TrxR protein activity. The initial experiments were done

[13] Q. Shen, S. Chada, C. Whitney, and P. E. Newburger, *Blood* **84**, 3902 (1994).

[14] C. H. Williams, Jr., L. D. Arscott, S. Müller, B. W. Lennon, M. L. Ludwig, P. Wang, D. M. Veine, K. Becker, and H. H. Schirmer, *Eur. J. Biochem.* **267**, 6110 (2000).

[15] I. Dreher, C. Schmutzler, F. Jakob, and J. Köhrle, *Trace Elements Med. Biol.* **11**, 83 (1997).

[16] N. Schütze, M. Bachthaler, A. Lechner, J. Köhrle, and F. Jakob, *Biofactors* **7**, 299 (1998).

[17] G. Shaw and R. Kamen, *Cell* **46**, 659 (1986).

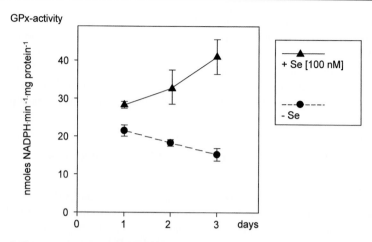

FIG. 2. Time course of stimulation of GPx activity by selenite in hFOB cells [reproduced from I. Dreher, N. Schütze, A. Baur, K. Hesse, D. Schneider, J. Köhrle, and F. Jacob, *Biochem. Biophys. Res. Commun.* **245**, 101 (1998) with permission]. hFOB cells were cultured in 10% (v/v) FCS-containing medium. From day 0 onward cells received serum-free medium plus/minus selenite at 100 nM. GPx enzyme activity was measured from cytosolic extracts prepared after 24, 48, and 72 hr. Data represent the means of two replicate measurements. GPx activity was measured from extracts of hFOB cells in a buffer consisting of 0.1 M Tris, 0.5 mM EDTA (pH 8.0), 200 μM NADPH, 2 mM glutathione, and glutathione reductase type IV from baker's yeast (1 U/ml). The reaction was started by the addition of the substrate *tert*-butyl hydroperoxide at 7 μM and the reaction was monitored after an initial incubation of 1 min at 340 nm for 2–3 min at 25°. Unspecific NADPH oxidation was measured by completely inhibiting GPx activity by the addition of 100 mM mercaptosuccinate before the addition of substrate and was subtracted from the results obtained.

in selenium-deficient media containing ∼5 nM selenium. Only after having cultured the cells in media supplemented with 100 nM selenium for several days could we observe the 1,25(OH)$_2$-vitamin D$_3$ stimulatory effect. Maximum TrxR activity after 1,25(OH)$_2$-vitamin D$_3$ stimulation of cells was found after 12 hr (Fig. 3A). Selenium supplementation (100 nM) per se did not result in any change in steady state mRNA levels. A corresponding enhanced protein expression was found in Western blots under similar conditions (Fig. 3B). Thus a biologically important event of stimulation of an immediate-early gene is obviously not effective in selenium deficiency in these hFOB cells. In contrast, in THP-1 monocytic leukemia cells 1,25(OH)$_2$-vitamin D$_3$ treatment resulted in a transient stimulation of TrxR activity at different levels in selenium deficiency as well as after 100 nM selenium supplementation.[11]

After identification of TrxRβ as a second TrxR localized in mitochondria we reprobed the Northern blots from hFOB cells with TrxRβ cDNA. There was low and constant expression in hFOB cells with no obvious response to 1,25(OH)$_2$-vitamin D$_3$ (data not shown). Thus the changes in TrxR activity measured in response to 1,25(OH)$_2$-vitamin D$_3$ can be attributed to TrxRα.

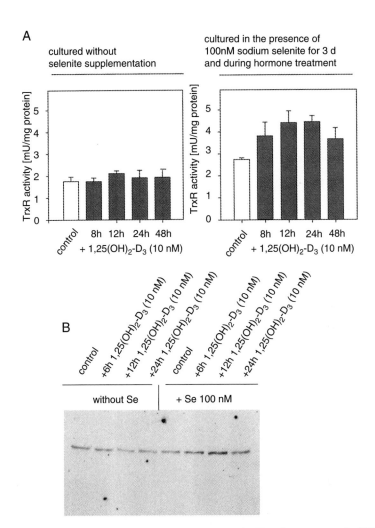

FIG. 3. (A) Induction of TrxRα activity by 1,25(OH)$_2$-D$_3$. hFOB cells were grown in 10% (v/v) FCS-containing medium. On day 0 cells received serum-free medium and were cultured with or without selenium supplementation of 100 nM for 3 days. From day 3 onward cells received 1,25(OH)$_2$-D$_3$ for various time periods as indicated and were cultured in serum-free medium with or without 100 nM selenite. Cytosolic extracts were analyzed for TrxRα activity, using the DTNB assay. Aliquots were analyzed in duplicate and the change in absorption at 412 nm was monitored for 2 min. The change in absorption from a reference cuvette containing the assay buffer and suspension buffer was monitored in parallel and subtracted from the results. Activity was defined as micromoles of TNB/2 generated ($\varepsilon_{412\,nm}$ is 13.6 mM^{-1} cm^{-1}) and results were expressed as activity per milligram of cellular protein determined by the Bio-Rad (Hercules, CA) protein assay. (B) Induction of TrxRα protein levels by 1,25(OH)$_2$-D$_3$. hFOB cells were cultured as described in (A). Equal amounts of cellular extracts of hFOB cells were separated by denaturing gel electrophoresis. Proteins were transferred to Optitran BA S85 membranes (Schleicher & Schuell, Dassel, Germany) and incubated with a rabbit TrxRα antiserum. The expression of TrxRα was detected with the ECL system (Amersham Pharmacia Biotech, Freiburg, Germany).

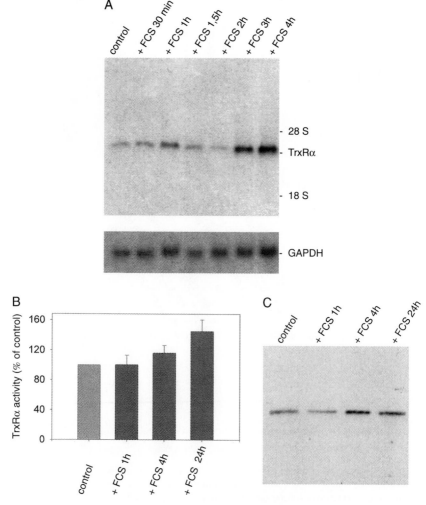

FIG. 4. (A) Regulation of TrxRα mRNA by serum. hFOB cells were cultured in serum-free medium for 3 days and then treated with 10% (v/v) FCS as indicated. Total RNA was isolated, processed for Northern analysis, and subsequently probed with a 2.3-kb TrxRα cDNA fragment and with a 1-kb GAPDH cDNA fragment. After hybridization and washing, the membrane was exposed to Kodak X-Omat X-ray film. (B) Regulation of TrxRα activity by serum. hFOB cells were cultured in serum-free medium for 3 days followed by the addition of 10% (v/v) FCS for various time periods as indicated. Cytosolic extracts were analyzed for TrxRα activity, using the DTNB assay as described in Fig. 3A. (C) Regulation of TrxRα protein levels by serum. hFOB cells were cultured in serum-free medium followed by the addition of 10% (v/v) FCS for various time periods as indicated. TrxRα protein levels were analyzed as described in Fig. 3B.

Response to Serum

We observed a significant early response of TrxRα mRNA production and activity after the addition of serum (10% fetal calf serum) to cell cultures that previously were serum depleted (Fig. 4A). TrxR activity was 30–60% enhanced. The response could be seen under conditions of selenium deficiency and after selenium supplementation at different levels (1.5- to 2-fold enhancement) (Fig. 4B). This comparably low sensitivity to selenium supply is due to the high efficiency of the TrxR selenocysteine insertion sequence (SECIS) element even in selenium deficiency.[18] In comparison, glutathione peroxidase activity in monocytic cells (THP-1) is stimulated up to 100-fold after selenium supplementation, probably because of enhanced translation of active cGPx (our unpublished results and Brigelius-Flohé[19]). Western blotting of cell extracts showed an approximately 2-fold stimulation of protein synthesis after the addition of serum (Fig. 4C).

Methodological Comparison of Different Assays for Thioredoxin Reductase Activity Measurement

Three assays for TrxR activity measurement were compared in human fetal osteoblasts: the measurement of DTNB reduction, insulin reduction, and direct reduction of thioredoxin from *Plasmodium falciparum*. In many reviews and discussions the problem of specificity and sensitivity of different assays has been elaborated. In our hands and cell systems the DTNB assay is not less specific or sensitive compared with the insulin assay. The application of recombinant thioredoxin from *P. falciparum* generated comparable results (see [35] and [36] in this volume [20,21]).

Expression of Glutathione Peroxidases and Thioredoxin Reductase α in Monocyte-Derived Cells

Glutathione peroxidase was reported to be enhanced during differentiation of promyeloic cells along the monocyte-macrophage pathway of differentiation.[13] We found several isoforms of glutathione peroxidases being expressed in cells of

[18] J. R. Gasdaska, J. W. Harney, P. Y. Gasdaska, G. Powis, and M. Berry, *J. Biol. Chem.* **274**, 25379 (1999).

[19] R. Brigelius-Flohé, *Free Radic. Biol. Med.* **27**, 951 (1999).

[20] S. M. Kangok, S. Raklfa, K. Becker, and R. Heiner Schirmer, *Methods Enzymol.* **347**, [35], 2002 (this volume).

[21] S. Gromer, H. Merkle, R. Heiner Schirmer, and K. Besker, *Methods Enzymol.* **347**, [36], 2002 (this volume).

monocyte origin, but no relevant amounts of pGPx. Whether glutathione peroxidases are expressed in osteoclasts remains to be elucidated.

We have previously shown that TrxRα is expressed in macrophage-like cells (THP-1) and in peripheral blood monocytes. It was as well regulated by 1,25(OH)$_2$-vitamin D$_3$ in terms of mRNA production and activity.[11] It is presently unknown whether mature osteoclasts express TrxRα and which cellular functions it might fulfill. However, a report on NO-mediated cytotoxicity in THP-1 cells confirmed our previous results of increased TrxR expression after differentiation induction and it was described that Trx expression was stimulated as well. Differentiated THP-1 cells were more resistant to NO-mediated cytotoxicity, as were cells transiently transfected with Trx.[22] This suggests that the TrxR/Trx system may be important for the prevention of NO-mediated cell damage in cells of the monocyte-macrophage differentiation pathway. Experiments exploiting mature osteoclasts remain to be performed.

Discussion

Selenium Deficiency and Bone Disease

A special form of osteoarthritis is associated with selenium deficiency and can be modulated by selenium supplementation. This disease, Kashin–Beck disease, occurs in regions of low selenium intake in Central Africa and in some provinces of China. The modulation of Kashin–Beck disease *in vivo* by selenium supplementation may indicate that selenoprotein expression in bone and possibly selenium compounds themselves are of biological relevance.[23]

Selenoproteins and Arthritis

Rheumatoid arthritis is an inflammatory autoimmune disease, in which destruction of synovia, cartilage, and bone, as well as locally enhanced bone resorption, take place at bone sites adjacent to the inflamed joints. Gold compounds have been used for several decades to treat this disease. The enzyme TrxR (in this case probably of mainly monocytic origin) is believed to be the main target for gold compounds *in vivo*. TrxR activity can effectively be inhibited *in vitro* by aurothioglucose. Inhibition of monocyte-macrophage TrxR activity may well affect both the inflammatory process as well as the locally enhanced process of bone resorption.[24]

---

[22] P. J. Ferret, E. Soum, O. Negre, E. E. Wollman, and D. Fradelizi, *Biochem. J.* **346,** 759 (2000).

[23] R. Moreno-Reyes, C. Suetens, F. Mathieu, F. Begaux, D. Zhu, M. T. Rivera, M. Boelaert, J. Néve, N. Perlmutter, and J. Vanderpas, *N. Engl. J. Med.* **339,** 1112 (1998).

[24] K. Becker, S. Gromer, H. H. Schirmer, and S. Müller, *Eur. J. Biochem.* **267,** 6118 (2000).

Redox Signaling and Reactive Oxygen Species and Bone

The present knowledge about the putative role of selenoproteins in bone is based on reports about the importance of oxidant conditions during bone remodeling. Reactive oxygen intermediates (ROIs) are produced during the process of bone remodeling, especially the process of bone resorption. Active osteoclasts produce superoxide, NO, and H_2O_2 by expressing NADPH-oxidase, inducible nitric oxide (iNO) synthase, and a 150-kDa superoxide dismutase-related protein, which produces H_2O_2. These processes appear to be regulated by osteoblast-derived secretory products, by 1,25(OH)$_2$-vitamin D_3 and phorbol esters. H_2O_2 is believed to be the principal stimulator of bone resorption (Evans and Ralston,[7] Fraser *et al.*,[8] Dreher *et al.*,[9] and references therein). Because NF-κB-mediated processes, such as the signaling of OPGL (osteoprotegerin ligand)/RANKL (receptor activator of NF-κB ligand) via RANK, are essential for osteoclast development, the modulation of NF-κB signaling by thioredoxin and H_2O_2 might represent an important means of regulation of bone metabolism.

Glutathione Peroxidases and Selenoprotein P in Bone

We have described selenium-dependent antioxidant systems in osteoblasts, such as the expression of cGPx and pGPx as well as selenoprotein P.[9] The coincident expression of various antioxidative systems in osteoblasts might represent a redundant rescue from cytotoxic ROIs, or the various systems might play individual roles in intracellular and extracellular compartments. Similarly, each enzyme system might *in vivo* show substrate preferences. Peroxynitrite, NO, and O_2^- are produced by osteoblasts as a consequence of tumor necrosis factor α (TNF-α) and interleukin 1β (IL-1β) signaling. Peroxynitrite was shown to inhibit whereas NO stimulated osteoblast-specific functions.[25] Given that one function of SeP is the neutralization of peroxynitrite[26] this would suggest that SeP represents an important factor to rescue osteoblast differentiation and function from the negative influence of the by-products of NO production.

Putative Functions of the Thioredoxin/Thioredoxin Reductase System in Bone

TrxR has been shown to be the only enzyme capable of reducing its main substrate, Trx. Thus the enzyme activity should modulate the effects of Trx itself.

---

[25] H. Hikiji, W. S. Shin, T. Koizumi, T. Takato, T. Susami, Y. Koizumi, Y. Okai-Matsuo, and T. Toyo-Oka, *Am. J. Physiol.* **278**, E1031 (2000).

[26] G. E. Arteel, V. Mostert, H. Oubrahim, K. Briviba, J. Abel, and H. Sies, *Biol. Chem.* **379**, 1201 (1998).

This protein has functions as an antioxidant, in DNA synthesis, and in gene transcription by reducing transcriptions factors such as NF-κB and steroid hormone receptors. Trx furthermore modulates cell growth and inhibits apoptosis.[14,27,28] All of these processes are important in bone differentiation and metabolism, especially apoptosis, which determines the fate of active osteoclasts.

Besides the putative regulatory role of the Trx/TrxR system in osteoclasts as well as in osteoblasts the cells of the microenvironment should have developed systems to defend themselves against cytotoxic effects of ROIs, which are locally produced in high quantities. This could be an important function of the Trx/TrxR system and other selenium-dependent proteins in osteoblasts, because it is known from experiments exploiting bone implant materials that osteoblast functions are altered by high concentrations of ROIs.

Trx and TrxR are intracellular compounds but both are secreted into the extracellular space, as described for TrxR.[29] As discussed above, there are a variety of extracellular matrix-associated signaling molecules in bone. Of these the family of cysteine-rich proteins (CCN family), such as CYR 61 and CTGFl,[4,30] comprising conserved cysteine amino acids, are candidates for sulfhydryl-dependent modulation of their biological activity as ligands for integrin receptors.

TrxR was reported to reduce lipid hydroperoxides, lipoic acid, NK-lysin, vitamin K, and dehydroascorbic acid.[28] Of these, vitamin K and ascorbate play a role in bone metabolism. Besides the fact that TrxR may reduce vitamin K during the regeneration cycle of γ-carboxylation, Trx could well be useful for reduction of a free cysteine in vitamin K-dependent carboxylase, necessary for vitamin K binding to the substrate box.[31] The vitamin K-dependent proteins bone gla protein (BGP, osteocalcin) and matrix gla protein (MGP) are important regulators of calcification. Thus there might be an association of TrxR activity (possibly as well in its secreted form) and mineralization, which has yet to be demonstrated.

Thus there are multiple potential situations in which the Trx/TrxR system could be of relevance in bone, for example, in intracellular and extracellular signaling as well as intracellular and extracellular antioxidative defense during processes in which the local burden of reactive oxygen intermediates (ROIs) is high.

[27] E. S. J. Arner and A. Holmgren, Eur. J. Biochem. 267, 6102 (2000).

[28] D. Mustacich and G. Powis, Biochem. J. 346, 1 (2000).

[29] A. Söderberg, B. Sahaf, and A. Rosén, Cancer Res. 60, 2281 (2000).

[30] N. Schütze, A. Lechner, C. Groll, H. Siggelkow, M. Hüfner, J. Köhrle, and F. Jakob, Endocrinology 139, 1761 (1998).

[31] B. Bouchard, B. Furie, and B. C. Furie, Biochemistry 38, 9517 (1999).

Summary

The expression of thioredoxin reductases and other selenoproteins in cells of the bone microenvironment may represent an important means of regulation of bone resorption and remodeling in health and disease. Selenoproteins and their substrates may influence intracellular and extracellular redox-dependent signaling, transcription factor activity, posttranslational modification of proteins, and general or compartmentalized scavenging from ROIs. However, the evaluation of their biological role in bone and their potential in terms of therapeutic approaches is just beginning.

[14] Selenoprotein W

By PHILIP D. WHANGER

A number of selenoenzymes and selenoproteins have been identified. They are the glutathione peroxidase (GPX) family containing at least four different seleno-enzymes,[1–4] the iodothyronine deiodinase family containing types I, II, and III deiodinases,[5–7] thioredoxin reductases,[8,9] and selenophosphate synthetase.[10] All of these selenoenzymes contain one selenium atom per polypeptide chain. Seleno-proteins P and W (SeW) are the two most studied selenoproteins without a known function. Selenoprotein P is the only known selenoprotein that contains multiple selenocysteine residues, which is 10.[11] SeW is a low molecular weight protein originally reported as a missing component in selenium-deficient lambs

[1] J. T. Rotruck, A. L. Pope, H. E. Ganther, A. B. Swanson, D. G. Hafeman, and W. G. Hoekstra, *Science* **179**, 588 (1973).
[2] K. Takahashi and H. J. Cohen, *Arch. Biochem. Biophys.* **256**, 677 (1986).
[3] F. Ursini, M. Maiorino, and C. Gregolin, *Biochem. Biophys. Acta* **839**, 62 (1985).
[4] F. F. Chu, J. H. Dorshow, and R. S. Esworth, *J. Biol. Chem.* **268**, 2571 (1993).
[5] D. Behne, A. Kyriakopoulus, H. Meinhold, and J. Kohrie, *Biochem. Biophys. Res. Commun.* **173**, 1143 (1990).
[6] J. C. Davey, K. B. Becker, M. J. Schneider, D. L. St. Germain, and V. A. Galton, *J. Biol. Chem.* **270**, 26786 (1995).
[7] W. Croteau, S. L. Whittemore, J. J. Schneider, and D. L. St. Germain, *J. Biol. Chem.* **270**, 16569 (1995).
[8] T. Tamura and T. C. Stadtman, *Proc. Natl. Acad. Sci. U.S.A.* **93**, 1006 (1996).
[9] D. Mustacich and G. Powis, *Biochem. J.* **246**, 1 (2000).
[10] I. Y. Kim and T. C. Stadtman, *Proc. Natl. Acad. Sci. U.S.A.* **92**, 7710 (1995).
[11] K. E. Hill, R. S. Lloyd, J. G. Yang, R. Read, and R. F. Burk, *J. Biol. Chem.* **266**, 10050 (1991).

suffering from white muscle disease.[12] Evidence has been presented for several other possible selenoproteins.[12a]

Procedures

Method Using Isotopes

Skeletal muscle tissue from 10 adult male rats is used to obtain sufficient quantities of SeW for quantitation and characterization studies. To monitor the purification, three of the rats are injected with 35 μCi of [^{75}Se]selenite each at 4 and 2 days before they are killed. The animals have been previously fed a diet with 1 mg of selenium per kilogram for 2 weeks to increase the SeW content. Skeletal muscle is removed as quickly as possible and frozen in liquid nitrogen. About 100 g of muscle is recovered from each animal.[13] All buffers used in the isolation are made under nitrogen, using degassed distilled water. The frozen tissue is placed in homogenization buffer [100 mM phosphate (pH 6.5), containing 0.1 mM phenylmethylsulfonyl fluoride (PMSF) and 0.02% (w/v) sodium azide] and processed under nitrogen at 4°. The tissues are partially thawed, chopped into fine pieces, and homogenized in an Omni-mixer for 3 min at maximum speed in a volume of buffer equivalent to four times the tissue weight. The homogenate is centrifuged at 10,000g for 30 min at 4°, and the supernatant is filtered through two layers of cheesecloth and recentrifuged at 100,000g for 90 min at 4° in an ultracentrifuge (L8-M; Beckman, Palo Alto, CA). Extracts from labeled and unlabeled muscle are combined and diluted to a final protein concentration of 8.4 mg/ml as determined by the Lowry assay. Finely ground ultrapure ammonium sulfate is added gradually to the extract while slowly stirring under nitrogen to a concentration of 2.0 M . The mixture is centrifuged at 2000g for 15 min and the pellet is discarded. Further additions of ammonium sulfate are made to the supernatant to a final concentration of 3.5 M. The precipitated protein is collected by centrifugation and the supernatant is discarded. The pelleted protein is redissolved in a minimum volume of homogenization buffer.

About 2400 mg of protein from the ammonium sulfate fraction is applied to a column (5.5 × 150 cm) of Sephadex G-50 and eluted with 100 mM phosphate buffer, pH 6.5, at a flow rate of 60 ml/hr, and approximately 18 ml is collected per fraction. The column eluant is monitored for radioactivity and absorbency at 280 nm. Two main radioactive peaks are obtained from this step. A pool of radioactive column eluant (about 2.5–3.0 liters) corresponding to a molecular mass

---

[12] N. D. Pedersen, P. D. Whanger, P. H. Weswig, and O. H. Muth, *Bioinorg. Chem.* **2**, 33 (1972).

[12a] V. N. Gladyshev, *in* "Selenium: Its Molecular Biology and Role in Human Health" (D. L. Hatfield, ed.). Kluwer Academic Publishers, Boston (2001).

[13] S. C. Vendeland, M. A. Beilstein, C. L. Chen, O. N. Jensen, E. Barofsky, and P. D. Whanger, *J. Biol. Chem.* **268**, 17103 (1993).

of 10,000 Da is concentrated in a stirred cell to 10 ml with an Amicon (Danvers, MA) YM3 ultrafiltration membrane. The concentrate is rechromatographed on a smaller Sephadex G-50 column (2.5 × 120 cm) with 100 mM phosphate buffer at a flow rate of 36 ml/hr. About 5 ml is collected per fraction. One main radioactive peak, which elutes slightly after the peak that contains cytochrome c, and a small one near the void volume are obtained. This main radioactive eluant peak is pooled to exclude significant contamination by cytochrome c and concentrated again by ultrafiltration. The G-50 pool is diluted with 50 mM sodium phosphate and chromatographed on a CM-Sephadex cation-exchange column (2.0 × 28 cm) with a sodium chloride gradient of 400 ml from 0 to 300 mM at a flow rate of 30 ml/hr. About 4 ml is collected per fraction. Radioactivity elutes from the column in three separated peaks at approximately 0.09, 0.13, and 0.17 M sodium chloride. The three radioactive peaks are pooled separately and designated as C-0, C-1, and C-2 respectively. The radioactivity from pool C-1 elutes as two peaks from a high-performance liquid chromatography (HPLC) column [Vydac (Hesperia, CA) C_{18}, 5 μm ODS 218 TP, 4.6 mm × 25 cm]. A gradient from 35 to 55% (v/v) acetonitrile in water with 0.1% (v/v) trifluoroacetic acid over 25 min is used at a flow rate of 1 ml/min. Column eluants are collected in 0.5-ml fractions. The two peaks are designated as C-11 and C-12 and elute, respectively, at approximately 42 and 48% (v/v) acetonitrile. The radioactivities from both pools of C-0 and C-2 elute as single peaks from the HPLC column, centered, respectively, at 42 and 48% (v/v) acetonitrile.

Method Using Antibodies

Because the antibody raised against the rat SeW peptide does not cross-react with primate SeW, other steps must be taken. For some unknown reason, antibodies raised against the primate peptide (residues 13 to 31) unfortunately do not perform in Western blots. Therefore, a histidine-tagged form of the human SeW cDNA is prepared.[14] The sense strand polymerase chain reaction (PCR) primer contains the first 15 codons of the cDNA, with a mutation that converts codon 13 from selenocysteine (TGA) to cysteine (TGT). This is done because TGA is a stop codon under these conditions. To facilitate ligation of the PCR product into the expression vector, this primer also incorporates an *Nco*I restriction site near the 5' end. The antisense PCR primer matches the terminal seven codons of the cDNA open reading frame followed by six histidine codons prior to the termination codon. This primer includes a *Bam*HI restriction site to facilitate ligation with the vector. The cloned human SeW cDNA[15] is used as a template for PCR. The PCR product and vector

[14] Q.-P. Gu, Y. Sun, L. W. Ream, and P. D. Whanger, *Mol. Cell. Biochem.* **204**, 49 (2000).
[15] Q.-P. Gu, M. A. Beilstein, S. C. Vendeland, A. Lugade, W. Ream, and P. D. Whanger, *Gene* **193**, 187 (1997).

pTrc99a are cleaved with the restriction enzymes *Nco*I and *Bam*HI, joined by T_4 DNA ligase, and transformed into *Escherichia coli* TGI. Transformants are screened for correct insert size by restriction analysis and selected clones are sequenced with vector primers. The recombinant plasmid is transformed into *E. coli* BL21DE pLyss for expression of the recombinant protein.

One liter of culture is induced with 0.1 mM isopropyl-β-D-thiogalactopyranoside (IPTG) for 3 hr. The harvested cells are resuspended in 100 ml of 100 mM Tris (pH 7.4) and lysed by freezing at $-20°$ and thawing two times. Thawed lysate is gently mixed for 30 min at $5°$ after addition of DNase I (10 U/ml), $MgCl_2$ (10 mM), NaCl (15 mM), PMSF (1 μg/ml), and Triton X-100 (0.1%, v/v). The lysate is centrifuged at 10,000g for 30 min at $5°$. The supernatant is applied to an Ni-NTA agarose column and washed extensively with 15 mM imidazole in phosphate-buffered saline (PBS), pH 6.8. The recombinant protein is eluted with a 40-ml gradient of 15 to 500 mM imidazole buffer. Protein in the eluate is pooled and concentrated to less than 0.25 ml with an Amicon Centricon-3 concentrator. The concentrates are chromatographed on C_{18} reversed-phase HPLC columns at a flow rate of 1 ml/min with a gradient from 30 to 60% (v/v) acetonitrile in water containing 0.1% (w/v) trifluoroacetate. Eluates from HPLC are collected in 0.5-ml fractions.

Rabbits are immunized for 9 weeks at 3-week intervals with histidine-tagged primate SeW conjugated to keyhole limpet hemocyanin in Freund's adjuvant (200 μg of protein per immunization). Blood is collected by cardiac puncture 1, 2, and 3 weeks after the last injection and the serum is stored frozen. The antibody is purified on an affinity column prepared from the bacterially expressed primate SeW bound to Sulfolink coupling gel (Pierce, Rockford, IL). After washing with 1 M NaCl and equilibration with PBS, rabbit sera are applied to the column. The column is washed with 6 column volumes of PBS and the antibody is eluted with 0.1 M glycine (pH 2.8), collected in 0.5-ml fractions, and neutralized by adding 50 μl of 1 M Tris (pH 9.5). The antibody-containing fractions are detected by absorbance at 280 nm, desalted, and changed to PBS buffer on a desalting column.

About 180 g of frozen monkey skeletal muscle is manually chopped with scissors in 4 volumes (w/v) of the homogenization buffer. The chopped muscle is homogenized and centrifuged, the supernatant is filtered and recentrifuged in an ultracentrifuge, and the centrifugal supernatant likewise is concentrated as described for rat muscle.

The concentrated cytosol is chromatographed on a large Sephadex G-50 column (5.5 × 150 cm), equilibrated and eluted with homogenization buffer. The column eluant is monitored for protein by absorbance at 280 nm and for SeW by slot blots developed with rabbit polyclonal antibody against the primate mutant protein.[14] A pool of low molecular weight immunoreactive column eluant is concentrated to 10 ml in a stirred cell, again with an Amicon YM3 ultrafiltration membrane. As for rats, the concentrate is rechromatographed on a smaller Sephadex

G-50 column (2.5 × 120 cm). The immunoreactive fractions are pooled to exclude significant contamination from the pink color, presumably cytochromes.

The sephadex G-50 pool is adjusted to a volume of 100 ml with homogenization buffer and chromatographed on a CM-Sephadex cation-exchange column (2.0 × 28 cm). The column is washed thoroughly with 200 ml of homogenization buffer and the selenoprotein is eluted with a 400-ml linear gradient of 0–0.3 M NaCl in the homogenization buffer. The immunoreactive fractions are pooled and concentrated to 250 μl, using a centifugable microporous concentrator (Centricon-3; Amicon). This preparation is chromatographed on a reversed-phase HPLC column (Vydac C_{18}, 5 μm ODS 218 TP column, 4.6 mm × 25 cm) with a 30–60% (v/v) acetonitrile gradient in water containing 0.1% (w/v) trifluoroacetic acid. Protein in the column eluant is monitored by absorbance at 280 nm. The fractionated column eluant is again assayed by slot blots for SeW. The protein concentration at each step is determined by the Lowry procedure. The protein concentration is multiplied by the volume to obtain the total protein content, which is applied to the next step.

Slot blots are performed with a commercial apparatus (Minifold II; Schleicher & Schuell, Keene, NH). Nitrocellulose membranes are cut to the size of the manifold and placed in the slot-blot apparatus. Samples are loaded on the slots and allowed to filter through membranes under low vacuum. After air drying, membranes are incubated with blocking solution [5% (w/v) nonfat dry milk in Tween Tris-buffered saline (TTBS)] for 1 hr, and incubated with polyclonal antibodies for 1.5 hr. After washing with TTBS, membranes are incubated with horseradish peroxidase-conjugated goat anti-rabbit IgG antibody. After washing with TTBS to eliminate excess secondary antibody, the membranes are incubated with enhanced chemiluminescence (ECL) reagents (Amersham Life Science, Arlington Heights, IL) and the signal is detected by exposure of ECL Hyperfilm (Amersham Life Science). Developed films are scanned with a personal densitometer SI (Molecular Dynamics, Sunnyvale, CA) and analyzed by the Image-QuaNT program (Molecular Dynamics).

There is a third possible method for purification of SeW by using immunoaffinity columns. Monclonal antibodies would be made against SeW and linked to the column, which would then retain this selenoprotein. After the solutions containing this SeW are entirely passed through, the column would be thoroughly washed. SeW would then be eluted from the column. This method is presently in development.

Characteristics of Selenoprotein W

The molecular weights of the four forms of protein have been determined by matrix-assisted laser desorption/ionization (MALDI) mass spectrometry to be 9549, 9592, 9853, and 9898. Thus the higher mass forms appear to be created by additions of moieties of about 42 and 306 Da to the lowest mass form. The 42-Da moiety still has not been identified but the 306-Da moiety has been determined to be reduced glutathione.[16] The 306-Da moiety was demonstrated to be glutathione by reductive release from the 9853-Da protein with a 1000-fold excess of dithiothreitol at 50°. Reduced glutathione could not be released at lower temperatures with this thio reagent, suggesting the protein must be denatured before glutathione could be removed. Evidence has been obtained to indicate that glutathione is bound to the amino acid at position 36, which is cysteine in SeW.[17]

The skeletal muscle SeW cDNA was isolated and sequenced.[18] A reverse transcription-coupled PCR product from rat muscle mRNA was used to screen a muscle cDNA library prepared from selenium-supplemented rats. The cDNA sequence confirmed the known protein primary sequence, including a selenocysteine residue encoded by TGA, and identified residues needed to complete the protein sequence. RNA-folding algorithms predict a stem–loop structure in the 3' untranslated region of the SeW mRNA that resembles selenocysteine insertion sequence (SECIS) elements identified in other selenocysteine-coding cDNAs. The cDNAs encoding skeletal muscle SeW from human, rhesus monkey, sheep, rat, and mouse contained highly similar SECIS elements that retained important features common to all known SECIS elements.[15] The rodent and sheep SeW mRNAs used UGA both as a stop codon and as a selenocysteine codon, indicating that this codon specified both selenocysteine incorporation and termination in a single mRNA. Even though UGA was used as the selenocysteine codon in primate SeW mRNAs, TAA was used instead as the stop codon.

SeW-coding sequences are highly conserved among the five species studied. Seventy-five (83%) of the amino acids in SeW are invariant in all five species, including the selenocysteine residue at position 13 and the cysteine residues at positions 10 and 37 (Table I).

Tissue Distribution and Effect of Selenium on Selenoprotein W Levels

Because the metabolic function of SeW is unknown, Western blots were used to assess the influence of selenium on its levels. The antibody raised against a peptide (amino acids 13 to 31) of the rodent SeW was used. These antibodies recognize SeW in muscles from rabbits, mice, guinea pigs, sheep, and cattle,[19] but not in tissues from monkeys and humans.[14] This is the reason antibodies were raised

[16] M. A. Beilstein, S. C. Vendeland, E. Barofsky, O. N. Jensen, and P. D. Whanger, *J. Inorg. Biochem.* **61**, 117 (1996).

[17] Q.-P. Gu, M. A. Beilstein, E. Barofsky, W. Ream, and P. D. Whanger, *Arch. Biochem. Biophys.* **361**, 25 (1999).

[18] S. C. Vendeland, M. A. Beilstein, J.-Y. Yeh, W. Ream, and P. D. Whanger, *Proc. Natl. Acad. Sci. U.S.A.* **92**, 8749 (1995).

[19] J.-Y. Yeh, M. A. Beilstein, J. S. Andrews, and P. D. Whanger, *FASEB J.* **9**, 392 (1995).

TABLE I

Deduced Selenoprotein W Amino Acid Sequences in Five Species of Animal[a]

Source	Sequence
	1 15
Primate	Met Ala Leu Ala Val Arg Val Val Tyr Cys Gly Ala Sec Gly Tyr Lys Ser Lys Tyr Leu
Rodent	Val Val
Sheep	Pro Pro
	30
Primate	Gln Leu Lys Lys Lys Leu Glu Asp Glu Phe Pro Gly Arg Leu Asp Ile Cys Gly Glu Gly
Rodent	Glu His
Sheep	Ser Cys
	45 60
Primate	Thr Phe Gln Ala Thr Gly Phe Phe Glu Val Met Val Ala Gly Lys Leu Ile His Ser Lys Lys
Rodent	Val Thr Val
Sheep	Val Phe Val
	75
Primate	Lys Gly Asp Gly Tyr Val Asp Thr Glu Ser Lys Phe Leu Lys Leu Val Ala Ala Ile Lys Ala
Rodent	Arg Arg
Sheep	Gly Thr
	88
Primate	Ala Leu Ala Gln Gly Gly
Rodent	Cys Gln
Sheep	Ala

[a] For the rodent and sheep proteins, only the amino acid residues that differ from the primate sequence are shown. The amino sequences for rats and mice (rodents), and monkeys and humans (primates), are identical.

against the primate mutant protein. Antibodies were raised against a peptide 13–31 of the primate SeW, but unfortunately these would not function in the Western blot. With adequate selenium intake SeW is highest in muscle, brain, spleen, testes, and skin of rats,[20] but in addition to these tissues this selenoprotein is also high in the hearts of sheep,[21] and monkeys and humans.[14] Thus, the SeW content in primate hearts is similar to that in sheep and suggests that rodents may not be a good model for studying cardiac SeW for primate application.

Differences in response of SeW in various tissue have been observed in the rat. SeW was not detectable in liver, thyroid, pancreas, pituitary, and eyes in rats regardless of the dietary level of selenium fed.[20] SeW was not detected in heart, lungs, prostate, esophagus, small intestine, tongue, skin, diaphragm, or skeletal muscle from rats fed selenium-deficient diets, but was present in these tissues with adequate or excessive selenium intakes. In other tissues such as the kidney and seminal vesicles, SeW was detected only in rats fed diets with excess selenium. The response of SeW to various levels of selenium is different in various tissues of the rat. For example, SeW did not start to increase in muscle until 0.06 mg of selenium per kilogram was fed in the diet, but a marked increase occurred in testes with additions of only 0.01 mg of selenium per kilogram of diet.[22] SeW levels appear to respond to selenium intake in humans. Fetal muscle and cardiac SeW content correlated with the selenium status of aborted fetuses from women living in selenium-deficient, selenium-adequate, and selenium-excessive areas of China.[23] Selenium deficiency resulted in the depletion of SeW in all tissues examined except the brain in both rats[20] and sheep,[21] indicating an affinity of this organ for this selenoprotein. Northern blots indicated that mRNA increased in muscle 4-fold[18] and 6-fold[22] in rats fed excess selenium as compared with deficient rats. Northern blots also indicated that mRNA levels were highest in monkey muscle and heart, which is similar to the pattern found with a human multiple tissue Northern blot.[14]

Selenium levels have been shown to affect the SeW content in tissue cultures, indicating that they can be used to study the relationship of this element to the content of this selenoprotein. SeW levels increased with selenium in the media for L8 muscle and for glial and neuroblastoma cells, and likewise decline, although at different rates, when selenium was removed from the media of these cells.[24,25] Northern blots indicated that the mRNA levels correlated with the selenium content of the media. Nuclear run-on experiments with isolated L8 nuclei showed the same

[20] Y. Sun, P.-C. Ha, J. A. Butler, B.-R. Ou, J.-Y. Yeh, and P. D. Whanger, *J. Nutr. Biochem.* **9,** 23 (1998).

[21] J.-Y. Yeh, Q.-P. Gu, M. A. Beilstein, N. E. Forsberg, and P. D. Whanger, *J. Nutr.* **127,** 2165 (1997).

[22] J.-Y. Yeh, S. C. Vendeland, Q.-P. Gu, J. A. Butler, B.-R. Ou, and P. D. Whanger, *J. Nutr.* **127,** 2165 (1997).

[23] J. A. Butler, Y. Xia, Y. Zhou, Y. Sun, and P. D. Whanger, *FASEB J.* **13,** A248 (1999).

[24] Y.-Y. Yeh, B.-R. Ou, N. E. Forsberg, and P. D. Whanger, *Biometals* **10,** 11 (1977).

[25] Y. Sun, J. Butler, N. Forsberg, and P. D. Whanger, *Nutr. Neurosci.* **2,** 227 (1999).

rate of SeW mRNA synthesis in cells cultured in either low selenium or selenium-supplemented medium, suggesting that the transcription rate of the SeW gene is independent of selenium. However, the estimated half-life of SeW mRNA was 57 hr for cells grown in low-selenium medium, but selenium supplementation increased this half-life 2-fold.[26] Therefore, the results indicate that selenium stabilizes SeW mRNA but has no effect on transcription.

[26] P. D. Whanger, *Cell. Mol. Life Sci.* **57**, 1846 (2000).

[15] Genetic and Functional Analysis of Mammalian Sep15 Selenoprotein

By EASWARI KUMARASWAMY, KONSTANTIN V. KOROTKOV, ALAN M. DIAMOND, VADIM N. GLADYSHEV, and DOLPH L. HATFIELD

Introduction

Approximately 20 vertebrate selenoproteins have been identified thus far as described in [6] and [8] in this volume.[1,1a] The functions of the majority of these proteins have not yet been determined. Identifying the biological roles of individual selenoproteins is essential to understanding the ways by which they contribute to better human health, as well as gaining a comprehensive appreciation of the overall beneficial effects of selenium. Selenium exerts its physiological function as a catalytically active constituent in many selenoenzymes, including glutathione peroxidases, thyroid hormone deiodinases, thioredoxin reductases, and selenophosphate synthetase 2 in mammals, and hydrogenases and formate dehydrogenases in bacteria and archaea.[1b,2] Selenium, in the form of the amino acid selenocysteine (Sec), is located at the enzyme active center of mammalian selenoproteins of known function and appears to be involved in various redox reactions.[1b,3]

Sep15 is selenoprotein with a molecular mass of 15 kDa, which was initially discovered in human T cells.[4] Genes encoding Sep15 have also been detected in mice and rats. The nucleotide sequence of Sep15 lacks significant homology

[1] G. V. Kryukov and V. N. Gladyshev, *Methods Enzymol.* **347**, [8], 2002 (this volume).

[1a] A. Lescure, D. Gautheret, and A. Krol, *Methods Enzymol.* **347**, [6], 2002 (this volume).

[1b] C. B. Allan, G. M. Lacourciere, and T. C. Stadtman, *Annu. Rev. Nutr.* **1**, 1 (1999).

[2] R. Wilting, S. Schorling, B. C. Persson, and A. Bock, *J. Mol. Biol.* **266**, 637 (1997).

[3] A. T. Diplock, *Mol. Aspects Med.* **15**, 293 (1994).

[4] V. N. Gladyshev, K. T. Jeang, J. C. Wootton, and D. L. Hatfield, *J. Biol. Chem.* **273**, 8910 (1998).

to previously characterized genes, suggesting that it may have evolved independently with a specialized function. The expression of Sep15 varies in human and mouse tissues, with the highest levels of its mRNA being apparent in liver, kidney, prostate, and testis.[5] The Sep15 gene spans 51 kb of the human genome and is organized in five exons and four introns. Its location on chromosome 1 at position p31 is a site mutated or deleted in many human cancers. Sep15 resides in the endoplasmic reticulum (ER) in association with UDP-glucose glycoprotein glucosyltransferase (UGTR),[6] an enzyme involved in the quality control of protein folding. Two polymorphic alleles of the Sep15 gene have been identified in the human population, differing at two positions that reside in the 3' untranslated region (3'-UTR). The polymorphic site located at position 1125 is located in the Sec insertion sequence (SECIS), an element required for Sec insertion. The identity of the nucleotide at that position influences Sec incorporation in a selenium-dependent manner. Allele frequencies have been assessed and shown to differ dramatically between African Americans and Caucasians.[7] In addition, allele frequencies have also been shown to differ in tumors of breast and head and neck origin, when compared with that of cancer-free individuals.[7] Collectively, these data raise the possibility that Sep15 is an important effector in the critical role that selenium plays in human biology.

The isolation and functional analysis of Sep15 are discussed in this article.

Materials and Methods

Purification of Sep15

Sep15 has been purified from human Jurkat T cells, rat prostate, and mouse liver.[4,5] Procedures for its purification and isolation from mouse liver are described herein (see Diagram 1).

General Considerations

All purification procedures are carried out at 4° and as rapidly as possible to avoid denaturation of Sep15. As selenoproteins can be specifically detected when labeled with ^{75}Se, mouse tissues are metabolically labeled with this isotope and mixed with unlabeled tissues before protein isolation. During purification,

---

[5] E. Kumaraswamy, A. Malykh, K. V. Korotkov, S. Kozyavkin, Y. Hu, S. Y. Kwon, M. E. Moustafa, B. A. Carlson, M. J. Berry, B. J. Lee, D. L. Hatfield, A. M. Diamond, and V. N. Gladyshev, *J. Biol. Chem.* **275**, 35540 (2000).

[6] K. V. Korotkov, E. Kumaraswamy, Y. Zhou, D. L. Hatfield, and V. N. Gladyshev, *J. Biol. Chem.* **276**, 15330 (2001).

[7] Y. T. Hu, K. V. Korotkov, R. Mehta, D. L. Hatfield, C. Rotimi, A. Luke, T. E. Prewitt, R. S. Cooper, W. Stock, E. E. Vokes, M. E. Dolan, V. N. Gladyshev, and A. M. Diamond, *Can. Res.* **61**, 2307 (2001).

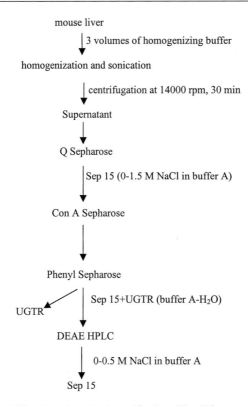

mouse liver

↓ 3 volumes of homogenizing buffer

homogenization and sonication

↓ centrifugation at 14000 rpm, 30 min

Supernatant

↓

Q Sepharose

↓ Sep 15 (0-1.5 M NaCl in buffer A)

Con A Sepharose

↓

Phenyl Sepharose

UGTR ← ↓ Sep 15+UGTR (buffer A-H₂O)

DEAE HPLC

↓ 0-0.5 M NaCl in buffer A

Sep 15

DIAGRAM 1. Flow chart depicting the purification of Sep15 from mouse liver.

chromatographic fractions containing peaks of radioactivity are analyzed by Coomassie blue staining of native and sodium dodecyl sulfate–polyacrylamide gel electrophoresis (SDS–PAGE) followed by PhosphorImager (Molecular Devices, Sunnyvale, CA) detection of radioactivity on gels. Because Sep15 is complexed with UGTR intracellularly, the complex migrates at ∼165 kDa on native gels and at 15 and ∼150 kDa on SDS gels.

Reagents and Materials

[^{75}Se]Selenious acid (University of Missouri Research Reactor, Columbia, MO)

Homogenizing buffer: 20 mM Tris-HCl (pH 7.5), 1 mM EDTA, 1 mM 4-(2-aminoethyl) benzenesulfonyl fluoride (AEBSF), 1 mM dithiothreitol (DTT), leupeptin (5 μg/ml), and aprotinin (5 μg/ml)

Buffer A: 20 mM Tris-HCl (pH 7.5), 1 mM EDTA, and 1 mM DTT

Buffer B: 20 mM Tris-HCl (pH 7.4), 0.5 M NaCl

High-performance liquid chromatograph (HPLC, model 1100; Hewlett-Packard, Palo Alto, CA)

Q-Sepharose and concanavalin A-Sepharose from Pharmacia (Piscataway, NJ)

Phenyl-Sepharose (TSK gel Phenyl-5PW) and DEAE-Sepharose (TSK gel DEAE 5PW) HPLC columns (Toso Haas, Tokyo, Japan)

Antibodies specific for Sep15 and UGTR

Purification from Mouse Liver

To obtain Sep15 of highest purity, a three-step purification procedure has been developed consisting of conventional chromatography on Q-Sepharose followed by HPLC on phenyl-Sepharose and DEAE-Sepharose columns. In addition, a concanavalin A-Sepharose (ConA-Sepharose) column is utilized as an intermediate step in the purification protocol. Inclusion of this step results in the isolation of UGTR to near homogeneity and identification of the Sep15–UGTR complex. ConA-Sepharose is a glycoprotein affinity column used previously in the purification of UGTR.[8] Sep15 is detected in chromatographic fractions by monitoring γ radioactivity. This procedure may also be used for purifying Sep15 from a rat prostate homogenate.

[75]Se Labeling. ^{75}Se-labeled mouse liver is obtained as follows[9]: 0.5 mCi of freshly neutralized [^{75}Se]selenious acid is injected intraperitoneally into the mouse, the mouse is killed after 48 hr, and ^{75}Se-labeled tissues are collected.

Step 1. Mouse liver (150 g) is mixed with ^{75}Se-labeled mouse liver (3 g) and homogenized with a Sorvall Omni mixer (Du Pont Instruments, Wilmington, DE) in 3 volumes of homogenizing buffer, containing 0.5 mM sodium orthovanadate. The homogenate is lysed with a 550 sonic dismembrator (Fisher Scientific, Pittsburgh, PA) (30-sec pulses for 3 min with 20-sec intervals) and clarified by centrifugation at 14,000 rpm for 30 min at 4°, and the supernatant is used for purification.

Step 2. The clear supernatant from step 1 is applied onto a Q-Sepharose (200-ml) column that has been previously equilibrated with buffer A. The column is washed with 2 volumes of buffer A, and the bound protein is eluted with a linear gradient of NaCl from 0 to 1.5 M in buffer A. UGTR and Sep15 are identified by Western analysis after gel electrophoresis of column fractions.

Step 3. Concanavalin A-Sepharose (ConA-Sepharose) column chromatography is used as the next purification step. Fractions containing labeled Sep15 are pooled, concentrated, and applied onto a ConA-Sepharose column (10 ml) equilibrated with buffer B. The column is washed with buffer B, and proteins are eluted by the application of a step gradient of 10, 50, 100, 200, and 300 mM

[8] S. E. Trombetta and A. J. Parodi, *J. Biol. Chem.* **267**, 9236 (1992).
[9] V. N. Gladyshev, V. M. Factor, F. Housseau, and D. L. Hatfield, *Biochem. Biophys. Res. Commun.* **251**, 488 (1998).

methyl-α-D-mannopyranoside in buffer B. The apparently homogeneous preparation of UGTR is then analyzed by Western blotting for the presence of Sep15.

Step 4. A phenyl-Sepharose HPLC column (15 cm × 21.5 mm) is equilibrated with 1 M NaCl in buffer A. The fractions containing Sep15 are pooled, concentrated, adjusted in 1 M NaCl, filtered (to remove fine particles before loading onto an HPLC column), and applied to the column. The column is washed and a linear gradient from 1 M NaCl in buffer A to buffer A without NaCl is applied. Sep15 is then eluted by a short gradient from buffer A to water. The protein, which is detected by Western blots or PhosphorImager analyses, is eluted in two overlapping peaks, the first peak containing both UGTR and Sep15, and the second peak containing only UGTR.

Step 5. Fractions containing Sep15 are pooled, concentrated, and subsequently applied onto a DEAE-HPLC column equilibrated in buffer A. After washing the column with buffer A, Sep15 is eluted by application of a gradient from buffer A to 0.5 M NaCl in buffer A.

Electrophoretic Analyses. Fractions containing peaks of radioactivity are concentrated and used for further native and SDS–PAGE. During purification, the protein migrates as noted above as an ~165-kDa molecular mass species on Coomassie blue staining of the native gel, whereas on SDS–PAGE, the protein appears as two species, running at ~150 and 15-kDa. The 150-kDa protein is UGTR and the 15-kDa protein is Sep15. Fractions containing Sep15 and/or UGTR are also assayed by Western blot analysis with antibodies specific for Sep15 or UGTR.

Characterization of Polymorphic Sites in Human Sep15 Gene

Homology

Protein and DNA homology analyses can be performed with BLAST programs.[10] Structural analysis of mRNA can be performed with the mfold[11] and SECISearch[12] programs. Our analyses revealed that both human and mouse Sep15 cDNAs encode an open reading frame (ORF) of 162 amino acid residues.[4] The ORF includes an in-frame TGA codon, encoding a selenocysteine residue at amino acid position 93.[13] An N-terminal signal peptide, spanning residues 1–27, is removed when the protein undergoes posttranslational processing. Sep15 is highly conserved among humans, mice, and rats and is not homologous to any known protein. A sequence comparison of Sep15 in mammals is presented in Fig. 1.

[10] T. L. Madden, R. L. Tatusov, and J. Zhang, *Methods Enzymol.* **266,** 131 (1996).

[11] D. H. Mathews, J. Sabina, M. Zuker, and D. H. Turner, *J. Mol. Biol.* **288,** 911 (1999).

[12] G. V. Kryukov, V. M. Kryukov, and V. N. Gladyshev, *J. Biol. Chem.* **274,** 33888 (1999).

[13] D. L. Hatfield, V. N. Gladyshev, J. Park, S. I. Park, H. S. Chittum, H. J. Baek, B. A. Carlson, E. S. Yang, M. E. Moustafa, and B. J. Lee, *in* "Comprehensive Natural Products Chemistry" (J. F. Kelly, ed.), Vol. 4, p. 353. Elsevier Science, New York, 1999.

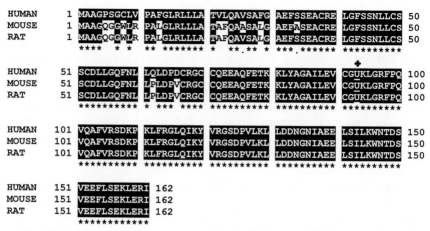

FIG. 1. Homology between human, mouse, and rat Sep15 sequences. The human, mouse, and rat Sep15 sequences are shown. Shaded sequences with asterisks (⋆) below indicate sequences that are 100% homologous in human, mouse, and rat. The sequences in mouse that are different from human and rat are indicated by dots (·). The cross (✚) indicates the position of the Sec residue (shown as U).

cDNA Polymorphism

The analysis of human expressed sequence tags (ESTs) indicated two polymorphic sites at nucleotide positions 811 and 1125 in the 3′-UTR of the human cDNA.[4] Further analysis of the human Sep 15 gene revealed that 811C is consistently associated with 1125G, and 811T with 1125A.

Selenocysteine Insertion Sequence Element

SECIS elements are stem–loop structures located in the mRNA of the 3′-UTR and are essential for insertion of selenocysteine into protein at UGA codons.[14] Both the human and mouse 3′-UTRs were found to contain SECIS-like structures ending approximately 60 nucleotides upstream of the polyadenylation signal sequence.

Functional Analysis of Selenocysteine Insertion Sequence Element

To determine the relative efficiencies of the SECIS element polymorphisms to support Sec insertion into protein, chimeric constructs are prepared in which the putative SECIS element encoding either 1125G or 1125A from the human Sep15 gene replaces the SECIS element in the deiodinase gene. HEK293 cells are transfected with the appropriate constructs, and the levels of deiodinase expression from these constructs are compared with that of the wild-type deiodinase gene, which is designated as 100%.[15] Data indicate that the human Sep15 gene

[14] S. C. Low and M. J. Berry, *Trends Biochem. Sci.* **21**, 203 (1996).
[15] M. J. Berry, L. Banu, and P. R. Larsen, *Nature (London)* **349**, 438 (1991).

apparently has only one functional SECIS element and it includes the 1125G/A polymorphic site.[5]

A more detailed analysis of SECIS element function may be achieved by using the double-reporter gene construct of Kollmus *et al.*[16] In this construct, the UGA codon is strategically placed between the genes for β-galactosidase (β-Gal) and luciferase (*luc*). The SECIS element for examination is cloned into the 3'-UTR of the *luc* gene and the ratio of *luc* to β-Gal expression determines the efficiency with which the SECIS element promotes translation of the UGA codon. In our studies, the 3'-UTR is amplified by polymerase chain reaction (PCR), using the human genomic DNA as templates, and the resulting PCR products are cloned into the double reporter gene construct. Assays and cloning manipulations are carried out as follows.

Reporter Gene Analyses

REAGENTS AND MATERIALS

Taq Polymerase
pBPLUGA vector
Mouse fibroblasts (NIH 3T3)
Dulbecco's modified Eagle's medium (DMEM)
Fetal bovine serum (FBS)
Lipofectin reagent

CLONING

1. Molecular cloning of sequences from the 3'-UTR of the Sep15 gene, including the SECIS element, is achieved by PCR with upstream primers containing the recognition sequence for *Spe*I (5'-AAAACTAGTGCTTTGTAACAGACTTGCG GTTAATTATGC-3') and downstream primers containing the recognition sequence for *Pst*I (5'-AAACTGCAGGGTCTTACAAATGATCACTTTTAAATGGAC-3'), using human genomic DNA as template.

2. The resulting 533-bp PCR product, including the polymorphic positions at 811 and 1125 (811T/1125A and 811C/1125G), are cleaved with *Spe*I and *Pst*I and directionally cloned into pBPLUGA.[16]

3. The construct is cotransfected into mouse fibroblasts (NIH 3T3) with the pSV$_2$neo plasmid at a 10 : 1 molar ratio, using Lipofectin reagent.

4. Cells are subjected to a single freeze–thaw cycle and suspended in reporter lysis buffer (Promega, Madison, WI), and the lysates are assayed for *luc* and β-Gal activities. *luc* activity is quantified with a Fentomaster FB12 luminometer from Zylux (Oak Ridge, TN) and β-Gal activity is assayed according to the manufacturer protocol.[16]

---

[16] H. Kollmus, L. Flohe, and J. E. McCarthy, *Nucleic Acids Res.* **24**, 1195 (1996).

5. For varying selenium supplementation conditions, the cells are grown in medium supplemented with 0, 30, 60, and 90 nM sodium selenite for 5 days before lysis. The tissue culture medium used for growing mammalian cells is considered to be selenium deficient.

If the SECIS is functional, then translation of the UGA would result in *luc* activity in the extracts as a result of readthrough into the *luc* gene. The efficiency with which the inserted SECIS sequence promotes UGA translation, as noted above, is measured by the ratio of *luc* to β-Gal activities.

Expression of Human 15-kDa Protein Gene

mRNA

Differential expression of Sep15 mRNA in humans and mice can be determined by Northern blot analyses.

HUMAN

1. Total RNA from several human tissues is electrophoresed on a 1.5% (w/v) formaldehyde agarose gel.
2. The RNA is transferred onto a nylon membrane.
3. The membrane is hybridized by standard hybridization techniques, using the human Sep15 cDNA as a probe.

MOUSE

The same procedure may be used to assess mouse Sep15 mRNA expression, except that mouse Sep15 cDNA is used as a probe. In addition, premade RNA blots of mouse tissues can be obtained from Clontech (Palo Alto, CA) and hybridized with the mouse cDNA probe.

The expression pattern of mRNAs from different tissues is similar in both humans and mice, with the highest levels observed in liver, kidney, brain, and testis whereas skeletal muscle, mammary gland, and trachea had virtually undetectable amounts.

Quantification of Sep15 Protein Levels

Sep15 protein levels are determined by Western analyses. Protein levels may be examined in a variety of normal and malignant tissues. However, we have focused on tumor and normal samples of prostate and liver, because these two tissues have been reported to be responsive to the chemopreventive effects of selenium.[17,18]

[17] K. Yoshizawa, W. C. Willett, S. J. Morris, M. J. Stampfer, D. Spiegelman, E. B. Rimm, and E. Giovannucci, *J. Natl. Cancer Inst.* **90,** 1219 (1998).
[18] L. C. Clark, G. F. Combs, Jr., B. W. Turnbull, E. H. Slate, D. K. Chalker, J. Chow, L. S. Davis, R. A. Glover, G. F. Graham, E. G. Gross, A. Krongrad, J. L. Lesher, Jr., H. K. Park, B. B. Sanders, Jr., C. L. Smith, and J. R. Taylor, *JAMA* **276,** 1957 (1996).

1. Protein samples (total protein) from liver tumors and surrounding normal tissues, mouse prostate adenocarcinoma cells, and normal mouse prostate are examined by SDS–PAGE. The isolated human Sep15 protein[5] is used as a control for immunoblot assays.

2. The samples are electrophoresed and transferred onto a polyvinylidene difluoride (PVDF) membrane.

3. The membrane is probed with antibodies specific for Sep15 and the levels of Sep15 in tumor tissues relative to those found in normal tissues are determined.

Intracellular Localization of Sep15

As noted above, Sep15 occurs as a complex with UGTR, an ER-resident protein,[19] which is involved in the recognition of misfolded proteins. UGTR glycosylates misfolded proteins for proper folding[20,21] or for degradation. To determine the intracellular localization of Sep15, several constructs have been made with Sep15 gene fused to the green florescent protein (GFP) gene. Generating the fusion protein facilitates tracking the expressed protein by monitoring green fluorescence in transfected cells, using confocal microscopy. The technique described below is designed to detect the cellular location of proteins in either the ER or the Golgi. However, because Sep15 is complexed with UGTR, which is known to reside in the ER, this technique can likely be used to demonstrate that Sep15 is located in the ER.

Reagents and Materials

Taq Polymerase
pEGFP-N1 and pEGFP-C3 vectors
Monkey CV-1 cells
LipofectAMINE
BODIPY TR ceramide (Molecular Probex, Eugene, OR)
Dulbecco's modified Eagle's medium (DMEM)
Fetal bovine serum (FBS)

Fusion Constructs

1. Human Sep15 cDNA is used as a template in all of the following constructions, where the 3′ untranslated region containing the SECIS element is removed, and the Sec UGA codon at position 93 is replaced with the Cys codon TGC. This construct is designated as U93C cDNA.

2. N-Sep15-C-GFP denotes the following: N is a 28-residue-long N-terminal signal peptide of Sep15; Sep15 is the Sep15 gene lacking its signal peptide and

[19] S. E. Trombetta, S. A. Ganan, and A. J. Parodi, Glycobiology 1, 155 (1991).

[20] A. J. Parodi, Biochem. J. 348, 1 (2000).

[21] A. J. Parodi, Annu. Rev. Biochem. 69, 69 (2000).

four C-terminal residues; C is a C-terminal tetrapeptide of Sep15; and GFP is the 239-residue-long green fluorescent protein.

3. Specific primers are used to PCR amplify fragments N, Sep15, and C and the resulting PCR products are cloned into expression vectors pEGFP-N1 to obtain N-Sep15-C-GFP and Sep15-C-GFP, or into pEGFP-C3 to obtain GFP-Sep15-C.

4. The N-terminal sequence of Sep15 is separately PCR amplified and the PCR product is cloned into GFP-Sep15-C and pEGFP-N1 to obtain N-GFP-Sep15-C and N-GFP, respectively.

5. N-GFP-Sep15 is obtained by mutagenesis of N-GFP-Sep15-C with specific primers.

6. Constructs are amplified by transforming into *Escherichia coli* strain Nova-Blue, and isolating the plasmids with a plasmid Maxi kit from Qiagen (Valencia, CA).

Transfection and Dual Fluorescence Imaging Confocal Microscopy

Monkey CV-1 cells are grown in 60-mm cell culture plates, transiently transfected[12] with 5 μg of the appropriate constructs and 30 μl of LipofectAMINE (GIBCO-BRL, Gaithersburg, MD), and incubated for 12 hr in a CO_2 incubator.

1. The green fluorescence of GFP is used to localize the expressed protein by confocal microscopy.

2. A fluorescent BODIPY TR ceramide that is known to label the ER and Golgi[22,23] is used as a reference maker for detecting perinuclear structures.[24]

3. The transfected cells are rinsed with serum-free DMEM containing 10 mM HEPES (DMEM–HEPES) and then incubated for 25 min at room temperature in the same medium containing 2 μM BODIPY TR ceramide.

4. The cells are washed twice in serum-free DMEM–HEPES and are immediately used for image collection. Double-labeled images of live cells are collected with a water immersion lens, using a dual excitation/emission and dual-channel mode on a Bio-Rad (Hercules, CA) MRC1024ES laser-scanning microscope.

Western Analyses

Western analyses for Sep15 and UGTR are carried out on cell extracts to determine the expression of fusion proteins. In addition, rabbit polyclonal antibodies specific for GFP are also used for the detection of GFP–Sep15 fusion proteins.

Using this system, we found that (1) the Sep15 N-terminal peptide is required for protein translocation (and is cleaved from the fusion proteins on translocation

[22] S.C. Ilgoutz, K. A. Mullin, B. R. Southwell, and M. J. McConville, *EMBO J.* **18**, 3643 (1999).

[23] M. Fukasawa, M. Nishijima, and K. Hanada, *J. Cell. Biol.* **144**, 673 (1999).

[24] L. W. Kok, T. Babia, K. Klappe, G. Egea, and D. Hoekstra, *Biochem. J.* **333**, 779 (1998).

to the ER); (2) the C-terminal sequence does not have a role in maintaining Sep15 in the ER; and (3) the internal selenoprotein sequence is responsible for complex formation with UGTR, which retains Sep15 in the ER.

Conclusions

Sep15 is one of the more recently identified selenoproteins. It has no homology to previously characterized proteins and its function is presently not known. We have described here the isolation of Sep15 from mouse liver, using procedures that involve minimal losses through denaturation.

Sep15 is located in the ER as a complex with UGTR, a 150-kDa protein that is involved in the protein-folding machinery, making it the first selenoprotein to be found in this cellular compartment. The association of Sep15 with UGTR suggests a role of Sep15 in protein folding.

Genetic analyses of the Sep15 gene revealed two single-nucleotide polymorphisms at positions 811 and 1125 in the human genome. The SECIS element in the 3'-UTR of the human Sep15 gene was shown to be functional by demonstrating it could replace the natural SECIS element of the thyroid hormone deiodinase 1 gene with comparable efficiency. It was also demonstrated that the identity of the nucleotide at position 1125 within the SECIS element influences the efficiency of Sec incorporation.

Data continue to accumulate in both animal experiments and human studies supporting a role for selenium in reducing the incidence of a wide variety of cancers. Several lines of evidence suggest that Sep15 may be an effector for this beneficial effect of selenium. These include differences in Sep15 expression in tumors compared with the corresponding nonmalignant tissue, its location in the human genome at a position frequently altered in cancers, and differences in polymorphic allele frequencies in human tumors as compared with those seen in cancer-free individuals. Collectively, these data raise the possibility that a better understanding of the biochemistry and biological roles of Sep15 will provide new insights into its function and the mechanism by which selenium exerts its protective effects.

Acknowledgment

Supported by NIH Grant CA080946 (to V.N.G).

[16] Selenocysteine Lyase from Mouse Liver

By Hisaaki Mihara and Nobuyoshi Esaki

Introduction

$$\text{L-Selenocysteine} \rightarrow \text{L-alanine} + \text{Se}$$

Selenocysteine lyase (SCL, EC 4.4.1.16) is a pyridoxal 5′-phosphate-dependent enzyme, which specifically catalyzes the decomposition of L-selenocysteine to L-alanine and elemental selenium. The enzyme was originally found in rat liver[1] and occurs widely in various organisms.[2] The enzymes from pig liver[3] and *Citrobacter freundii*[4] are described elsewhere in this series.[5,6] In this article, methods utilized for cDNA cloning, expression, and characterization of SCL from mouse liver are described.

Cloning of cDNA for Mouse Selenocysteine Lyase

Amino acid sequences of pig liver SCL are obtained from the proteolysate of the purified enzyme by sequence analysis with an automated protein sequencer PPSQ-10 (Shimadzu, Kyoto, Japan).[7] Mouse cDNA sequences (GenBank accession numbers AA107712 and MUS94C09) encoding peptides resembling the sequences of pig SCL are found in the expressed sequence tag (EST) database, using the BLAST[8] program. These sequences are used as probes to isolate full-length cDNA for mouse liver SCL as follows. Primers 5′-CCGACAGTGCGCTCCCTT CAA-3′ and 5′-GTGAACCATGTATCCCTTCAG-3′ are used to amplify a 340-bp fragment of AA107712, and primers 5′-CAGGATCGGTGCTCTGTATGT-3′ and 5′-GGCTGTTCAAATGGATTCTCT-3′ are used to amplify a 250-bp fragment of MUS94C09. The polymerase chain reaction (PCR) products are gel purified, labeled with digoxigenin (DIG), and used as probes to obtain an SCL cDNA clone from a mouse liver λZAP cDNA library constructed with the ZAP-cDNA synthesis kit (Stratagene, La Jolla, CA). Immunodetection is performed with anti-DIG, Fab fragment–Ap conjugate (Roche Diagnostics, Basel, Switzerland), nitroblue

---

[1] N. Esaki, T. Nakamura, H. Tanaka, T. Suzuki, Y. Morino, and K. Soda, *Biochemistry* **20,** 4492 (1981).

[2] P. Chocat, N. Esaki, T. Nakamura, H. Tanaka, and K. Soda, *J. Bacteriol.* **156,** 455 (1983).

[3] N. Esaki, T. Nakamura, H. Tanaka, and K. Soda, *J. Biol. Chem.* **257,** 4386 (1982).

[4] P. Chocat, N. Esaki, K. Tanizawa, K. Nakamura, H. Tanaka, and K. Soda, *J. Bacteriol.* **163,** 669 (1985).

[5] N. Esaki and K. Soda, *Methods Enzymol.* **143,** 415 (1987).

[6] N. Esaki and K. Soda, *Methods Enzymol.* **143,** 493 (1987).

[7] H. Mihara, T. Kurihara, T. Watanabe, T. Yoshimura, and N. Esaki, *J. Biol. Chem.* **275,** 6195 (2000).

[8] S. F. Altschul, W. Gish, W. Miller, E. W. Myers, and D. J. Lipman, *J. Mol. Biol.* **215,** 403 (1990).

tetrazolium, and 5-bromo-4-chloro-3-indolylphosphate *p*-toluidine salt. Positive clones are isolated, and their cDNA inserts are analyzed by sequencing. If necessary, 5′-rapid amplification of cDNA ends[9] and CapFinder (Clontech, Palo Alto, CA) techniques are employed. The total length of SCL cDNA, termed *Scly*, is 2172 bp, containing an open reading frame of 1296 bp encoding a polypeptide chain of 432 amino acid residues (GenBank accession number AF175407). The 3′ untranslated region contains a poly(A) tail and two potential overlapping polyadenylation signals, AATTAA and ATTAAA, located 13 and 12 bp, respectively, upstream from the poly(A) tail.

Expression of Selenocysteine Lyase in *Escherichia coli*

The entire coding sequence of *Scly* is obtained by reverse transcriptase (RT)-PCR with the primers 5′-GGGGAATT*CATATG*GACGCGGCGCGAAATGGC GCG-3′ and 5′-CCCC*AAGCTT*CTAGAGCCGCCCTTCCAGTTGGGCC-3′, and mouse liver total RNA as a template. The initiation codon is underlined, and the restriction enzyme sites for *Nde*I and *Hin*dIII are shown in italics. The 1.3-kbp fragment of amplified cDNA is ligated into the *Nde*I and *Hin*dIII sites of pET21a(+) to yield pESL. *Escherichia coli* BL21(DE3) is used as a host strain. The functional enzyme amounts to approximately 20% of the total protein in the crude extract of the recombinant strain.

Assay Method

Principle

Elemental selenium, eliminated from selenocysteine by the enzyme, is reduced to selenide by dithiothreitol (DTT) in the reaction mixture. Selenide is determined with lead acetate as colloidal PbSe in an acidic solution.

Reagents

Tricine–NaOH buffer, 0.5 M, pH 9.0
L-Selenocystine,[10] 10 mM, suspended in distilled water
Pyridoxal 5′-phosphate (PLP), 10 mM
DTT, 1 M
Lead acetate, 5 mM, dissolved in 0.1 M HCl

Procedure

The standard assay system contains 5 mM L-selenocysteine, 50 mM DTT, 0.2 mM PLP, 120 mM Tricine–NaOH (pH 9.0), and the appropriate amount of

[9] M. A. Frohman, M. K. Dush, and G. R. Martin, *Proc. Natl. Acad. Sci. U.S.A.* **85,** 8998 (1988).
[10] H. Tanaka and K. Soda, *Methods Enzymol.* **143,** 240 (1987).

enzyme sample, in a final volume of 0.1 ml. The reaction is allowed to proceed at 37° for 5 to 30 min and is terminated by the addition of 0.7 ml of a lead acetate solution. A nonenzyme blank is used for background subtraction for each assay set. Turbidity of the brown PbSe colloid is measured at 400 nm within 15 min. The molar turbidity coefficient of PbSe at 400 nm is $1.18 \times 10^4 \, M^{-1} \, cm^{-1}$.

Definition of Specific Activity

Specific activity is expressed as units per milligram of protein, with 1 unit of SCL defined as the amount of enzyme that catalyzes the formation of 1 μmol of selenide in 1 min.

Purification Procedure

Unless otherwise stated, all steps are performed at 0–4°, and a potassium phosphate buffer (KPB), pH 7.4, is used as the standard buffer. *Escherichia coli* BL21(DE3) cells harboring pESL are grown at 28° in 500 ml of Luria–Bertani (LB) medium containing ampicillin (0.1 mg/ml). When the A_{600} of the culture reaches 1.5, the gene expression is induced by 1 mM isopropyl-β-D-thiogalactopyranoside. After 16 hr, the cells are harvested and suspended in 50 mM KPB containing 2 mM phenylmethylsulfonyl fluoride, 5 mM EDTA, and pepstatin A (2 μg/ml), and then disrupted by sonication. The crude extract is fractionated with ammonium sulfate (1.0–3.0 M) and applied to a butyl-Toyopearl column (3 × 9.5 cm) equilibrated with 10 mM KPB containing 1.0 M ammonium sulfate. The enzyme is eluted with a 0.8-liter linear gradient of 1.0–0 M ammonium sulfate in the buffer, and fractions are collected. The fractions containing SCL are pooled and concentrated by 3.0 M ammonium sulfate. The enzyme is dissolved in 10 mM KPB containing 1 mM DTT, 0.5 mM phenylmethylsulfonyl fluoride, 1 mM EDTA, and pepstatin A (1 μg/ml), and then loaded onto a Sephadex G-25 column (2 × 24 cm) equilibrated with 50 mM KPB. The active fractions are collected and applied to a Q-Sepharose column (3 × 10 cm) equilibrated with 50 mM KPB. The enzyme is eluted with a 0.8-liter linear gradient of 0–0.2 M NaCl in the buffer, and the active fractions are pooled and concentrated with ammonium sulfate as described above. The enzyme is dialyzed against 10 mM KPB containing 1 mM DTT, 0.5 mM EDTA, and 20 μM PLP. Starting from 10 g of *E. coli* cell paste, approximately 60 mg of bright yellow-colored protein is obtained. The purified enzyme can be stored at −80° for a few weeks without loss of activity.

Properties

Physicochemical Properties

The N-terminal sequence of the purified enzyme, MDAARNGALG, agrees with that deduced from the nucleotide sequence of *Scly*. The molecular weight of the enzyme as predicted by cDNA sequence is 47,201. The molecular weight as determined by sodium dodecyl sulfate (SDS)–polyacrylamide gel electrophoresis is about 47,000. In the native state, the molecular weight based on gel filtration with Superose 12 10/30 is approximately 105,000, indicating that the enzyme is a homodimer. The absorption spectrum of the pure enzyme presents an absorption maximum at approximately 420 nm, which is compatible with the presence of PLP as a cofactor. Reduction of the enzyme with sodium borohydride results in irreversible inactivation and disappearance of the absorption band at 420 nm with a concomitant increase in absorbance at 330 nm. After incubation with 1 mM hydroxylamine, the activity decreases to 25% of the original activity. The addition of 0.2 mM PLP to the dialyzed enzymes restores 90% of the original activity.

Kinetic Properties and Substrate Specificity

The enzyme catalyzes the formation of alanine and selenide in a 1 : 1 stoichiometric ratio from L-selenocysteine in the presence of DTT. The selenide product is produced by the nonenzymatic reduction of elemental selenium by DTT. Maximum activity is obtained at pH about 9.0 when measured in Tricine–NaOH (pH 7.0–9.5) and glycine–NaOH (pH 8.5–11) buffers. The enzyme exhibits an extremely high, although not absolute, specificity for L-selenocysteine. The k_{cat}/K_m for L-selenocysteine is 4.6 mM^{-1} sec^{-1} and is approximately 100 and 4200 times higher than that for L-cysteine sulfinate and L-cysteine, respectively (Table I).

Tissue Distribution and Intracellular Localization

Reverse Transcriptase-Polymerase Chain Reaction

The tissue distribution of the enzyme is determined by RT-PCR with the primers specific to the *Scly* transcript: 5'-TGGGCAGTGTGGAGAG-3' and 5'-GTGCCCCAGAAGTGAAGATGATGT-3'. Tissues are excised from a BALB/c mouse (6 weeks, male), and total RNA isolated with Sepasol-RNAI (Nacalai Tesque, Kyoto, Japan) is subjected to reverse transcription. A 270-bp fragment is amplified in brain, heart, lung, stomach, liver, kidney, spleen, and

TABLE I
KINETIC CONSTANTS FOR SELENOCYSTEINE LYASE FROM MOUSE LIVER

Substrate	K_m (mM)	k_{cat} (sec^{-1})
L-Selenocysteine	9.9	46
L-Cysteine sulfinate	8.6	0.35
L-Cysteine	5.2	0.0058

testis. In the control reaction using the mouse genomic DNA as a template, a fragment of about 4.5 kbp is amplified, indicating that this region of the genomic DNA contains at least one intron.

Western Blot Analysis

The tissue distribution of the enzyme is semiquantitatively determined by Western blot analysis. Polyclonal antibodies raised against SCL are used to detect the amount of the enzyme in various tissue homogenates from a mouse. An immunoreactive protein of 47 kDa is detected and analyzed with the public domain NIH Image program (National Institutes of Health, Bethesda, MD). Liver, kidney, spleen, and testis have significant amounts of SCL (Fig. 1).

The subcellular localization of the enzyme is also determined by Western blot analysis. Mouse liver is homogenized in ice-cold 0.25 M sucrose containing 3 mM Tris-HCl (pH 7.4) and 0.1 mM EDTA. The homogenate is centrifuged at $700g$ for 10 min. The supernatant is centrifuged at $7000g$ for 10 min to obtain

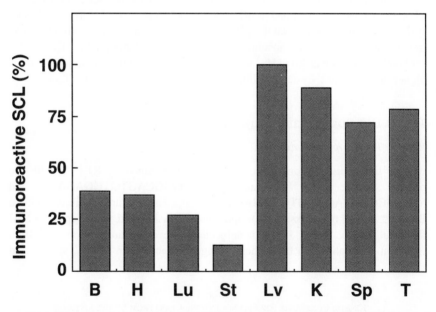

FIG. 1. Expression of selenocysteine lyase (SCL) in mouse tissues. Extracts from eight tissues of mouse (B, brain; H, heart; Lu, lung; St, stomach; Lv, liver; K, kidney; Sp, spleen; T, testis) were subjected to 12.5% (w/v) SDS–polyacrylamide gel electrophoresis, blotted onto a polyvinylidene difluoride membrane, and analyzed with a rabbit antiserum raised against mouse SCL. The amount of immunoreactive bands was estimated with the NIH Image program (National Institutes of Health). Amounts of SCL in the tissues are shown relative to the amount in liver, which is set at 100%.

crude mitochondrial pellets. The supernatant is centrifuged at 105,000g for 60 min to obtain a microsomal fraction (precipitate) and a soluble cytosolic fraction. The immunoreactive 47-kDa enzyme band is detected mainly in the cytosolic fraction.

[17] Selenocysteine Methyltransferase

By BERNHARD NEUHIERL and AUGUST BÖCK

Introduction

Several plant species have been known for their ability to accumulate large amounts of organoselenium compounds.[1] This accumulation may be the result of a detoxification reaction that becomes necessary when these plants grow on "seleniferous" soils and produce toxic selenium compounds via sulfur biosynthetic pathways.[1] As a consequence, accumulator plants store selenium analogs of compounds from normal sulfur biosynthetic pathways, such as selenocystathionine or relatively simple derivatives such as Se-methyl-selenocysteine, the latter being stored in especially high concentrations by selected members of the genus *Astragalus* (Fabaceae).[1]

An enzyme producing Se-methyl-selenocysteine, selenocysteine methyltransferase, was first purified from cultured cells of *Astragalus bisulcatus*,[2] an accumulator species. It catalyzes Se-methylation of selenocysteine and selenohomocysteine[2,3]:

$$\text{Selenocysteine} + \text{methyl-R} \rightarrow \text{Se-methylselenocysteine} + \text{R} \qquad (1)$$

$$\text{Selenohomocysteine} + \text{methyl-R} \rightarrow \text{Selenomethionine} + \text{R} \qquad (2)$$

Methyl donors (methyl-R) can be S-methylmethionine or S-adenosylmethionine; R is methionine or S-adenosylhomocysteine, respectively.

Enzyme Assays

There are two methods to test for activity of selenocysteine methyltransferase: A semiquantitative assay uses the ninhydrin reaction to visualize reaction products after separation by thin-layer chromatography. A quantitative assay is similar but

---

[1] T. A. Brown and A. Shrift, *Biol. Rev.* **57**, 59 (1982).
[2] B. Neuhierl and A. Böck, *Eur. J. Biochem.* **239**, 235 (1996).
[3] B. Neuhierl, M. Thanbichler, F. Lottspeich, and A. Böck, *J. Biol. Chem.* **274**, 5407 (1999).

uses [^{14}C]methyl-S-adenosylmethionine to allow quantification of the product, Se-[^{14}C]methylselenocysteine.

Semiquantitative Assay

A control reaction without enzyme is of special importance in this assay, because a slow but measurable chemical reaction will occur under the conditions used. Through the addition of dithiothreitol (DTT), selenocysteine is sufficiently protected from oxidation to allow reaction times of up to 60 min.

Reagents

L-Selenocystine, 5 mM
Sodium borohydride, 100 mM
Sodium citrate buffer (pH 6.0), 500 mM
Magnesium acetate, 100 mM
Dithiothreitol, 50 mM
S-Methylmethionine (or S-adenosylmethionine), 5 mM

The selenocystine solution is first mixed with the freshly prepared sodium borohydride solution at a ratio of 1 : 1 (v/v) and kept at room temperature for at least 30 min to ensure complete reduction. For the assay, combine the following (per reaction).

Component	Volume (μl)	Final concentration (mM)
Citrate buffer	5	50
Magnesium acetate	5	10
DTT	5	5
Water	20	
Selenocysteine (reduced)	5	0.5
Enzyme	5	

Tubes are equilibrated in a 30° water bath for 5 min, then the reaction is started by addition of 5 μl of methyl donor (0.5 mM final concentration). At the appropriate time points (usually 0, 10, 20, and 30 min), 5-μl portions are spotted directly onto silica gel thin-layer chromatography plates. Plates are developed in *n*-butanol–acetic acid–water (4 : 1 : 1, v/v/v), dried, and stained by spraying with an acidic ninhydrin reagent[2] and heating to 80° until spots can be detected. Se-Methylselenocysteine can be identified by its R_f of 0.59. If S-methylmethionine is used as a methyl donor, the second product, methionine, is also visible.

Quantitative Assay

For the quantitative assay, the following changes are made as compared with the semiquantitative assay: the reaction volume is reduced to 15 μl;

[^{14}C]methyl-S-adenosylmethionine at a concentration of 15 μM is used as methyl-donor; at the appropriate time points, 2.5-μl samples are withdrawn, pipetted into 2.5 μl of glacial acetic acid, and spotted onto silica gel plates; and developed and dried plates are exposed to phosphoimager screens and the radioactivity is quantified.

Purification from Plant Cell Cultures

Selenocysteine methyltransferase can be purified from cultured cells of *Astragalus bisulcatus.*[2] However, this purification is tedious and the yields are low (approximately 50 μg of pure enzyme per 100g of cultured cells). For this reason, purification of large amounts should be accomplished after overexpression of the *smtA* gene (encoding selenocysteine methyltransferase) from *A. bisulcatus* in *Escherichia coli,*[3] in this way up to 1 mg of pure enzyme protein an be obtained from 1g of cells.

Overproduction

The plasmid p7AMTT contains the *smtA* open reading frame from *A. bisulcatus* followed by the *E. coli trxA* gene, both under the control of the T7 promoter. *Escherichia coli* BL21(DE3) cells harboring this plasmid plus the plasmid pUBS520[4] are grown in 8 liters of overproduction medium [3% (w/v) tryptone, 1% (w/v) yeast extract, and 1% (w/v) NaCl] containing ampicillin (100 μg/ml) and kanamycin sulfate (50 μg/ml) in a 10-liter laboratory fermentor (Braun, Melsungen, Germany) at 37° and under maximal aeration. At an OD_{600} of 1.5, overproduction is induced by the addition of isopropyl β-D-thiogalactopyranoside to a final concentration of 50 μM. Incubation is continued until mass production in the culture ceases (approximately 3–4 hr); the cells are then harvested by centrifugation, washed with buffer A (25 mM Tris-HCl, 10 mM magnesium acetate, 1 mM EDTA, and 2 mM DTT; pH 7.5 at 4°), frozen in liquid nitrogen, and stored at −20°. If the enzyme is to be produced in a smaller volume, for example, in Erlenmeyer flasks, overexpression should be induced at an OD_{600} of 0.5, followed by a production period of 2 to 3 hr.

Purification from Overexpressing Escherichia coli Cells

All steps of the purification are performed at 0–4°. After each step, fractions are analyzed by sodium dodecyl sulfate (SDS)–polyacrylamide gel electrophoresis followed by Coomassie staining; those fractions containing selenocysteine methyltransferase are pooled.

[4] U. Brinkmann, R. E. Mattes, and P. Buckel, *Gene* **85,** 109 (1989).

Crude Extract and Ammonium Sulfate Precipitation

For the preparation of a crude extract, 20 g of cells is resuspended in 40 ml of buffer A containing 1 mM phenylmethylsulfonyl fluoride and lysed by three passages through a French pressure cell at 13,000 lb/in^2. The soluble (S100) fraction is obtained by centrifugation (30 min at 30,000g, supernatant = S30, and 2 hr at 100,000g, supernatant = S100) and adjusted to 35% (w/v) ammonium sulfate saturation. The supernatant of the following centrifugation (20 min at 25,000g) is readjusted to 60% (w/v) ammonium sulfate saturation. Selenocysteine methyltransferase is recovered in the sediment of the following centrifugation and dissolved in 25 ml of buffer containing 1 M ammonium sulfate.

Column Chromatography

The dissolved enzyme preparation is loaded onto a 1.6 × 10 cm Phenyl-Sepharose column equilibrated with buffer containing 1 M ammonium sulfate and eluted with a gradient from 1 to 0 M ammonium sulfate over 5 column volumes. Fractions containing the enzyme are pooled and proteins are precipitated from 70% (w/v) ammonium sulfate overnight. Pelleted protein is dissolved in a minimal volume and chromatographed on a Superdex 75 gel-filtration column (1.6 × 60 cm) equilibrated with buffer A containing 50 mM KCl. Fractions containing selenocysteine methyltransferase are chromatographed on a 1.6 × 10 cm Q-Sepharose column, using a gradient from 0 to 500 mM KCl in buffer A over 20 column volumes. Apparently pure fractions are pooled, whereas fractions containing contaminating proteins are further purified on a Mono Q column, using the same gradient as described previously. However, when elution of selenocysteine methyltransferase starts, the gradient is held at the same KCl concentration until all protein is eluted.

Storage and Stability

Pure fractions from Mono Q chromatography are combined with the pool from Q-Sepharose chromatography and selenocysteine methyltransferase is precipitated from 70% (w/v) ammonium sulfate overnight. The enzyme is stable as an ammonium sulfate pellet for several months. Alternatively, it can be dialyzed against 500 ml of buffer containing 50% (v/v) glycerol and kept at −20°.

Properties

Selenocysteine methyltransferase is a monomeric protein of approximately 38 kDa. Its pH optimum was determined to be 7.5; reactions, however, are preferentially performed at pH 6 to increase stability of the methyl donor substrate and to reduce the background (chemical) reaction. The temperature optimum is at 30°. K_m values are 15 μM for S-adenosylmethionine and 0.35 mM for DL-selenocysteine.[2]

The most intriguing feature of selenocysteine methyltransferase is its pronounced specificity for selenium. Reaction with the sulfur analog of selenocysteine, cysteine, is catalyzed with an approximately 1000-fold lower efficiency. Accordingly, this enzyme has been proposed to play an important role in selenium detoxification and accumulation by selenium-accumulating plants: Selenocysteine is the primary organoselenium compound produced by the cells when selenium intrudes on sulfur metabolic pathways[1] and this toxicity is believed to be mediated, at least in part, by the free selenol function of the molecule. Methylation would then mask the selenol and could mark the molecule for storage. This model was further corroborated by experiments in which cultured cells from the selenium nonaccumulating plant *Astragalus cicer* were long-term selected for selenium tolerance. Tolerant cells were found to produce an enzyme similar to selenocysteine methyltransferase, which is absent from nontolerant cells.[5]

Relation to Homocysteine Methyltransferases

When the cDNA encoding selenocysteine methyltransferase was cloned, it became obvious that several selenium-non-accumulating organisms such as baker's yeast (*Saccharomyces cerevisiae*) or *E. coli* contain one or more apparent homologs. These were later shown to be homocysteine methyltransferases,[3,6] which catalyze the following reactions:

$$\text{Selenohomocysteine} + \text{methyl-R} \rightarrow \text{selenomethionine} + \text{R} \qquad (3)$$

$$\text{Homocysteine} + \text{methyl-R} \rightarrow \text{methionine} + \text{R} \qquad (4)$$

Methyl-R can be the same methyl donor compounds as for Eqs. (1) and (2). With methyl-R being S-methylmethionine, the reaction according to Eq. (4) will yield 2 mol of methionine. This methionine-producing activity of homocysteine methyltransferase appears to be the physiological role of the enzyme, as it was shown for *E. coli*[6] and *Arabidopsis thaliana*.[7]

[5] Y. Wang, A. Böck, and B. Neuhierl, *BioFactors* **9**, 3 (1999).

[6] M. Thanbichler, B. Neuhierl, and A. Böck, *J. Bacteriol.* **181**, 662 (1999).

[7] P. Ranocha, F. Bourgis, M. J. Ziemak, D. Rhodes, D. A. Gage, and A. D. Hanson, *J. Biol. Chem.* **275**, 15962 (2000).

[18] Phospholipid–Hydroperoxide Glutathione Peroxidase in Sperm

By Antonella Roveri, Leopold Flohé, Matilde Maiorino, and Fulvio Ursini

Introduction

The phospholipid hydroperoxide glutathione peroxidase (PHGPx) gene (*gpx-4*) was found to be highly expressed in postpubertal mammalian testis.[1] Mature testis was reported to have the highest PHGPx activity of all mammalian tissues investigated.[2] There it is found associated, in part at least, with mitochondria,[3] which complies with a tissue-specific transcription of the PHGPx gene into an mRNA encoding a protein with a mitochondrial leader peptide.[4,5] Biosynthesis of PHGPx in testis occurs predominantly in round spermatids.[6,7] Correspondingly, specific PHGPx activity in testis appears to parallel the thickness of the spermatid layer.[6] The hormonal regulation of testicular PHGPx activity[1] was shown to be indirect; Leydig cell-derived testosterone stimulates the seminiferous epithelium and thereby augments the number of spermatids that produce PHGPx.[6]

Despite the abundance of PHGPx in spermatids, PHGPx activity is practically undetectable in mature spermatozoa by conventional activity assays.[8] Surprisingly, however, PHGPx protein could easily be detected by Western blotting in reduced and sodium dodecyl sulfate (SDS)-solubilized sperm proteins.[9] This enzymatically inactive PHGPx protein was found to make up at least 50% of the keratinous material embedding the mitochondrial helix in the midpiece of sperm mitochondria.[9] It accounts for most, if not all, of the selenium content in the midpiece of spermatozoa and thus is the real "mitochondrial capsule selenoprotein," a term that had

[1] A. Roveri, A. Casasco, M. Maiorino, P. Dalan, A. Calligaro, and F. Ursini, *J. Biol. Chem.* **267,** 6142 (1992).

[2] A. Roveri, M. Maiorino, and F. Ursini, *Methods Enzymol.* **233,** 202 (1994).

[3] M. Maiorino, A. Roveri, and F. Ursini, in "Free Radicals: From Basic Science to Medicine" (G. Poli, E. Albano, and M. U. Dianzani, eds.), p. 412. Birkhäuser Verlag, Basel, 1993.

[4] T. R. Pushpa-Rekha, A. L. Burdsall, L. M. Oleksa, G. M. Chisolm, and D. M. Driscoll, *J. Biol. Chem.* **270,** 26993 (1995).

[5] M. Arai, H. Imai, D. Sumi, T. Imanaka, T. Takano, N. Chiba, and Y. Nakagawa, *Biochem. Biophys. Res. Commun.* **227,** 433 (1996).

[6] M. Maiorino, J. B. Wissing, R. Brigelius-Flohé, F. Calabrese, A. Roveri, P. Steinert, F. Ursini, and L. Flohé, *FASEB J.* **12,** 1359 (1998).

[7] K. Mizuno, S. Hirata, K. Hoshi, A. Shinohara, and M. Shiba, *Biol. Trace Elem. Res.* **74,** 1112 (2000).

[8] R. Brigelius-Flohé, K. Wingler, and C. Müller, *Methods Enzymol.* **347,** [9], 2002 (this volume).

[9] F. Ursini, S. Heim, M. Kiess, M. Maiorino, A. Roveri, J. Wissing, and L. Flohé, *Science* **285,** 1393 (1999).

misleadingly been introduced for the "sperm mitochondria-associated cysteine-rich protein" (SMCP).[10]

The process transforming the enzymatically active selenoperoxidase into a structural protein during late spermatogenesis remains largely elusive. Obviously, the transformation results from oxidative cross-linking of PHGPx with itself and/or other proteins via Se–S and/or S–S bonds, because PHGPx protein can be solubilized and PHGPx activity can be recovered from the capsule material by drastic reductive procedures.[9] The change in enzymatic activity and physical properties of PHGPx during spermatogenesis is a typical example of "moonlighting," as has also been described, for example, for proteins making up the eye lens.[11,12] This moonlighting certainly indicates a dual role of PHGPx in male reproduction; the enzyme may protect the rapidly dividing spermatogenic cells from potentially harmful oxidants, whereas it later becomes part of a structural element essential for the function of mature spermatozoa. The latter biological role of PHGPx appears at least as important for male fertility as its presumed antioxidant activity; interestingly, it is precisely the morphological alteration and mechanical instability of the midpiece material that lead to the impaired fertilization potential of selenium-deficient spermatozoa.[13,14]

Measurement of Phospholipid-Hydroperoxide Glutathione Peroxidase in Human Spermatozoa

As outlined in the introduction, PHGPx is found in spermatozoa mainly in the midpiece, where it is part of the "mitochondrial capsule." The enzyme is catalytically inactive in this form. The measurement of specific activity therefore requires a "rescue" procedure. The solubilization–reactivation is efficiently carried out in the presence of high concentrations of thiols and guanidine. The accuracy and reproducibility of specific activity measurements require (1) the complete and reproducible solubilization of all proteins of the sperm and (2) the careful removal of guanidine and of the thiols used for solubilization–reactivation, because they compete with glutathione in the peroxidase reaction, thus leading to an underestimation of the activity.

Specimen Collection

Human sperm are collected by masturbation between days 3 and 7 of abstinence. On each sample sperm count, morphological and functional parameters are immediately evaluated. Sperm are centrifuged at 300g for 10 min at 4°. Pellets are

---

[10] L. Cataldo, K. Baig, R. Oko, M. A. Mastrangelo, and K. C. Kleene, *Mol. Reprod. Dev.* **45**, 320 (1996).

[11] J. Piatigorski, *Prog. Ret. Eye Res.* **17**, 145 (1998).

[12] C. J. Jefferey, *Trends Biol. Sci.* **24**, 8 (1999).

[13] M. Maiorino, L. Flohé, A. Roveri, P. Steinert, J. Wissing, and F. Ursini, *BioFactors* **10**, 251 (1999).

washed twice in phosphate-buffered saline (PBS). Washed pellets can be stored at $-20°$ for a few weeks without any apparent loss of activity.

Sample Preparation

The pellet is solubilized in 0.1 M Tris-HCl, 6 M guanidine hydrochloride, 0.1% (v/v) Triton X-100, 0.1 M 2-mercaptoethanol, pH 7.4 (solubilization buffer; SB). Sperm count in SB is $15–25 \times 10^6$ cell/ml. Solubilization is carried out by vortex mixing. In the standardized procedure three periods of 30 sec of vortexing are used with a pause in ice of 1 min between each of them. The absence of any visible pellet after centrifugation ensures a thorough solubilization. The protein concentration thus obtained in the SB ranges between 0.5 and 1.0 mg/ml. Solubilized samples can be stored in this buffer at $-20°$ and, under these conditions, PHGPx activity is stable for several days without any obvious loss of activity.

Sample Preparation for Activity Measurement

NAP 10 desalting columns (Amersham Pharmacia Biotech, Uppsala, Sweden) are equilibrated just before use in 0.1 M Tris-HCl, 5 mM glutathione (GSH), 0.1% (v/v) Triton X-100, 5 mM EDTA, pH 7.4 (elution buffer; EB). A sample containing 0.30–0.40 mg of solubilized sperm proteins is diluted to exactly 1 ml with SB and loaded on the NAP 10 column. After the sample has entered the gel bed, elution is carried out with 1.5 ml of EB according to the manufacturer instructions.

The buffer exchange procedure usually needs to be repeated twice in order to assure a complete removal of 2-mercaptoethanol.

Activity Measurement

Activity is measured in 2.2 ml of EB to which glutathione reductase (GSSG reductase, 0.6 IU/ml) and 0.030 mM NADPH are added. The mixture is incubated for 5 min at 25° and, after recording the basal rate of NADPH oxidation, the reaction is started by adding 30 μM phosphatidylcholine hydroperoxide (PC-OOH) prepared as previously described.[15] PHGPx activity is measured from the time course of absorbance decrease at 340 nm ($\varepsilon = 6.22$ mM^{-1} cm^{-1} for NADPH). The basal rate, although negligible, is subtracted from the enzyme activity rate.

PHGPx activity is calculated from the regression curve obtained from three different amounts of sample.[8] A good sample volume-to-activity correlation (R, 0.995 ± 0.004) is reproducibly found.

---

[14] L. Flohé, R. Brigelius-Flohé, M. Maiorino, A. Roveri, J. Wissing, and F. Ursini, *in* "Selenium: Its Molecular Biology and Role in Human Health" (D. Hatfield, ed.), p. 273. Kluwer Academic Publisher, 2001.

[15] M. Maiorino, C. Gregolin, and F. Ursini, *Methods Enzymol.* **186,** 448 (1990).

Specific Activity

Protein measurement is carried out by the Lowry procedure on the sample after elution from the NAP 10 column. Protein concentration is calculated by linear regression of activity data obtained with increasing amounts of sample (from 0.02 to 0.12 ml). Care is always taken that the amount of protein recovered corresponds to that loaded onto the column. The PHGPx measurements by this procedure of 30 normal subjets resulted in 200.5 ± 50.2 nmol/min × mg protein.

Comments on Procedure

Homogeneous PHGPx from pig heart was used to test the effect of the procedure on the activity of the enzyme. A constant loss of 25–30% of activity was observed to take place during the procedure. This is apparently due to the presence of guanidine. Because no further loss of activity was observed both over a 30-min incubation in the presence of guanidine at room temperature and after storage for weeks at $-20°$, it is apparent that guanidine does not progressively denature the protein but simply slows down, for an unknown reason, the catalytic activity. Because guanidine is useful to ensure a complete and reproducible solubilization of sperm, which is crucial for precision of the measurement, the decrease in activity was considered acceptable and not affecting the precision of the analysis when different samples are compared.

1. To optimize solubilization of sperm a relatively large dilution of the sperm samples must be used ($<25 \times 10^6$ sperm/ml).
2. The amount and concentration of solubilized proteins to be loaded on buffer exchange column must be kept relatively constant ($\pm 20\%$) in order to optimize the recovery from the column and the reproducibility of the results.
3. The removal of 2-mercaptoethanol should be performed as carefully as possible. It is advisable to use the 5,5′-dithio-bis(2-nitrobenzoic acid) (DTNB) reaction for calibrating the column elution volumes. Futhermore, a linear correlation between different volumes of sample and measured activities gives a valuable index of the absence of thiols interfering with the measurement of activity.

Discussion

Determination of PHGPx in sperm may be considered a possible way to characterize hitherto poorly understood disturbances of male fertility. The idea that PHGPx is of pivotal importance for sperm architecture and function[9] has been supported by partially successful gene disruption in mice.[16] Nine chimeric male

[16] M. Conrad, U. Heinzelmann, W. Wurst, G. W. Bornkamm, and M. Brielmeier, *in* "7th International Symposium on Selenium in Biology and Medicine," October 1–5, 2000, Venice, Italy. [Abstract]

mice having at least 50% PHGPx[+/−] cells produced 190 offspring homozygous for PHGPx. It is clear that the PHGPx[+/−] cells did not contribute to the germ line. Interestingly, the testes of the chimeric mice displayed mosaic-like disturbances, with a few spermatozoa displaying morphological alterations reminiscent of those observed in selenium-deficient rats, for example, fuzzy or broken midpieces, disoriented tails, or even isolated heads and tails.[17,18] Reduced PHGPx content in sperm can therefore be expected to result in disturbed sperm morphology and function, as does selenium deficiency. Whether selenium shortage may affect PHGPx synthesis and sperm function in humans remains to be demonstrated. In rodents, male fertility was observed to be affected only on long-lasting selenium deprivation for several generations.[17–20] These findings are not surprising if PHGPx is indeed the selenoprotein responsible for sperm integrity, because (1) testis tends to retain selenium even under severe selenium restriction[21] and (2) PHGPx responds poorly to selenium deprivation.[22–24] It can therefore hardly be anticipated that minor variations of selenium intake affect testicular PHGPx synthesis and fertility in men. It can nevertheless be envisaged that a reduced content of functional PHGPx in human testes due to genetic, hormonal, or severe nutrional deficiencies is etiologically linked to certain forms of male fertility problems.

Acknowledgments

The preparation of this article was supported by the Deutsche Forschungsgemeinschaft (Grant F161/12-1) and the Volkswagenstiftung (ZN548).

[17] A. S. H. Wu, J. E. Oldfield, P. D. Whanger, and P. H. Weswig, *Biol. Reprod.* **8,** 625 (1973).

[18] A. S. H. Wu, J. E. Oldfield, L. R. Shull, and P. R. Cheeke, *Biol. Reprod.* **20,** 625 (1979).

[19] E. Wallace, G. W. Cooper, and H. I. Calvin, *Gamete Res.* **4,** 389 (1983).

[20] E. Wallace, H. I. Calvin, and G. W. Cooper, *Gamete Res.* **4,** 377 (1983).

[21] D. Behne, T. Hofer, R. von Bersworat-Wallrabe, and W. Egler, *J. Nutr.* **112,** 1682 (1982).

[22] F. Weitzel, F. Ursini, and A. Wendel, *Biochim. Biophys. Acta* **1036,** 88 (1990).

[23] R. Brigelius-Flohé, *Free Radic. Biol. Med.* **27,** 951 (1999).

[24] L. Flohé, E. Wingender, and R. Brigelius-Flohé, "Oxidative Stress and Signal Transduction" (H. J. Forman and E. Cadenas, eds.), p. 415. Chapman & Hall, New York, 1997.

[19] *In Vivo* Antioxidant Role of Glutathione Peroxidase: Evidence from Knockout Mice

By XIN GEN LEI

Introduction

Cellular glutathione peroxidase (glutathione : H_2O_2 oxidoreductase, EC 1.11.1.9, GPX1) was initially found in erythrocytes by Mills[1] in 1957, when Schwarz and Foltz[2] showed the nutritional essentiality of selenium (Se) in rats. These two important findings were merged in 1973 by the identification of GPX1 as a selenium-dependent enzyme.[3,4] This milestone discovery signified the beginning of a new era of selenium biology, followed by biochemical characterizations of the GPX1 enzyme.[5–7] After the first GPX1 gene was cloned by Chambers *et al.*[8] in 1986 from murine erythroblasts, intensive research efforts were focused on mechanisms of selenium incorporation into the GPX1 peptide[9] and selenium regulation of GPX1 gene expression, in comparison with those of other selenoprotein genes.[10,11]

However, the physiological function of GPX1 has remained unclear for many years, although its activity in tissues is widely used to assess body status and nutritional needs for selenium. Early empirical observations of selenium deficiency symptoms did not help in distinguishing the roles of selenium, let alone of GPX1, from those of vitamin E because selenium deficiency alone, without depriving vitamin E, does not produce these symptoms. Deprivation of body selenium causes GPX1 activity to fall to extremely low levels in liver and other tissues, but growth and well-being of experimental animals remain apparently normal if adequate vitamin E is provided and the selenium deficiency does not last to the second generation.[11] There is no specific inhibitor of any of the multiple selenoproteins

[1] G. C. Mills, *J. Biol. Chem.* **229**, 189 (1957).
[2] K. Schwarz and C. M. Foltz, *J. Am. Chem. Soc.* **79**, 3292 (1957).
[3] J. T. Rotruck, A. L. Pope, H. E. Ganther, A. B. Swanson, D. G. Hafeman, and W. G. Hoekstra, *Science* **179**, 585 (1973).
[4] L. Flohé, W. A. Günzler, and H. H. Schock, *FEBS Lett.* **32**, 132 (1973).
[5] L. Flohé, G. Loschen, W. A. Günzler, and E. Eichole, *Physiol. Chem.* **353**, 987 (1972).
[6] J. W. Forstrom, J. J. Zakowski, and A. L. Tappel, *Biochemistry* **17**, 2639 (1972).
[7] R. J. Kraus, S. J. Foster, and H. E. Ganther, *Biochemistry* **22**, 5853 (1983).
[8] I. Chambers, J. Frampton, P. Goldfarb, N. Affara, W. McBain, and P. R. Harrison, *EMBO J.* **5**, 1221 (1986).
[9] Q. Shen, F.-F. Chu, and P. E. Newburger, *J. Biol. Chem.* **268**, 11463 (1993).
[10] R. F. Burk, *FASEB J.* **5**, 2274 (1991).
[11] R. A. Sunde, *in* "Selenium in Biology and Human Health" (R. F. Burk, ed.), p. 47. Springer-Verlag, New York, 1994.

0076-6879/02 $35.00

identified in mammalian tissues.[12] Thus, it is difficult, if not impossible, to assess metabolic roles of GPX1 using genetically intact animals and cells, without confounding effects of other selenoproteins.

Numerous studies[13–23] have shown that expression of GPX1 mRNA, protein, and activity in tissues or cells is affected more by changes in selenium supply and is saturated at higher levels of selenium than that of other selenoproteins. On the basis of mass, selenium content, and GPX1 activity, Behne and Wolters[24] estimated that GPX1 binds 63% of liver total selenium in selenium-adequate adult female rats. This abundance and the apparent low rank in the hierarchy[25] of selenium partitioning for selenoprotein synthesis have resulted in questions concerning an important *in vivo* antioxidant role of GPX1.[10,11] However, overexpression of GPX1 enhances the resistance to clastogenic oxidant stress in T47 D human breast cells,[26] FL5.12 cells,[27] and NIH 3T3 and MCF-7 cells.[28] In contrast, suppression of GPX1 expression, by transfecting antisense cDNA of GPX1, decreases the protection of DG-44 cells from pro-oxidant toxicity of paraquat and Adriamycin.[29] Although these GPX1-transfected cells are indeed more specific than the conventional, selenium-deficient animal models for the study of GPX1 roles, it is arguable that responses of such cells do not necessarily represent physiological reactions of the whole body. Therefore, an ideal model is required to bear the specificity of the GPX1-altered cell and the physiological condition of the whole animal. GPX1 knockout [GPX1(–/–)] mice, along with GPX1 overexpression [GPX1(+)] mice, satisfy both requirements for the task.

[12] L. Flohé, J. R. Andreesen, R. Brigelius-Flohé, M. Maiorino, and F. Ursini, *Life* **49**, 411 (2000).

[13] N. Avissar, E. A. Kerl, S. S. Baker, and H. J. Cohen, *Arch. Biochem. Biophys.* **309**, 239 (1994).

[14] R. D. Baker, S. S. Baker, K. LaRosa, C. Whitney, and P. E. Newburger, *Arch. Biochem. Biophys.* **304**, 53 (1993).

[15] G. Bermano, F. Nicol, A. A. Dyer, R. A. Sunde, G. J. Beckett, J. R. Arthur, and J. E. Hesketh, *Biochem. J.* **311**, 425 (1995).

[16] M. J. Christensen, P. M. Cammack, and C. D. Wray, *Nutr. Biochem.* **6**, 367 (1995).

[17] M. Gross, M. Oertel, and J. Köhrle, *Biochem. J.* **306**, 851 (1995).

[18] S. A. B. Knight and R. A. Sunde, *J. Nutr.* **117**, 732 (1987).

[19] X. G. Lei, J. K. Evenson, K. T. Thompson, and R. A. Sunde, *J. Nutr.* **125**, 1438 (1995).

[20] J. H. Mitchell, F. Nicol, G. J. Beckett, and J. R. Arthur, *J. Mol. Biol.* **16**, 259 (1996).

[21] S. Vadhanavikit and H. E. Ganther, *J. Nutr.* **123**, 1124 (1993).

[22] F. Weitzel, F. Ursini, and A. Wendel, *Biochim. Biophys. Acta* **1036**, 88 (1990).

[23] J. G. Yang, K. E. Hill, and R. F. Burk, *J. Nutr.* **119**, 1010 (1989).

[24] D. Behne and W. Wolters, *J. Nutr.* **113**, 456 (1983).

[25] R. Brigelius-Flohé, *Free Radic. Biol. Med.* **27**, 951 (1999).

[26] M.-E. Mirault, A. Tremblay, N. Beaudoin, and M. Tremblay, *J. Biol. Chem.* **266**, 20752 (1991).

[27] D. M. Hockenbery, Z. N. Oltvai, X.-M. Yin, C. L. Milliman, and S. J. Korsmeyer, *Cell* **75**, 241 (1993).

[28] M. J. Kelner, R. D. Bagnell, S. F. Uglik, M. A. Montoya, and G. T. Mullenbach, *Arch. Biochem. Biophys.* **323**, 40 (1995).

[29] S. D. Taylor, L. D. Davenport, M. J. Speranza, G. T. Mullenbach, and R. E. Lynch, *Arch. Biochem. Biophys.* **306**, 600 (1993).

Development and Characterization of GPX1(+) and GPX1(–/–) Mouse Models

Several groups have developed the GPX1(–/–) and/or GPX1(+) mice.[30–34] The two lines we have used in our study were provided by Y.-S. Ho at Wayne State University (Detroit, MI) and were the first reported of this kind. Procedures in generating these GPX1(–/–) and GPX1(+) mice have been described in detail previously.[31,34–36] Briefly, the GPX1(–/–) mice were derived from 129/SVJ × C57B L/6 mice by microinjecting C57BL/6 blastocysts with recombinant embryonic stem cells carrying a target mutation in the GPX1 gene. The GPX1(+) mice were derived from the B6C3 (C57BL × C3H) hybrid line. The GPX1 gene was isolated from a bacteriophage FIX II genomic library prepared with DNA from a 129/SVJ mouse. Initially, suppression or elevation of the GPX1 gene expression was verified by Northern analysis of various tissues of founders fed adequate selenium and vitamin E. No difference was detected in activities of glutathione reductase (EC 1.6.4.2), catalase (EC 1.11.1.6), glucose-6-phosphate dehydrogenase, and Cu,Zn- and Mn-superoxide dismutases (SOD, EC 1.15.1.1) in various tissues between these GPX1-altered lines and their respective wild-type controls.[31,34]

The specificity of GPX1(–/–) and GPX1(+) mice for the functional study of GPX1 was validated by us in four experiments.[35–37] GPX1(+) mice, GPX1(–/–) mice, and their respective wild-type mice (3 weeks to 2 months old) were fed with a selenium-deficient, *Torula* yeast basal diet (selenium at <0.02 mg/kg) supplemented with selenium at 0, 0.2, 0.5, or 3.0 mg/kg as sodium selenite (Sigma, St. Louis, MO; all chemicals were from Sigma unless otherwise indicated) and all-*rac*-α-tocopheryl acetate at 0, 15, or 50 mg/kg for 5–13 weeks. At the end of the feeding trials, all mice were anesthetized with carbon dioxide and killed by exsanguination via heart puncture, using a heparinized syringe to collect tissues for assays.

[30] O. Mirochnitchenko, U. Palnitkar, M. Philbert, and M. Inouye, *Proc. Natl. Acad. Sci. U.S.A.* **92,** 8120 (1995).

[31] Y.-S. Ho, J.-L. Magnenat, R.T. Bronson, J. Cao, M. Gargano, M. Sugawara, and C. D. Funk, *J. Biol. Chem.* **272,** 16644 (1997).

[32] J. B. de Haan, C. Bladier, P. Griffiths, M. Kelner, R. D. O'Shea, N. S. Cheung, R. T. Bronson, M. J. Silvestro, S. Wild, S. S. Zheng, P. M. Beart, P. J. Hertzog, and I. Kola, *J. Biol. Chem.* **273,** 22528 (1998).

[33] L. A. Esposito, J. E. Kokoszka, K. G. Waymire, B. Cottrell, G. R. MacGregor, and D. C. Wallace, *Free Radic. Biol. Med.* **28,** 754 (2000).

[34] A. Spector, Y. Yang, Y.-S. Ho, J.-L. Magnenat, R.-R. Wang, W. Ma, and W.-C. Li, *Exp. Eye Res.* **62,** 521 (1996).

[35] W. H. Cheng, Y.-S. Ho, D. A. Ross, Y. M. Han, G. F. Combs, Jr., and X. G. Lei, *J. Nutr.* **127,** 675 (1997).

[36] W. H. Cheng, Y.-S. Ho, D. A. Ross, B. A. Valentine, G. F. Combs, Jr., and X. G. Lei, *J. Nutr.* **127,** 1445 (1997).

[37] W. H. Cheng, G. F. Combs, Jr., and X. G. Lei, *Biofactors* **7,** 311 (1998).

Compared with the wild-type controls, knockout of GPX1 resulted in almost complete abolishment of GPX1 mRNA and 80 to 99% reduction in total GPX (H_2O_2) activity in various tissues, including testis, intestine, and stomach, in the selenium-adequate animals.[36] In contrast, overexpression of GPX1 enhanced total GPX activity by 1-to 6-fold ($p < 0.0001$) in kidney, lung, heart, intestine, stomach, and muscle of both selenium-deficient and -adequate animals, compared with the wild-type controls.[35] In liver, however, such increases were observed only in the selenium-deficient mice. Higher levels of GPX1 mRNA were also detected in kidney (81%) and lung (7-fold) of the selenium-deficient GPX1(+) mice than those of the wild-type mice. Total liver selenium concentrations were not different between the GPX1(–/–) and wild-type mice fed selenium at 0 mg/kg, but were reduced by 61 and 64% in the GPX1(–/–) mice fed selenium at 0.5 and 3.0 mg/kg, respectively.[36,37] Meanwhile, mRNA and/or activity expression of other selenoproteins, including GPX3, GPX4 (EC 1.11.1.12), thioredoxin reductase (TR, EC 1.6.4.5) and selenoprotein P, was not altered by GPX1 knockout or overexpression in assayed tissues of mice fed selenium-deficient, -adequate, or -excess diets.[35–40] Plasma selenium concentration was unaffected in these animals as well. Activities of several selenium-independent antioxidant enzymes, including glutathione transferase (EC 2.5.1.18), catalase, and superoxide dismutase (SOD) were not different between the nonstressed GPX1(–/–) and wild-type mice.

Clearly, GPX1 gene knockout in the GPX1(–/–) mice and GPX1 gene overexpression in the GPX1(+) mice are specific. Because this scenario maintains across a wide range of dietary selenium levels, expression of GPX1 is apparently independent of that of other selenoproteins and does not seem to affect body selenium channeling into syntheses of these proteins. The 60% reduction in total liver selenium in the GPX1(–/–) mice fed adequate selenium confirms the early estimate[24] of the proportion of GPX1 in the total liver selenium of adult female rats, indicating that this enzyme is probably the most abundant selenoprotein, at least in rodents. The relatively substantial reductions in total GPX activities in testis, stomach, and intestines of the GPX1(–/–) mice[36,37] suggest a need for reconsideration of the activity contribution and physiological importance of GPX1 in these tissues.

Protection by Glutathione Peroxidase against Acute, Lethal Oxidative Stress

Pro-oxidants paraquat and diquat have been used to study the *in vivo* antioxidant functions of selenium and selenoproteins because both compounds can

[38] Y. X. Fu, W.-H. Cheng, J. M. Porres, D. A. Ross, and X. G. Lei, *Free Radic. Biol. Med.* **27**, 605 (1999).

[39] Y. X. Fu, W.-H. Cheng, D. A. Ross, and X. G. Lei, *Proc. Soc. Exp. Biol. Med.* **222**, 164 (1999).

[40] W. H. Cheng, Y. X. Fu, J. M. Porres, D. A. Ross, and X. G. Lei, *FASEB J.* **13**, 1467 (1999).

generate superoxide radicals ($O_2 \cdot ^-$) through the redox cycling as a futile reductive attempt.[41,42] Under the catalysis of SOD, the superoxide radicals are converted to hydrogen peroxide (H_2O_2) that subsequently serves as a substrate of GPX1 and catalase. If the generation of hydrogen peroxide exceeds the reducing capacity of peroxidases, it undergoes a series of iron-catalyzed reactions to produce hydroxyl-free radicals ($\cdot$ OH). These radicals are extremely reactive with macromolecules, resulting in multiple organ injuries leading to death. The primary target organ is lung for paraquat and liver for diquat.[43] Several groups[43–48] have shown that selenium is protective against the toxicity of these compounds in rats, mice, and chicks, but there is no consensus on the role of GPX1 as the putative mediator of selenium protection. Although many experimental factors might have contributed to this controversy, the most apparent reason has been the difficulty in distinguishing the contribution of GPX1 from that of other selenoproteins in the conventional animal models. Thereby, the GPX1(–/–) and GPX1(+) mice offer us an unprecedented, powerful model to clarify the role of GPX1 in protecting against various sources of oxidative stress and its relative contribution to the total protection of selenium.

Impacts of Glutathione Peroxidase Expression on Paraquat and Diquat Lethality

To produce an acute, lethal oxidative stress *in vivo,* we challenged selenium-deficient and selenium-adequate GPX1(–/–) and wild-type mice with an intraperitoneal injection of paraquat[49] or diquat[38] (50 or 24 mg/kg of body weight, respectively). Before the injections, mice (3 weeks old) were fed the *Torula* yeast basal diet without selenium (selenium deficient) or supplemented with selenium at 0.3 to 0.4 mg/kg as sodium selenite (selenium adequate) for 5 weeks. All diets were supplemented with 20 (paraquat study) or 75 (diquat study) mg of all-*rac*-α-tocopheryl acetate per kilogram. Both paraquat (methyl viologen, 1,1′-dimethyl-4,4′-bipyridinium dichloride) and diquat (dibromide monohydrate; Chemical Service, West Chester, PA) were dissolved in isotonic saline and filtered sterilized. Injection volume were controlled at 10 ml/kg of body weight.

After the intraperitoneal injection of paraquat, the selenium-deficient and selenium-adequate GPX1(–/–) mice survived 4.9 and 6.4 hr, respectively, whereas

[41] J. S. Bus, S. D. Aust, and J. E. Gibson, *Biochem. Biophys. Res. Commun.* **58**, 749 (1974).
[42] L. L. Smith, *Hum. Toxicol.* **6**, 31 (1987).
[43] R. F. Burk, R. A. Lawrence, and J. M. Lane, *J. Clin. Invest.* **65**, 1024 (1980).
[44] G. F. Combs, Jr. and F. J. Peterson, *J. Nutr.* **113**, 538 (1983).
[45] J. A. Awad, R. F. Burk, and L. J. Roberts, *J. Pharmacol. Exp. Ther.* **270**, 858 (1994).
[46] S. D. Mercurio and G. F. Combs, Jr., *J. Nutr.* **116**, 1726 (1986).
[47] S. D. Mercurio and G. F. Combs, Jr., *Biochem. Pharmacol.* **35**, 4505 (1986).
[48] S. Z. Cagen and J. E. Gibson, *Toxicol. Appl. Pharmacol.* **40**, 193 (1977).
[49] W. H. Cheng, Y.-S. Ho, B. A. Valentine, D. A. Ross, G. F. Combs, Jr., and X. G. Lei, *J. Nutr.* **128**, 1070 (1998).

the selenium-deficient and selenium-adequate wild-type mice survived 4.6 and 69 hr, respectively.[49] Apparently, dietary selenium supplementation in the wild-type mice prolonged their survival time by 14-fold, but had only a marginal effect in the GPX1(–/–) mice. On the other hand, the GPX1 knockout caused a 10-fold reduction in survival time only in the presence of adequate dietary selenium that supported the full expression of GPX1 in the wild-type mice. In fact, the survival time of mice was mainly dependent on tissue GPX1 activity rather than body selenium status. The diquat injection also produced 100% mortality in all mice except for the selenium-adequate wild-type group, which survived until day 7 when all the mice were terminated to collect tissues for analysis.[38] The selenium-deficient wild-type and selenium-adequate GPX1(–/–) mice had similar survival times (4.1 and 3.9 hr, respectively), which were longer ($p < 0.05$) than that of the selenium-deficient GPX1(–/–) mice (2.4 hr). These three GPX1-deficient groups had higher levels of carbonyl (an indicator of protein oxidation)[50] and/or plasma alanine transaminase (ALT, EC 2.6.1.2; an indicator of liver injury and detected with a kit from Sigma) activities than in selenium-adequate wild-type mice. Similar impacts of GPX1 knockout on mouse survival were shown by another group.[32] It is apparent that GPX1 is not only protective against this type of acute lethal oxidative stress induced by high levels of paraquat or diquat, but is also the major metabolic mediator of body selenium for the protection. The contribution of other selenium-dependent proteins or factors to this protection, as shown by the differences in survival time and oxidative injury between the selenium-adequate and the selenium-deficient GPX1(–/–) mice, is relatively small. Overall, GPX1 plays an antioxidant role *in vivo* and its normal expression is essential for mice to cope with drastic oxidative attack. Furthermore, results from our GPX1(+) mouse study[49] indicate that overexpression of GPX1 above the physiological level is beneficial in this regard. After the selenium-adequate GPX1(+) and wild-type mice were given an intraperitoneal injection of paraquat (125 mg/kg), the survival time of the former was 10 times as long as for the latter (59 vs. 5.8 hr, $p < 0.001$). Thiobarbituric acid-reacting substances (TBARS) in postmortem liver of GPX1(+) mice was only 15% of that in wild-type mice. In a comparative sense, GPX1 overexpression helps mice protect against paraquat toxicity as does ebselen, a GPX mimetic, in chicks.[47]

Biochemical Mechanism of GPX1 Protection

There was no distinct histopathology of typical paraquat toxicity in the GPX1 (–/–) and the selenium-deficient wild-type mice that died shortly after challenge.[49] Instead, severe acute pulmonary interstitial necrosis was found only in the GPX1(+) and the selenium-adequate wild-type mice that had an extended survival time. In

---

[50] R. L. Levine, J. A. Williams, E. R. Stadtman, and E. Shacter, *Methods Enzymol.* **233**, 346 (1994).

addition, there was no simple correlation between hepatic TBARS content and mouse survival time. Presumably, these mice were killed by paraquat through metabolic disruptions at an early stage before the development of tissue lesions. As a sharp decrease in ratios of NADPH to NAPD in the lung of rats was seen 60 min after an intravenous injection of paraquat (40 mg/kg),[51] we compared the time course of the paraquat-induced depletion of NADPH and NADH between GPX1(−/−) and wild-type mice (3 weeks old) fed the selenium-deficient (<0.02 mg/kg) or the selenium-adequate (0.4 mg/kg) diet for 7 weeks.[40] After an intraperitoneal injection of paraquat (50 mg/kg), mice were killed to collect tissues at 0 (saline-injected controls), 0.5, 1, 2, 3, or 4 hr ($n = 3$). Concentrations of NADPH, NADH, NADP, and NAD in lung and liver were assayed by high-performance liquid chromatography (a 501 pump, a 712 wisp, and a 490E UV detector with an interface module; Waters, Milford, MA).[52] NADPH/NADP and NADH/NAD ratios in lung were abruptly reduced by 50–75% only 0.5 hr after the injection in all groups, and thereafter exhibited no further decrease or group difference. Hepatic NADH/NAD ratios in the selenium-adequate wild-type mice were 20–30% higher than those of the GPX1(−/−) and the selenium-deficient wild-type mice between 2 and 4 hr after the injection. While the hepatic NADPH/NADP ratios remained 70–80% of the initial level in the selenium-adequate wild-type mice from 1 to 4 hr, the other three groups showed linear decreases to only 20% of the initials over this period. Thus, our data provide direct evidence of protection by GPX1 against the paraquat-mediated oxidation of NADPH and NADH. Consistently, the depletion profiles of NADPH and NADH in liver paralleled the survival time pattern of mice given the same stress. Most likely, GPX1 deficiency aggravates the paraquat-induced depletion of cellular reducing equivalents and disrupts the redox status and many NADPH-dependent pathways, causing sudden death before the formation of tissue lesions. Nonetheless, a defined role of GPX1 in protecting against oxidation of NADPH in lung, the primary target of paraquat, needs further clarification with appropriate doses of the compound other than 50 mg/kg, which produced a drastic oxidative stress exceeding the total defense of even selenium-adequate wild-type mice.

Because total F_2-isoprostane content is considered a reliable and sensitive marker of *in vivo* lipid peroxidation,[53] and carbonyl content is used as an indicator of protein oxidation,[50] we detected the time course of the formation of these oxidative products, along with plasma ALT activity, after paraquat injection.[40] Pulmonary F_2-isoprostanes and hepatic carbonyl formation peaked at 1 hr in the GPX1(−/−) and the selenium-deficient wild-type mice. No such induction by paraquat was shown in any tissue of the selenium-adequate wild-type mice

[51] H. Witschi, S. Kacew, K.-I. Hirai, and M. G. Côté, *Chem. Biol. Interact.* **19**, 143 (1977).
[52] T. F. Kalhorn, K. E. Thummel, S. D. Nelson, and J. T. Slattery, *Anal. Biochem.* **151**, 343 (1985).
[53] J. D. Morrow and L. J. Roberts II, *Methods Enzymol.* **233**, 163 (1994).

throughout. Whereas F_2-isoprostane formation was accelerated by both GPX1 knockout and selenium deficiency in liver, it was not significantly elevated by paraquat treatment in the brain of any group. In a similar study[38] in which mice were killed 0, 1, and 3 hr after an intraperitoneal injection of diquat (24 mg/kg), formation of hepatic F_2-isoprostanes peaked at 1 hr in the GPX1(−/−) and the selenium-deficient wild-type mice and remained high at 3 hr. The peak of hepatic lipid peroxidation preceded the rise in plasma ALT in the selenium-adequate GPX1(−/−) mice. Diquat did not cause any rise in hepatic F_2-isoprostane or carbonyl content in the selenium-adequate wild-type mice, so that both measures were significantly different between these animals and the three groups of GPX1-deficient mice. In summary, GPX1 protects against pro-oxidant-induced oxidative destruction of lipids and protein *in vivo*, although it is still unclear to us whether these changes are primarily responsible for the distinct differences in survival rate and time between the GPX1(−/−) and the wild-type mice after the challenge.[54,55]

Notably, the paraquat injection resulted in a linear increase from 0 to 2 hr and then a decrease in total GPX activity in the lungs of the selenium-adequate wild-type mice.[40] Western analysis with an antibody against the human GPX1 protein and GPX4 activity assay indicated that these GPX activity changes[32] were mainly due to fluctuation of pulmonary GPX1 expression. However, the diquat treatment did not affect liver total GPX or TR activity, but was associated with a linear increase over time in liver GPX4 activity in the selenium-adequate GPX1(−/−) mice.[38] Paraquat caused a decrease in lung total SOD activity[40] whereas diquat produced an increase in liver total SOD activity in the GPX1(−/−) and the selenium-deficient wild-type mice,[38] compared with that of the selenium-adequate wild-type mice. Hepatic catalase activities were not affected by either the GPX1 knockout, the selenium deficiency, or the diquat stress.[38] As compensatory responses of GPX1 to oxidant stress have been seen in Cu,Zn-SOD-overexpressing cells,[56] the inconsistent changes in total SOD activity by paraquat and diquat stresses in GPX1-deficient mice warrants further study.

Interactions of GPX1 and Vitamin E

Because GPX1 is the most abundant selenoenzyme in mice, the GPX1(−/−) mice are an excellent model with which to study the metabolic relationship between selenium and vitamin E. Initially, we found that feeding the GPX1(−/−) and the wild-type mice with or without all-*rac*-α-tocopheryl acetate (20 mg/kg) had no impact on mortality or survival time after the intraperitoneal injection of paraquat (50 mg/kg), despite a significant effect on liver TBARS content.[49] To determine whether high levels of dietary vitamin E (up to 100-fold of nutritional

[54] C. Steffen and K. J. Netter, *Toxicol. Appl. Pharmacol.* **47**, 593 (1979).
[55] H. Shu, R. E. Talcott, S. A. Rice, and E. T. Wei, *Biochem. Pharmacol.* **28**, 327 (1979).
[56] P. Amstad, R. Moret, and P. Cerutti, *J. Biol. Chem.* **269**, 1606 (1994).

needs) could act in place of GPX1 against this lethal oxidative stress, we fed the GPX1(–/–) and the wild-type mice with the selenium-adequate (0.4-mg/kg) diet plus 0, 75, 750, or 7500 mg of all-*rac*-α-tocopheryl acetate for 5 weeks before an intraperitoneal injection of paraquat (50 mg/kg).[57] Once again, the stress resulted in 8- to 15-fold differences in survival times between these two types of mice, independent of dietary vitamin E levels. Although increasing tocopheryl acetate from 0 to 750 mg/kg extended the survival time of the GPX1(–/–) mice for 2 hr, the highest tocopheryl acetate level resulted in a decrease in survival time in the wild-type mice. In a following time-course study, mice were fed the diet plus 0 or 750 mg of all-*rac*-α-tocopheryl acetate per kilogram for 5 weeks and were killed 1 or 3 hr after an intraperitoneal injection of diquat at 12, 24, or 48 mg/kg.[57] Whereas the diquat-induced formation of hepatic F_2-isoprostanes was attenuated by the vitamin E supplementation, the diquat-induced increases in plasma ALT activities were enhanced by the GPX1 knockout irrespective of the dietary vitamin E levels. Therefore, high levels of dietary vitamin E in both studies did not replace the protection of GPX1 against the pro-oxidant-induced lethality or tissue injury in mice, despite a potent inhibition of hepatic lipid peroxidation. It is inferable that GPX1 played roles other than in preventing the pro-oxidant-induced lipid peroxidation to extend the survival time of the wild-type mice. The inhibition on the pro-oxidant-induced rise of plasma ALT activity is one of those roles of GPX1 that is not shared by vitamin E. Limited protection of vitamin E against diquat-induced cytotoxicity[58,59] or acute oral paraquat lethality in chicks[44] has been previously reported. It is worth mentioning that liver GPX4 activities were greater at 750 and 7500 mg than at 0 and 75 mg of all-*rac*-α-tocopheryl acetate per kilogram in both GPX1(–/–) and wild-type mice killed by paraquat at 50 mg/kg.[57] Moreover, hepatic GPX4 activities were significantly decreased by various doses of diquat only in the vitamin E-deficient GPX1(–/–) mice. Previously, addition of tocopherol to peroxidizing microsomes was shown to enhance GPX4 activity.[60] Our results indicate that high levels of dietary vitamin E in GPX1(–/–) mice or normal expression of GPX1 in vitamin E-deficient wild-type mice may "spare" liver GPX4 activity under oxidative stress. No such changes in liver TR activity were observed.

Protection by Glutathione Peroxidase against Moderate and Metabolic Oxidative Stress

    In contrast to the remarkable effects of GPX1 expression on lethal oxidative stress, results concerning the necessity for GPX1 in protecting against moderate or metabolic oxidative stress are highly variable. While most groups noticed

[57] W. H. Cheng, B. A. Valentine, and X. G. Lei, *J. Nutr.* **129**, 1951 (1999).
[58] L. Eklow-Lastbom, L. Rossi, H. Thor, and S. Orrenius, *Free Radic. Res. Commun.* **2**, 57 (1986).
[59] M. S. Sandy, D. Di Monte, and M. T. Smith, *Toxicol. Appl. Pharmacol.* **93**, 288 (1988).
[60] M. Maiorino, M. Coassin, A. Roveri, and F. Ursini, *Lipids* **24**, 721 (1989).

little effect of GPX1 knockout or overexpession on the growth, reproduction, and well-being of adult mice,[31,32,35–37] one group, which developed their own GPX1(–/–) line, reported a 20% reduction in body weight in these mice compared with the wild-type controls.[33] They suggested that the growth retardation was presumably due to increased mitochondrial oxidative stress that decreases energy production. No increased sensitivity to hyperoxia was observed by Ho et al.[31] in GPX1(–/–) mice, whereas the same mice were shown to be more susceptible to myocardial ischemia reperfusion injury[61] and induced apoptosis.[62] Likewise, two groups, using the same GPX1(–/–) and GPX1(+) mouse lines, came to different views on the role of GPX1 in protecting against peroxide-induced damage in lenses.[34,63,64] Several groups demonstrated clear impacts of GPX1 knockout on neuron susceptibility to pro-oxidant drugs,[32,65,66] but one group failed to observe such impacts on dopaminergic neuronal responses to glutathione depletion.[67] Despite being the sole GPX in erythrocytes, GPX1 knockout had little effect on the defense of red blood cells against exogenous peroxides, even at high concentrations.[68] However, the GPX1 protection of red blood cell membrane integrity became important when severe hypotonic stress was employed after an in vivo oxidative insult.[69] In our early studies,[35,36] the GPX1-altered mice exhibited no change in susceptibility to dietary selenium and vitamin E deficiency. However, the selenium-adequate GPX1(–/–) mice developed myocarditis, just as did the selenium-deficient wild-type mice, after infection with benign coxsackievirus B3, indicating an association between host GPX1 expression and viral genome mutations.[70] Also, the GPX1(–/–) mice were more susceptible to neutrophil-mediated hepatic parenchymal cell injury during endotoxemia.[71]

Obviously, there is no simple explanation for the above-described discrepancies. One possibility lies in the existence of a critical threshold of oxidative stress that turns on the action of GPX1.[72] To determine such a threshold for the GPX1

[61] T. Yoshida, N. Maulik, R. M. Engelman, Y. S. Ho, J.-L. Magnenat, J. A. Rousou, J. E. Flack III, D. Deaton, and D. K. Das, Circulation 96, II-20 (1997).

[62] N. Maulik, T. Yoshida, and D. K. Das, Mol. Cell. Biochem. 196, 13 (1999).

[63] V. N. Reddy, L. R. Lin, Y. S. Ho, J.-L. Magnenat, N. Ibaraki, F. J. Giblin, and L. Dang, Ophthalmologica 211, 192 (1997).

[64] A. Spector, W. Ma, R.-R. Wang, Y. Yang, and Y. S. Ho, Exp. Eye Res. 64, 477 (1997).

[65] P. Klivenyi, O. A. Andreassen, R. J. Ferrante, A. Dedeoglu, G. Mueller, E. Lancelot, M. Bogdanov, J. K. Andersen, D. Jiang, and M. F. Beal, J. Neurosci. 20, 1 (2000).

[66] J. Zhang, D. G. Graham, T. J. Montine, and Y. S. Ho, J. Neuropathol. Exp. Neurol. 59, 53 (2000).

[67] K. Nakamura, D. A. Wright, T. Wiatr, D. Kowlessur, S. Milstien, X. G. Lei, and U. J. Kang, J. Neurochem. 74, 2305 (2000).

[68] R. M. Johnson, G. Goyette, Jr., Y. S. Ho, Blood 96, 1985 (2000).

[69] T. M. Mullins and X. G. Lei, FASEB J. 14, A538 (2000).

[70] M. A. Beck, R. S. Esworthy, Y. S. Ho, and F. F. Chu, FASEB J. 12, 1143 (1998).

[71] H. Jaeschke, Y. S. Ho, M. A. Fisher, J. A. Lawson, and A. Farhood, Hepatology 29, 443 (1999).

[72] O. Toussaint, A. Houbion, and J. Remacle, Toxicology 81, 89 (1993).

function, we injected (intraperitoneally) selenium-adequate GPX1(–/–) and wild-type mice with diquat at 6, 12, 24, or 48 mg/kg of body weight.[39] Whereas the wild-type mice survived all four doses, the GPX1(–/–) mice survived only 6 mg of diquat per kilogram and died at diquat doses of 12, 24, and 48 mg/kg at 52, 4.4, and 3.9 hr, respectively. Compared with those of surviving mice that were terminated on day 7, the dead GPX1(–/–) mice had diquat dose-dependent increases in plasma ALT activity and higher liver carbonyl content. Thus, GPX1 protection is essential only when the selenium-adequate mice are exposed to oxidative stress induced by diquat at 12 mg/kg and higher. As shown previously,[73] the diquat challenge also produced a dose-dependent reduction in total GPX activity and GPX1 protein in liver of the wild-type mice. However, it had no consistent effect on liver GPX4, TR, or catalase activity.

Similarly, both the selenium-adequate GPX1(–/–) and wild-type mice survived for at least 2 weeks after an intraperitoneal injection of paraquat (12.5 mg/kg).[49] When these animals were terminated for tissue collection, there was no rise in plasma ALT activity or overt injuries in liver, lung, and kidney in either group. Again, the selenium-adequate GPX1(–/–) mice are able to tolerate this mild oxidative stress induced by the low level of paraquat, consistent with the results of another group.[32]

Strikingly, the impacts of GPX1 knockout are highly visible in selenium-deficient mice exposed to moderated oxidative stress. After weanling GPX1(–/–) and wild-type mice were fed a selenium-deficient diet (<0.02 mg/kg) or selenium-adequate diet (0.40 mg/kg) for 5 weeks, an intraperitoneal injection of paraquat at 12.5 mg/kg of body weight produced a higher mortality rate (67 vs. 33%) in the selenium-deficient GPX1(–/–) (10 of 15) than that in the selenium-deficient wild-type mice (5 of 15).[74] All the selenium-adequate mice, regardless of their GPX1 levels, survived as described above.[49] Postmortem histology indicated that the selenium-deficient GPX1(–/–) mice had much more pronounced acute centrilobular vacuolar necrosis of liver, acute renal necrosis of proximal convoluted tubules, and apoptosis than the selenium-deficient wild-type mice.[75] As there was no difference in liver GPX4 activities and presumably in expression of other selenoproteins between these two groups, their mortality and histopathological differences were directly related to their difference in total GPX activity. In subsequent experiments,[76] both these selenium-deficient groups were injected with selenium at 50 μg/kg 3, 6, and 12 hr before the paraquat injection. Depending on the selenium injection time, it reduced the mortality of only the wild-type mice and/or

[73] P. Garberg and M. Thullberg, *Chem. Biol. Interact.* **99**, 165 (1996).

[74] W. H. Cheng, Y. X. Fu, and X. G. Lei, *FASEB J.* **14**, A538 (2000).

[75] X. G. Lei and W. H. Cheng, unpublished data (2001).

[76] X. G. Lei, W. H. Cheng, Y. X. Fu, J. M. Porres, T. M. Mullins, and C. A. Roneker, "Selenium 2000, 7th International Symposium on Selenium in Biology and Medicine," Venice, Italy, October 1–5, p. 92, 2000.

extended their survival time much longer than that of the GPX1(–/–) mice (5 to 56 vs. 5 to 12 hr). The selenium injection resulted in a similar rise in liver GPX4 activity between both groups, but raised total liver GPX activity in the wild-type mice to be significantly higher than in the GPX1(–/–) mice. Consequently, the selenium injection postponed the onset of apoptosis and reduced the severity in wild-type mice compared with GPX1(–/–) mice. Amazingly, residual GPX1 activity in the selenium-deficient or the partially selenium-replete wild-type mice, even at as low as only 3% of the normal physiological level in the selenium-adequate animals, confers significant protection against apoptosis[75] and death induced by 12.5 mg of paraquat per kilogram. Comparatively, this type of GPX1 role is not absolutely indispensable or easily detectable in the selenium-adequate animals.

Unanswered Questions

Although convincing direct evidence of an *in vivo* antioxidant role for GPX1 has been generated by us and others using GPX1(+) and GPX1(–/–) mice, the physiological importance of its role varies with particular circumstances. When animals are exposed to high levels of oxidative stress, normal expression of GPX1 is essential and overexpression of GPX1 is beneficial to protect against the induced lethality and oxidative destruction of NADPH, NADH, lipid, and protein. Other selenoproteins and antioxidant enzymes or high levels of dietary vitamin E cannot replace GPX1 in these functions. In selenium-inadequate animals with compromised antioxidant defense, minute residual GPX1 activity confers significant protection against lethality and apoptosis induced by mild oxidative stress. In contrast, selenium-adequate animals do not seem to require GPX1 expression to cope with "normal" metabolic or mild oxidative stress. However, GPX1 knockout attenuates body defense against viral infection[70] and endotoxin[71] in animals fed adequate dietary selenium and vitamin E.

Undoubtedly, these GPX1-altered mice have enabled us to make significant progress in understanding the metabolic role of GPX1. However, there are many unanswered questions. First, what is the exact mechanism by which GPX1 reduces mortality and extends, survival time of mice challenged with high doses of paraquat and diquat? Clearly, GPX1 is not redundant under these circumstances, and it would be useful to find out whether the knockout of other selenoproteins such as GPX4 or selenoprotein P exerts impacts similar to those of GPX1. The comparisons will teach us whether the role of GPX1 is special or just a part of the total body defense in antioxidation. Second, full functions of GPX1 are clearly dependent on relative balances of oxidative stress and antioxidant defense in the body. The question is, how does the body sense a need for GPX1 function and turn it on? We are studying the impacts of GPX1 knockout on stress-related signaling,[77] and hope to relate the signal cascades of cell death and GPX1 actions. Third, reactive nitrogen species (RNS), like reactive oxygen species (ROS), are major players in many physiological and pathological processes, but research on the role of GPX1 in detoxifying RNS has been scarce.[78] Using primary hepatocytes isolated from GPX1(–/–) and wild-type mice, we have shown that knockout of GPX1 rendered the cells more resistant to apoptosis induced by peroxynitrite.[77] This unexpected positive effect of GPX1 knockout has also been shown in mouse brain resistance to induced epileptic seizure.[79] Experiments should be designed to study the interaction of GPX1, ROS, and RNS under physiological conditions. Finally, the long-term impacts of GPX1 knockout, such as on aging and chronic diseases, should be investigated.

The development of GPX1(–/–) and GPX1(+) mice has created an exciting new field of GPX1 functional genomics, and the successful illustration of the *in vivo* antioxidant function of GPX1 bears significant biomedical implications. Nutritionally, these data justify the current use of tissue GPX1 activity for the assessment of dietary selenium needs and body selenium status.[80] The remarkable benefit of GPX1 overexpression in protecting oxidative stress indicates a potential of GPX1 mimetic, such as ebselen,[81] in clinical therapy. Understanding the relationships of GPX1 function, signaling, and metabolic disorders will help us in developing novel diagnosis and treatment for certain diseases.

Acknowledgment

The research conducted in the author's laboratory is supported by NIH Grant DK53018.

[77] Y. X. Fu, H. Sies, and X. G. Lei, *J. Biol. Chem.* **276,** in press (2001).

[78] H. Sies, V. S. Sharov, L. O. Klotz, and K. Briviba, *J. Biol. Chem.* **272,** 27812 (1997).

[79] D. Jiang, G. Akopian, Y. S. Ho, J. P. Walsh, and J. K. Andersen, *Exp. Neurol.* **164,** 257 (2000).

[80] O. A. Levander and P. D. Whanger, *J. Nutr.* **126,** 2427S (1996).

[81] G. Tiegs, S. Kusters, G. Kunstle, H. Hentze, A. K. Kiemer, and A. Wendel, *J. Pharmacol. Exp. Ther.* **287,** 1098 (1998).

[20] Recombinant Expression of Mammalian Selenocysteine-Containing Thioredoxin Reductase and Other Selenoproteins in *Escherichia coli*

By ELIAS S. J. ARNÉR

Introduction

Heterologous expression of recombinant proteins in *Escherichia coli* for use in basic research studies, as therapeutics, in diagnostics, or as research reagents, is a much utilized technique in molecular biology with high protein yield and a well-characterized production system. Recombinant protein production will also play a significant role in the analysis of the vast number of candidate proteins (the "proteomes") predicted from the many diverse ongoing genomic sequencing programs. Many organisms of all branches of life carry proteins that contain selenocysteine, the 21st amino acid. Examples of these selenoproteins are the three formate dehydrogenase isoenzymes of *E. coli* or the mammalian selenoprotein P, glutathione peroxidases, thyroid hormone deiodinases, and thioredoxin reductases; about 20 mammalian selenoproteins have been characterized.[1–5] The list of newly discovered naturally occurring selenoproteins will continue to increase, either through database searches and predictions or as the result of experimental screening approaches.[6–8] The selenoproteins have until more recently generally been excluded from conventional production based on recombinant expression in bacteria, because of highly species-specific machineries for cotranslational insertion of the reactive selenocysteine residue. Yet, if the species barriers are circumvented, the advantages and potential applications of targeted selenocysteine (Sec) insertion into recombinant proteins would include (1) use of conventional recombinant methodology for studies of naturally occurring selenoproteins, (2) the possibility of introducing selenocysteine into proteins for improving phase determination in X-ray crystallography or as a method for studying protein folding

---

[1] E. S. J. Arnér and A. Holmgren, *Eur. J. Biochem.* **267,** 6102 (2000).

[2] A. Böck, K. Forchhammer, J. Heider, W. Leinfelder, G. Sawers, B. Veprek, and F. Zinoni, *Mol. Microbiol.* **5,** 515 (1991).

[3] A. Hüttenhofer and A. Böck, *in* "RNA Structure and Function" (R. W. Simons, and M. Grunberg-Manago, eds.), p. 603. Cold Spring Harbor Laboratory Press, Cold Spring Harbor, New York, 1998.

[4] S. C. Low and M. J. Berry, *Trends Biochem. Sci.* **21,** 203 (1996).

[5] T. C. Stadtman, *Annu. Rev. Biochem.* **65,** 83 (1996).

[6] A. Lescure, D. Gautheret, P. Carbon, and A. Krol, *J. Biol. Chem.* **274,** 38147 (1999).

[7] G. V. Kryukov, V. M. Kryukov, and V. N. Gladyshev, *J. Biol. Chem.* **274,** 33888 (1999).

[8] G. V. Kryukov and V. N. Gladyshev, *Genes Cells* **5,** 1049 (2000).

0076-6879/02 $35.00

(reviewed in Besse *et al.*[9]), (3) targeting of Sec into proteins, making possible the insertion of high-energy selenium isotopes and constituting an alternative method of radiolabeling proteins with high specific activity, and (4) the possible use of selenium isotope-labeled proteins in positron emission tomography (PET) studies (discussed in Bergmann *et al.*[10]).

The intricate synthesis machineries of selenoproteins in diverse organisms have been described in detail in several reviews[2–5] as well as in other articles in this volume of *Methods in Enzymology*. The reader is therefore referred to those reviews and the mechanisms of selenoprotein synthesis shall not be repeated here at length. However, we should note that selenocysteine is not stable in its free form because of its high reactivity and the residue is in all organisms inserted cotranslationally at the position of an opal (UGA) codon, which normally confers termination of translation. The UGA codon is encoded as selenocysteine by highly complex translation machineries, which differ between gram-negative[3,5] and gram-positive bacteria,[11] archaea[12] and higher eukaryotes.[13–15] The synthesis machinery in *E. coli* has been characterized in detail mainly through the work of Böck and co-workers, using synthesis of formate dehydrogenase H as the main model system.[2,3,5] The mRNA for the *E. coli* selenoprotein contains a specific sequence immediately following the UGA codon, with the nucleotides in this sequence both providing the information for translation of the amino acids following the selenocysteine residue, and forming a stem–loop secondary structure—a SECIS (selenocysteine insertion sequence) element. The SECIS element binds the SelB protein, the *selB* gene product. SelB is the selenocysteine-specific elongation factor, which is homologous to elongation factor EF-Tu but, in addition, has a carboxy-terminal domain that binds the SECIS element of the selenoprotein mRNA. SelB also binds a selenocysteine-specific tRNA (tRNA[Sec]), the *selC* gene product, in its selenocysteinylated form. In analogy with EF-Tu, SelB may then catalyze selenocysteine insertion at the ribosome, which occurs under hydrolysis of GTP and insertion of selenocysteine at the specific position of the selenocysteine-encoding UGA codon. The tRNA[Sec] is originally charged with a seryl residue which by selenocysteine synthase, the *selA* gene product, is converted to selenocysteinyl, utilizing selenophosphate as the selenium donor. The selenophosphate, in turn, is provided by

[9] D. Besse, N. Budisa, W. Karnbrock, C. Minks, H. J. Musiol, S. Pegoraro, F. Siedler, E. Weyher, and L. Moroder, *Biol. Chem.* **378,** 211 (1997).

[10] R. Bergmann, P. Brust, G. Kampf, H. H. Coenen, and G. Stocklin, *Nucl. Med. Biol.* **22,** 475 (1995).

[11] T. Gursinsky, J. Jäger, J. R. Andreesen, and B. Söhling, *Arch. Microbiol.* **174,** 200 (2000).

[12] M. Rother, R. Wilting, S. Commans, and A. Böck, *J. Mol. Biol.* **299,** 351 (2000).

[13] P. R. Copeland, V. A. Stepanik, and D. M. Driscoll, *Mol. Cell. Biol.* **21,** 1491 (2001).

[14] M. T. Nasim, S. Jaenecke, A. Belduz, H. Kollmus, L. Flohé, and J. E. McCarthy, *J. Biol. Chem.* **275,** 14846 (2000).

[15] D. Fagegaltier, N. Hubert, K. Yamada, T. Mizutani, P. Carbon, and A. Krol, *EMBO J.* **19,** 4796 (2000).

selenophosphate synthetase, the *selD* gene product. Taken together, selenocysteine insertion during selenoprotein translation in *E. coli* involves a SelB-compatible SE-CIS element next to the UGA codon in the selenoprotein mRNA, and the *selA, selB, selC,* and *selD* gene products.

A SECIS element is also present in mammalian selenoprotein mRNA but has other secondary structural features than those found in *E. coli* and, moreover, is situated in the 3' untranslated region several hundred nucleotides downstream of the selenocysteine-encoding UGA.[4,16,17] Also archaea[18] and gram-positive bacteria[11] have characteristic SECIS elements, with the latter being somewhat reminiscent of the *E. coli* SECIS element.[11] It should therefore be emphasized that although the specific secondary structures of mRNA directing selenocysteine insertion may be called SECIS elements in all organisms, in analogy to the nomenclature of the mammalian system,[4] the location and secondary structures of the different SECIS elements in diverse organisms are generally not similar to, or compatible with, each other. The differences between SECIS elements are the basis for species barriers in recombinant selenoprotein synthesis. In this article the critical factors for successfully by-passing the species barrier in order to enable heterologous selenoprotein production in *E. coli* are discussed with a focus on the possible general use of the technique.

tRNA^{Sec} Defining the Selenoprotein World

The specific tRNA for selenocysteine insertion, tRNA^{Sec} with it UCA anti-codon, may be viewed as the common denominator of the selenoprotein world. Homozygous tRNA^{Sec}-deficient mice display early embryonic death,[19] indicating vital functions of one or several mammalian selenoproteins. With the number of canonical tRNAs identified in sequence-determined genomes ranging from 36 for *Methanococcus jannaschii* to 497 for humans or 584 for *Caenorhabditis elegans* it is interesting to note that these organisms (human, *C. elegans, Drosophila melanogaster, M. Jannaschii,* and *E. coli*) all seem to rely on a single tRNA^{Sec} gene for selenoprotein synthesis.[20] Moreover, as a common denominator, tRNA^{Sec} from several species can complement an *E. coli* mutant deficient in tRNA^{Sec}.[11,21–23]

[16] L. Zhong, E. S. J. Arnér, J. Ljung, F. Åslund, and A. Holmgren, *J. Biol. Chem.* **273**, 8581 (1998).

[17] R. Walczak, E. Westhof, P. Carbon, and A. Krol, *RNA* **2**, 367 (1996).

[18] M. Rother, A. Resch, W. L. Gardner, W. B. Whitman, and A. Böck, *Mol. Microbiol.* **40**, 900 (2001).

[19] M. R. Bosl, K. Takaku, M. Oshima, S. Nishimura, and M. M. Taketo, *Proc. Natl. Acad. Sci. U.S.A.* **94**, 5531 (1997).

[20] International Human Genome Sequencing Consortium, *Nature (London)* **409**, 860 (2001).

[21] J. Heider, W. Leinfelder, and A. Böck, *Nucleic Acids Res.* **17**, 2529 (1989).

[22] P. Tormay, R. Wilting, J. Heider, and A. Böck, *J. Bacteriol.* **176**, 1268 (1994).

[23] C. Baron, C. Sturchler, X. Q. Wu, H. J. Gross, A. Krol, and A. Böck, *Nucleic Acids Res.* **22**, 228 (1994).

As long as a tRNA^{Sec} species is expressed at a balanced level, its function or specificity thereby is not generally a factor, making heterologous selenoprotein synthesis in *E. coli* a challenging task.

Species Barriers in Heterologous Selenoprotein Synthesis

It was recognized early on that the TGA in a mammalian glutathione peroxidase cDNA could not direct selenocysteine incorporation at the corresponding UGA when it was expressed in *E. coli*[24] and the same incompatibility was found for glycine reductase of *Clostridium sticklandii*.[25] *Methanobacterium formicicum* has a cysteine homolog of the selenocysteine-containing *E. coli* formate dehydrogenase H. In 1992, Heider and Böck used this gene to demonstrate successful targeted selenocysteine insertion in *E. coli* by changing the UGC, encoding cysteine in the native *M. formicicum* gene, to UGA and introducing an *E. coli* SECIS element from the homologous formate dehydrogenase gene immediately following the UGA.[26] They had at that time identified features of the *E. coli* formate dehydrogenase SECIS element necessary for directing selenocysteine insertion,[27,28] which was the basis for those studies. Details of the barriers to heterologous expression of a selenoprotein gene in bacteria were subsequently analyzed by Tormay and Böck, demonstrating that not only must the heterologous SECIS element be compatible with the host selenoprotein synthesis machinery; the barrier can also not be overcome by concomitantly introducing additional heterologous selenocysteinyl-tRNA and SelB, most likely due to incompatibility between the heterologous SelB and the *E. coli* ribosome, as demonstrated by expressing the *hydV* gene from *Desulfomicrobium baculatum* in *E. coli*.[29] The general capacity for synthesizing recombinant selenoproteins in *E. coli* was, however, demonstrated in an overproduction of the endogenous *E. coli* formate dehydrogenase H.[30] That study demonstrated that the bacteria indeed possessed an inherent capacity for high-level selenoprotein production, but the production was found only under anaerobic conditions. The reason for strict anaerobic expression should have depended on use of the endogenous formate dehydrogenase H promoter kept in the plasmid construct[30] because subsequent findings show that the capacity for selenoprotein production in *E. coli* is indeed also efficient under aerobic conditions.

[24] C. Rocher, C. Faucheu, F. Herve, C. Benicourt, and J. L. Lalanne, *Gene* **98**, 193 (1991).

[25] G. E. Garcia and T. C. Stadtman, *J. Bacteriol.* **174**, 7080 (1992).

[26] J. Heider and A. Böck, *J. Bacteriol.* **174**, 659 (1992).

[27] J. Heider, C. Baron, and A. Böck, *EMBO J.* **11**, 3759 (1992).

[28] F. Zinoni, J. Heider, and A. Böck, *Proc. Natl. Acad. Sci. U.S.A.* **87**, 4660 (1990).

[29] P. Tormay and A. Böck, *J. Bacteriol.* **179**, 576 (1997).

[30] G. T. Chen, M. J. Axley, J. Hacia, and M. Inouye, *Mol. Microbiol.* **6**, 781 (1992).

Introduction of Selenocysteine Insertion Sequence Element Compatible with Bacterial Selenoprotein Synthesis Machinery as Method for Recombinant Selenoprotein Production

When mammalian thioredoxin reductase was found to be a selenoprotein with a selenocysteine residue positioned in a unique carboxy-terminal -Gly-Cys-Sec-Gly motif,[16,31] we decided to analyze whether it was possible to engineer a bacterial-type SECIS element in the rat cDNA for the enzyme, thereby enabling targeted insertion of selenocysteine at the correct position in the recombinant selenoprotein. As is discussed below, the penultimate position of the Sec residue is particularly suitable for this strategy. The study, producing recombinant selenocysteine-containing rat thioredoxin reductase, showed that *E. coli* also has a high inherent capacity for selenoprotein production under aerobic conditions; this was the first evidence that the bacteria could be utilized for production of a recombinant mammalian selenoprotein.[32] Other groups have since utilized the approach of engineering a bacterial-type SECIS element in a heterologous gene in order to enable expression in *E. coli,* demonstrating production of human thioredoxin reductase[33,34] or of a cysteine to selenocysteine-substituted plant glutathione peroxidase.[35] The critical factors in this production strategy are discussed in further detail below.

Critical Factors for Recombinant Selenoprotein Production in *Escherichia coli*

Functional Selenocysteine Insertion Sequence Element

It is clear that the SECIS element that is to be introduced into a gene for expression as a selenoprotein in *E. coli* must be functionally compatible with the *E. coli* SelB elongation factor. The determinants for this interaction have been analyzed in detail by Böck and co-workers, as reviewed,[3] and also further studied by others.[35–39] To summarize, native *E. coli* SECIS elements are found in the three genes encoding the only selenoproteins of *E. coli,* namely formate dehydrogenase H (FDH-H, encoded by the *fdhF* gene), N (*fdnG*), and O (being a formate

[31] V. N. Gladyshev, K.-T. Jeang, and T. C. Stadtman, *Proc. Natl. Acad. Sci. U.S.A.* **93,** 6146 (1996).

[32] E. S. J. Arnér, H. Sarioglu, F. Lottspeich, A. Holmgren, and A. Böck, *J. Mol. Biol.* **292,** 1003 (1999).

[33] S. Bar-Noy, S. N. Gorlatov, and T. C. Stadtman, *Free Radic. Biol. Med.* **30,** 51 (2001).

[34] R. Koishi, T. Nakamura, T. Takazawa, C. Yoshimura, and N. Serizawa, *J. Biochem. (Tokyo)* **127,** 977 (2000).

[35] S. Hazebrouck, L. Camoin, Z. Faltin, A. D. Strosberg, and Y. Eshdat, *J. Biol. Chem.* **275,** 28715 (2000).

[36] Z. Liu, M. Reches, and H. Engelberg-Kulka, *J. Mol. Biol.* **294,** 1073 (1999).

[37] Z. Liu, M. Reches, I. Groisman, and H. Engelberg-Kulka, *Nucleic Acids Res.* **26,** 904 (1998).

[38] K. E. Sandman and C. J. Noren, *Nucleic Acids Res.* **28,** 755 (2000).

[39] C. Li, M. Reches, and H. Engelberg-Kulka, *J. Bacteriol.* **182,** 6302 (2000).

oxidase and probably identical to FDH-Z encoded by *fdoG*[40,41]). Normally, FDH-H is expressed only under anaerobic conditions, FDH-N is induced anaerobically in the presence of nitrate, whereas the 110-kDa FDH-O is the only selenoprotein endogenously expressed in *E. coli* grown under aerobic conditions, during which it is expressed at low levels under conventional culture conditions.[40] The SECIS elements of FDH-O and FDH-N are identical whereas that of FDH-H (*fdhF*) differs slightly from these, with the latter being the most studied in terms of the functional determinants for interaction with the SelB elongation factor. The major determinants of the *fdhF* SECIS necessary for maintained function as an SECIS element in *E. coli* involve preservation of the loop region binding SelB, which must be positioned at a distance of about 11 nucleotides downstream of the Sec UGA codon. These determinants are schematically shown in Fig. 1, where it is also shown how a variant of this SECIS element could be engineered to maintain function yet enable production of rat thioredoxin reductase when it was fused to the open reading frame of that enzyme and expressed in *E. coli*.[32]

SECIS elements that can be introduced into a gene tailored for heterologous selenoprotein production in *E. coli* other than those suggested from Fig. 1 may also be possible. Such novel SECIS elements may be characterized and found either by general screening of UGA suppression or function in SelB binding, using randomized sequence elements,[42,43] or by combining the analysis of mutated SECIS elements with random mutagenesis of the SelB factor itself.[39,44] Moreover, in the production of a selenocysteine-containing variant of plant glutathione peroxidase, it was reported that the proximal base pairs of the conserved loop region, around the bulged U nucleotide next to the loop (see Fig. 1), may possibly be substituted and still have maintained function.[35] Use of such novel SECIS elements that deviate from the hitherto well-characterized consensus motifs (Fig. 1) in applications to achieve heterologous selenoprotein production should, however, be performed with great care to make certain that the supposedly improved system does not lead to suppression of the UGA codon by the insertion of other amino acids than selenocysteine (see also discussion on suppression below).

Importance of Correct Stoichiometry

The SelB elongation factor must form a quarternary complex with GTP, selenocysteinyl-tRNA[Sec], and the SECIS element of the mRNA, and must interact

[40] G. Sawers, J. Heider, E. Zehelein, and A. Böck, *J. Bacteriol.* **173**, 4983 (1991).
[41] H. Abaibou, J. Pommier, S. Benoit, G. Giordano, and M. A. Mandrand-Berthelot, *J. Bacteriol.* **177**, 7141 (1995).
[42] S. J. Klug, A. Hüttenhofer, M. Kromayer, and M. Famulok, *Proc. Natl. Acad. Sci. U.S.A.* **94**, 6676 (1997).
[43] S. J. Klug, A. Hüttenhofer, and M. Famulok, *RNA* **5**, 1180 (1999).
[44] M. Kromayer, B. Neuhierl, A. Friebel, and A. Böck, *Mol. Gen. Genet.* **262**, 800 (1999).

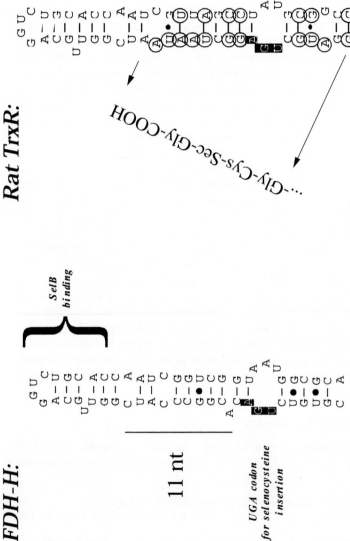

Rat TrxR:

FDH-H:

SelB
binding

11 nt

UGA codon
for selenocysteine
insertion

...-Gly-Cys-Sec-Gly-COOH

fdhF: ...-NNN-**TGA**-NNN-NNN-NNN-NNG-GTT-GCA-GGT-CTG-CAC-CNN-...
(fdnG: ...-NNN-**TGA**-NNN-NNN-NNN-NNG-GTA-GCA-AGT-CTT-GCT-CCN-...)

with the ribosome in order to catalyze a completed peptidyltransferase reaction with insertion of selenocysteine.[3] It is clear that the stoichiometry between the separate factors involved in this final step is of importance for resulting yield and efficiency in bacterial selenoprotein synthesis. Hence it was shown that sole overexpression of either SelB or tRNA[Sec], or of an mRNA carrying an SECIS element, may reduce the UGA readthrough capacity as a result of disturbances in this stoichiometry.[45] In essence, SelB may be tethered to nonfunctional binary or tertiary complexes with either tRNA[Sec] or the mRNA SECIS, thereby not forming the functional complete quarternary complex.[3,45] We also found that the total yield of the selenocysteine-containing rat thioredoxin reductase expressed from a strong T7 promoter-driven pET vector could be increased about 8-fold (from 0.6 to 5 mg of selenoprotein produced per liter of *E. coli* culture) when the bacteria were cotransformed with a plasmid carrying accessory *selA*, *selB*, and *selC* genes under the control of their endogenous promoters.[32] Use of titrable promoters guiding the expression level of the heterologous selenoprotein mRNA in order to achieve optimal stoichiometry and thereby total yield will possibly be shown as one method to further optimize this production system.

Selenium Metabolism and Non-Selenocysteine-Mediated UGA Suppression

Selenium is needed for selenoprotein production. However, selenium incorporation may also be unspecific, with selenium entering the cysteine or methionine pathways. This unspecific incorporation must be avoided in recombinant selenoprotein production and can be blocked by addition of excess L-cysteine or L-cystine to the growth medium.[46] It is also important to note that an adequate selenium supply is essential for correct selenoprotein synthesis. In the absence of selenium in the growth medium the translation will either be terminated at the position of the

---

FIG. 1. Scheme of the determinants for a functional *E. coli* SECIS element. *Top left:* Native SECIS element of formate dehydrogenase H (encoded by the *fdhF* gene). The determinants for function in directing targeted cotranslational selenocysteine insertion at the UGA codon are indicated. The corresponding DNA sequence necessary to preserve these functional determinants is given below the SECIS element, in essence illustrating a SECIS cassette that may be introduced into a gene in order to achieve heterologous expression as a recombinant selenoprotein in *E. coli*. *Top right:* An example of such an engineered SECIS variant, which was utilized for production of rat thioredoxin reductase (TrxR), with the carboxy-terminal motif being -Gly-Cys-Sec-Gly-COOH [E. S. J. Arnér, H. Sarioglu, F. Lottspeich, A. Holmgren, and A. Böck, *J. Mol. Biol.* **292**, 1003 (1999)]. It should also be noted that the stem of the SECIS element may be shortened or elongated with a few nucleotides and yet maintain function [J. Heider, C. Baron, and A. Böck, *EMBO J.* **11**, 3759 (1992)], and that the "stem" in fact does not need to base pair as long as the SelB-binding part is kept within a functional distance [Z. Liu, M. Reches, I. Groisman, and H. Engelberg-Kulka, *Nucleic Acids Res.* **26**, 904 (1998)]. Below the DNA sequence corresponding to the SECIS determinants of the *fdhF* gene is also that of the slightly different *fdnG* gene, which probably also would be functional for use in heterologous selenoprotein production, although this has not yet been demonstrated. See text for further details.

UGA or, more detrimental in the case of recombinant selenoprotein production, be suppressed by insertion of other amino acids than selenocysteine. It has been shown that the nucleotide immediately following the UGA codon may influence the extent of UGA suppression, with tryptophan most often suppressing the UGA when selenocysteine is not inserted,[38] although the native SECIS element or the sequence immediately upstream of the UGA in fact seems to prevent tryptophan suppression.[36] For practical applications in heterologous selenoprotein production it is important that as long as the selenium supply is sufficient and a proper SECIS element follows the UGA little or no non-selenocysteine-mediated UGA suppression can be noted.[36,38] The resulting recombinant selenoprotein expressed under optimal conditions thereby either contains selenocysteine at the position of the UGA or, alternatively, becomes truncated at the site of the UGA. This was illustrated in the detailed characterization by mass spectrometry of tryptic digests of the rat thioredoxin reductase produced in *E. coli,* in which no peptide masses corresponding to non-selenocysteine-mediated UGA suppression could be detected.[32]

Inherent Lack of Efficiency in Bacterial Selenocysteine Insertion

The UGA-directed selenocysteine insertion in *E. coli* is by nature inefficient, with a significant translational pause resulting from SelB binding to the SECIS element.[47] This inefficiency may possibly affect the yield of recombinant selenoproteins produced in *E. coli*. Using the same plasmid carrying accessory *selA, selB,* and *selC* genes as utilized by us in the production of rat thioredoxin reductase (see above), the amount of full-length selenocysteine-containing product was found to be 11% (with 89% truncated protein formed) compared with 4% without overproduction of the accessory *sel* genes, when a native SECIS element was located within the open reading frame of an analyzed selenoprotein mRNA.[47] This should be compared with the production of thioredoxin reductase, where termination codons were introduced into the nonconserved stem of the SECIS element in order to encode the carboxy-terminal -Sec-Gly-COOH motif of the enzyme (Fig. 1), resulting in about 25% full-length product using the accessory *sel* genes, compared with about 3% without such cotransformation.[32] It is possible that this case, which in essence located the SelB-binding motif to an untranslated region of the mRNA (Fig. 1), increased the yield as a result of a less pronounced competition between SelB and EF-Tu. Inefficiencies in targeted selenocysteine insertion within an open reading frame may also be one of several possible factors leading to a low yield in heterologous production of the Sec-containing plant[35] or human (our unpublished observations, 2000) glutathione peroxidases, using this technique. The

[45] P. Tormay, A. Sawers, and A. Böck, *Mol. Microbiol.* **21,** 1253 (1996).
[46] S. Muller, J. Heider, and A. Böck, *Arch. Microbiol.* **168,** 421 (1997).
[47] S. Suppmann, B. C. Persson, and A. Böck, *EMBO J.* **18,** 2284 (1999).

positioning of a bacterial-type SECIS element within an open reading frame also imposes limitations to the use of codons corresponding to the SelB-binding loop region, which must be conserved (Fig. 1). Even if these limitations exist, the strategy nonetheless can also yield functional recombinant enzyme species when the selenocysteine residue is located at internal positions of a protein and the whole SECIS element therefore must be translated.[35]

Other Factors Influencing Bacterial Recombinant Protein Production

Naturally, all the factors affecting yield in heterologous expression of recombinant proteins in *E. coli,* such as promoter or codon usage, growth conditions, inclusion body formation, proteolysis, and mRNA stability,[48] also apply to production of recombinant selenoproteins. Such other factors, not depending on the fact that a selenoprotein is produced, may explain the lower yield reported for production of human thioredoxin reductases[33,34] compared with the production of the corresponding enzyme from rat.[32]

Conclusions

Escherichia coli has a significant capacity for heterologous production of recombinant selenoproteins, far exceeding the natural level of synthesis of the endogenous bacterial selenoproteins. Targeted insertion of selenocysteine at specific UGA codons in recombinant proteins requires, however, that attention be given to a number of critical factors as summarized herein. This production system may nonetheless be utilized in studies of naturally occuring selenoproteins as well as for the production of selenocysteine-containing variants of nonselenoproteins; the applications of such selenosubstituted species may be several, considering the high-energy irradiation of selenium isotopes that can be introduced or the specific physical and biochemical characteristics of the selenocysteine moiety.

[48] G. Hannig and S. C. Makrides, *Trends Biotechnol.* **16,** 54 (1998).

[21] Mammalian Thioredoxin Reductases as Hydroperoxide Reductases

By LIANGWEI ZHONG and ARNE HOLMGREN

Introduction

Mammalian thioredoxin reductases (TrxRs) differ fundamentally from the enzyme from bacteria, fungi, or plants, as has been known since purification and characterization was started more than two decades ago.[1–3] Mammalian TrxRs are much larger, with two subunits of 55 kDa or higher as compared with 35 kDa for the well-characterized smaller *Escherichia coli* enzyme.[4] In addition, the mammalian enzymes have a surprisingly wide range of different substrates apart from carrying out the reduction of oxidized thioredoxin (Trx-S_2) by NADPH yielding reduced thioredoxin [Trx-$(SH)_2$)] [reaction (1)]:

$$\text{Trx-S}_2 + \text{NADPH} + \text{H}^+ \xrightarrow{\text{TrxR}} \text{Trx-(SH)}_2 + \text{NADP}^+ \tag{1}$$

$$\text{Trx-(SH)}_2 + \text{protein-S}_2 \longrightarrow \text{Trx-S}_2 + \text{protein-(SH)}_2 \tag{2}$$

Trx-$(SH)_2$ is the cell major protein disulfide reductase [reaction (2)] with a large number of functions in cell growth, redox regulation by thiol redox control, and defense against oxidative stress (for a review see Arnér and Holmgren[5]).

In a previous article[6] we described the substrate specificity of mammalian TrxR. Substrates involve thioredoxin from different species, protein disulfide isomerases (PDI), NK-lysin, alloxan, vitamin K, dehydroascorbic acid, lipoic acid, S-nitrosylated glutathione, etc.

Of particular interest for the theme of this article was the discovery that selenium compounds such as selenite[7] and selenodiglutathione[8] were efficiently reduced directly by the mammalian TrxR but not by the *E. coli* enzyme. Anaerobically, selenite is reduced to selenide by consumption of 3 mol of NADPH, whereas in the presence of oxygen selenite is redox cycling, with the enzyme leading to a large nonstoichiometric oxidation of NADPH.[7] The reduction of selenite by the enzyme, which is stimulated by the presence of thioredoxin, will provide a source of selenide for synthesis of bioactive selenium compounds including selenocysteine

[1] A. Holmgren, *J. Biol. Chem.* **252**, 4600 (1977).
[2] M. Luthman and A. Holmgren, *Biochemistry* **21**, 6628 (1982).
[3] A. Holmgren, *Annu. Rev. Biochem.* **54**, 237 (1985).
[4] C. H. J. Williams, *FASEB J.* **9**, 1267 (1995).
[5] E. S. Arnér and A. Holmgren, *Eur. J. Biochem.* **267**, 6102 (2000).
[6] E. S. J. Arnér, L. Zhong, and A. Holmgren, *Methods Enzymol.* **300**, 226 (1999).
[7] S. Kumar, M. Björnstedt, and A. Holmgren, *Eur. J. Biochem.* **207**, 435 (1992).
[8] M. Björnstedt, S. Kumar, and A. Holmgren, *J. Biol. Chem.* **267**, 8030 (1992).

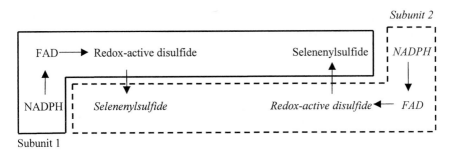

FIG. 1. Proposed mode of intersubunit electron transfer in mammalian TrxRs. Each subunit of the enzyme consists of a glutathione reductase equivalent carrying a redox-active disulfide. Electrons from its thiols are transferred to the selenenyl sulfide found in a 16-residue C-terminal extension of the other subunit. The selenenyl sulfide formed from the conserved -Gly-Cys-SeCys-Gly is thereby reduced to a selenolthiol, which is the active site [L. Zhong and A. Holmgren, *J. Biol. Chem.* **275**, 18121 (2000); L. Zhong, E. S. J. Arnér, J. Ljung, F. Åslund, and A. Holmgren, *J. Biol. Chem.* **273**, 8581 (1998); T. Sandalova, L. Zhong, Y. Lindqvist, A. Holmgren, and G. Schneider, *Proc. Natl. Acad. Sci. U.S.A.* **98**, 9533 (2001); and L. Zhong, E. S. J. Arnér, and A. Holmgren, *Proc. Natl. Acad. Sci. U.S.A.* **97**, 5854 (2000)].

(SeCys).[7] Selenodiglutathione is also an excellent direct substrate for mammalian TrxR, in sharp contrast to oxidized glutathione (GSSG), which is not reduced.[8] Surprisingly, mammalian TrxRs exhibit NADPH-dependent peroxidase activity and are capable of reducing organic hydroperoxides and lipid hydroperoxides,[9] as well as H_2O_2,[10] having a function similar to that of the known selenoproteins of the glutathione peroxidase family. For all these reactions free selenocysteine or selenoglutathione strongly stimulated the reductions by catalytically generated selenols.[9] The discoveries of mammalian TrxR as a selenoprotein[11,12] provided a clue to its ability both to reduce selenite and to act as a hydroperoxide reductase. Mammalian TrxRs comprise a glutathione reductase equivalent elongated by 16 residues and a C-terminal -Gly-Cys-SeCys-Gly sequence. Mammalian TrxRs contain at least three redox centers per subunit, that is, FAD, a redox-active disulfide identical to that of glutathione reductase, and a selenenyl sulfide (Fig. 1).[10,12,13,18] An internal "electron shuttle" is an important part of the overall electron transport process. Theoretically, electron transfer in proteins occurs down a potential "well," the depth of which is determined by the reduction potentials of the centers. Two other factors, the distance between the centers and the intervening medium, determine the efficiency of the electron transfer. On the basis of a three-dimensional

[9] M. Björnstedt, M. Hamberg, S. Kumar, J. Xue, and A. Holmgren, *J. Biol. Chem.* **270**, 11761 (1995).
[10] L. Zhong and A. Holmgren, *J. Biol. Chem.* **275**, 18121 (2000).
[11] T. Tamura and T. C. Stadtman, *Proc. Natl. Acad. Sci. U.S.A.* **93**, 1006 (1996).
[12] L. Zhong, E. S. J. Arnér, J. Ljung, F. Åslund, and A. Holmgren, *J. Biol. Chem.* **273**, 8581 (1998).
[13] T. Sandalova, L. Zhong, Y. Lindqvist, A. Holmgren, and G. Schneider, *Proc. Natl. Acad. Sci. U.S.A.* **98**, 9533 (2001).

structure of rat SeCys498Cys TrxR in complex with NADP[13] and a number of biochemical studies, the following set of electron transfer reactions can be envisaged.

1. Binding of NADPH is proposed to involve a conformational change to allow the dihydropyridine ring to take a position close to the FAD. Occurrence of the NADPH-induced conformational change is implied from Western blot results using monoclonal antibodies generated with native human placenta TrxR, which did not recognize NADPH-prereduced human TrxR (L. Zhong and A. Holmgren, unpublished result, 2000).

2. NADPH reduces the cofactor FAD.

3. Reduction of the redox-active disulfide by the reduced flavin results in flavin returning to the oxidized state, and in the disulfide being reduced. One of the nascent cysteine residues may exist as a thiolate anion that may form a covalent bond at flavin C(4a), as suggested from spectroscopic data.[10,14,15]

4. The other cysteine residue may be involved in a nucleophilic attack of the selenenyl sulfide to form the Cys497-thiol and SeCys498-selenol. This has been demonstrated by chemical modification studies.[12,16,17] The SeCys498 should be the residue involved in attacking the active site disulfide of Trx or reacting with H_2O_2.[18]

The electron transfer from the redox-active disulfide to the selenenyl sulfide will occur between the two subunits, rather than between domains from the same subunit. Association of two subunits is "head to tail,"[18] as schematically represented in Fig. 1.

A number of reports have highlighted the crucial role played by SeCys in catalytic activities of mammalian TrxRs.[10,19,20] We have examined the result of replacing the penultimate SeCys residue in rat TrxR by cysteine or serine or of removing it together with the C-terminal glycine residue.[10] Only the SeCys498Cys TrxR showed activity in reduction of thioredoxin, with a 100-fold lower k_{cat} and 10-fold lower K_m compared with the wild-type rat enzyme.[10] All three mutant proteins showed undetectable activity with H_2O_2 by themselves.[10] However, addition of selenocystine and thioredoxin resulted in considerable activity in H_2O_2 reduction

[14] L. D. Arscott, S. Gromer, R. H. Schirmer, K. Becker, and C. H. J. Williams, *Proc. Natl. Acad. Sci. U.S.A.* **94**, 3621 (1997).

[15] T. W. Gilberger, R. D. Walter, and S. Muller, *J. Biol. Chem.* **272**, 29584 (1997).

[16] J. Nordberg, L. Zhong, A. Holmgren, and E. S. J. Arnér, *J. Biol. Chem.* **273**, 10835 (1998).

[17] S. N. Gorlatov and T. C. Stadtman, *Proc. Natl. Acad. Sci. U.S.A.* **95**, 8520 (1998).

[18] L. Zhong, E. S. J. Arnér, and A. Holmgren, *Proc. Natl. Acad. Sci. U.S.A.* **97**, 5854 (2000).

[19] J. R. Gasdaska, J. W. Harney, P. Y. Gasdaska, G. Powis, and M. J. Berry, *J. Biol. Chem.* **274**, 25379 (1999).

[20] S. R. Lee, S. Bar-Noy, J. Kwon, R. L. Levine, T. C. Stadtman, and S. G. Rhee, *Proc. Natl. Acad. Sci. U.S.A.* **97**, 2521 (2000).

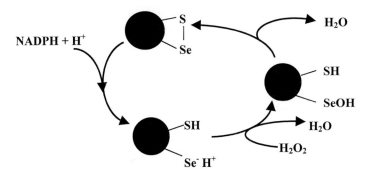

FIG. 2. Proposed mechanism of mammalian TrxR and its C-terminal selenolthiol couple formed from the conserved -Gly-Cys-SeCys-Gly-OH sequence in reducing H_2O_2 and other hydroperoxides [L. Zhong and A. Holmgren, *J. Biol. Chem.* **275**, 18121 (2000); L. Zhong, E. S. J. Arnér, J. Ljung, F. Åslund, and A. Holmgren, *J. Biol. Chem.* **273**, 8581 (1998); and L. Zhong, E. S. J. Arnér, and A. Holmgren, *Proc. Natl. Acad. Sci. U.S.A.* **97**, 5854 (2000)].

catalyzed by the SeCys498Cys TrxR.[10] Thus, SeCys is involved in the peroxidase mechanism of mammalian TrxRs. Two redox states of the SeCys residue have been unambiguously identified as a selenol[12] and a selenenyl sulfide,[18,20] which allow us to propose the mechanism for H_2O_2 reduction by mammalian TrxRs[18] (Fig. 2).[10,12,18] First the selenenyl sulfide receives electrons from NADPH via the FAD and redox-active dithiol of the opposite subunit to produce a Cys497-SH and SeCys498-SeH in a second subunit. Because of the low pK_a value of the selenol, selenolate ($-Se^-$) should be a predominant form under physiologic conditions. Being a strong nucleophile, selenolate is more susceptible to oxidation by H_2O_2 than thiols, yielding selenenic acid ($-SeOH$). One cysteine thiol (most likely Cys-497) reacts with the selenenic acid to produce water and reform the selenenyl sulfide. A second thiol (most likely Cys-59 from the other subunit, the number referring to rat TrxR1[10,18]) would attack the bridge to regenerate the selenol. Therefore, the selenenyl sulfide serves as either a catalytically essential redox center or transient intermediate during hydrogen peroxide reduction. According to its high apparent K_m value (2.5 mM) for H_2O_2 and K_{cat} of $100\times$ min^{-1}, this antioxidant defense function of mammalian TrxRs may be expected to play a role only with elevated H_2O_2 concentration. However, the relative efficiency with other, different hydroperoxides requires further studies, as does the role of selenium compounds to stimulate this reaction.

Purification of Mammalian Thioredoxin Reductase from Wet Tissue

Sources

Mammalian cytosolic TrxRs have been purified to homogeneity from calf thymus and liver,[1] rat liver,[2] human placenta,[14,21] human lung adenocarcinoma cells,[11] a human Jurkat T cell line,[22] and mouse ascites cells.[23] For enzyme assays there is full cross-reactivity between cytosolic TrxR from all these sources and the respective cytosolic thioredoxins from the heterogeneous mammalian species. There have been several modifications of basic purification procedures.[2,6] A protocol described here, using calf liver as starting material, is also applicable to rat liver or human placenta. A novel method using an engineered *E. coli* strain capable of synthesizing recombinant selenocysteine containing rat TrxR with 25%, of specific activity of homogenous rat liver TrxR has been described.[24] Pure bovine liver enzyme is available from IMCO (Stockholm, Sweden; www.imcocorp.se).

Preparation of Crude Extract

Calf liver is obtained from a local slaughterhouse and stored frozen at $-70°$. The tissue is allowed to thaw at $4°$ and all subsequent procedures are performed at $4°$. The liver is homogenized in 50 mM Tris-HCl, 1 mM EDTA, pH 7.5 (TE buffer) with 1 mM phenylmethylsulfonyl fluoride (PMSF) as protease inhibitor (one drop of octanol is added to reduce foam formation), and then the homogenate is diluted with TE buffer to five times the tissue volume and allowed 1 hr to extract the proteins. The diluted homogenate is centrifuged in a GSA rotor (Sorvall, Newtown, CT) at 10,000 rpm for 30 min, and the supernatant is recovered as crude extract.

Acidic Acid Treatment

The crude extract is acidified to pH 5.0 by dropwise addition of 1 M acidic acid with gentle stirring, and then centrifuged in the GSA rotor at 13,000 rpm for 30 min. The recovered supernatant is neutralized to pH 7.5 by dropwise addition of 1 M NH$_4$OH as soon as possible.

Ammonium Sulfate Fractionation

To the neutralized supernatant, solid ammonium sulfate is slowly added to 40% (w/v) saturation with gently stirring, holding at pH 7.5. After 1 hr, precipitate is removed by centrifugation under the above-described conditions to keep the supernatant, which is then saturated with solid ammonium sulfate to 80% (w/v) under gentle stirring for 2 hr. The precipitate is recovered by centrifugation under the above-described conditions and kept at $-20°$ until use.

---

[21] J. E. Oblong, P. Y. Gasdaska, K. Sherrill, and G. Powis, *Biochemistry* **32,** 7271 (1993).
[22] V. N. Gladyshev, K. T. Jeang, and T. C. Stadtman, *Proc. Natl. Acad. Sci. U.S.A.* **93,** 6146 (1996).
[23] S. Gromer, R. H. Schirmer, and K. Becker, *FEBS Lett.* **412,** 318 (1997).
[24] E. S. J. Arnér, H. Sarioglu, F. Lottspeich, A. Holmgren, and A. Böck, *J. Mol. Biol.* **292,** 1003 (1999).

DEAE-Cellulose Chromatography

The protein precipitate is dissolved in TE buffer in as small a volume as possible, and then dialyzed against TE buffer to completely remove ammonium sulfate. The resulting protein solution is loaded on a DEAE-cellulose column (2.5 × 20 cm) equilibrated with TE buffer. Unbound proteins are washed away with TE buffer until the $A_{280\,nm}$ reaches baseline, and then TrxR is eluted with a linear gradient of 0 to 0.3 M NaCl in TE buffer (1000 ml each). TrxR activity is determined by NADPH-co⋯d reduction of 5,5′-dithiobis (2-nitrobenzoic acid) (DTNB) assay,[1] and appeaɪ at about 0.12 M NaCl. Fractions containing TrxR activity are pooled and extensively dialyzed against TE buffer.

2′,5′-ADP-Sepharose Chromatography

The dialyzed sample is loaded on a 2′,5′-ADP Sepharose 4B column (1.5 × 7 cm) equilibrated with TE buffer. The column should be carefully washed with TE buffer to remove unbound proteins. TrxR is eluted with a linear gradient of 0 to 0.3 M NaCl in TE buffer (400 ml each). Fractions containing TrxR activity are pooled and dialyzed against TE buffer to remove NaCl, and then applied to a second 2′,5′-ADP-Sepharose column, repeating the same procedures described above. The fractions with DTNB reductase activity are pooled and concentrated in an ultrafiltration system (Diaflo; Millipore) using a YM30 membrane, and then washed with TE buffer three times to completely remove salt. The obtained TrxR exhibits a single band on a sodium dodecyl sulfate. (SDS)–polyacrylamide gel stained with Coomassie Brilliant Blue.

Assay of Mammalian Thioredoxin Reductase Activity

The DTNB reduction assay[2] is used during preparation of the enzyme. The thioredoxin-coupled insulin reduction assay[2,6] is used in measuring activity of the pure enzyme.

Assay of Hydroperoxide Reductase Activity

Reduction of H_2O_2

Principle of Assay. Mammalian TrxRs catalyze the reduction of H_2O_2 by NADPH, which is determined as the decrease in absorbance at 340 nm due to oxidation of NADPH [reaction (3)]. To determine kinetic parameters, H_2O_2 concentrations are varied from 0 to 550 μM.

$$H_2O_2 + NADPH + H^+ \xrightarrow{TrxR} 2H_2O + NADP^+ \qquad (3)$$

Reagents

H_2O_2, 10 mM
Potassium phosphate (pH 7.5), 0.5 M

EDTA (pH 8.0), 0.2 M
Human placenta TrxR, 3 μM
Assay mixture: 0.1 M Potassium phosphate (pH 7.5), 2 mM EDTA, and 0.2 mM NADPH.

Procedure. To the cuvettes with 500 μl of assay mixture, H_2O_2 is added and mixed well. The reaction is started by adding human placenta TrxR to the sample cuvette (a final concentration of 60 nM) and an equal volume of distilled H_2O to the reference cuvette. The reduction is monitored at 340 nm and at 20°, using a millimolar extinction coefficient of 6.2.

Reduction of Lipid Hydroperoxide

Principle of Assay. Mammalian TrxRs catalyze the conversion of lipid hydroperoxide into water and corresponding alcohols [reaction (4)], which is monitored as the decrease in absorbance at 340 nm, owing to oxidation of NADPH.

$$ROOH + NADPH + H^+ \xrightarrow{\text{TrxR}} ROH + H_2O + NADP^+ \tag{4}$$

Reagents

(15S)-Hydroperoxyeicosatetraenoic acid (HPETE) dissolved in 99.5% (v/v) ethanol, 13 mM
Potassium phosphate (pH 7.0), 0.5 M
EDTA (pH 8.0), 0.2 M
Human placenta TrxR, 3 μM
Assay mixture: 50 mM Potassium phosphate (pH 7.0), 2 mM EDTA, and 0.5 mM NADPH

Procedure. To the cuvettes with 500 μl of assay mixture, HPETE is added to a final concentration of 0.15 μM and mixed well. The reaction is started by adding human placenta TrxR to the sample cuvette (a final concentration of 50 nM) and an equal volume of 99.5% (v/v) ethanol to the reference cuvette. The reduction is monitored at 340 nm and at 20°, using a millimolar extinction coefficient of 6.2.

Discussion

Oxidative stress occurs when cells are exposed to elevated levels of reactive oxygen species[25] such as superoxide ($O_2{}^{\cdot-}$), hydrogen peroxide (H_2O_2), and alkyl hydroperoxides (ROOH) such as cumene hydroperoxide and *tert*-butyl hydroperoxide. Oxidative stress can lead to DNA damage and mutations, lipid peroxidation, the disassembly of iron–sulfur clusters, disulfide bond formation, and

[25] B. Halliwell and J. M. Gutteridge, *in* "Free Radicals in Biology and Medicine," 3rd Ed. Oxford University Press, New York, 1999.

other types of protein oxidation.[25] It is known that thioredoxins and glutaredoxins via reduced glutathione (GSH) are required for disulfide bond reduction and for glutathione- and thioredoxin-dependent peroxidase activities, and the GSH- and thioredoxin-based systems have been proposed to protect cells against oxidative stress.[26] Lipid hydroperoxides can accumulate in tissues and exert harmful effects. Specifically, (15S)-hydroperoxyeicosatetraenoic acid (HPETE) can oxidatively modify low-density lipoprotein and has been implicated in atherosclerosis.[27] The conversion of lipid hydroperoxides to corresponding alcohols in the presence of mamma' in TrxRs and NADPH serves as an important alternative to glutathione peroxidase.[9] Selenol groups of mammalian TrxRs have low redox potentials and potential redox-sensing activity, and their importance in redox signaling requires further studies. Of particular interest is the unique direct coupling to NADPH by the mammalian enzyme without any soluble intermediates.

Comments

The SeCys residue is essential to maintain the hydroperoxide reductase activity of mammalian TrxRs,[10] because when the SeCys was mutated to a cysteine residue, the hydroperoxide reductase function of mammalian TrxRs was severely impaired.[10] The truncated enzyme lacking the C-terminal SeCys-Gly residues is folded but inactive.[10] Therefore, this enzyme activity is correlated with selenium content and selenium status, which may vary from tissue sources or from different enzyme preparations.[12,17,23] To prepare highly active enzyme, fresh tissues should always be used as starting material. A gentle manipulation of the protein should always be considered in the whole purification procedure and during storage because selenium can be easily lost.[17]

Acknowledgments

This research was supported by grants from the Swedish Medical Research Council (13X-3529), the Swedish Cancer Society (961), and the Karolinska Institute. L.Z. acknowledges the support of a fellowship from the David and Astrid Hegeléns Foundation.

[26] A. Holmgren, *Antioxid. Redox. Signal.* **2**, 811 (2000).
[27] S. Ylä-Herttuala, M. E. Rosenfeld, S. Parthasarathy, C. K. Glass, E. Sigal, J. L. Witztum, and D. Steinberg, *Proc. Natl. Acad. Sci. U.S.A.* **87**, 6959 (1990).

[22] Tryparedoxin and Tryparedoxin Peroxidase

By Leopold Flohé, Peter Steinert, Hans-Jürgen Hecht, and Birgit Hofmann

Introduction

Trypanothione [N^1,N^8-bis(glutathionyl)spermidine, T(SH)$_2$], discovered in 1985 by Fairlamb and Cerami,[1] is the main low molecular weight thiol in trypanosomatids of the genera *Crithidia, Trypanosoma,* and *Leishmania,* the latter two comprising causative agents of life-threatening or disabling diseases such as African sleeping sickness, Chagas' disease, *kala azar,* and *espundia* or oriental sore. Like reduced glutathione (GSH), T(SH)$_2$ can be oxidized by H$_2$O$_2$ or other hydroperoxides. Like oxidized glutathione (GSSG), the trypanothione disulfide (TS$_2$) is specifically reduced by a member of the FAD-containing disulfide reductases called trypanothione reductase.[2] Unlike GSH, T(SH)$_2$ does not, however, serve as an immediate substrate of a thiol-dependent peroxidase. The selenoprotein glutathione peroxidase (GPx), typical of mammalian hydroperoxide metabolism,[3] is not found in trypanosomatids, nor could any homologous enzyme be detected in this family of parasites. Instead, the reduction of hydroperoxides by T(SH)$_2$ is catalyzed by two independent catalytic events[4]: T(SH)$_2$ reduces a thioredoxin-related protein called tryparedoxin (TXN) and reduced TXN in turn reduces a peroxiredoxin-type peroxidase, tryparedoxin peroxidase (TXNPx).

This "trypanothione peroxidase system" was first discovered to operate in the insect pathogen *Crithidia fasciculata.*[4] Components of the system have been demonstrated in many of the species pathogenic to humans and live stock, for example, in *Trypanosoma brucei rhodesiense,*[5] *T. brucei brucei,*[6] *T. cruzi,*[7] *Leishmania major,*[8] and *L. donovani.*[9] It is likely that the peculiar pathway is characteristic

[1] A. H. Fairlamb and A. Cerami, *Mol. Biochem. Parasitol.* **14,** 187 (1985).

[2] L. Flohé, H.-J. Hecht, and P. Steinert, *Free Radic. Biol. Med.* **27,** 966 (1999).

[3] L. Flohé, *in* "Selenium: Its Molecular Biology and Role in Human Health" (D. Hatfield, ed.). Kluwer Academic, Boston, 2001.

[4] E. Nogoceke, D. U. Gommel, M. Kiess, H. M. Kalisz, and L. Flohé, *Biol. Chem.* **378,** 827 (1997).

[5] N. M. A. El-Sayed, GenBank accession No. U26666 (1995).

[6] H. Lüdemann, M. Dormeyer, C. Sticherling, D. Stallmann, H. Follmann, and R. L. Krauth-Siegel, *FEBS Lett.* **431,** 381 (1998).

[7] J. A. Lopez, T. U. Carvalho, W. de Souza, L. Flohé, S. A. Guerrero, M. Montemartini, H. M. Kalisz, E. Nogoceke, M. Singh, J. M. Alves, and W Colli, *Free Radic. Biol. Med.* **28,** 767 (2000).

[8] M. P. Levick, E. Tetaud, A. H. Fairlamb, and J. M. Blackwell, *Mol. Biochem. Parasitol.* **96,** 125 (1998).

[9] S. Kansal-Kalavar, P. Steinert, M. Singh, and L. Flohé, GenBank accession number AF225212 (1999).

0076-6879/02 $35.00

of the whole trypanosomatid family. In *C. fasciculata*[10] and *T. cruzi*,[7] TXN and TXNPx were shown to be colocalized in the cytosolic compartment as was TR.[11] The detoxification of hydroperoxides at the expense of NADPH by the concerted action of TR, TXN, and TXNPx can thus proceed *in vivo*, as has been deduced from *in vitro* reconstitution of the system from its components having either been isolated from *C. fasciculata*[4] or heterologously expressed in *E. coli*.[12]

With three oxidoreductases plus the unique redox metabolite trypanothione and two auxiliary enzymes to synthesize the latter,[13] the trypanosomatids have apparently developed the most complex metabolic pathway to detoxify hydroperoxides (Fig. 1). The pathway appears to be of vital importance to trypanosomatids, because conditioned knockout of trypanothione reductase in *T. brucei* resulted in increased sensitivity to H_2O_2-induced killing, arrest of proliferation, and loss of virulence.[14]

Tryparedoxins

General Characteristics

TXNs are small proteins of about 16 kDa. They belong to the thioredoxin superfamily but are about 50% larger than typical thioredoxins. Also, their active site motif is WCPPCR, whereas in thioredoxins the motif usually is WCGPCK. Two distinct isoforms, TXN1 and TXN2, exist in *C. fasciculata*.[4,15,16] Further, microheterogeneities may occur, because TXNs are encoded by chromosomally located multicopy genes,[10,16] as is frequently observed in trypanosomatids. Sequence similarity with thioredoxins is pronounced near the active site motif, but otherwise restricted to a few amino acid residues, notably an aspartate and a *cis*-proline (positions 71 and 90 in TXN1 of *C. fasciculata; Cf*TXN1), which are typical of most thioredoxin-related proteins. Similarly, X-ray structures of *Cf*TXN1[17,18] and

[10] P. Steinert, K. Dittmar, H. M. Kalisz, M. Montemartini, E. Nogoceke, M. Rohde, M. Singh, and L. Flohé, *Free Radic. Biol. Med.* **26**, 844 (1999).

[11] D. Meziane-Cherif, M. Aumercier, I. Kora, C. Sergheraert, A. Tartar, J. F. Dubremetz, and M. A. Ouaissi, *Exp. Parasitol.* **79**, 536 (1994).

[12] E. Tetaud and A. H. Fairlamb, *Mol. Biochem. Parasitol.* **96**, 111 (1998).

[13] K. Koening, U. Menge, M. Kiess, V. Wray, and L. Flohé, *J. Biol. Chem.* **272**, 11908 (1997).

[14] S. Krieger, W. Schwarz, M. R. Ariyanagam, A. Fairlamb, R. L. Krauth-Siegel, and C. E. Clayton, *Mol. Microbiol.* **35**, 542 (2000).

[15] M. Montemartini, H. M. Kalisz, M. Kiess, E. Nogoceke, M. Singh, P. Steinert, and L. Flohé, *Biol. Chem.* **379**, 1137 (1998).

[16] S. A. Guerrero, L. Flohé, H. M. Kalisz, M. Montemartini, E. Nogoceke, H.-J. Hecht, P. Steinert, and M. Singh, *Eur. J. Biochem.* **259**, 789 (1998).

[17] M. S. Alphey, G. A. Leonard, D. G. Gourley, E. Tetaud, A. H. Fairlamb, and W. N. Hunter, *J. Biol. Chem.* **274**, 25613 (1999).

[18] B. Hofmann, H. Budde, K. Bruns, S. A. Guerrero, H. M. Kalisz, U. Menge, M. Montemartini, E. Nogoceke, P. Steinert, J. Wissing, L. Flohé, and H. J. Hecht, *Biol. Chem.* **382**, 459 (2001).

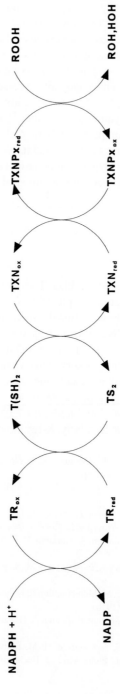

Fig. 1. NADPH-dependent alkyl hydroperoxide reduction in trypanosomatids. TR, Trypanothione reductase; T(SH)$_2$ and TS$_2$, reduced and oxidized trypan-othione, respectively; TXN, trypanoredoxin; TXNPx, tryparedoxin peroxidase; ROOH, alkyl hydroperoxide; ROH, alcohol.

CfTXN2[18] display a folding pattern similar to that of thioredoxins, an almost identical conformation of the active site motif with Cys-40 (in CfTXN1; corresponding to Cys-41 in CfTXN2) exposed to the solvent, but considerable deviations in more remote parts, an extra helix corresponding to a large insertion, and some minor insertions[18] (Fig. 2).

Catalytic Mechanism

TXNs may systematically be described as trypanothione:protein disulfide oxidoreductases. In the context of trypanosomal peroxide metabolism the disulfide substrate is the peroxiredoxin-type TXNPx. It can, however, be anticipated that in trypanosomes TXNs might exert additional activities that are catalyzed by thioredoxins in other species. The TXN of $T.$ $brucei$ $brucei,$ for instance, was reported to serve as reducing substrate of ribonucleotide reductase.[6,19,20]

TXNs appear to be specifically reduced by $T(SH)_2$. GSH is a poor substrate[21] and, unlike thioredoxins, TXN is not efficiently reduced directly by thioredoxin reductase, or by other members of the disulfide reductase family such as trypanothione reductase or glutathione reductase.[4,21,22] The molecular basis of this specificity has become evident from molecular modeling,[18] site-directed mutagenesis,[18,22] and X-ray crystallography[17,18](Fig. 3). If the thiol group of the N^1-glutathionyl residue in $T(SH)_2$ is to attack the sulfur of Cys-41 in oxidized CfTXN2, its negatively charged carboxylate function must be bound to the positively charged Arg-129. In this orientation fixation of the glutathionyl residue is further supported by hydrogen bridges to the main-chain atoms near the conserved cis-proline bond. Out of this complex, the Cys^{41}–Cys^{44} disulfide bond is split, whereby likely a CfTXN2 derivative is formed having trypanothione linked to Cys-41 via a disulfide bridge. To terminate the reduction of CfTXN2, the remote N^8-glutathionyl residue must bend back to the active site. This is probably achieved by electrostatic attraction of the basic secondary amine of the spermidine moiety to Glu-73 and of the second carboxylate to Arg-45. In this intermediate, then, the fully reduced CfTXN2 and oxidized trypanothione are readily formed by thiol/disulfide exchange reactions.[18]

In reduced CfTXN2, the thiol of Cys-41 (Cys-40 in CfTXN1, respectively) is activated by a network of hydrogen bridges.[18] Thereby this solvent-exposed cysteine residue is dissociated, as is also evident from selective alkylation by iodoacetamide.[21] The thiolate function of Cys-41 must therefore be considered as the most likely candidate to attack the disulfide bridge of the protein substrates.

[19] M. Domeyer, N. Reckenfelderbäumer, H. Lüdemann, and R. L. Krauth-Siegel, $J.$ $Biol.$ $Chem.$ **276**, 10602 (2001).

[20] R. L. Krauth-Siegel and H. Schmidt, $Methods$ $Enzymol.$ **347**, [23], 2002 (this volume).

[21] D. U. Gommel, E. Nogoceke, M. Morr, M. Kiess, H. M. Kalisz, and L. Flohé, $Eur.$ $J.$ $Biochem.$ **248**, 913 (1997).

[22] P. Steinert, K. Plank-Schumacher, M. Montemartini, H.-J. Hecht, and L. Flohé, $Biol.$ $Chem.$ **381**, 211 (2000).

Fig. 2.

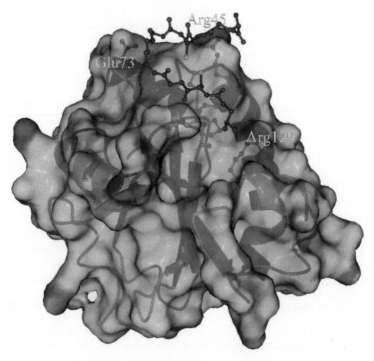

Fig. 3.

Needless to say, the mechanism, as proposed, is an enzyme substitution mechanism, in which the reducing and the oxidizing substrate independently react with the enzyme. Accordingly, a typical ping–pong pattern is observed with TXNs in steady state kinetic analyses.[16,21,22]

The initial velocity v is adequately described by the Dalziel[23] Eq. (1):

$$\frac{[E_0]}{v} = \phi_0 + \frac{\phi_1}{[\text{TXNPx}]} + \frac{\phi_2}{[\text{T(SH)}_2]} \tag{1}$$

In Eq. (1), ϕ_0 equals $1/k_{\text{cat}}$, the coefficient ϕ_1 describes the TXNPx-dependent (i.e., the oxidative) part of the catalytic cycle, and ϕ_2 characterizes the reduction of oxidized enzyme. The catalytic cycle of the TXNs may be schematically described in Scheme 1, wherein E means the reduced TXN, F the oxidized TXN, and (EA) and (FB) represent the pertinent enzyme–substrate complexes.[21]

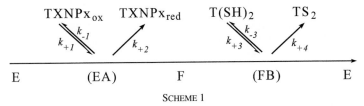

SCHEME 1

If backward reactions involving products are neglected, steady state treatment of the partial reactions comprised in Scheme 1 yields the definitions of the Dalziel coefficients in terms of individual rate constants[23]:

$$\phi_0 = \frac{1}{k_{+2}} + \frac{1}{k_{+4}} \tag{2}$$

[23] K. Dalziel, *Acta Chem. Scand.* **11**, 1706 (1957).

FIG. 2. Ribbon diagram of *Cf* TXN2 with the core structure similar to thioredoxin shown in green, less similar parts (high RMS deviation) in gray, and sequence stretches deleted in thioredoxin in red. The N terminus differing between *Cf* TXN2 and *Cf* TXN1 is shown in yellow. Active site cysteines are shown as balls and sticks. α Helices and β strands are labeled sequentially [adapted from B. Hofmann, H. Budde, K. Bruns, S. A. Guerrero, H. M. Kalisz, U. Menge, M. Montemartini, E. Nogoceke, P. Steinert, J. Wissing, L. Flohé, and H. J. Hecht, *Biol. Chem.* **382**, 459 (2001)].

FIG. 3. Interaction of trypanothione with tryparedoxin. *Cf* TXN1 is shown as a ribbon representation analogous to the one in Fig. 2, encased, however, in a semitransparent molecular surface and color-coded according to electrostatic charge (red, negative; blue, positive). The orientation of the molecule is similar to that in Fig. 2. The model is based on the experimentally determined structures of TXNs, mutagenesis studies, and the structure of a covalent reaction intermediate between the Cys44Ser variant of *Cf* TXN2 and glutathionylspermidine. Trypanothione and selected residues of tryparedoxin are shown as ball and sticks. For experimental details supporting this model see B. Hofmann, H. Budde, K. Bruns, S. A. Guerrero, H. M. Kalisz, U. Menge, M. Montemartini, E. Nogoceke, P. Steinert, J. Wissing, L. Flohé, and H. J. Hecht, *Biol. Chem.* **382**, 459 (2001).

TABLE I
DALZIEL COEFFICIENTS AND DERIVED KINETIC CONSTANTS FOR TXNs FROM *Crithidia fasciculata.*

TXN	ϕ_0 (min)	$\phi_{TXNPx}{}^b$ ($10^{-3}\mu M$ min)	$\phi_{T(SH)2}$ ($10^{-3}\mu M$ min)	$V_{max}[E]$ (min^{-1})	$K_{m\,TXNPx}{}^b$ (μM)	$K_{mT(SH)2}$ (μM)	Ref.
TXN1	2.55×10^{-3}	5.5	330	392	2.2	130	21
TXN1H$_6{}^a$	0.63×10^{-3}	3.0	255	1587	4.8	406	16
TXN2H$_6$	1.05×10^{-3}	4.4	35	952	4.2	33	15
TXN2H$_6$	1.68×10^{-3}	75 ± 14	60 ± 6	595	45	36	22
TXN2H$_6$-W40F	1.31×10^{-3}	47 ± 6	145 ± 4	762	36	110	22
TXN2H$_6$-W40Y	4.16×10^{-3}	52 ± 5	95 ± 3	240	13	23	22
TXN2H$_6$-P42G	3.79×10^{-3}	69 ± 11	160 ± 3	264	18	42	22

[a] H$_6$, Polyhistidine tail at the C terminus of TXNs.
[b] Authentic [M. Montemartini, H. M. Kalisz, M. Kiess, E. Nogoceke, M. Singh, P. Steinert, and L. Flohé, *Biol. Chem.* **379,** 1137 (1998); S. A. Guerrero, L. Flohé, H. M. Kalisz, M. Montemartini, E. Nogoceke, H. J. Hecht, P. Steinert, and M. Singh, *Eur. J. Biochem.* **259,** 789 (1998); and D. U. Gommel, E. Nogoceke, M. Morr, M. Kiess, H. M. Kalisz, and L. Flohé, *Eur. J. Biochem.* **248,** 913 (1997)] or His-tagged [P. Steinert, K. Plank-Schumacher, M. Montemartini, H. J. Hecht, and L. Flohé, *Biol. Chem.* **381,** 211 (2000)] TXNPx.

$$\phi_1 = \frac{k_{-1} + k_{+2}}{k_{+1}k_{+2}} \tag{3}$$

$$\phi_2 = \frac{k_{-3} + k_{+4}}{k_{+3}k_{+4}} \tag{4}$$

Numeric values for kinetic constants of TXNs are compiled in Table I.[15,16,21,22]

Assay Procedure

The choice of assay conditions requires consideration of several peculiarities of the tryparedoxins: (1) Because of the kinetic mechanism the apparent K_m values vary with the cosubstrate concentrations[21]; (2) limiting K_m values for T(SH)$_2$ are in the medium micromolar range whereas those for TXNPx are in the low micromolar range[15,16,21,22] (Table I); (3) substrate inhibition by T(SH)$_2$ can be observed[22]; (4) the oxidizing protein substrate TXNPx is not particularly stable; TXNPx of *L. donovani* is, for example, denatured by an excess of hydroperoxides (our unpublished data, 2000).

Determination of the trypanothione:peroxiredoxin oxidoreductase activity is most conveniently achieved by a coupled test that continuously measures trypanothione reductase (TR)-catalyzed NADPH oxidation by reduction of TS$_2$, which is becoming oxidized by TXN, while the TXNPx is kept oxidized by an excess of hydroperoxide. The sequence of events is illustrated in Fig. 1. Apart from the usual precautions that must be considered in coupled test systems, three points are of crucial importance: (1) because of the sensitivity of TXNPx to excess hydroperoxides, in particular to H_2O_2, the reaction must be started by the peroxide substrate; (2) because working under substrate saturation is neither feasible nor advisable owing to substrate inhibition and usually limited availability of substrates, the initial rate obtained depends on the concentration of both substrates, $T(SH)_2$ and oxidized TXNPx, although the initial turnover seems to follow zero-order kinetics due to substrate regeneration; (3) because spontaneous turnover of $T(SH)_2$ by ROOH proceeds by higher level kinetics than the pseudo-zero-order kinetics of the enzymatic reaction, blanks are not easily considered just by subtraction of slopes. It is therefore recommended that TXN activity be maximized in the assay and that the spontaneous reaction be minimized by choice of a hydroperoxide displaying marginal reactivity with SH groups, such as *tert*-butyl hydroperoxide. A typical test procedure is as follows: The sample containing TXN in a range of 100–1000 nM corresponding to 1.6 to 16 μg ml^{-1} is preincubated in 50 mM HEPES (or Tris)–1 mM EDTA (pH 7.6) at the selected temperature (25 or 37) for 15 min with (final concentrations)

NADPH, 150 μM
TR, 1.0 U ml^{-1}
$T(SH)_2$, 20 μM
TXNPxH_6, 10 μM

where H_6 represents a polyhistidine tail. The reaction is then started with 70 μM *tert*-butyl hydroperoxide and the initial slope, which should be linear, is monitored in a spectrophotometer at 340 nm for 5 min. If (partially) oxidized trypanothione is used in this assay, the NADPH concentration must be adjusted to assure an OD_{340} of at least 0.8. Blanks are run identically but without TXN. Activities are calculated as usual from differences of slopes and given as $U \equiv \Delta \mu$mol NADPH/min $= \Delta E \cdot \varepsilon^{-1} \cdot$ min$^{-1} \cdot$ dilution factor $= 159 \cdot \Delta OD_{340} \cdot$ min$^{-1} \cdot$ dilution factor.

If activities measured under different conditions are to be compared, the rate equation and kinetic parameters listed in Table I may be used. Under the conditions recommended, the rate is primarily limited by the reduction of oxidized TXN by $T(SH)_2$ and should, accordingly, be roughly proportional to the $T(SH)_2$ concentration, at least up to 30 μM. Substrate inhibition (as, e.g., observed with CfTXN2) may, however, render such estimates invalid. It is therefore recommended that the optimal testing conditions be reevaluated if new TXNs are to be determined.

Tryparedoxin Peroxidases

General Characteristics

Tryparedoxin peroxidase (TXNPx) belongs to the peroxiredoxins, a family of redox-active proteins widely spread in nature.[24] The first member of the family

[24] S. G. Rhee, S. W. Kang, L. E. Netto, M. S. Seo, and E. R. Stadtman, *BioFactors* **10**, 207 (1999).

to be discovered was the "thiol-specific antioxidant protein (TSA)" of yeast,[25] which was later reclassified as thioredoxin peroxidase.[26] Peroxiredoxins appear to be common in bacteria, where they constitute the alkyl hydroperoxide reductase (AhpC) system,[27-29] but they are also present in protozoa, plants, lower animals, and mammals, in which their precise biological role remains largely elusive.[2,24] The general enzymatic activity of peroxiredoxins appears to be the reduction of hydroperoxides at the expense of thiols.[30] They all appear to be homo-oligomeric proteins with subunit sizes in the 20-kDa range. Catalysis is mediated by one or two activated cysteines. TXNPx of *C. fasciculata*,[31] *T. brucei*,[5] *T. cruzi*,[7,32] *L. major*,[8] and *L. donovani*[9] proved to be typical two-cysteine peroxiredoxins, in which the active cysteines are embedded in VCP motifs, as often found in peroxiredoxins.[2] TXNPx of *C. fasciculata* (*Cf*TXNPx) is active only as an oligomer with a molecular mass beyond 200 kDa,[4] which according to a low-resolution X-ray crystallographic analysis is a ring composed of five dimers of inversely oriented monomers.[33] Like TXN, TXNPx appears also to be encoded by multiple gene copies as demonstrated for *C. fasciculata*[31] and *L. major*.[8]

Catalytic Mechanism

The present view of the catalytic mechanism, as illustrated in Fig. 4, was originally deduced from previous investigations of the homologous yeast thioredoxin peroxidase[34] and the first established X-ray structure of a peroxiredoxin, that of hORF6,[35] now reclassified as a nonselenium glutathione peroxidase.[36] In the yeast peroxiredoxin, an exchange of the cysteine in the N-proximal VCP motif against serine led to complete loss of activity. In contrast, the analogous mutation in the distal VCP motif did not abrogate the peroxidatic activity when measured with

[25] K. Kim, I. H. Kim, K.-Y. Lee, S. G. Rhee, and E. R. Stadtman, *J. Biol. Chem.* **263**, 4704 (1988).

[26] H. Z. Chae, S. J. Chung, and S. G. Rhee, *J. Biol. Chem.* **269**, 27670 (1994).

[27] L. A. Tartaglia, G. Storz, M. H. Brodsky, A. Lai, and B. N. Ames, *J. Biol. Chem.* **265**, 10535 (1990).

[28] Y. Niimura, L. B. Poole, and V. Massey, *J. Biol. Chem.* **270**, 25645 (1995).

[29] D. R. Sherman, K. Mdluli, M. J. Hickey, M. C. E. Berry, and C. K. Stover, *BioFactors* **10**, 211 (1999).

[30] L. E. S. Netto, H. Z. Chae, S.-W. Kang, S. G. Rhee, and E. R. Stadtman, *J. Biol. Chem.* **271**, 15315 (1996).

[31] M. Montemartini, E. Nogoceke, M. Singh, P. Steinert, L. Flohé, and H. M. Kalisz, *J. Biol. Chem.* **273**, 4864 (1998).

[32] S. A. Guerrero, J. A. Lopez, P. Steinert, M. Montemartini, H. M. Kalisz, W. Colli, M. Singh, M. J. M. Alves, and L. Flohé, *Appl. Microbiol. Biotechnol.* **53**, 410 (2000).

[33] M. S. Alphey, C. S. Bond, E. Tetaud, A. H. Fairlamb, and W. N. Hunter, *J. Mol. Biol.* **300**, 903 (2000).

[34] H. Z. Chae, T. B. Uhm, and S. G. Rhee, *Proc. Natl. Acad. Sci. U.S.A.* **91**, 7022 (1994).

[35] H.-J. Choi, S. W. Kang, C.-H. Yang, S. G. Rhee, and S.-E. Ryu, *Nat. Struct. Biol* **5**, 400 (1998).

[36] J. W. Chen, C. Dodia, S. I. Feinstein, M. K. Jain, and A. B. Fisher, *J. Biol. Chem.* **275**, 28421 (2000).

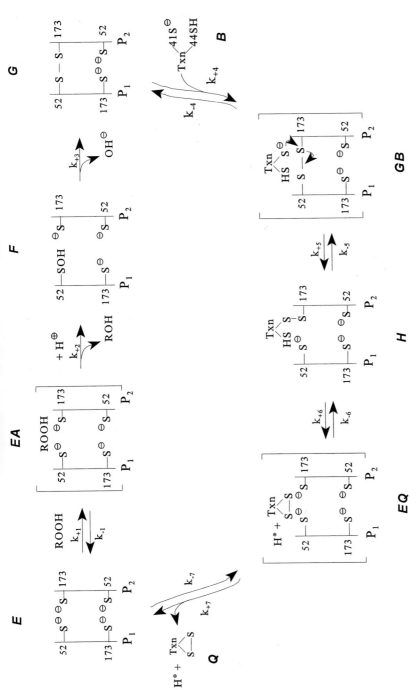

FIG. 4. Scheme of the reaction mechanism of *Cf*TXNPx. TXN, Tryparedoxin (sequence number of *Cf*TXN2); ROOH, hydroperoxide; ROH, alcohol; E, ground state enzyme; F, G, H, intermediates; P₁ and P₂, first and second subunit of the dimeric protein. The sulfhydryl group of Cys-41 in *Cf*TXN2 is shown to be dissociated to emphasize its presumed nucleophilic attack on the disulfide bridge of oxidized TXNPx. Identical reactions may occur with the second pair of SH groups of the intersubunit active site (not shown).

dithiothreitol (DTT) as donor substrate but blocked the reaction with thioredoxin.[34] The three-dimensional (3D) structure of hORF6 revealed the subunits of the dimeric molecule to be inversely oriented.[35] Taken together, the findings favored the following assumptions[34]: (1) the proximal cysteine should be the component reacting with the peroxide and therefore is indispensable for any kind of peroxidatic activity; (2) a sulfenic acid derivative formed by the reaction of the proximal cysteine could then react with the distal cysteine of an inversed second subunit to form an intersubunit disulfide bridge; and (3) the specific donor substrate thioredoxin can attack the oxidized peroxidase only at the sulfur of the distal cysteine. These ideas are likely valid also for TXNPx. Exchange of the Cys-52 versus serine in CfTXNPx abolishes activity, whereas the mutation of Cys-173 only reduces activity,[37] and the 3D structure[33] in fact shows the subunits of TXNPx inversely oriented in such a way that the proximal and distal VCP motifs could interact.

Despite the multiple steps implicated in the catalysis of TXNPx, the kinetic pattern and the rate equation look surprisingly simple[4]: It is a ping–pong pattern with infinite maximum velocities and infinite Michaelis constants, as is also observed with the selenium-containing glutathione peroxidases.[38] In terms of a Dalziel equation[23] the initial velocities of TXNPx are described by Eq. (5):

$$\frac{[E_0]}{v} = \phi_0 + \frac{\phi_1}{[ROOH]} + \frac{\phi_2}{[TXN]} \qquad (5)$$

where ϕ_1 and ϕ_2 are the reciprocal values of the net forward apparent oxidative and reductive rate constants, respectively. The ϕ_0 term of the general equation for two-substrate ping–pong mechanisms [Eq. (1)] is usually zero. According to the scheme shown in Fig. 4, this implies that the formation of enzyme–substrate complexes, that is, the formation of EA and GB, is slower than any consecutive reaction within such complexes. In consequence, no enzyme–substrate complex can accumulate to become rate limiting. The rate-limiting steps, depending on substrate concentrations, consist of either the one yielding EA from E and A or that forming GB from G and B. In other words, the empirical Dalziel coefficients ϕ, being the reciprocal values of rate-limiting apparent constants, are more precisely defined as: Follows $\phi_1 = 1/(k_{+1} - k_{-1})$ and $\phi_2 = 1/(k_{+4} - k_{-4})$. Numeric values of pertinent constants are compiled in Table II. Deviations from the kinetic pattern were observed when TXNPx was investigated with a heterologous TXN[21] or when the distal reaction center was altered by site-directed mutagenesis.[37] In such cases saturation kinetics were observed, indicating that one or the other of the reactions within the complexes has become rate limiting (Table II[4,31,32,37,38]).

---

[37] M. Montemartini, H. M. Kalisz, H.-J. Hecht, P. Steinert, and L. Flohé, *Eur. J. Biochem.* **264,** 516 (1999).
[38] R. Brigelius-Flohé, K. Wingler, and C. Müller, *Methods Enzymol.* **347,** [9], 2002 (this volume).

TABLE II

KINETIC CONSTANTS OF TRYPAREDOXIN PEROXIDASE

Type of TXNPx	Oxidizing substrate (A)	Reducing substrate (B)	k'_1 $(10^6 M^{-1} sec^{-1})$	k'_2 $(10^6 M^{-1} sec^{-1})$	k_{cat} (sec^{-1})	K_{mA} (μM)	K_{mB} (μM)	Ref.
Cf TXNPx	H_2O_2	Cf TXN1	0.1	3.5	∞	∞	∞	4
	tert-bOOH	Cf TXN1	0.2	1.9	∞	∞	∞	4
	LOOH	Cf TXN1	0.1	2.0	∞	∞	∞	4
	PCOOH	Cf TXN1	0.04	1.6	∞	∞	∞	4
Cf TXNPxH6	tert-bOOH	Cf TXN1	0.08	0.9	∞	∞	∞	31
Cf TXNPxH6-C173S	tert-bOOH	Cf TXN1	0.008	3.8	0.2	29.9	0.06	37
TcH6TXNPx	tert-bOOH	Cf TXN1H6	0.03	1.0	1.7	51.8	1.65	32
GPx	H_2O_2	GSH	59	0.5	∞	∞	∞	38

[a] tert-bOOH, tert-Butyl hydroperoxide; LOOH, linoleic acid hydroperoxide; PCOOH, phosphatidylcholine hydroperoxide; Cf TXNPx, authentic TXNPx isolated from C. fasciculata; Cf TXNPxH6, C-terminally His-tagged recombinant Cf TXNPx; Cf TXNPxH6-C173S, mutein thereof; TcH6TXNPx, N-terminally His-tagged recombinant TXNPx of T. cruzi; Cf TXN1, authentic TXN1 isolated from C. fasciculata; Cf TXN1H6, C-terminally His-tagged recombinant Cf TXN1; values for the selenoprotein cytosolic glutathione peroxidase (GPx) are listed for comparison. For other related values see R. Brigelius-Flohé, K. Wingler, and C. Müller, Methods Enzymol. 347, [9], 2002 (this volume).

There remains the question concerning what renders the proximal cysteine residue active enough to make TXNPx a peroxidase. In fact, the net forward rate constant k'_{+1} of TXNPx, which is $k_{+1} - k_{-1}$, is small when compared with those of selenium- or heme-containing peroxidases (Table II). It is nevertheless close to that of sulfur-containing homologs of glutathione peroxidases.[39,40] This implies that Cys-52 must be located in an environment that enforces dissociation of its thiol group, because only under such circumstances can a satisfying reaction rate with a hydroperoxy group be envisaged. On the basis of homology considerations and site-directed mutagenesis, activation of Cys-52 of Cf TXNPx, that is, the cysteine of the first VCP motif, by Arg-128 and Trp-87 had been suggested.[2,37] Although an electrostatic activation of Cys-52 by Arg-128 was compatible with molecular modeling,[2,37] and later supported by X-ray analysis,[33] the supposed hydrogen bonding of the ring nitrogen of Trp-87 to the sulfur of Cys-52 remained questionable.[33,37] Instead, Thr-49 appears to contribute the dissociation of Cys-52 by hydrogen bonding. Mutagenesis studies with TXNPx of L. donovani (Ld TXNPx) showed that Thr-49 may be replaced by serine without impairment of activity, whereas an apolar residue in this position dramatically reduces activity

[39] M. Maiorino, K.-D. Aumann, R. Brigelius-Flohé, D. Doria, J. van den Heuvel, J. McCarthy, A. Roveri, F. Ursini, and L. Flohé, Biol. Chem. 376, 651 (1995).
[40] H. Sztajer, B. Gamain, K.-D. Aumann, C. Slomianny, K. Becker, R. Brigelius-Flohé, and L. Flohé, J. Biol. Chem. 276, 7397 (2001).

(our unpublished data, 2001). Thr-49 is strictly conserved in all known TXNPx species, is highly conserved in most peroxiredoxins, and is replaced by a functionally equivalent serine in some of them. It is likely, therefore, that the first catalytic step in TXNPx is achieved by means of a catalytic triad in which the essential cysteine is activated by the positive charge of Arg-128 and hydrogen bonding by Thr-49. This conclusion is consistent with mutagenesis studies and molecular modeling,[2,37] it appears compatible with the X-ray analysis of CfTXNPx,[33] although the coordinates of the structures have not yet been made available, and is further supported by the structural analysis of a human peroxiredoxin showing the homologous residues in ideal coordination to fulfill the presumed catalytic role (Fig. 5).[41,42] How the second conserved cysteine (Cys-173) is activated and precisely interacts with the first one in the consecutive steps of the catalytic cycle (Fig. 4) remains to be established.

Assay Procedure

TXNPx, like other peroxiredoxins, is active in the assay system measuring protection of glutamine synthetase against oxidative inactivation by iron and DTT.[25] More conveniently, TXNPx activity is measured in the coupled test system applied for TXN determination (see above and Fig. 1) with appropriate modifications. Most importantly, instant regeneration of reduced TXN must be guaranteed by increasing the concentration of T(SH)$_2$. The most convenient TXN-type substrate is the easily prepared[16] histidine-tagged TXN of *C. fasciculata* (CfTXN1H$_6$). As the condition of "substrate saturation" cannot be achieved because of the kinetic mechanism (see above), the concentrations of TXN and peroxide can, in principle, be chosen arbitrarily, but must be clearly stated to guarantee comparability of results.

Our standard assay conditions are as follows: TXNPx in a range of 1 to 4 μg/ml is preincubated in 50 mM HEPES (pH 7.6)–1 mM EDTA at 25° with 150 μM NADPH (300 μM, if TS$_2$ is used), TR (1 U/ml), 130 μM T(SH)$_2$, and 20 μM CfTXN1H$_6$ (final concentrations). After 15 min the reaction is started with *tert*-butyl hydroperoxide at an initial test concentration of 73 μM. The rate of hydroperoxide reduction is calculated from the initial rate of NADPH consumption as described above. Blanks are processed identically with omission of TXNPx. The activity is defined as $U \equiv \Delta$ μmol NADPH/min $= \Delta$ μmol ROOH/ min $= \Delta E \cdot \varepsilon^{-1} \cdot$ min$^{-1} \cdot$ dilution factor $= 159 \cdot \Delta OD_{340} \cdot$ min$^{-1} \cdot$ dilution factor.

Some precautions should be considered: (1) If a TXNPx less characterized than CfTXNPx is to be measured, make sure that the regeneration of the reduced TXN

[41] E. Scröder, J. A. Littlechild, A. A. Lebedev, N. Errington, A. A. Vagin, and M. N. Isupov, *Struct. Fold Des.* **15,** 605 (2000).
[42] P. J. Kraulis, *J. Appl. Crystallogr.* **24,** 946 (1991).

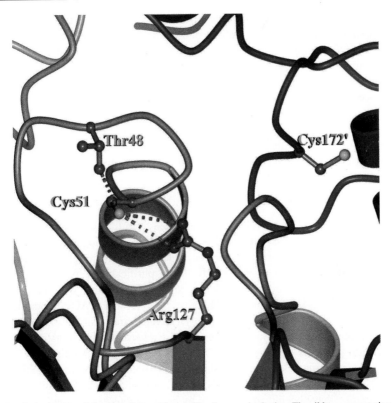

FIG. 5. Activation of the essential cysteine residue in peroxiredoxins. The ribbon presentation reproduces the structure of a peroxiredoxin from human erythrocytes (pdb-id1QMV) [E. Scröder, J. A. Littlechild, A. A. Lebedev, N. Errington, A. A. Vagin, and M. N. Isupov, *Struct. Fold Des.* **15,** 605 (2000)], which in its N-terminal domain (brown) is similar to TXNPx. The sulfur of Cys-51, which here is modeled in reduced form, is coordinated to the OH of Thr-48, and to the positively charged guanidino group of Arg-127. The Cys-51, Thr-48, and Arg-127 of the structure shown are homologous to the residues presumed to build up the N-terminal catalytic triad in the tryparedoxin peroxidases, for example, Cys-52, Thr-49, and Arg-128 in *Cf* TXNPx. The distal conserved Cys-172 in the inversed second subunit (green) is located in the homologous position of Cys-173 in *Cf* TXNPx, which there is presumed to interact with Cys-52 during catalysis (see Fig. 4). To facilitate such interaction substantial structural changes evidently occur during catalysis. (Prepared with Molscript [P. J. Kraulis, *J. Appl. Crystallogr.* **24,** 946 (1991)] and POV-ray, www.povray.org.)

does not become rate limiting. This can hardly be achieved by increasing the T(SH)$_2$ concentration, because a T(SH)$_2$ concentration really saturating TXN (>1 mM) may be rated as commercially not affordable and may not be helpful because of substrate inhibition of TXN. Dilution of the TXNPx sample to be investigated is the more appropriate solution; (2) some of the TXNPx (e.g., *Ld* TXNPx) tend to become readily inactivated by hydroperoxides, in particular by H$_2$O$_2$. It is therefore advised to rely on initial velocities only; (3) the kinetics of TXNPx imply that the velocities measured depend on the concentrations of both substrates, TXNH$_2$ and ROOH. The units of activity are therefore valid only for the conditions applied. Recalculations of activities from measurements performed at other substrate concentrations by means of rate Eq. (2) are justified only if the ϕ or k' values for the particular TXNPx have been established; and (4) purified TXNPX of *C. fasciculata,* as likely those of other species, tends to become inactivated if diluted to less than 300 μg ml^{-1}. Even when stored on ice, dilution of a 300-μg ml^{-1} solution to 150 μg ml^{-1} results in a 50% loss of activity within 1hr. It is therefore recommended to transfer undiluted aliquots of the TXNPx solution into the test cuvette by means of micropipettes.

Sources of Purified Tryparedoxin and Tryparedoxin Peroxidase

Authentic TXN1 and TXNPx can be isolated from *C. fasciculata* as described by Nogoceke *et al.*[4] In essence, *Cf*TXNPx is purified to homogeneity from lysates of *C. fasciculata* by chromatography on S-Sepharose, hydroxylapatite, and Resource Q. *Cf*TXN1 is obtained by the same purification procedure from the flowthrough of the S-Sepharose column, there still coeluting with TR. Separation of *Cf*TXN1 and TR is achieved by absorbing TR to 2,5-ADP-Sepharose.

More conveniently, *Cf*TXN1[16] and *Cf*TXNPx[31] are obtained as polyhistine-tagged proteins heterologously expressed in *Escherichia coli* transformed with expression plasmids based on pET22 or pET24 (Novagen, Madison, WI). The His-tagged proteins can be readily isolated by His-bind resin (Novagen). Similarly, TXN2 of *C. fasciculata*[15] and the TXN from *T. brucei*[6] and TXNPx from *T. cruzi*[32] and *L. major*[8] have been made available by heterologous expression.

Conclusions

Tryparedoxins and tryparedoxin peroxidases together with trypanothione and trypanothione reductase constitute the peroxidase system typical of trypanosomatids. It is obviously of vital importance to trypanosomes. In consequence, the enzymes constituting the system are considered attractive targets in the search for trypanocidal drugs. The assays presented may easily be adapted for high-throughput screening for suitable inhibitors. The knowledge on the 3D structures of TR, TXN, and TXNPx provides an ideal basis for rational inhibitor design.

Acknowledgments

Our more recent work related to this article was supported by the Deutsche Forschungsgemeinschaft (Grants FL61/8–3, FL61/11-1, and He2554/2-1).

[23] Trypanothione and Tryparedoxin in Ribonucleotide Reduction

By R. Luise Krauth-Siegel and Heide Schmidt

Introduction

Trypanosomatids are the causative agents of South American Chagas' disease (*Trypanosoma cruzi*), African sleeping sickness (*T. brucei rhodesiense* and *T. brucei gambiense*), Nagana cattle disease (*T. congolense and T. brucei brucei*), and the three manifestations of leishmaniasis. All these parasitic protozoa have in common that the ubiquitous glutathione–glutathione reductase system is replaced by a trypanothione–trypanothione reductase system.

Monoglutathionylspermidine (Gsp) and trypanothione [N^1,N^8-bis(glutathionyl) spermidine; $T(SH)_2$] are the main low molecular mass thiols and are responsible for the redox balance of the cell.[1-3] These glutathionylspermidine conjugates are kept reduced by the flavoenzyme trypanothione reductase (TR, EC 1.6.4.8): TS_2 + NADPH + H^+ → $T(SH)_2$ + NADP (TS_2, trypanothione disulfide), an essential enzyme of the parasite.[2,4]

Trypanothione is responsible for the detoxification of hydrogen peroxide in trypanosomatids, the reaction being catalyzed by an enzyme cascade composed of trypanothione, trypanothione reductase, tryparedoxin, and a tryparedoxin peroxidase (see Nogoceke *et al.*[5] and [22] in this volume[6]). Tryparedoxin is a thiol protein with a molecular mass of 16 kDa and an active site WCPPC motif. The protein was first isolated from the insect parasite *Crithidia fasciculata*[7] but sequences encoding tryparedoxins have been detected in a wide variety of trypanosomatid genoms. *Tpx* genes have been cloned and overexpressed from *T. brucei*[8] and *C. fasciculata*.[9-11]

[1] A. H. Fairlamb and A. Cerami, *Annu. Rev. Microbiol.* **46**, 695 (1992).
[2] R. L. Krauth-Siegel and G. Coombs, *Parasitol. Today* **15**, 404 (1999).
[3] R. L. Krauth-Siegel, E. M. Jacoby, and R. H. Schirmer, *Methods Enzymol.* **251**, 287 (1995).
[4] S. Krieger, W. Schwarz, M. R. Ariyanayagam, A. H. Fairlamb, R. L. Krauth-Siegel, and C. Clayton, *Mol. Microbiol.* **35**, 542 (2000).
[5] E. Nogoceke, D. U. Gommel, M. Kiess, H. M. Kalisz, and L. Flohé, *Biol. Chem.* **378**, 827 (1997).
[6] L. Flohé, P. Steinert, H.-J. Hecht, and B. Hofmann, *Methods Enzymol.* **347**, [22], 2002 (this volume).
[7] D. U. Gommel, E. Nogoceke, M. Morr, M. Kiess, H. M. Kalisz, and L. Flohé, *Eur. J. Biochem.* **248**, 913 (1997).
[8] H. Lüdemann, M. Dormeyer, C. Sticherling, D. Stallmann, H. Follmann, and R. L. Krauth-Siegel, *FEBS Lett.* **431**, 381 (1998).
[9] M. Montemartini, H. M. Kalisz, M. Kiess, E. Nogoceke, M. Singh, P. Steinert, and L. Flohé, *Biol. Chem.* **379**, 1137 (1998).
[10] S. A. Guerrero, L. Flohé, H. M. Kalisz, M. Montemartini, E. Nogoceke, H. J. Hecht, P. Steinert, and M. Singh, *Eur. J. Biochem.* **259**, 789 (1999).

Tryparedoxin functions as a trypanothione-dependent thiol–disulfide oxidoreductase with catalytic properties intermediate between those of classic thioredoxins and glutaredoxins.[8]

African trypanosomes possess a typical eukaryotic class I ribonucleotide reductase (EC 1.17.4.1). The genes encoding the *Trypanosoma brucei* R1 and R2 proteins have been cloned and overexpressed in *Escherichia coli*.[12–14] *Trypanosoma brucei* ribonucleotide reductase is regulated via the R2 subunit. Whereas the R1 protein is present throughout the life cycle of the parasite, the R2 protein is not found in cell cycle-arrested short stumpy trypanosomes.[15]

Reduction of the 2′-OH group of ribonucleoside diphosphates to the corresponding deoxynucleotides requires external electron donors. For class I enzymes, small thiol proteins with an active site CXXC motif, such as thioredoxin (CGPC) and glutaredoxin (CPYC), are well-known hydrogen donors.[16–18] Oxidized thioredoxin is subsequently reduced by thioredoxin reductase at the expense of NADPH.[18] The dithiol form of glutaredoxin is spontaneously regenerated by glutathione. Glutathione disulfide formed in the reaction is then reduced by NADPH and glutathione reductase.[19,20] The trypanosomatid-specific dithiol trypanothione—in contrast to the monothiol glutathione—is a direct donor of reducing equivalents for *T. brucei* ribonucleotide reductase and the reaction is catalyzed by tryparedoxin (Fig. 1).[21]

Methods

Materials

[³H]GDP is purchased from Amersham Pharmacia (Braunschweig, Germany), GDP and dTTP from Sigma (Deisenhofen, Germany), trypanothione disulfide (TS$_2$) and glutathionylspermidine disulfide (Gsp$_{ox}$) from Bachem (Heidelberg, Germany), and NaBH$_4$ from Fluka (Wiesbaden, Germany). The reduced forms of the glutathionylspermidine conjugates are prepared as described below.

[11] E. Tetaud and A. H. Fairlamb, *Mol. Biochem. Parasitol.* **96,** 111 (1998).
[12] A. Hofer, P. P. Schmidt, A. Gräslund, and L. Thelander, *Proc. Natl. Acad. Sci. U.S.A.* **94,** 6959 (1997).
[13] M. Dormeyer, R. Schöneck, G. A. G. Dittmar, and R. L. Krauth-Siegel, *FEBS Lett.* **414,** 449 (1997).
[14] A. Hofer, J. T. Ekanem, and L. Thelander, *J. Biol. Chem.* **273,** 34098 (1998).
[15] T. Breidbach, R. L. Krauth-Siegel, and D. Steverding, *FEBS Lett.* **473,** 212 (2000).
[16] A. Jordan and P. Reichard, *Annu. Rev. Biochem.* **67,** 71 (1998).
[17] A. Holmgren, *J. Biol. Chem.* **264,** 13963 (1989).
[18] A. Holmgren, *Annu. Rev. Biochem.* **54,** 237 (1985).
[19] A. Holmgren, *J. Biol. Chem.* **254,** 3664 (1979).
[20] A. Holmgren, *J. Biol. Chem.* **254,** 3672 (1979).
[21] M. Dormeyer, N. Reckenfelderbäumer, H. Lüdemann, and R. L. Krauth-Siegel, *J. Biol. Chem.* **276,** 10602 (2001).

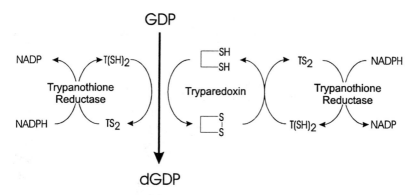

FIG. 1. Hydrogen donors of *T. brucei* ribonucleotide reductase. *Left:* Spontaneous reduction of ribonucleotide reductase by trypanothione. *Right:* Tryparedoxin-catalyzed reduction of ribonucleotide reductase by trypanothione.

$2'$-Deoxyadenosine $5'$-(γ-4-aminophenyl)triphosphate, sodium salt is purchased from Amersham Life Sciences or synthesized as described by Berglund and Eckstein.[22] The dATP-Sepharose is prepared according to published procedures.[22] C_{18} cartridges are obtained from Millipore (Bedford, MA), the Aminex A-9 anion-exchange resin is from Bio-Rad (München, Germany), and the PD-10 columns are from Pharmacia (Freiburg, Germany).

Recombinant *T. brucei* tryparedoxin[8] and *T. cruzi* trypanothione reductase[23] are purified according to published procedures. Alkaline phosphatase from calf intestine is purchased from Roche Molecular Biochemicals (Mannheim, Germany).

Expression and Purification of Trypanosoma brucei Ribonucleotide Reductase

Expression of R1 Gene. An overnight culture of recombinant *E. coli* BL21(DE3) pLysS cells with the pET3*Tb*R1 plasmid[12] is grown at 37° in Terrific Broth (TB) medium containing carbenicillin (100 μg/ml). Fifty milliliters of overnight culture is diluted into 5 liters of TB medium with carbenicillin (100 μg/ml) and the culture is shaken at 37°. At an OD_{600} of 0.6, expression is induced by adding 1 mM isopropyl-β-D-thiogalactopyranoside (IPTG) and the suspension is incubated for 18 hr at 16°. The cells are harvested by centrifugation and stored at $-80°$.

Purification of R1 Protein. All steps are carried out at 4° and the buffers are purged with argon before use.

Cells from 2.5 liters of *E. coli* culture are suspended in 40 ml of 50 mM HEPES, pH 7.3, and disintegrated by sonication. After centrifugation (20 min, 30,000g), the supernatant is kept on ice and the pellet is again suspended in 40 ml of 50 mM

[22] O. Berglund and F. Eckstein, *Eur. J. Biochem.* **28**, 492 (1972).
[23] F. X. Sullivan and C. T. Walsh, *Mol. Biochem. Parasitol.* **44**, 145 (1991).

HEPES, pH 7.3, sonicated, and centrifuged. The supernatants are combined and ammonium sulfate is added to 50% saturation. The suspension is kept on ice for 60–90 min and then centrifuged (20 min, 30,000g). The pellet is dissolved in 50–100 ml of 50 mM HEPES, pH 7.3, and cleared by centrifugation (the conductivity of the solution should be <30 mS/cm). The supernatant is applied onto a 5-ml dATP-Sepharose column. The column is washed with 50 mM HEPES, pH 7.3, at a flow rate of ≤0.5 ml/min until the ΔA_{280} of the runthrough is ≤0.1, followed by addition of each 20 ml of 0.5 mM ADP in 50 mM HEPES, pH 7.3, 0.2 M KCl in 50 mM HEPES, pH 7.3, and 1 mM ATP in 50 mM HEPES, pH 7.3. The R1 protein is eluted with 15 ml of 10 mM ATP in 50 mM HEPES, pH 7.3, and precipitated by 80% ammonium sulfate. After 2 hr on ice the suspension is centrifuged (20 min, 30,000g), the protein pellet is suspended in 80% ammonium sulfate in 50 mM HEPES, pH 7.3, and again centrifuged. The pellet is dissolved in 500–700 μl of 50 mM Tris-HCl, 0.1 M KCl, 10% (v/v) glycerol, pH 7.6 and desalted on a PD-10 column equilibrated in the same buffer. The concentration of the R1 protein is adjusted to 3.0 mg/ml and aliquots are stored at −80°. The total yield of the R1 protein is about 3 mg per liter of culture. The protein solution is thawed only once.

Expression of R2 Gene. An overnight culture of *E. coli* BL21(DE3) cells containing the pET3*Tb*R2 plasmid is grown at 37° in Luria–Bertani (LB) medium with carbenicillin (100 μg/ml).[12] Fifty milliliters of overnight culture is added to 5 liters of LB medium with carbenicillin (100 μg/ml) and incubated at 37°. At an OD$_{600}$ of 0.6, 1 mM IPTG is added and the culture is shaken for another 4 hr at 37°. The cells are harvested by centrifugation (15 min, 4000g) and stored overnight at −80°.

Purification of Trypanosoma brucei R2 Protein. The purification is carried out at 4° and the buffers are purged with argon.

The cell pellet of a 5-liter culture is suspended in 40 ml of 50 mM Tris-HCl, pH 7.6, and the cells are disintegrated by sonication. After centrifugation (20 min, 30,000g), the supernatant is stored on ice and the pellet is suspended in another 40 ml of buffer and sonicated. After centrifugation, both supernatants are combined and streptomycin sulfate is added to 2.5% (w/v). The solution is stirred for 15 min and centrifuged (20 min, 30,000g). The R2 protein is precipitated by a 40–48% ammmonium sulfate fractionation. The protein pellet is dissolved in 1–1.5 ml of 10 mM potassium phosphate, pH 7.0, and desalted on a PD-10 column. The solution is applied onto a 10-ml DEAE-Sepharose (DE52; Whatman, Clifton, NJ) column equilibrated with 10 mM potassium phosphate, pH 7.0. The column is washed with 80 mM KCl in 10 mM potassium phosphate, pH 7.0, until the ΔA_{280} of the runthrough is ≤0.1. The R2 protein is eluted with 160 mM KCl in 10 mM potassium phosphate, pH 7.0. The volume (about 10 ml) is reduced to 1 ml in Centriprep 10 concentrators (Millipore) and the protein solution is desalted on a PD-10 column equilibrated with 50 mM Tris-HCl, pH 7.6. The protein concentration is adjusted to 5–10 mg/ml and fractions are stored at −80°. The final yield is about 2.5 mg of R2 protein per liter of culture. The protein solution is thawed only once.

Ribonucleotide Reductase Assay

The assay described here uses GDP as substrate and contains dTTP as positive effector of GDP reduction by *T. brucei* ribonucleotide reductase.[14] The conditions are derived from those described for CDP reduction by calf thymus ribonucleotide reductase.[24]

In a final volume of 200 μl of 50 mM HEPES–KOH, pH 7.6, the assay mixture contains 500 μM GDP (including 1.25 μCi of [^{3}H]GDP), 100 μM dTTP, 100 mM KCl, 6.4 mM MgCl$_2$, and variable concentrations of thiols (see below). The reaction is started by adding the R1 and R2 proteins and allowed to proceed at 37° for 10 to 30 min. The R1 concentration is chosen so that not more than 15% of the GDP is consumed during the assay. The R2 protein is present in an at least 3-fold molar excess. Because of the low activity of isolated ribonucleotide reductases the specific activity refers to nanomoles per minute per milligram of R1. The reaction is stopped by boiling for 10 min, precipitated protein is removed by centrifugation, and the nucleotides are dephosphorylated by adding 10 U of alkaline phosphatase. After incubation at 37° for 45 min the sample is again boiled for 5 min and centrifuged. The supernatant is used for nucleoside analysis.[25] Guanosine, deoxyguanosine, and guanine are separated by high performance liquid chromatography (HPLC) on a thermostatted Aminex A-9 cation exchange column (250 × 4 mm) at 37°. Samples (40 μl) are injected and the components are eluted isocratically with 100 mM ammonium borate buffer, pH 8.2, at a flow rate of 1 ml/min. Six 3-ml fractions are directly collected in scintillation vials. Each two vials correspond to guanosine, deoxyguanosine, and guanine (if any) (Fig. 2). Ribonucleotide reductase activity is calculated from the percent radioactivity in the deoxyguanosine fractions in relation to the total radioactivity in the assay.

Trypanothione as Reductant of Ribonucleotide Reductase

Reduction of nucleoside diphosphates to the respective deoxyribonucleotides catalyzed by class I ribonucleotide reductases requires external hydrogen donors. The thioredoxin–NADPH–thioredoxin reductase and the glutaredoxin–glutathione–NADPH–glutathione reductase systems are well-known electron donors. *In vitro* the dithiol DTE (dithioerythritol) can replace the physiological systems,[24,25] but monothiols such as glutathione are unable to deliver reducing equivalents directly.

Trypanothione, T(SH)$_2$, the main low molecular mass thiol in trypanosomatids, is a direct donor of reducing equivalents for ribonucleotide reduction. The dithiol form of trypanothione is generated *in situ* to ensure a constant thiol concentration during the reaction and to prevent accumulation of the disulfide. The assay mixture contains 2 mM trypanothione disulfide, 200 mU of *T. cruzi* TR, 5 mM NADPH,

[24] Y. Engström, S. Eriksson, L. Thelander, and M. Åkerman, *Biochemistry* **18**, 2941 (1979).

[25] A. Willing, H. Follmann, and G. Auling, *Eur. J. Biochem.* **170**, 603 (1988).

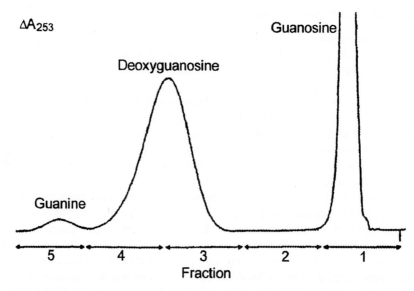

FIG. 2. Separation of guanosine, deoxyguanosine, and guanine by HPLC on Aminex A-9 in 100 mM ammonium borate, pH 8.2, at 37°.

and all components of the ribonucleotide reductase assay except R1, R2, and radiolabeled GDP. The mixture is incubated for 15 min at 37° and the reaction is then started by adding R1, R2, and [^{3}H]GDP. V_{max} is calculated using a K_m value of 2.1 ± 0.4 mM of *T. brucei* ribonucleotide reductase for trypanothione.[21]

Tryparedoxin-Mediated Ribonucleotide Reduction

Tryparedoxin is readily reduced by trypanothione and then delivers reducing equivalents to *T. brucei* ribonucleotide reductase. The assay mixture contains all components for GDP reduction. The thiol system is composed of 2.5 mM trypanothione disulfide, 6 mM NADPH, 200 mU of trypanothione reductase, and 4 μM tryparedoxin. As described in the previous section, the reduced form of trypanothione is generated before starting the reaction by adding R1, R2, and [^{3}H]GDP. The apparent K_m value of *T. brucei* ribonucleotide reductase for tryparedoxin in the presence of 2.5 mM trypanothione is 3.7 ± 0.5 μM.[21]

Chemical Reduction of Trypanothione Disulfide

Trypanothione disulfide (10 mM) in 1 ml of water is incubated on ice with 100 mM NaBH$_4$ for 1 hr. The pH of the solution is adjusted to pH 3.0 with 1 M HCl in order to prevent reoxidation of the thiol after decomposition of excess hydrid. A C_{18} cartridge is washed with 4 ml of acetonitrile followed by 10 ml of water. The reaction mixture is applied and the cartridge is washed with 3 ml of 0.1% (v/v) trifluoroacetic acid (TFA). Reduced trypanothione is eluted with 1.5 ml of 80% (v/v) acetonitrile in 0.1% (v/v) TFA, lyophilized, dissolved in 50 mM HEPES, pH 7.6, to a final concentration of 25 mM, and immediately used. The thiol concentration is determined by reaction with 5,5'-dithiobis(2-nitrobenzoic acid) (DTNB).[26]

Abolished Reduction of Ribonucleotide Reductase by Carboxamidomethylation of Tryparedoxin

In a final volume of 100 μl of 50 mM HEPES, pH 7.6, 50 μM tryparedoxin is incubated with 1 mM T(SH)$_2$ for 15 min under an argon atmosphere and 3 μl of 200 mM iodoacetamide in water is added. After 60 min of incubation at room temperature in the dark, the reaction is stopped by adding 15 μl of 200 mM DTE in water. A control reaction contains tryparedoxin and iodoacetamide but no T(SH)$_2$ and is not stopped by DTE. Low molecular mass reaction components are removed by centrifugation in a Centricon 3 concentrator (Millipore) and the protein is washed several times with 50 mM HEPES, pH 7.6. This procedure results in a homogeneous sample that represents tryparedoxin specifically modified at the first cysteine residue (Cys-40) of the WCPPC motif[8] as described for C. fasciculata tryparedoxin.[7]

The alkylated T. brucei tryparedoxin is not able to catalyze the reduction of ribonucleotide reductase by trypanothione, which shows that catalysis requires the active site dithiol of the protein.

Inhibition of Tryparedoxin Activity by Trypanothione Disulfide

Trypanothione disulfide, TS$_2$, is a strong inhibitor of the tryparedoxin-mediated reaction but shows only a minor direct effect on ribonucleotide reductase.

In the presence of 1 mM trypanothione, 2.5 mM TS$_2$ lowers the rate of GDP reduction by ribonucleotide reductase by about 40%. In contrast, the tryparedoxin-mediated reaction is much more sensitive. At 1 mM T(SH)$_2$ and 10 μM tryparedoxin, 2.5 mM TS$_2$ inhibits deoxyribonucleotide formation by 90%. The 50% inhibitory concentration (IC$_{50}$) value of tryparedoxin for trypanothione disulfide under these conditions is about 50 μM.[21]

The pronounced sensitivity of tryparedoxin toward trypanothione disulfide becomes evident when adding NADPH and trypanothione reductase to the ribonucleotide reductase assays. With trypanothione alone the rate of dGDP formation increases by only 10% whereas with trypanothione and tryparedoxin, ribonucleotide reductase activity is doubled. The sample of chemically reduced trypanothione

---

[26] G. L. Ellman, *Arch. Biochem. Biophys.* **82,** 70 (1959).

contains still about 4% disulfide as revealed by an end-point determination in a trypanothione reductase assay. This corresponds to a concentration of 40 μM TS$_2$ at the beginning, in addition to trypanothione disulfide formed during the reaction, and explains the pronounced effect of trypanothione reductase–NADPH.

Conclusions

The parasite-specific dithiol trypanothione—but not glutathione—is a direct reductant of *T. brucei* ribonucleotide reductase. The different behavior of the thiols is not related to their redox potentials, which are similar [−242 and −230 mV for T(SH)$_2$ and GSH, respectively[1]]. In contrast, the pK values of the thiols differ significantly. A pK value of 7.4 has been reported for trypanothione, which is more than 1 pH unit lower than the pK of 8.7 for GSH.[27] Because second-order rate constants for thiol–disulfide exchanges exhibit an optimum when the thiol pK is equal to the pH of the solution, T(SH)$_2$ is expected to be much more reactive than glutathione (GSH) under physiological conditions. In addition, as reductants for intramolecular disulfides as in the R1 protein of ribonucleotide reductase, dithiols are kinetically superior to monothiols.[28] Trypanothione is the first example of a natural low molecular mass dithiol directly delivering reducing equivalents for ribonucleotide reduction. At submillimolar concentrations, the reaction is strongly accelerated by tryparedoxin, a small parasite protein with a WCPPC active site motif.

The disulfide form of trypanothione is a powerful inhibitor of the tryparedoxin-mediated reaction, which may represent a physiological control mechanism. For *E. coli* glutaredoxin a similar behavior has been observed. Glutaredoxin is strongly inhibited by oxidized glutathione (GSSG),[20] indicating a regulation of deoxyribonucleotide synthesis by the redox state of the cell. The trypanothione–tryparedoxin system is a new system providing electrons for a class I ribonucleotide reductase, in addition to the well-known thioredoxin and glutaredoxin systems described in other organisms.

Acknowledgment

Our work is supported by the Deutsche Forschungsgemeinschaft (SFB 544: Kontrolle tropischer Infektionskrankheiten and Graduiertenkolleg: Pathogene Mikroorganismen: Molekulare Mechanismen und Genome).

[27] M. Moutiez, D. Meziane-Sherif, M. Aumercier, C. Sergheraert, and A. Tartar, *Chem. Pharm. Bull.* **42,** 2641 (1994).
[28] H. F. Gilbert, *Adv. Enzymol. Relat. Areas Mol. Biol.* **63,** 69 (1990).

[24] Selenium- and Vitamin E-Dependent Gene Expression in Rats: Analysis of Differentially Expressed mRNAs

By ALEXANDRA FISCHER, JOSEF PALLAUF, and GERALD RIMBACH

Introduction

Selenium (Se) and vitamin E (VE) have been shown to affect gene expression in mammalian cells.[1-5] Until recently, researchers have used mostly Northern blot analysis, RNase protection, or reverse transcription-polymerase chain reaction (RT-PCR) to detect or quantitate differentially regulated genes. These methods have the disadvantage of being inherently serial, involving measurement of a single mRNA at a time, and of being difficult to automate.[6] A key feature of genomics is that it allows simultaneous holistic analysis of a large number of genes and how their activities are orchestrated. Global gene expression data at the mRNA level can be produced by a set of different technologies including cDNA microarrays[7,8] amplified fragment length polymorphism (AFLP),[9] differential display,[10] serial analysis of gene expression (SAGE),[11] and others (for review see Steiner and Anderson).[12] Array-based methods involve the spotting of nucleic acids (either PCR products or oligonucleotides) onto nylon membranes or modified glass microscope slides. These techniques have the advantage of a high throughput, direct and rapid readout of the hybridization results, and immediate information about the gene of interest.[13] There are two major application forms for the DNA microarray

[1] A. Azzi, I. B. M. Feher, M. Pastori, R. Ricciarelli, S. Spycher, M. Staffieri, A. Stocker, S. Zimmer, and J.-M. Zingg, *J. Nutr.* **130,** 1649 (2000).

[2] G. Bermano, F. Nicol, J. A. Dyer, R. A. Sunde, G. J. Beckett, J. R. Arthur, and J. E. Hesketh, *Biochem. J.* **311,** 425 (1995).

[3] L. G. F. Combs, Jr., *Med. Klin.* **94,** 18 (1999).

[4] L. Flohé, J. R. Andreesen, R. Brigelius-Flohé, M. Maiorino, and F. Ursini, *IUBMB Life* **49,** 411 (2000).

[5] L. Packer, S. U. Weber, and G. Rimbach, *J. Nutr.* **131,** 3698 (2001).

[6] D. J. Lockhart, H. Dong, M. C. Byrne, M. T. Follettie, M. V. Gallo, M. S. Chee, M. Mittmann, C. Wang, M. Kobayashi, H. Horton, and E. L. Brown, *Nat. Biotechnol.* **14,** 1675 (1996).

[7] M. B. Eisen and P. O. Brown, *Methods Enzymol.* **303,** 179 (1999).

[8] E. A. Winzeler, M. Schena, and R. W. Davis, *Methods Enzymol.* **306,** 3 (1999).

[9] L. Y. Wong, V. Belonogoff, V. L. Boyd, N. M. Hunkapiller, P. M. Casey, S. N. Liew, K. D. Lazaruk, and S. Baumhueter, *Biotechniques* **28,** 776 (2000).

[10] E. D. Harris, *Nutr. Rev.* **54,** 287 (1996).

[11] V. E. Velculescu, L. Zhang, B. Vogelstein, and K. W. Kinzler, *Science* **270,** 484 (1995).

[12] S. Steiner and N. L. Anderson, *Toxicol. Lett.* **112-113,** 467 (2000).

[13] S. Backert, M. Gelos, U. Kobalz, M. L. Hanski, C. Bohm, B. Mann, N. Lovin, A. Gratchev, U. Mansmann, M. P. Moyer, E. O. Riecken, and C. Hanski, *Int. J. Cancer* **82,** 868 (1999).

0076-6879/02 $35.00

technology: (1) polymorphism analysis and genotyping and (2) determination of the expression level (abundance) of genes, that is, comparing patterns of expression in different tissues and developmental stages, in normal and disease states, or in response to different treatments.[14] Methods for large-scale measurement of gene expression are also becoming powerful tools in the field of free radical research.[6] Importantly, cDNA arrays can help to discover redox-regulated genes, transcription factors, and potential biomarkers of oxidative stress. Furthermore, genomics can be applied to obtain more insight into the molecular functions of antioxidants, thereby addressing them more specifically, as well as for screening and developing of new antioxidants.

This article focuses on the analysis of differentially expressed mRNAs in selenium and vitamin E deficiency.[15] Herein is described how to produce combined selenium and vitamin E deficiency in laboratory rats and how to analyze differentially expressed mRNAs in rat liver by a cDNA expression array.

Production of Selenium and Vitamin E Deficiency in Rats

Selenium- and Vitamin E-Deficient Diet

Feeding a selenium- and VE-deficient diet plays a key role in the production of selenium- and VE-deficient animals. Because rats can conserve their selenium stores when deprived of the element, we began depletion in animals weighing about 35 g. Male rats are generally preferred, because they have a higher selenium requirement than females.[16]

Powdered and purified diets are composed of *Torula* yeast, sucrose, vitamin E-stripped corn oil, cellulose, and DL-methionine as shown in Table I. Both the selenium content of the yeast and assay of the complete diet yielded a value of <30 μg of selenium and <0.10 mg of α-tocopherol per kilogram. The mineral and vitamin mix given in Table I is modified from AIN-93G[17] and is suitable for use with *Torula* yeast, which contains naturally about 6.5 g of calcium, 13 g of potassium, 14 g of phosphorus, and 1.2 g of magnesium per kilogram. VE and selenium are omitted from these premixes and can be added separately according to the experimental design.

For our experiment, four groups of eight male albino rats (Wistar Unilever, Harlan/Winkelmann, Paderborn, Germany) received either the control diet (group I) or the control diet supplemented with 75 mg of DL-α-tocopheryl acetate per kilogram (group II) or a 200 μg/kg concentration of selenium as sodium selenate (group III) or both supplements (group IV). Animals were housed

[14] J. D. Rockett and D. J. Dix, *Xenobiotica* **30**, 155 (2000).

[15] A. Fischer, J. Pallauf, K. Gohil, S. U. Weber, L. Packer, and G. Rimbach, *Biochem. Biophys. Res. Commun.* **281**, 470 (2001).

[16] R. F. Burk, *Methods Enzymol.* **143**, 307 (1987).

[17] P. G. Reeves, F. H. Nielsen, and G. C. Fahey, Jr., *J. Nutr.* **123**, 1939 (1993).

TABLE I

COMPOSITION OF SELENIUM- AND VITAMIN E DEFICIENT DIET FOR RATS

Component	Percentage by weight
Torula yeast[a]	30
Cellulose[b]	5
Sucrose	10
Vitamin E-stripped corn oil[c]	5
Mineral mix[d]	3.5
Vitamin mix[e]	1
DL-Methionine	0.3
Choline chloride	0.2
Corn starch[f]	45

[a] Attisholz *Torula utilis* dried yeast (Cellulose Attisholz, Luterbach, Switzerland).
[b] Vitacol L600-30 (Rettenmaier & Söhne, Rosenberg, Germany).
[c] ICN Biomedicals (Eschwege, Germany).
[d] Contains (in milligrams per kilogram diet): $CaCO_3$, 7490; potassium citrate, 1940; $MgSO_4 \cdot 7H_2O$, 2310; NaCl, 2590; ferric citrate, 153; $MnSO_4 \cdot H_2O$, 31; $ZnSO_4 \cdot 7H_2O$, 132; $CuSO_4 \cdot 5H_2O$, 24; Kl, 0.26; ammonium heptamolybdate, 0.28; $KCr(SO_4)_2 \cdot 12H_2O$, 9.6; NaF, 2.2; $Na_2SiO_3 \cdot 9H_2O$, 51; LiCl, 0.61; H_3BO_3, 2.9; $NiSO_4 \cdot 6H_2O$, 2.2; NH_4VO_3, 0.23.
[e] Contains (in milligrams per kilogram diet): nicotinic acid, 30; calcium pantothenate, 15; pyridoxine hydrochloride, 6; thiamin hydrochloride, 5; riboflavin, 6; folic acid, 2; menadione, 0.9; biotin, 0.2; vitamin B_{12}, 0.025; retinyl palmitate to provide 4000 IU of vitamin A per kilogram diet; cholecalciferol to provide 1000 IU of vitamin D per kilogram diet.
[f] Roquette (Roquette Frères, Lestrem, France).

individually in metal-free metabolic cages under standard conditions (22°, 55% humidity, 12-hr light : dark cycle) and had free access to diets and doubly distilled water.

Development of Selenium Deficiency

Selenium and VE deficiency is readily produced by feeding the diet described above. Table II[18–22] shows various selenium and VE status parameters derived from animals after 7 weeks on trial. These animals were further used for the analysis of differently expressed mRNAs.

[18] B. Welz and M. Melcher, *Anal. Chim.* **165,** 131 (1984).
[19] J. G. Bieri, T. J. Tolliver, and G. L. Catignani, *Am. J. Clin. Nutr.* **32,** 2143 (1979).
[20] R. A. Lawrence and R. F. Burk, *Biochem. Biophys. Res. Commun.* **71,** 952 (1976).
[21] W. H. Habig and W. B. Jakob, *Methods Enzymol.* **77,** 398 (1981).
[22] S. Gromer, L. D. Arscott, C. H. Williams, Jr., R. H. Schirmer, and K. Becker, *J. Biol.Chem.* **273,** 20096 (1998).

TABLE II

SELENIUM AND VITAMIN E CONCENTRATIONS, GLUTATHIONE PEROXIDASE,
GLUTATHIONE TRANSFERASE, AND THIOREDOXIN REDUCTASE[a]

	Group			
Parameter	I (−Se, −VE)	II (−Se, +VE)	III (+Se, −VE)	IV (+Se, +VE)
Selenium (μg/kg FW)[b]	25.8 ± 3.7	31.8 ± 5.3	892.3 ± 78.4	881.5 ± 41.4
α-Tocopherol (μg/g FW)[c]	1.03 ± 0.2	29.6 ± 3.8	1.15 ± 0.2	32.9 ± 3.3
cGPx (mU/mg protein)[d]	5.45 ± 0.96	10.9 ± 1.6	154.2 ± 11.9	161.0 ± 15.2
TrxR (mU/mg protein)[d]	1.32 ± 0.17	1.16 ± 0.11	9.46 ± 1.05	9.25 ± 0.89
GST (mU/mg protein)[d]	209 ± 16.9	226 ± 16.8	141 ± 10.8	162 ± 8.2

Abbreviations: VE, vitamin E; FW, fresh weight; cGPx, hepatic glutathione peroxidase; TrxR, thioredoxin reductase; GST, glutathione transferase.

[a] In liver of growing rats fed diets containing different levels of VE and selenium (means ± SD).

[b] Determination of selenium content of the liver was determined by hydride-generation atomic absorption spectrometry after microwave-supported wet digestion [B. Welz and M. Melcher, *Anal. Chim.* **165**, 131 (1984)].

[c] Vitamin E was analyzed by C_{18} reversed-phase HPLC (RP-HPLC) with fluorescence detection [J. G. Bieri, T. J. Tolliver, and G. L. Catignani, *Am. J. Clin. Nutr.* **32**, 2143 (1979)].

[d] Assays performed on 10,000g supernatant fluid. The substrate used for the glutathione peroxidase assay was H_2O_2 [R. A. Lawrence and R. F. Burk, *Biochem. Biophys. Res. Commun.* **71**, 952 (1976)], that for the glutathione transferase assay was 1-chloro-2,4-dinitrobenzene [W. H. Habib and W. B. Jakob, *Methods Enzymol.* **77**, 398 (1981)], and that for the thioredoxin reductase assay was DTNB [S. Gromer, L. D. Arscott, C. H. Williams, Jr., R. H. Schirmer, and K. Becker, *J. Biol. Chem.* **273**, 20096 (1998)].

Feed intake, live weight gain, and feed conversion efficiency were not significantly different in the four experimental groups. Hepatic glutathione peroxidase (cGPx) activity had fallen to 5% of control by 7 weeks. Feeding selenium-deficient diets led furthermore to a significant decrease in hepatic selenium concentrations as well as in hepatic thioredoxin reductase activities as compared with rats fed selenium-adequate diets. Glutathione transferases (GSTs), which are believed to compensate for the depletion of cGPx under selenium deficiency, were increased in the selenium-deficient animals. Rats fed diets enriched with VE had 30-fold higher α-tocopherol liver concentrations as compared with nonsupplemented control animals.

Analysis of Differentially Expressed mRNAs by Atlas cDNA Expression Arrays

Atlas cDNA expression arrays (Clontech, Heidelberg, Germany) include hundreds of cDNAs spotted on positively charged nylon membranes. On the larger arrays, the cDNAs are arrayed into functional classes. We use Atlas rat toxicology array II, which includes 450 rat cDNAs of genes that carry out critical functions such as DNA damage and repair, cell cycle, xenobiotic metabolism, apoptosis, inflammation, and stress response, 9 housekeeping control cDNAs for normalizing mRNA abundance; and plasmid and bacteriophage DNAs as negative controls to confirm hybridization specificity. The cDNAs are spotted in duplicate dots, whereby each dot contains 10 ng of a PCR-amplified fragment. Using the reagents provided, expression of mRNAs present at 0.005–0.01% of poly(A)$^+$ RNA—a sensitivity corresponding to approximately 10–20 target transcripts per cell—can be detected.

Target Labeling and Hybridization of cDNA Array

The quality of the RNA used to make probes is the most important factor influencing the sensitivity and reproducibility of the hybridization pattern. A poor-quality RNA preparation leads to high background on the membrane and/or an inaccurate hybridization pattern. These problems are typically caused by residual RNase and genomic DNA contamination. To avoid this, we check for the existence of genomic DNA and RNA degradation on an agarose gel. Any remaining DNA, that may still exist but not be visible on the gel, does not seem to interfere with the production of the array probe. When using arrays to compare mRNA populations, it is recommended that the same method be used to isolate total or poly(A)$^+$ RNA from each sample. It is essential that all RNA samples used for analysis be of the same quality, otherwise it will be impossible to determine whether to differences in the pattern of hybridization are due to differential gene expression or to differences in RNA quality. For hybridization to the Atlas array, we isolate total RNA from rat liver with an RNeasy minikit (Qiagen, Hilden, Germany). Three samples within one group with the same diet treatment are pooled to prepare complex cDNA probes.

In a 10-μl reaction, 2–5 μg of total RNA can be converted into ^{32}P-labeled first-strand cDNA. ^{32}P gives an approximately four times stronger signal than ^{33}P, thereby increasing assay sensitivity. ^{33}P, however, tends to improve spatial resolution and quality of the resulting images, which facilitates image analysis and signal quantification. To verify that the cDNA synthesis system works, a control RNA should always be included. For each experimental RNA sample, 2 μl of the 5× reaction buffer [250 mM Tris-HCl (pH 8.3), 375 mM KCl, 15 mM MgCl$_2$], 1 μl of 10× dNTP mix (dCTP, dGTP, and dTTP, 5 mM each), 3.5 μl of [α-^{32}P]dATP (3000 Ci/mmol, 10 μCi/μl; Amersham, Arlington Heights, IL), and 0.5 μl of dithiothreitol (DTT, 100 mM) must be combined at room temperature in a 0.5-ml microcentrifuge tube to give a total volume of 7 μl. It is recommended that a master mix for all labeling reactions be prepared. One to 2 μl of RNA (which gives a concentration of 2–5 μg), 1 μl of CDS primer mix (a mixture of primers specific for each different type of Atlas array, which ensures that cDNAs are synthesized only for the genes on the particular Atlas array), and, if necessary, doubly distilled H$_2$O to give a total volume of 3 μl are mixed well by pipetting, spun down briefly in a microcentrifuge, and subsequently incubated at 70° for 2 min. Thereafter, the temperature is reduced to 50° and the tubes are incubated for a further 2 min. During this incubation, 1 μl of moloney murine leukemia virus (Mo-MuLV) reverse transcriptase per reaction is added to the master mix, mixed by pipetting, and kept at room temperature. After completion of the 2-min incubation at 50°, 8 μl of master mix is added to each reaction tube. The contents of the tubes are mixed by pipetting, immediately returned to 50°, and incubated for 25 min. The reaction is stopped by adding 1 μl of 10× termination mix [0.1 M EDTA (pH 8.0), glycogen (1 mg/ml)]. To purify the labeled cDNA from unincorporated ^{32}P-labeled nucleotides and small (<0.1-kb) cDNA fragments, it is recommended that the provided Atlas NucleoSpin extraction kit be used. Samples can be stored at −20°.

Before hybridizing labeled cDNA probes to Atlas arrays, the quality of each probe should be checked by hybridizing it to the supplied control (blank) nylon membrane. This will allow the level of nonspecific background resulting from impurities in the RNA samples to be estimated. If a high level of background occurs on the control membrane, total RNA samples should be treated with DNase I and probes should be remade.

To prepare the prehybridization solution, 5 ml of supplied ExpressHyb should be prewarmed at 68°. Sheared salmon testis DNA (0.5 mg at 10 mg/ml; Sigma, St, Louis, MO) must be heated at 95–100° for 5 min and then chilled quickly on ice. The heat-denatured sheared salmon testis DNA needs to be mixed with prewarmed ExpressHyb and kept at 68° until use. Subsequently the hybridization bottle is filled with doubly distilled H_2O and the Atlas array is soaked in a dish of doubly distilled H_2O and then placed into the hybridization bottle. All water is poured off from the bottle, whereby the membrane should adhere to the inside walls of the container without creating air pockets. Five milliliters of the hybridization solution is added so that the solution is evenly distributed over the membrane. This step needs to be performed quickly to prevent the array membrane from drying. The membrane is prehybridized for 30 min with continuous agitation at 68°.

To prepare the probe for hybridization, 200 μl of labeled probe (entire pool; $0.5–20 \times 10^6$ cpm) is mixed with 10× denaturing solution (1 M NaOH, 10 mM EDTA) to give a total volume of 222 μl and incubated at 68° for 20 min. Five microliters of C_0t-1 DNA (1 mg/ml) and 225 μl of 2× neutralizing solution (1 M NaH$_2$PO$_4$, pH 7.0) are added to give a total volume of 450 μl and incubated at 68° for 10 min. The mixture is added carefully to the array (avoid pouring the concentrated probe directly on the surface of the membrane) and the prehybridization solution. We perform hybridization overnight with continuous agitation at 68°.

The next day, wash solution 1 [2× SSC (1× SSC is 0.15 M NaCl plus 0.015 M sodium citrate), 1% (w/v) sodium dodecyl sulfate (SDS)] and wash solution 2 (0.1× SSC, 0.5% SDS) are prewarmed at 68°. Carefully the hybridization solution is removed, discarded in an appropriate radioactive waste container, and replaced with 200 ml of prewarmed wash solution 1. The Atlas array needs to be washed four times for 30 min with continuous agitation at 68°. Another 30-min wash is performed in 200 ml of prewarmed wash solution 2 with continous agitation at 68° and one final 5-min wash in 200 ml of 2× SSC with agitation at room temperature. Using forceps, the Atlas array is removed from the container and the excess wash solution is shaken off without allowing the membrane to dry. If the membrane dries even partially, subsequent removal of the probe (stripping) from the Atlas array will be difficult. The damp membrane is sealed in plastic to prevent drying and exposed to a PhosphorImaging screen at room temperature. The expression profiles are obtained by PhosphorImaging for 1 week.

Image Analysis

The expression profiles are evaluated by using Atlas software. Each array is aligned to the AtlasImage grid template, which allows AtlasImage to determine the location of all the genes on the array.

For background calculation different options are provided. We use the default external background calculation, in which the background is set at the median intensity of the "blank space" between the different panels of the array. In areas of intense signal, there is the possibility of signal bleed to neighboring genes. In this case, we adjust the background level for this spot individually. If the background pattern on the blot is uneven or splotchy, it is recommended that the custom external option be used, in which the background is set at the median intensity of a user-defined area of the array membrane.

In most cases, it is useful to normalize the signal intensity between the two arrays being compared. For this step, different methods are available. Included on the Atlas arrays are several housekeeping genes (e.g., glyceraldehyde-3-phosphate dehydrogenase, tubulin α_1, cytoplasmic β-actin) that can serve as positive controls. When normalizing the results of the closely related cells or tissues, the ratios of the intensities of the hybridization signals for most of the housekeeping control cDNAs will be approximately identical. In this case, it might be advisable to choose one or two of the least variable housekeeping control cDNAs to normalize the results from both membranes being compared. However, when comparing unrelated tissues or cells, it is often easiest to use the signals of cDNAs surrounding a target cDNA for normalization. The advantage of this approach is that the cDNAs being compared are in close proximity on the filter. We use the average intensity of all genes on the array to normalize the results from the membranes compared. This method is best suited when only a few genes are expected to be differentially expressed (e.g., for the comparison of similar tissues). By using the sum method, the values of signal over background for all genes on the arrays are added to calculate the normalization coefficient. This normalization coefficient is used together with the background level to calculate the adjusted intensity for a gene, which equals the intensity minus the background value, multiplied by the normalization coefficient.

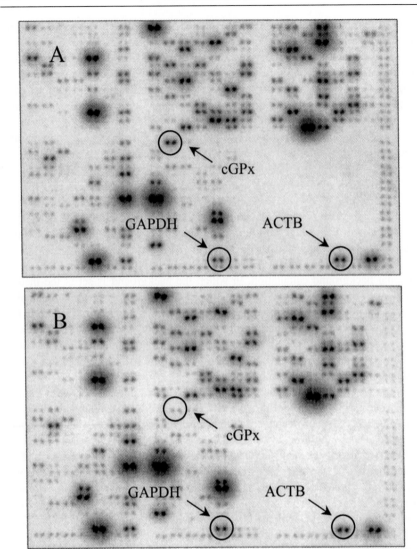

FIG. 1. Gene expression profiles of ^{32}P-labeled cDNA probes, prepared from total RNA from selenium-adequate rats (A) and selenium-deficient rats (B), on an Atlas rat cDNA toxicology array.

To set the limit above which a signal from a gene will be considered "real" and not background noise, it is important to define the signal threshold. This will help to filter out weak signals that are not attributable to actual gene expression or that are too weak for meaningful interpretation. We use the background-based signal threshold, whereby any gene whose adjusted intensity is at least twice the background value is considered a genuine signal.

To obtain significant differences while comparing adjusted intensities of different membranes, the ratio of intensities on both arrays is calculated for each gene. We consider significant a difference in gene expression with a ratio threshold of two or more.

Typical Results

In Fig. 1 gene expression profiles of ^{32}P-labeled cDNA probes from selenium-adequate rats (Fig. 1A) and selenium-deficient rats (Fig. 1B) on an Atlas rat cDNA toxicology array are shown.[15] Each gene is represented in duplicate dots, whose intensity is calculated as the average of the total signal from the left and right spots.

Expression levels of the housekeeping genes did not change between the different treatments, which is shown for glyceraldehyde-3-phosphate dehydrogenase (GAPDH) and cytoplasmic β-actin (ACTB) in Fig. 1. Selenium deficiency was characterized by significant inductions in the expression of genes encoding proteins involved in cell adhesion, inflammation, and the acute-phase response. In addition, a downregulation was evident for genes important in the inhibition of apoptosis, cell cycling, and antioxidant defense. A selection of downregulated changes in

TABLE III
SELECTION OF DOWNREGULATED CHANGES IN GENE EXPRESSION INDUCED
BY SELENIUM AND VITAMIN E DEFICIENCY IN LIVER

GenBank accession no.	Δ −Se −VE^a (fold)	Gene	Function
Stress response			
X12367	↓ 18.8	Cellular glutathione peroxidase I	Peroxide detoxification
J05181	↓ 3.4	Glutamate–cysteine ligase catalytic subunit	Glutathione synthesis
PO4800	↓ 2.5	Cytochrome P-450 3A1	Xenobiotic metabolism
Cell cycle			
J03969	↓ 2.9	Nucleophosmin (NPM)	Stimulation of normal cell growth
D14014	↓ 3.1	G_1/S-specific cyclin D_1 (CCND1)	Initiation of cell cycle, oncogene
Apoptosis			
Y13336	↓ 2.0	Defender against cell death 12 protein (DAD1)	Protection against apoptosis
AF081503	↓ 2.6	Inhibitor of apoptosis protein 1	Protection against apoptosis
U72350	↓ 3.2	Bcl2-L1	Promotes cell survival
Inflammation			
U22424	↓ 2.2	11β-Hydroxysteroid dehydrogenase 2	Conversion of corticosterone into 11-dehydrocorticosterone
L49379	↓ 2.3	cMOAT(canalicular multispecific organic anion transporter)	Detoxification, export of leukotriene C_4

^{a} Change in selenium and vitamin E deficiency relative to control.

gene expression induced by selenium and vitamin E deficiency compared with controls in liver as well as their related function is given in Table III. The largest differential expression between control and deficient animals was observed for cGPx (18.8-fold change), which is clearly shown by less intensive dots representing cDNAs for cGPx in Fig. 1B compared with Fig. 1A.

The accuracy of an expression profile depends mainly on the type of array technology used to generate the data. Although DNA chips have been shown to have a higher sensitivity than cDNA arrays, a close correlation between the expression results obtained with both techniques has been observed.[23] After significant changes in the mRNA levels of selected genes have been discovered by the cDNA array, results should always be validated either by RT-PCR or Northern blotting. Finally, Western blotting or two-dimensional polyacrylamide gel electrophoresis analysis can clarify whether differences in the mRNA level are also apparent in terms of protein expression.

[23] P. Cohen, M. Bouaboula, M. Bellis, V. Baron, O. Jbilo, C. Poinot-Chazel, S. Galiégue, E.-H. Hadibi, and P. Casellas, *J. Biol. Chem.* **275,** 11181 (2000).

Section II

Thioredoxin

[25] Overview

By HIROSHI MASUTANI and JUNJI YODOI

Introduction

Thiol reduction by the thioredoxin system and the glutathione (GSH) system plays key roles in the regulation of a variety of biological functions such as apoptosis, cell cycle, and growth control. The thioredoxin system is composed of several related molecules interacting through the active site cysteine residues. Thioredoxin couples with thioredoxin-dependent peroxidases (peroxiredoxin) to scavenge hydrogen peroxide. In addition, thioredoxin acts not simply as a scavenger of reactive oxygen species (ROS) but also as an important regulator of oxidative stress response by protein–protein interaction. Thioredoxin also interacts with other related molecules to exert biological functions: For example, thioredoxin interacts with redox factor 1 (Ref-1) and affects the function of various transcription factors. The interaction of thioredoxin and thioredoxin-binding proteins such as thioredoxin-binding protein 2 (TBP-2)/vitamin D_3-upregulated protein 1 (VDUP-1) and apoptosis-signaling kinase 1 (ASK-1) suggested unpredicted functions of thioredoxin and a novel mechanism of the redox regulation. Our studies of the regulation of the thioredoxin gene have shown that the redox and detoxifying enzymes have a common regulatory mechanism and may have coordinate roles against environmental stressors. This article is an overview of thioredoxin-interacting proteins and the role of the thioredoxin system in redox signaling.

Thioredoxin and Related Molecules

The thioredoxin system in concert with the glutathione system constitutes a cellular reducing environment. Thioredoxin is a small protein with two redox-active cysteine residues in an active center (Cys-Gly-Pro-Cys) and operates together with NADPH and thioredoxin reductase as an efficient reducing system of exposed protein disulfides (see [26] in this volume[1]). We have identified and cloned a human homolog of thioredoxin.[2] Several cytokine-like factors are identical or related to thioredoxin. Thioredoxin has been reported to have chemokine-like activity (see [31] in this volume[3]). Mammalian thioredoxin 2 (thioredoxin 2) has high homology with thioredoxin and has an active site Cys-Gly-Pro-Cys with

---

[1] A. Vlamis-Gardikas and A. Holmgren, *Methods Enzymol.* **347**, [26], 2002 (this volume).

[2] Y. Tagaya, Y. Maeda, A. Mitsui, N. Kondo, H. Matsui, J. Hamuro, N. Brown, K. Arai, T. Yokota, H. Wakasugi, and J. Yodoi, *EMBO J.* **8**, 757 (1989).

[3] Y. Nishinaka, H. Nakamura, and J. Yodoi, *Methods Enzymol.* **347**, [31], 2002 (this volume).

METHODS IN ENZYMOLOGY, VOL. 347

thiol-reducing activity and is specifically localized in mitochondria.[4] Thioredoxin 2 plays roles in protection against oxidative stress in mitochondria.[4a] Thioredoxin reductase has a selenium-containing active center in the C terminus. There exist several isoforms of thioredoxin reductase.[1] Thioredoxin-dependent peroxidases (peroxiredoxin) are considered to be members of a family of proteins reducing intracellular hydrogen peroxide (see [41] in this volume[5]). In addition, several thioredoxin-related proteins have been reported. Nucleoredoxin is a nuclear protein.[6] TRP32 is a cytosolic 32-kDa protein that contains a thioredoxin domain with a conserved thioredoxin active site and has thioredoxin-like reducing activity.[7] The function of such thioredoxin-related proteins should be further investigated. Collectively, the thioredoxin system is composed of several related molecules forming a network of recognition and interaction with its active site cysteine residues.

Thioredoxin-Binding Proteins

The interaction of thioredoxin and thioredoxin-binding proteins is a novel mechanism of the redox regulation. We isolated vitamin D_3-upregulated protein 1 (VDUP1) as thioredoxin-binding protein 2(TBP-2).[8] VDUP1 was originally reported as an upregulated gene in HL-60 cells stimulated by $1\alpha,25$-dihydroxyvitamin D_3.[9] TBP-2/VDUP1 serves as a negative regulator of the function and expression of thioredoxin. In $1\alpha,25$-dihydroxyvitamin D_3-treated HL-60 cells, TBP-2/VDUP1 expression was enhanced, whereas thioredoxin expression and the reducing activity were downregulated. Thioredoxin–TBP-2 VDUP1 interaction may play an important role in the redox regulation of growth and differentiation of cells sensitive to $1\alpha,25$-dihydroxyvitamin D_3. Thioredoxin was also isolated as an apoptosis-signaling kinase 1 (ASK1)-binding protein by a yeast two-hybrid system.[10] When thioredoxin is oxidized by reactive oxygen species (ROS), the binding between thioredoxin and ASK-1 is dissociated and ASK-1 is activated to transduce the signal of apoptosis. Interestingly, interaction of glutathione

[4] G. Spyrou, E. Enmark, V. A. Miranda, and J. Gustafsson, *J. Biol. Chem.* **272**, 2936 (1997).

[4a] T. Tanaka, Y. Yamaguchi-Iwai, F. Hosoi, H. Nakamura, H. Masutani, S. Ueda, A. Nishiyama, S. Takeda, H. Wada, G. Spyrou, and J. Yodoi, submitted (2001).

[5] H. Nakamura, A. Matsui, and J. Yodoi, *Methods Enzymol.* **347**, [41], 2002 (this volume).

[6] H. Kurooka, K. Kato, S. Minoguchi, Y. Takahashi, J. Ikeda, S. Habu, N. Osawa, A. M. Buchberg, K. Moriwaki, H. Shisa, and T. Honjo, *Genomics* **39**, 331 (1997).

[7] K. K. Lee, M. Murakawa, S. Takahashi, S. Tsubuki, S. Kawashima, K. Sakamaki, and S. Yonehara, *J. Biol. Chem.* **273**, 19160 (1998).

[8] A. Nishiyama, M. Matsui, S. Iwata, K. Hirota, H. Masutani, H. Nakamura, Y. Takagi, H. Sono, Y. Gon, and J. Yodoi, *J. Biol. Chem.* **274**, 21645 (1999).

[9] K. S. Chen and H. F. DeLuca, *Biochim. Biophys. Acta* **1263**, 1 (1995).

[10] M. Saito, H. Nishitoh, M. Fujii, K. Takeda, K. Tobiume, Y. Sawada, M. Kawabata, K. Miyazono, and H. Ichijyo, *EMBO J.* **17**, 2596 (1998).

S-transferase p (GSTp) with Jun kinase has been reported. Therefore, direct interaction between a redox-regulating proteins and its binding protein may be a basic mechanism of the redox regulation of cellular processes.

Redox Regulation of Transcriptional Factors by Thioredoxin

The activity of several transcription factors is posttranslationally altered by redox modification(s) of specific cysteine residue(s). The redox regulation of the glucocorticoid receptor was one of the earliest studies to concern the redox regulation of transcription factors. Free sulfhydryl groups of iron-responsive element (IRE)-binding protein (IRE-BP) are required for the specific interaction between IRE-BP and IRE.[11] The importance of conserved cysteine residues of transcription factors was revealed by a report showing that DNA binding of the Fos–Jun heterodimer was modulated by reduction–oxidation (redox) of a single conserved cysteine residue in the DNA-binding domains of the two proteins.[12] Redox factor 1 (Ref-1) was identified as a factor to facilitate activator protein 1 (AP-1) DNA-binding activity.[13] Interestingly, Ref-1 is identical to formally described apurinic/apyrimidinic (AP) endonuclease, although the redox and DNA repair activities of Ref-1 are encoded by nonoverlapping domains.[14] DNA binding of AP-1 is modified by Ref-1 and the activity of Ref-1 is in turn modulated by thioredoxin.[15] DNA binding of NF-κB to the κB site was also proved to be regulated by the redox status (see [33] in this volume[16]). In the cytoplasm, thioredoxin interferes with the signals to IκB (inhibitor of NF-κB) kinase and blocks the degradation of IκB, whereas in the nucleus, thioredoxin enhances NF-κB transcriptional activities by enhancing its ability to bind DNA.[17] DNA binding of glucocorticoid receptor and estrogen receptor is also redox regulated by thioredoxin. Thioredoxin and Ref-1 efficiently promote the DNA-binding activity of the RUNX family of transcription factors.[18] We previously showed that thioredoxin and Ref-1 enhance p53-dependent p21 activation for DNA repair.[19] The Ref-1-mediated activation of p53 is also shown by other reports.[20,21] In a yeast system, the importance of the thioredoxin system in p53 activity has also been reported, in which deletion of the thioredoxin reductase gene

[11] M. W. Hentze, T. A. Rouault, J. B. Harford, and R. D. Klausner, *Science* **244,** 357 (1989).

[12] C. Abate, L. Patel, F. J. Rauscher, and T. Curran, *Science* **249,** 1157 (1990).

[13] S. Xanthoudakis and T. Curran, *EMBO J.* **11,** 653 (1992).

[14] S. Xanthoudakis, G. G. Miao, and T. Curran, *Proc. Natl. Acad. Sci. U.S.A.* **91,** 23 (1994).

[15] K. Hirota, M. Matsui, S. Iwata, A. Nishiyama, K. Mori, and J. Yodoi, *Proc. Natl. Acad. Sci. U.S.A.* **94,** 3633 (1997).

[16] T. Okamato, K. Samitsu, and T. Tetauka, *Methods Enzymol.* **347,** [33], 2002 (this volume).

[17] K. Hirota, M. Murata, Y. Sachi, H. Nakamura, J. Takeuchi, K. Mori, and J. Yodoi, *J. Biol. Chem.* **274,** 27891 (1999).

[18] Y. Akamatsu, T. Ohno, K. Hirota, H. Kagoshima, J. Yodoi, and K. Shigesada, *J. Biol. Chem.* **272,** 14497 (1997).

[19] M. Ueno, H. Masutani, R. J. Arai, A. Yamauchi, K. Hirota, T. Sakai, T. Inamoto, Y. Yamaoka, J. Yodoi, T. Sakai, and T. Nikaido, *J. Biol. Chem.* **274,** 35809 (1999).

inhibited p53-dependent reporter gene expression. Thioredoxin-dependent redox regulation of p53 activity indicates coupling of the oxidative stress response and p53-dependent repair mechanism. Interaction of hypoxia-inducible factor (HIF) and coactivators may also be a target of redox regulation. The C-terminal activation domain of HIF-1α and its related factor has a specific cysteine. Expression of thioredoxin and Ref-1 enhanced interaction of these factors with a coactivator, CREB-binding protein (CBP)/p300.[22] Thioredoxin translocation from cytosol to nucleus was induced by a wide variety of oxidative stresses including UV irradiation, hypoxia, treatment with cis-diamminedichloroplatinum(II) (cisplatin, CDDP), and X-ray irradiation. Nuclear localization of thioredoxin is often observed in pathological tissue,[23,24] although the significance and mechanism of these observations remain unclear. The mechanism of thioredoxin translocation should be further clarified.

Redox regulation appears to be involved in various steps in the activation of transcription factors. In yeast, Yap1 and Skn7 cooperate on the yeast thioredoxin 2 promoter to induce transcription in response to oxidative stress. Cysteine residues in the nuclear export signal (NES)-like sequence of Yap1p are essential for regulated nuclear localization, suggesting that the Yap1p regulatory domain can confer the oxidative stress sensor function.[25] In the thioredoxin-deficient mutant, Yap1p was constitutively concentrated in the nucleus and the level of expression of the Yap1 target genes was high.[26] Taken together, thioredoxin seems to be a key component of redox signaling (Fig. 1).

Cytoprotective Action of Thioredoxin

Thioredoxin (TRX) plays a cytoprotective role against various oxidative stresses in a variety of systems.[27] Thioredoxin has a radical-scavenging activity. Thioredoxin can protect cells from tumor necrosis factor (TNF) or anti-Fas antibody, hydrogen peroxide, activated neutrophils, and ischemic reperfusion injury. TRX can effectively reduce lens-soluble protein disulfide bonds generated by

[20] L. Jayaraman, K. G. Murthy, C. Zhu, T. Curran, S. Xanthoudakis, and C. Prives, Genes Dev. 11, 558 (1997).

[21] C. Gaiddon, N. C. Moorthy, and C. Prives, EMBO J. 18, 5609 (1999).

[22] M. Ema M, K. Hirota, J. Mimura, H. Abe, J. Yodoi, K. Sogawa, L. Poellinger, and Y. Fujii-Kuriyama, EMBO J. 18, 1905 (1999).

[23] S. Fujii, Y. Nanbu, H. Nonogaki, I. Konishi, T. Mori, H. Masutani, and J. Yodoi, Cancer 68, 1583 (1991).

[24] T. Tanaka, Y. Nishiyama, K. Okada, K. Hirota, M. Matsui, J. Yodoi, H. Hiai, and S. Toyokuni, Lab. Invest. 77, 145 (1997).

[25] S. Kuge, T. Toda, N. Iizuka, and A. Nomoto, Genes Cells 3, 521 (1998).

[26] S. Izawa, K. Maeda, K. Sugiyama, J. Mano, Y. Inoue, and A. Kimura, J. Biol. Chem. 274, 28459 (1999).

[27] H. Masutani, M. Ueno, S. Ueda, and J. Yodoi, in "Antioxidant and Redox Regulation of Genes" (C. K. Sen, H. Sies, and P. A. Baeuerle, eds.). Academic Press, San Diego, California, 1999.

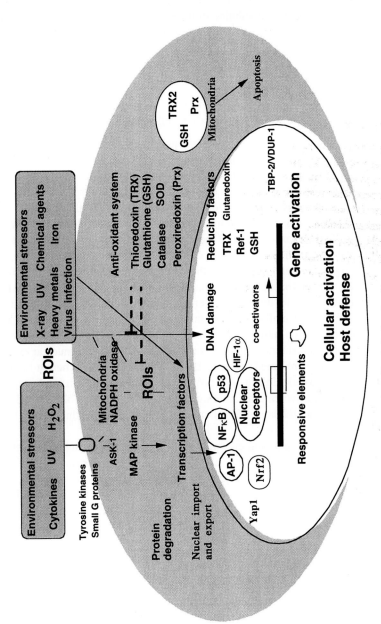

FIG. 1. Redox signaling.

H_2O_2 and act as a free radical quencher (see [40] in this volume[28]). Homozygous mutants carrying a targeted disruption of the thioredoxin gene were lethal shortly after implantation, suggesting that thioredoxin expression is essential for early differentiation and morphogenesis of the mouse embryo.[29] Studies using thioredoxin transgenic mice show that thioredoxin protects cells against oxidative stress (see [41] in this volume[5]).

Environmental Stressors and Redox Signal

Thioredoxin expression is induced by a variety of stresses, including virus infection, mitogens, phorbol myristate acetate (PMA), X-ray and ultraviolet irradiation, hydrogen peroxide, and ischemic reperfusion.[27] We analyzed the 5' upstream region of the thioredoxin gene and identified an oxidative responsive element, which was inducible by various oxidative agents.[30] In an erythroleukemic cell line K562, thioredoxin is induced transcriptionally by hemin (ferriprotoporphyrin IX).[31] In the 5' upstream sequence of the human thioredoxin gene, there exist putative binding sites such as AP-1, the cyclic AMP responsive element (CRE), the antioxidant responsive element (ARE), the xenobiotic responsive element (XRE), and SP-1. We showed that hemin activates the thioredoxin gene through the ARE. We also reported that the thioredoxin gene is regulated through the ARE by the binding of NF-E2p45/small Maf under unstimulated conditions, Nrf2/small Maf under hemin stimulation, and Jun/Fos families of proteins under PMA stimulation. The binding of these factors to the ARE correlated well with their nuclear expression pattern. We also presented evidence that Nrf2 plays a role in the hemin-induced activation of the thioredoxin gene. We proposed a novel mechanism for the regulation of the ARE by a switch of its binding factors including CNC-bZIP/small Maf transcription factors and the Jun/Fos families of proteins, depending on different stimuli.[32]

Groups of redox-regulating enzymes such as γ-glutamylcysteine synthetase, NAD(P)H:quinone oxidoreductase, and glutathione S-transferase Ya genes are known to contain the antioxidant responsive element/electrophile response element (ARE/ EpRE) for response to electrophile-targeting xenobiotics. The importance of Nrf2 in oxidative stress-induced gene activation has been shown by several studies.[33] It is interesting, therefore, that thioredoxin and these redox enzymes all

[28] K. C. Bhuyan, P. G. Reddy, and D. K. Bhuyan, *Methods Enzymol.* **347**, [40], 2002 (this volume).
[29] M. Matsui, M. Oshima, H. Oshima, K. Takaku, T. Maruyama, J. Yodoi, and M. M. Taketo, *Dev. Biol.* **178**, 179 (1996).
[30] Y. Taniguchi, U. Y. Taniguchi, K. Mori, and J. Yodoi, *Nucleic Acids Res.* **24**, 2746 (1996).
[31] S. Leppa, L. Pirkkala, S. C. Chow, J. E. Eriksson, and L. Sistonen, *J. Biol. Chem.* **272**, 30400 (1997).
[32] Y.-C. Kim, H. Masutani, Y. Yamaguchi, K. Itoh, M. Yamamoto, and J. Yodoi, *J. Biol. Chem.* **276**, 18399 (2001).
[33] T. Ishii, K. Itoh, and M. Yamamoto, *Methods Enzymol.* **348**, in press (2002).

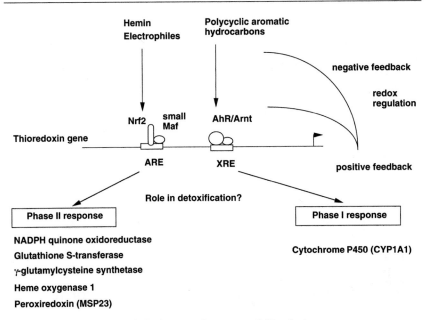

FIG. 2. Environmental stressors and thioredoxin.

share common regulatory mechanisms and may play coordinated roles against a subset of oxidative stress in which the ARE-mediated activation pathway is involved in the response.

Polycyclic aromatic hydrocarbons (PAHs) including 2,3,7,8-tetrachlorodibenzo-p-dioxin (TCDD) are ubiquitous environmental pollutants that have many untoward effects in humans and wildlife, such as immune suppression, thymic involution, endocrine disruption, wasting syndrome, birth defects, and carcinogenesis.[34] Most of the biological effects of PAHs are mediated by the aryl hydrocarbon receptor (AhR), which binds to the cognate enhancer sequence, XRE. The XRE sequence is found in the promoter regions of phase I detoxification enzymes including members of the cytochrome P-450 family and aldo–keto reductases (AKRs). The thioredoxin gene has a sequence similar to XRE and is induced by PAHs.[35] Therefore, thioredoxin and other redox-regulating enzymes may also plays role against xenobiotic-induced stress response. We here propose a model in which thioredoxin is a redox-censoring system against environmental stressors and plays a role in the regulation of the redox signaling (Fig. 2).

[34] D. W. Nebert, A. Puga, and V. Vasiliou, *Ann. N.Y. Acad. Sci.* **685**, 624 (1993).
[35] Y.-W. Kwon, S. Ueda, M. Veno, J. Yodoi, and H. Masutani, submitted (2001).

Concluding Remarks and Perspectives

Thioredoxin target proteins seem to be divided into several categories. Thioredoxin interacts with (1) members of the thioredoxin system; (2) thioredoxin interacting proteins such as Ref-1; (3) thioredoxin binding proteins; (4) substrates of thioredoxin. Thioredoxin seems to exert its function by these multiple ways of interaction. Further elucidation of molecular mechanism of thioredoxin function is required to understand how cells integrate and differentiate redox signals.

Acknowledgments

We thank members of the laboratory for helpful discussion, and Ms. Y. Kanekiyo for secretarial help. This work was supported by a grant-in-aid for scientific research from the Ministry of Education, Culture, Sports, Science, and Technology of Japan and a grant-in-aid for research for the future from the Japan Society for the Promotion of Science, Japan.

[26] Thioredoxin and Glutaredoxin Isoforms

By ALEXIOS VLAMIS-GARDIKAS and ARNE HOLMGREN

Introduction

Intracellular environments are characterized by reducing conditions generally due to the presence of millimolar concentrations of glutathione (GSH) or other low molecular weight thiols. Intracellular proteins contain free thiols rather than disulfides, which exist on extracellular proteins or proteins of the cell surface.[1] The reduction of intracellular disulfides is mediated by NADPH via the thioredoxin and glutaredoxin systems.[2-4] Thioredoxins and glutaredoxins are ubiquitous proteins with redox-active cysteines and a growing number of isoforms in different species including viruses. Sequences of many novel thioredoxins, glutaredoxins, and homologous proteins from different species are now available, particularly through the sequencing of different genomes. The role of many of these proteins is currently being examined. In this review we focus only on characterized thioredoxins/glutaredoxins from the bacterium *Escherichia coli*, the unicellular eukaryote *Saccharomyces cerevisiae*, and humans. We do not deal with proteins with thioredoxin/glutaredoxin-like domains (e.g., protein disulfide-isomerase[5]) or other members of the thioredoxin fold superfamily.[6]

[1] H. F. Gilbert, *Adv. Enzymol. Relat. Areas Mol. Biol.* **63**, 69 (1990).

[2] A. Holmgren, *Annu. Rev. Biochem.* **54**, 237 (1985).

[3] A. Holmgren, *J. Biol. Chem.* **264**, 13963 (1989).

[4] A. Holmgren and F. Åslund, *Methods Enzymol.* **252**, 283 (1995).

Thioredoxins

The thioredoxin system consists of thioredoxin (Trx), thioredoxin reductase (TrxR), and NADPH.[2] For all thioredoxin systems, oxidized Trx (Trx-S_2) is reduced by TrxR to reduced Trx [Trx-(SH_2)], using electrons from NADPH [Reaction (1)]. Trx-(SH)$_2$ is the major disulfide reductase of intracellular protein disulfides [Reaction (2)]:

$$\text{Trx-S}_2 + \text{NADPH} + \text{H}^+ \xrightarrow{\text{TrxR}} \text{Trx-(SH}_2) + \text{NADP}^+ \qquad (1)$$

$$\text{Trx-(SH}_2) + \text{protein-S}_2 \longrightarrow \text{Trx-S}_2 + \text{protein-(SH)}_2 \qquad (2)$$

There is a major difference in TrxR from mammalian cells and the enzyme from bacteria, fungi, and plants.[7] The latter, exemplified by the well-characterized *E. coli* enzyme, is composed of two 35-kDa subunits each with an FAD molecule and a redox-active disulfide.[7] These smaller TrxRs are generally highly specific to the Trx from the same species. In contrast, TrxR from mammalian cells is composed of two subunits of 55 kDa or higher, and has broad substrate specificity and an entirely different structure and catalytic mechanism.[7,8] The mammalian enzymes are built from a glutathione reductase equivalent with a 16-amino acid residue elongation containing the C-terminal conserved sequence Gly-Cys-SeCys-Gly, where SeCys is selenocysteine.[9] Selenium is essential for the enzyme and a unique selenenyl-sulfide bridge from the Cys-SeCys sequence is present in the oxidized enzyme. Reduction of this bridge to a selenolthiol generates the active site in TrxR, able to reduce oxidized Trx and a large variety of substrates.[10] In contrast to the remarkable evolutionary divergence in thioredoxin reductases, thioredoxins from different species are generally similar (Fig. 1).

Escherichia coli Thioredoxins

Escherichia coli has two thioredoxins, Trx1 and Trx2 (Fig. 1). Trx1, the first discovered, heat-stable 12-kDa thioredoxin,[2] has a well-described structure[11,12] (thioredoxin fold[6]) and WCGPC as the redox-active sequence of its active site. Trx1 reduces the disulfide formed in ribonucleotide reductase (RR), methionine

---

[5] D. M. Ferrari and H.-D. Söling, *Biochem. J.* **33**, 91 (1999).

[6] J. L. Martin, *Structure* **3**, 245 (1995).

[7] C. H. Williams, L. D. Arscott, S. Muller, B. W. Lennon, M. L. Ludwig, P. F. Wang, D. M. Veine, K. Becker, and R. H. Schirmer, *Eur. J. Biochem.* **267**, 6110 (2000).

[8] E. S. J. Arnér and A. Holmgren, *Eur. J. Biochem.* **267**, 6102 (2000).

[9] L. Zhong, E. S. J. Arnér, J. Ljung, F. Åslund, and A. Holmgren, *J. Biol. Chem.* **273**, 8581 (1998).

[10] L. Zhong, E. S. J. Arnér, and A. Holmgren, *Proc. Natl. Acad. Sci. U.S.A.* **97**, 5854 (2000).

[11] S. K. Katti, D. M. LeMaster, and H. Eklund, *J. Mol. Biol.* **212**, 167 (1990).

[12] M. F. Jeng, A. P. Cambell, T. Begley, A. Holmgren, D. A. Case, P. E. Wright, and H. J. Dyson, *Structure* **2**, 853 (1994).

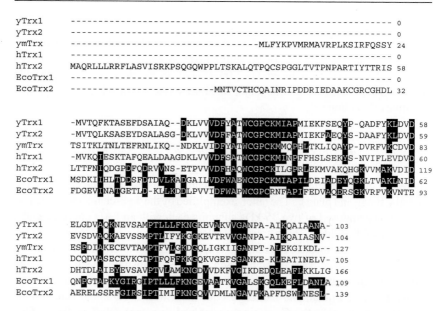

FIG. 1. Thioredoxin isoforms: Sequence alignments of cytosolic (yTrx1, yTrx2) and mitochondrial (ymitTrx) thioredoxins from *Saccharomyces cerevisiae,* thioredoxins from *Escherichia coli* (Eco Trx1 and Eco Trx2), and human cytosolic (hTrx1) and mitochondrial (hTrx2) thioredoxins. All amino acid numberings include the initiating methionine. Residues identical to *E. coli* Trx1 are printed on a black background. All alignments of thioredoxin or glutaredoxin species were performed online with the CLUSTAL W Multiple Sequence Alignment program at http://www.clustalw.genome.ad.jp/.

sulfoxide reductase, and 3'-phosphoadenosyl sulfate (PAPS) reductase.[8,13] Reduced Trx1 together with gene 5 protein constitute T7 DNA polymerase.[14] Trx1 is accordingly essential for the life cycle of phage T7 and is also essential for the assembly of phages M13 and f1.[15] Trx2 was discovered after completion of the sequencing of the *E. coli* genome.[16] Trx1 and Trx2 have 29% sequence identity, with the greatest difference being a 32-amino acid extension of Trx2 at its amino terminus. The two thioredoxins have similar steady state kinetic properties with TrxR.[16] Trx2 posesses additional cysteine thiols apart from those of the active site, which when oxidized downregulate its activity in the reduction of insulin disulfides.[16] Trx2 is also less heat stable than Trx1 and does not reduce

[13] C. H. Lillig, A. Prior, J. D. Schwenn, F. Aslund, D. Ritz, A. Vlamis-Gardikas, and A. Holmgren, *J. Biol. Chem.* **274,** 7695 (1999).

[14] D. F. Mark and C. C. Richardson, *Proc. Natl. Acad. Sci. U.S.A.* **74,** 789 (1976).

[15] M. Russel, *Methods Enzymol.* **252,** 264 (1995).

[16] A. Miranda-Vizuete, A. E. Damdimopoulos, J. Gustafsson, and G. Spyrou, *J. Biol. Chem.* **272,** 30841 (1997).

RR as efficiently as Trx1.[16] Trx2 participates in the OxyR-orchestrated antioxidant response.[17] Trx1 expression is not controlled by OxyR, but by ppGpp in the stationary phase.[18]

Saccharomyces cerevisiae Thioredoxins

Yeast has three thioredoxins, two cytosolic (yTrx1 and yTrx2)[19] and one mitochondrial[20] (ymTrx). Null mutants for yTrx1 and yTrx2 cannot live on sulfate under aerobic conditions because yeast thioredoxins are electon donors to PAPS reductase.[21] Double null mutants for yTrx1 and yTrx2 have a prolonged S phase and shortened G_1 phase although their deoxyribonucleotide pools are not affected, consistent with an unknown effect on the cell cycle.[22] yTrx2 is involved in the response against H_2O_2. A null mutant for yTrx1 and yTrx2 and yeast glutaredoxins 1 and 2 is not viable, showing that the mitochondrial thioredoxin (ymTrx) cannot substitute for the functions of the other two thioredoxins.[23] The transcription of yTrx2 is regulated by yAP-1[24] and Skn7.[25]

Human Thioredoxins

Two thioredoxins have been described for humans, one cytosolic[26] (hTrx1) and one mitochondrial[27] (hTrx2). hTrx1 can modulate the DNA-binding activity of many transcription factors including NF-κB[28] and AP-1.[29] Reduced thioredoxin prevents apoptosis by complexing apoptosis-signaling kinase 1[30] (ASK1). Secreted

---

[17] D. Ritz, H. Patel, B. Doan, M. Zheng, F. Åslund, G. Storz, and J. Beckwith, *J. Biol. Chem.* **275**, 2505 (2000).

[18] C. J. Lim, T. Daws, M. Gerami-Nejad, and J. A. Fuchs, *Biochim. Biophys. Acta* **1491**, 1 (2000).

[19] Z.-R. Gan, *J. Biol. Chem.* **266**, 1692 (1991).

[20] J. R. Pedrajas, A. Miranda-Vizuete, N. Javanmardy, J.-Å. Gustafsson, and G. Spyrou, *J. Biol. Chem.* **275**, 16296 (2000).

[21] E. G. D. Muller, *J. Biol. Chem.* **266**, 9194 (1991).

[22] E. G. D. Muller, *J. Biol. Chem.* **269**, 24466 (1994).

[23] T. Draculic, I. W. Dawes, and C. M. Grant, *Mol. Microbiol.* **36**, 1167 (2000).

[24] S. Kuge and N. Jones, *EMBO J.* **13**, 655 (1994).

[25] B. A. Morgan, G. R. Banks, W. M. Toone, D. Raitt, S. Kuge, and L. H. Johnston, *EMBO J.* **16**, 1035 (1997).

[26] Y. Tagaya, Y. Maeda, A. Mitsui, N. Kondo, H. Matsui, J. Hamuro, N. Brown, K. Arai, T. Yokota, H. Wakasugi, and J. Yodoi, *EMBO J.* **8**, 757 (1989).

[27] G. Spyrou, E. Enmark, A. Miranda-Vizuete, and J.-Å. Gustafsson, *J. Biol. Chem.* **272**, 2936 (1997).

[28] K. Hirota, M. Murata, Y. Sachi, H. Nakamura, J. K. M. Takeuchi, and J. Yodoi, *J. Biol. Chem.* **274**, 27891 (1999).

[29] H. Schenk, M. Klein, W. Erdbugger, W. Droge, and K. Schulze-Osthoff, *Proc. Natl. Acad. Sci. U.S.A.* **91**, 1672 (1994).

[30] M. Saitoh, H. Nisitoh, M. Fujii, K. Takeda, K. Tobiume, Y. Sawada, M. Kawabata, K. Miyazono, and H. Ichio, *EMBO J.* **17**, 2596 (1998).

thioredoxin is a cocytokine[31] and chemokine[32] whereas if secreted from glial cells it promotes neuronal survival after ischemia–reperfusion.[33] Thioredoxin participates in the defense against oxidative stress via the reduction of peroxiredoxins[34] (or thioredoxin peroxidases).

Structural Similarities

With the exception of the leader peptides of mitochondrial thioredoxins, all thioredoxins share the same fold[6] originally discovered for *E. coli* Trx1.[35] Thus, it is possible to classify all thioredoxins in one family. The major difference concerns mammalian thioredoxins, which have two or three additional structural cysteine residues. The reversible oxidation of the structural cysteines can regulate the enzyme activity.[36]

Assays for Thioredoxin Activity

These assays employ the ability of thioredoxin to reduce the disulfides of insulin or enzymes such as RR and have been described in detail previously in this series.[37,38] The most common assay of thioredoxin uses insulin as a substrate.

Insulin Assay

1. Insulin disulfides are good substrates for thioredoxin. Reduction of the disulfides can be monitored directly by the change in A_{340}, using a reaction mixture with thioredoxin, thioredoxin reductase, insulin, and NADPH.[38] The sensitivity of this assay is in the area of 10 pmol.

2. A more sensitive indirect assay for the reduction of insulin disulfides measures the free thiols of insulin after reduction by thioredoxin. This method uses 5,5'-dithiobis(2-nitrobenzoic acid) (DTNB) to measure generated thiols at A_{412} under denaturing conditions. The sensitivity of the assay is in the low picomolar range.

[31] H. Nakamura, K. Nakamura, and J. Yodoi, *Annu. Rev. Immunol.* **15,** 351 (1997).

[32] R. Bertini, O. M. Howard, H. F. Dong, J. J. Oppenheim, C. Bizzarri, R. Sergi, G. Caselli, S. Pagliei, B. Romines, J. A. Wilshire, M. Mengozzi, H. Nakamura, J. Yodoi, K. Pekkari, R. Gurunath, A. Holmgren, L. A. Herzenberg, L. A. Herzenberg, and P. Ghezzi, *J. Exp. Med.* **189,** 1783 (1999).

[33] K. Hori, M. Katayama, N. Sato, K. Ishii, S. Waga, and J. Yodoi, *Brain Res.* **652,** 304 (1999).

[34] S. W. Kang, H. Z. Chae, M. S. Seo, K. Kim, I. C. Baines, and S. G. Rhee, *J. Biol. Chem.* **273,** 6297 (1998).

[35] A. Holmgren, B.-O. Söderberg, H. Eklund, and C.-I. Branden, *Proc. Natl. Acad. Sci. U.S.A.* **72,** 2305 (1975).

[36] X. Ren, M. Björnstedt, B. Shen, M. L. Ericson, and A. Holmgren, *Biochemistry* **32,** 9701 (1993).

[37] A. Holmgren, *Methods Enzymol.* **107,** 295 (1984).

[38] E. S. J. Arnér, L. Zhong, and A. Holmgren, *Methods Enzymol.* **300,** 226 (1999).

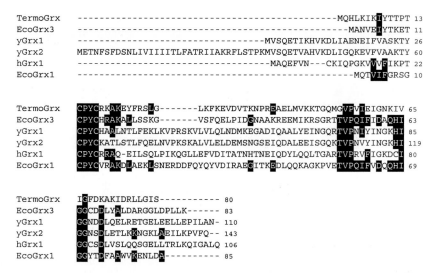

Fig. 2. Isoforms of dithiol glutaredoxins with CPYC active site, excluding *Escherichia coli* glutaredoxin 2. Sequences include the glutaredoxin from *Thermotoga maritima* (TermoGrx), glutaredoxins 1 (yGrx1) and 2 (yGrx2) from *Saccharomyces cerevisiae,* the human glutaredoxin (hGrx1), and glutaredoxins 1 (EcoGrx1) and 3 (EcoGrx3) from *Escherichia coli.* All amino acid numberings include the initiating methionine irrespective of the sequence of the mature protein. Residues identical to *E. coli* Grx1 are printed against a black background.

3. Reduced insulin B-chain precipitates readily by aggregation. Dithiothreitol (DTT) reduces Trx-S_2 more than two orders of magnitude faster than the direct reduction of insulin, whereas Trx-$(SH)_2$ reduces insulin four orders of magnitude faster than DTT.[39] The net effect is that Trx catalyzes the reduction of insulin by DTT. Therefore it is convenient to determine the activity of thioredoxin in a reaction mixture with 1 mM DTT and 160 μM insulin by recording the turbidity generated at 650 nm. The sensitivity of this assay is in the area of 100 pmol.

Glutaredoxins

Glutaredoxins (Grx) (Figs. 2–4) catalyze reductions of disulfides or GSH mixed disulfides in a coupled system with GSH, NADPH, and glutathione reductase (GR).[4] Glutaredoxins can be envisaged as reductants of disulfides via GSH controlling at the same time the levels of GSH-mixed disulfides, particularly under conditions of oxidative stress.[4,40]

[39] A. Holmgren, *J. Biol. Chem.* **254,** 9113 (1979).

[40] M.-J. Prieto-Alamo, J. Juardo, R. Gallardo-Madueno, F. Monje-Casas, A. Holmgren, and C. Pueyo, *J. Biol. Chem.* **275,** 13398 (2000).

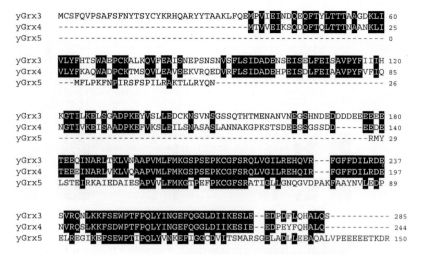

```
yGrx3  MCSFQVPSAFSFNYTSYCYKRHQARYYTAAKLFQEMPVIEINDQEQFTYLTTTAAGDKLI  60
yGrx4  ---------------------------------MTVVEIKSQDQFTQLTTTNAANKLI  25
yGrx5  ----------------------------------------------------------  0

yGrx3  VLYFHTSWAEPCKALKQVFEAISNEPSNSNVSFLSIDADENSEISELFEISAVPYFIIIH  120
yGrx4  VLYFKAQWADPCKTMSQVLEAVSEKVRQEDVRFLSIDADEHPEISDLFEIAAVPYFVFIQ  85
yGrx5  ---MFLPKFNPIRSFSPILRAKTLLRYQN-----------------------------  26

yGrx3  KGTILKELSGADPKEYVSLLEDCKNSVNSGSSQTHTMENANVNEGSHNDEDDDDEEEEE  180
yGrx4  NGTIVKEISAADPKEFVKSLEILSNASASLANNAKGPKSTSDEESSGSSDD-----EEDE  140
yGrx5  ---------------------------------------------------RMY  29

yGrx3  TEEQINARLTKLVNAAPVMLFMKGSPSEPKCGFSRQLVGILREHQVR---FGFFDILRDE  237
yGrx4  TEEEINARLVKLVQAAPVMLFMKGSPSEPKCGFSRQLVGILREHQIR---FGFFDILRDE  197
yGrx5  LSTEIRKAIEDAIESAPVVLFMKGTPEFPKCGFSRATIGLLGNQGVDPAKFAAYNVLEDP  89

yGrx3  SVRQNLKKFSEWPTFPQLYINGEFQGGLDIIKESLE--EDPDFLQHALQS----------  285
yGrx4  NVRQSLKKFSDWPTFPQLYINGEFQGGLDIIKESLE--EDPEYFQHALQ-----------  244
yGrx5  ELREGIKEFSEWPTIPQLYVNKEFIGGCDVITSMARSGELADLLEEAQALVPEEEEETKDR  150
```

FIG. 3. Yeast (*Saccharomyces cerevisiae*) monothiol glutaredoxins 3–5 (yGrx3, -4, and -5) with CGFS as the active site. The alignment is based on the sequence of yGrx3. Identical residues in the other two sequences are printed on a black background. All sequences include the initiating methionine residue irrespective of the sequence obtained finally in the mature protein.

```
GSTO1    MSGASARSLGKGSAPPGPVPEGQIRVYSMRFCPFAQRTLMVLKAKGIRHEIININLKNKP  60
p28      MSGESARSLGKGSAPPGPVPEGSIRIYSMRFCPFAERTRLVLKAKGIRHEVININLKNKP  60
CLIC1    MAEEQPQVELFVKAGSDGAKIG--------NCPFSQRLFMVLWLKGVTFNVTTVDTKRRT  52
EcoGrx2  ------MKLYIYDH---------------CPYCLKARMIFGLKNIPVELHVLLN-DDA  36

GSTO1    EWFFEKNPFGLVPVLENTQGHLITESVITCEYLDEAYPEKKLFPDDPYEKACQKMTFELF  120
p28      EWFFKKNPFGLVPVLENSQGQLIYESAITCEYLDEAYPGKKLLPDDPYEKACQKMILELF  120
CLIC1    ETVQKLCPGGQLPFLLYG-TEVHTDTNKIEEFLEAVLCPPRYPKLAALNPESNTAGLDIF  111
EcoGrx2  ETPTRMVGQKQVPILQKDDSRYMPESMDIVHYVDKLDGKPLLTGKRSPAIEEWLRKVNGY  96

GSTO1    SKVPSLVTSFIRAKRKEDHPGIKEELKKEFSKL-----------EEAMANKRTAFFGGN  168
p28      SKVPSLVGSFIRSQNKEDYAGLKEEFRKEFTKL-----------EEVLTNKKTTFFGGN  168
CLIC1    AKFSAYIKNSNPALNDNLEKGLLKALKVLDNYLTSPLPEEVDETSAEDEGVSQRKFLDGN  171
EcoGrx2  ANKLLLPRFAKSAFDEFSTPAARKYFVDKKEAS-----------AGNFADLLAHSDGLI  144

GSTO1    SLSMIDYLIWPWFQRLEALELNECIDHTPKLKLWMATMQEDPVASSHFIDAKTYRDYLSL  228
p28      SISMIDYLIWPWFERLEAMKLNECVDHTPKLKLWMAAMKEDPTVSALLTSEKDWQGFLEL  228
CLIC1    ELTLADCNLLPKLHIVQVVCKKYRGFTIPEAFRGVHRYLSNAYAREEFASTCPDDEEIEL  231
EcoGrx2  KNISDDLRALDKLIVKPNAVNGELSEDDIQLFPLLRNLTLVAGINWPSRVADYRDNMAKQ  204

GSTO1    YLQDSPEACDYGL  241
p28      YLQNSPEACDYGL  241
CLIC1    AYEQVAKALK---  241
EcoGrx2  TQINLLSSMAI--  215
```

FIG. 4. Isoforms of large glutaredoxins with similarity to *Escherichia coli* glutaredoxin 2 (EcoGrx2). Sequences shown here are human glutathione transferase omega 1 (GSTO1), mouse glutathione transferase theta-like stress response protein (p28), and human chloride intracellular channel 1 (CLIC1). Residues identical to EcoGrx2 are printed on a black background.

Catalytic Mechanism of Glutaredoxin/Thioredoxin

Glutaredoxins can reduce disulfides by a dithiol or a monothiol mechanism.[41] In the dithiol mechanism[42] the thiolate of the active site (typically CPYC) closer to the N terminus attacks and reduces the substrate disulfide via a transient mixed disulfide [Reaction (3)].

$$Grx\text{-}(SH)_2 + protein\text{-}S_2 \longrightarrow Grx\text{-}S_2 + protein\text{-}(SH)_2 \qquad (3)$$

In the monothiol mechanism (e.g., active site CGFS), the thiolate of glutaredoxin initiates a nucleophilic attack on the mixed disulfide of a protein thiol with GSH (protein-S-SG) [Reaction (4)]. A New disulfide between glutaredoxin and GSH is formed, while the protein substrate is released in the reduced form. The mixed disulfide between glutaredoxin and GSH can be reduced by GSH, which will make a nucleophilic attack on the disulfide of the glutaredoxin, to give as final products GSSG and reduced monothiol glutaredoxin [Reaction (5)]. The preference of glutaredoxin to form a disulfide with GSH instead of the protein in the initial nucleophilic attack is explained by the properties of the GSH-binding site of glutaredoxins.[43,44] Note that both Reactions (4) and (5) are reversible. The GSSG produced is finally reduced by GR [Reaction (6)].

$$Grx\text{-}SH + protein\text{-}S\text{-}SG \longrightarrow Grx\text{-}S\text{-}SG + protein\text{-}SH \qquad (4)$$

$$Grx\text{-}S\text{-}SG + GSH \longrightarrow Grx\text{-}SH + GSSG \qquad (5)$$

$$GSSG + NADPH + H^+ \xrightarrow{\text{GR}} 2GSH + NADP^+ \qquad (6)$$

The monothiol/dithiol mechanism of catalytic action of glutaredoxins determines the substrate specificity of the enzymes. Monothiol glutaredoxins (usually CGFS) will catalyze the reduction only of thiol–glutathione mixed disulfides, whereas dithiol glutaredoxins (mostly CPYC) will be able to reduce both protein disulfides (e.g., RR) and GSH-mixed disulfides. Glutaredoxins are central in the response against oxidative stress as the biological activity of many proteins is modified by the formation of GSH-mixed disulfides (glutathionylation).[45] Glutaredoxins exhibit a much greater variability of isoforms compared with thioredoxins (Figs. 2–4).

Escherichia coli Glutaredoxins

Escherichia coli thus far contains three characterized glutaredoxins.[41] The first (Grx1, 9 kDa) was discovered as a glutathione-dependent hydrogen donor for class

[41] A. Holmgren, E. S. J. Arnér, and K. D. Berndt, "Encyclopedia of Molecular Biology" (T. E. Creighton, ed.), p. 1020. John Wiley & Sons, New York, 1999.

[42] J. H. Bushweller, F. Åslund, K. Wüthrich, and A. Holmgren, *Biochemistry* **31**, 9288 (1992).

[43] J. H. Bushweller, M. Billeter, A. Holmgren, and K. Wüthrich, *J. Mol. Biol.* **235**, 1585 (1994).

[44] K. Nordstrand, F. Åslund, A. Holmgren, G. Otting, and K. Berndt, *J. Mol. Biol.* **286**, 541 (1999).

[45] I. A. Cotgreave and R. G. Gerdes, *Biochem. Biophys. Res. Commun.* **242**, 1 (1998).

1a ribonucleotide reductase (RR1a) in null mutants for thioredoxin.[46] The other two glutaredoxins, glutaredoxin 2 (Grx2, 24.3 kDa) and glutaredoxin 3 (Grx3, 9 kDa), were purified from an *E. coli* null mutant for glutaredoxin 1 and thioredoxin 1.[47] Grx1 and Grx3 show 33% sequence identity (Fig. 2) and have a similar three-dimensional structure, being members of the thioredoxin fold superfamily.[43,44] Both proteins can reduce RR1a, with Grx1 being the most efficient.[47] The larger Grx2 has limited amino acid sequence identity to Grx1 or Grx3[48] and cannot reduce RR1a.[47] Grx2 (Fig. 4) is radically different in terms of molecular weight, amino acid sequence, specificity, and lack of a consensus GSH-binding site.[48] In wild-type cells, Grx2 provides as much as 80% of the total GSH oxidoreductase activity for low molecular weight mixed disulfides with GSH.[47] Steady state kinetics analysis using 2-hydroxyethyl disulfide (HED) as substrate showed that Grx2 has the highest apparent k_{cat} of all known glutaredoxins.[48] In a system using the mixed disulfide of arsenate reductase (ArsC) with glutathione as a substrate, Grx2 was 100-fold more active than the other two glutaredoxins.[49] ArsC catalyzes the reduction of arsenate to arsenite.[49] Expression of Grx1 is regulated by OxyR (exponential phase).[40]

Saccharomyces cerevisiae Glutaredoxins

Five glutaredoxins have been described so far in yeast.[50,51] The first two glutaredoxins are typical CPYC proteins[50] (Fig. 2) whereas the other three (Fig. 3) have monothiol active sites (CGFS).[51] All yeast glutaredoxins participate in the response against oxidative stress.[50,51] yGrx1 participates in the response against the superoxide anion, and yGrx2 against hydrogen peroxide.[50] Lack of yGrx5 renders cells growth deficient and sensitive to oxidative damage (carbonylation) or hyperosmotic treatment.[51] Yeast cells without yGrx2 and yGrx5 or without yGrx3, yGrx4, and yGrx5 are not viable, suggesting that a specific minimal monothiol glutaredoxin activity is essential for yeast viability.[51]

Human Glutaredoxin

One classic human glutaredoxin of 12 kDa (CPYC active site, Fig. 2) has been described so far.[52,53] The protein has three additional structural cysteine residues

[46] A. Holmgren, *Proc. Natl. Acad. Sci. U.S.A.* **73,** 2275 (1976).

[47] F. Åslund, B. Ehn, A. Miranda-Vizuete, C. Pueyo, and A. Holmgren, *Proc. Natl. Acad. Sci. U.S.A.* **91,** 9813 (1994).

[48] A. Vlamis-Gardikas, F. Åslund, G. Spyrou, T. Bergman, and A. Holmgren, *J. Biol. Chem.* **272,** 11236 (1997).

[49] J. Shi, A. Vlamis-Gardikas, F. Åslund, A. Holmgren, and B. P. Rosen, *J. Biol. Chem.* **274,** 36039 (1999).

[50] S. Luikenhuis, G. Perrone, I. W. Dawes, and C. M. Grant, *Mol. Biol. Cell* **9,** 1081 (1998).

[51] M. T. Rodriguez-Manzaneque, J. Ros, E. Cabiscol, A. Sorribas, and E. Herrero, *Mol. Cell. Biol.* **19,** 8180 (1999).

and can reduce recombinant mouse RR.[52] Different glutaredoxins generally exhibit specificity for ribonucleotide reductases (e.g., viral glutaredoxins). A novel human glutaredoxin of 18 kDa with mitochondrial and nuclear isoforms obtained by differential splicing has been isolated.[54]

Classification of Glutaredoxins

In terms of their structure and catalytic mechanism, glutaredoxins can now be classified in three categories. The first consists of classic glutaredoxins, which are 9- to 12-kDa proteins with the sequence CPYC at their active site (Fig. 2). Typical examples are Grx1 and Grx3 from *E. coli,* glutaredoxins 1 and 2 from yeast, and human and phage T4 glutaredoxins.[55] The second category (Fig. 3) is exemplified by the monothiol yeast glutaredoxins (yGrx3, yGrx4, and yGrx5).[51] Their glutaredoxin activity has been confirmed by comparing the activity of total cell extracts from null mutants and the wild type. Further studies are required to determine the *in vitro* activity and structure of these proteins. The third class of glutaredoxins (Fig. 4) is defined by *E. coli* Grx2, a protein of 24.3 kDa with close homologs in *Actinobacillus actinomycetemcomitans* (87% amino acid identity), *Neisseria meningitidis* (58%), and *Vibrio cholerae* (42%).[56] The three-dimensional structure of Grx2 shows that the protein has extended structural similarities to the glutathione transferase (GST) family of proteins.[57] *Escherichia coli* Grx2 is like a GST that has become a glutaredoxin after introducing the CPYC sequence in the glutaredoxin domain of GST. Grx2 should therefore be considered as a link between glutaredoxins and GSTs. Other proteins that are likely to belong to this category are the human glutathione transferase omega 1 (GSTO1),[58] the mouse glutathione transferase theta-like stress response protein (p28),[59] and the human chloride intracellular channel 1 (CLIC1).[60] Common structural characteristics are a two-domain structure, the first domain being a thioredoxin domain in which the active site resides and the second domain an α-helical structure. The

[52] C. A. Padilla, E. Martinez-Galisteo, T. Bárcena, G. Spyrou, and A. Holmgren, *Eur. J. Biochem.* **227,** 27 (1995).

[53] C. Sun, M. J. Berardi, and J. H. Bushweller, *J. Mol. Biol.* **280,** 687 (1998).

[54] M. Lundberg, C. Johansson, J. Chandra, M. Enoksson, G. Jacobsson, J. Ljung, M. Johansson, and A. Holmgren, *J. Biol. Chem.* **276,** 26269 (2001).

[55] H. Eklund, M. Ingelman, B. O. Soderberg, T. Uhlin, P. Nordlund, M. Nikkola, U. Sonnerstam, T. Joelson, and K. Petratos, *J. Mol. Biol.* **228,** 596 (1992).

[56] BLAST search.

[57] B. Xia, A. Vlamis-Gardikas, A. Holmgren, P. E. Wright, and H. J. Dyson, *J. Mol. Biol.* **310,** 907 (2001).

[58] P. G. Board, M. Coggan, G. Chelvanayagam, S. Easteal, L. S. Jermiin, G. K. Schulte, D. E. Danley, L. R. Hoth, M. C. Griffor, A. V. Kamath, M. H. Rosner, B. A. Chrunyk, D. E. Perregaux, C. A. Gabel, K. F. Geoghegan, and J. Pandit, *J. Biol. Chem.* **275,** 24798 (2000).

[59] R. Kodym, P. Calkins, and M. Story, *J. Biol. Chem.* **274,** 5131 (1999).

[60] A. Dulhunty, P. Cage, S. Curtis, G. Chelvanayagam, and P. Board, *J. Biol. Chem.* **276,** 3319 (2001).

Grx2-like family consists of both mono-(e.g., GLIC; p28, and GSTO1) and dithiol active site proteins (*E. coli* Grx2).

Assays for Glutaredoxin Activity

Reduction of Low Molecular Weight Disulfides

Low molecular weight substrate disulfides are those of hydroxyethyl, L-cystine, coenzyme A, and S-sulfocysteine.[61] It is believed that the actual substrate is not the disulfide itself but the mixed disulfide forming spontaneously between GSH and the thiol part of the disulfide substrate. Low molecular weight disulfides such as those of hydroxyethyl (hydroxyethyl disulfide), cysteine (cystine), GSH (GSSG), and cystamine are relatively poor substrates for thioredoxin.[39] The assay for low molecular weight disulfides is therefore a specific and sensitive indicator of glutaredoxin activity. The sensitivity using HED as substrate is, for *E. coli* Grx1, 0.1 ng/ml.[48]

Reduction of High Molecular Weight Disulfides

Reduction of high molecular weight disulfides corresponds to the reduction of the disulfide of RR, which is followed by the reduction of [^{3}H]CDP to dCDP.[62] Reducing equivalents for the reduction of CDP can be provided by 1 mM NADPH, 4 mM GSH, and GR (10 μg/ml).

Reduction of High Molecular Weight Mixed Disulfides

Reduction of high molecular weight mixed disulfides can be monitored spectrophotometrically (A_{340}) for RNase-SG,[63] ArsC-SG,[49] bovine serum albumin (BSA)-Cys and papain-Cys[64] or by radioimmunoassay (RIA) for radiolabeled BSA, methemoglobin (metHb), oxyHb, papain, and BSA forming mixed disulfides with cysteine or glutathione.[64]

Reduction of Dehydroascorbic Acid

Reduction of dehydroascorbic acid (DHA) is a spectrophotometric assay measuring changes in A_{265}.[4] Care should be taken that DHA solutions are fresh.

Acknowledgments

This research was supported by grants from the Swedish Cancer Society (961), the Swedish Medical Research Council (13X-3529), and the Karolinska Institute.

[61] J. J. Mieyal, D. W. Starke, S. A. Gravina, and B. A. Hocevar, *Biochemistry* **30,** 8883 (1991).
[62] A. Holmgren, *Methods Enzymol.* **113,** 525 (1985).
[63] J. Lundstrom-Ljung, A. Vlamis-Gardikas, F. Åslund, and A. Holmgren, *FEBS Lett.* **443,** 85 (1999).
[64] S. Gravina and J. J. Mieyal, *Biochemistry* **32,** 3368 (1993).

[27] Mammalian Thioredoxin Reductases

By TAKASHI TAMURA and THRESSA C. STADTMAN

Introduction

Thioredoxin reductase (TrxR) is an NADPH-dependent, FAD-containing disulfide reductase that plays an important role in cell proliferation.[1] Unlike the well-characterized homologs from yeast and prokaryotes, the larger mammalian enzyme is a selenoprotein that contains a selenocysteine (Secys) residue[2] in the sequence-Cys-Secys-Gly (end) at the C terminus of each subunit.[3-5] Catalysis of electron transfer from NADPH to thioredoxin, which in turn is linked to critical components of cell metabolism such as ribonucleotide reductase,[6] AP-1 and NF-κB transcription factors,[7-10] vitamin K epoxide reductase,[11] thiol peroxidase,[12] and plasma glutathione peroxidase,[13] illustrates the diversity of processes that depend on this selenium-containing TrxR. The provision of reduced thioredoxin for two important cell processes, DNA synthesis and gene transcription, implicates TrxR as a key enzyme in the control of cell growth.

The selenocysteine residue in mammalian thioredoxin reductase (TrxR) was first identified in the ^{75}Se-labeled protein isolated from a human lung adenocarcinoma cell line.[2] The unexpected discovery of a selenocysteine residue in a protein that proved to be mammalian TrxR originated from experiments designed to characterize a putative selenoprotein produced by nonsense mutants of a cytochrome *P*-450 isozyme.[14] A ^{75}Se-labeled protein was purified to apparent homogeneity from the human lung adenocarcinoma cells but its physicochemical properties did not match those of a cytochrome *P*-450 species. Instead of a cytochrome

---

[1] A. Holmgren, *Annu. Rev. Biochem.* **54,** 237 (1985).

[2] T. Tamura and T. C. Stadtman, *Proc. Natl. Acad. Sci. U.S.A.* **93,** 1006 (1996).

[3] V. N. Gladyshev, K.-T. Jeang, and T. C. Stadtman, *Proc. Natl. Acad. Sci. U.S.A.* **93,** 6146 (1996).

[4] S.-Y. Liu and T. C. Stadtman, *Proc. Natl. Acad. Sci. U.S.A.* **94,** 6138 (1997).

[5] P. Y. Gasdaska, J. R. Gasdaska, S. Cochran, and G. Powis, *FEBS Lett.* **373,** 5 (1995).

[6] L. Thelander and P. Reichard, *Annu. Rev. Biochem.* **48,** 133 (1979).

[7] G. Spyrou, M. Bjornstedt, S. Kumar, and A. Holmgren, *FEBS Lett.* **368,** 59 (1995).

[8] M. L. Handel, C. L. Watts, A. DeFazio, R. O. Day, and R. L. Sutherland, *Proc. Natl. Acad. Sci. U.S.A.* **92,** 4497 (1995).

[9] V. Makropoulos, T. Bruning, and K. Schulze-Osthoff, *Arch. Toxicol.* **70,** 277 (1996).

[10] I. Y. Kim and T. C. Stadtman, *Proc. Natl. Acad. Sci. U.S.A.* **94,** 12904 (1997).

[11] R. B. Silverman and D. L. Nandi, *Biochem. Biophys. Res. Commun.* **155,** 1248 (1988).

[12] H. Z. Chae, S. J. Chung, and S. G. Rhee, *J. Biol. Chem.* **269,** 27670 (1994).

[13] M. Bjornstedt, J. Xue, W. Huang, B. Akesson, and A. Holmgren, *J. Biol. Chem.* **269,** 29382 (1994).

[14] S. Yamano, P. T. Nhamburo, T. Aoyama, U. A. Meyer, T. Inaba, W. Kalow, H. V. Gelboin, O. W. McBride, and F. J. Gonzalez, *Biochemistry* **28,** 7340 (1989).

0076-6879/02 $35.00

chromophore the dimeric 57-kDa subunit protein contained FAD and the flavin was reduced specifically by NADPH. Unfortunately this selenocysteine-containing protein was N-blocked and N-terminal amino acid sequence information was not obtained. However, the total amino acid compositional analysis indicated similarities to the amino acid content of rat liver thioredoxin reductase.[15] Indeed, the ^{75}Se-labeled enzyme was shown to catalyze the NADPH-dependent reduction of 5,5'-dithiobis(2-nitrobenzoate) (DTNB) and also thioredoxin as substrates, thus indicating its identity as a thioredoxin reductase. Subsequently selenium-containing TrxRs from other mammalian cells and tissues were identified.[3,4,16] Mammalian TrxRs now appear to have three isozymes designated as TrxR1, TrxR2,[17] and TrxR3.[18] It seems widely accepted to designated TrxR1 as the dominant cytosolic enzyme, whereas TrxR2 is referred to as the mitochondrial type. TrxR3 was first identified in a gene sequence but its expression seems to be lower than that of TrxR1. Alternatively, TrxR3 expression may be organ specific or produced under certain physiological conditions in cells.

The present article focuses on skills and techniques for studies of mammalian thioredoxin reductase and also for general selenium biochemistry. The radioisotope ^{75}Se is a useful tool for the detection and identification of selenoproteins on sodium dodecyl sulfate (SDS)–polyacrylamide gels, and the ^{75}Se-labeled selenocysteine residue can be chemically identified by amino acid analysis after chemical derivatization. Technical precautions in the purification of selenoproteins are also described for researchers seeking undiscovered selenoenzymes.

Labeling Proteins with Selenium-75

The radioisotope selenium-75 is a powerful isotope used in selenium biochemistry. It emits γ rays with 0.265 eV (59%) and 0.280 eV (25%), and decays with a half-life time of 120.4 days. ^{75}Se-labeled proteins can be detected by this moderately strong radioactivity, yet this half-life time gives us an opportunity to diminish the radioactivity of the biological wastes in just a few months. The presence of ^{75}Se is readily detected with a portable Geiger–Muller counter, and the radioactivity is determined in a γ-ray counter such as the Beckman (Fullerton, CA) γ-5500 or the Wallace 1470 Wizard automatic γ counter. [^{75}Se]Selenite can be purchased from the Research Reactor Facility, University of Missouri (Columbia, MO). The radioisotope is delivered in the form of selenious acid (H_2SeO_3) in a small volume of 7–30% nitric acid. Its specific activity is high enough to allow us to

[15] M. Luthman and A. Holmgren, *Biochemistry* **21**, 6628 (1982).
[16] J. Nordberg, L. Zhong, A. Holmgren, and E. S. Arner, *J. Biol. Chem.* **273**, 10835 (1998).
[17] S. R. Lee, J. R. Kim, K. S. Kwon, H. W. Yoon, R. L. Levine, A. Ginsburg, and S. G. Rhee, *J. Biol. Chem.* **274**, 4722 (1999).
[18] Q. A. Sun, Y. Wu, F. Zappacosta, K.-T. Jeang, B. J. Lee, D. L. Hatfield, and V. N. Gladyshev, *J. Biol. Chem.* **274**, 24552 (1999).

ignore the original content of selenium. Because selenite is one of the more effective chemical forms in labeling the selenocysteine residue with ^{75}Se, the sodium [^{75}Se]selenite is directly added to the culture broth and incubated for the desired period of time. In most cases the small amount of nitric, acid added with the selenite has no effect on the growth of cells. It is strongly recommended that a small amount of "cold" or nonradioactive sodium selenite be added before use to avoid radiocolloid formation. In experiments with the human adenocarcinoma cell line NCI-H441,[2] 0.1 μM sodium selenite containing [75]Se (368 Ci/mmol) was added to the medium and cultures were routinely incubated for 4 days. To determine the optimum time of harvest cells were collected at 6, 24, 48, 72, and 96 hr and directly analyzed by SDS–Polyacrylamide gel electrophoresis (PAGE) followed by PhosphorImager (Molecular Devices, Sunnyvale, CA) detection of radioactivity.

Sodium Dodecyl Sulfate-Polyacrylamide Gel Electrophoresis Analysis and Autoradiography

PhosphoImager analysis provides a convenient and sensitive method for detecting proteins labeled with ^{75}Se after separation by SDS–PAGE and drying of gels. A cell suspension (about 5 mg wet weight in 30 μl of phosphate-buffered saline is mixed with 30 μl of SDS–PAGE sampling buffer, and boiled for 10 min. The selenocysteine residue can survive the heat treatment because of the high content of 2-mercaptoethanol. Then 5- to 20-μl portions of the heated sample are loaded on an SDS–polyacrylamide gel. Selenoproteins can be developed by SDS–PAGE without any obvious degradation unless ammonium peroxodisulfate, a radical polymerization initiator, remains in the polyacrylamide gel. Oxidizing reagents and radical species are reactive with selenocysteine residues, and ^{75}Se may be totally eliminated from the selenoproteins during the course of electrophoresis. This can be avoided by running 10 ml of buffer containing 10% (w/v) thioglycolate through the gel before the sample is loaded, or alternatively by including 5 mM 2-mercaptoethanol or dithiothreitol (DTT) in the running buffer. Autoradiography is more conveniently and better performed by a PhosphorImager technique. Authors have noted that 3000 cpm of ^{75}Se is sufficient for development of a clear image when the gel is exposed to an imaging plate for only 2 hr. For routine analysis, the exposure is usually performed overnight (Fig. 1).

Identification of Selenocysteine Residues

Identification of ^{75}Se-labeled selenocysteine is the critical evidence for characterization of a selenocysteine-dependent selenoprotein.[19,20] Some proteins can

[19] R. Read, T. Bellew, J.-G. Yang, K. E. Hill, I. S. Palmer, and R. F. Burk, *J. Biol. Chem.* **268**, 17899 (1990).

[20] J. E. Cone, R. M. del Rio, J. N. Davis, and T. C. Stadtman, *Proc. Natl. Acad. Sci. U.S.A.* **73**, 2659 (1976).

1 2 3 4 5   1 2 3 4 5

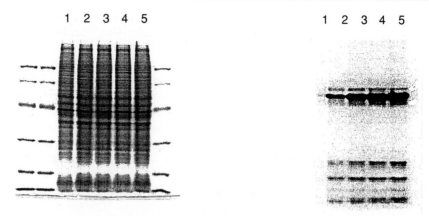

FIG. 1. SDS–PAGE and autoradiography of ^{75}Se-labeled proteins in the human lung adenocarcinoma cell line NCI-H441. Cells were grown in RPMI 1640 medium containing 10% (v/v) fetal bovine serum and 680 kBq of ^{75}Se-labeled selenite for (lane 1) 6, (lane 2) 24, (lane 3) 48, (lane 4) 72, and (lane 5) 96 hr. The cells were harvested by trypsin–EDTA treatment, boiled in SDS–PAGE sample buffer, and subjected to SDS–PAGE analysis (*left;* Coomassie Brilliant Blue staining). The gel was dried and subjected to autoradiography (right).

bind elemental ^{75}Se or [^{75}Se]selenite with high affinity, microbial samples may contain significant amounts of [^{75}Se]selenomethionine,[21] and the resulting radioactive proteins may be confused with specific selenoproteins.[22,23] Selenocysteine can be identified on an amino acid analyzer but it requires chemical derivatization before the ^{75}Se-labeled protein is subjected to acid hydrolysis. Selenocysteine can be almost completely decomposed when heated at 110° in 6 N HCl in the presence of trace amounts of oxygen (survival rate, 6%). Iodoacetate is frequently used for the protective derivatization of selenocysteine residues, and the resulting *Se*-carboxymethylselenocysteine (CM-Secys) can survive the entire procedure. A homologous derivatization can be carried out with 3-bromopropionate, which yields *Se*-carboxyethylselenocysteine (CE-Secys). A combination of CM-Secys and CE-Secys can illustrate that the selenium moiety of the ^{75}Se-labeled protein is in the form of selenocysteine as described in the following procedure.

Purified ^{75}Se-labeled protein (40 μg) is washed three times with distilled water and concentrated to 40 μl with a Centricon-10 microcencentrator (Amicon, Danvers, MA). The solution is mixed with 60 μl of 100 mM NaBH$_4$ in 20 mM NaOH aqueous solution, and incubated under argon at room temperature for 30 min. Sodium iodoacetate (or sodium 3-bromopropionate) is added to the mixture

[21] M. G. M. Hartmanis and T. C. Stadtman, *Proc. Natl. Acad. Sci. U.S.A.* **79,** 4912 (1982).
[22] M. P. Bansal, C. J. Oborn, K. G. Danielson, and D. Medina, *In Vitro* **3,** 167 (1989).
[23] R. Sinha, M. P. Bansal, H. Ganther, and D. Medina, *Carcinogenesis* **14,** 1895 (1993).

to a final concentration of 50 μM, and the mixture is incubated under argon at room temperature for 40 min. Then, 2-mercaptoethanol is added to quench the alkylation reaction. The protein is washed three times with distilled water by ultrafiltration, taken to dryness, and then hydrolyzed in 6 M HCl at 155° under argon. The hydrolysate is dried, treated with small amount of NaBH$_4$, mixed with authentic CM-Secys and CE-Secys (each at 1.3 μmol), and chromatographed on an amino acid analyzer.[20,21] The eluate from the analyzer column is collected in 1-min fractrions, and the radioactivity contained in these fractions is determined with a Beckman model 5500 γ counter. On the amino acid analyzer, CM-Secys is usually eluted earlier than CE-Secys. When the labeled protein is alkylated with iodoacetate, the radioactive elution profile of the hydrolysate coincides exactly with CM-Secys (Fig. 2A). When 3-bromopropionate is used for the alkylation,

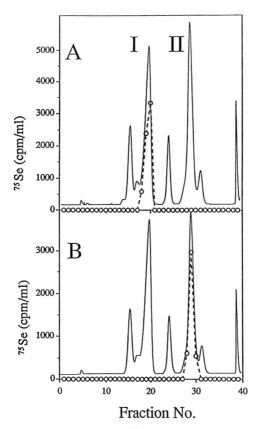

Fraction No.

FIG. 2. Amino acid analyzer chromatogram of ^{75}Se-labeled compounds from an acid hydrolysate of carboxymethylated (A) and carboxyethylated (B) selenoprotein. The hydrolysate was mixed with CM-Secys (I) and CE-Secys (II) before the chromatography. Solid lines represent amino acid elution and open circles represent ^{75}Se radioactivity.

[75]Se in the hydrolysate coincides with that of CE-Secys (Fig. 2B). Throughout the procedure of alkylation, hydrolysis, and chromatography, the recovery of ^{75}Se is usually 47 and 65% for CM-Secys and CE-Secys formation, respectively.

Precautions for Purifying Selenocysteine-Containing Proteins

Purification can be carried out by conventional protein purification methods, but certain precautions are taken. First, selenocysteine has a lower redox potential and lower pK_a than cysteine, and it requires some reducing reagent to maintain the selenol state. In the absence of such a reducing reagent, it may undergo oxidation to seleninate ($-SeO_2H$) and decompose to a dehydroalanine residue through α,β-elimination under alkaline conditions. Dithiothreitol at 2 mM is a favorite reagent of the present authors and it is always accompanied by EDTA-Mg,K complex at 0.1 mM to prevent metal-catalyzed thiol radical formation that is also harmful to selenocysteine residues. In addition, the selenocysteine residue is also reactive toward various nucleophiles. Therefore sodium azide should be omitted from buffers used for gel-filtration column chromatography. ^{75}Se-labeled protein can be located in the column bed if a Geiger–Muller counter is held close to the column tube during the course of purification. This physicochemical property as a radiant γ-ray emitter is particularly useful in estimating the elution of selenoproteins during gel filtration or other type of protein purification in open columns.

Separation of Thioredoxin Reductase 1 from Thioredoxin Reductase 2 in Rat Liver Homogenates

A convenient and efficient method of separation of rat liver TrxR1 from TrxR2 by adjustment of pH has been described by S. G. Rhee and co-workers.[17] Rat livers (1 kg) are homogenized in 4 liters of 20 mM Tris-HCl (pH 7.8) containing 1 mM EDTA, 1 mM dithiothreitol, 0.05 mM 4-(2-aminoethyl)benzenesulfonyl fluoride hydrochloride (AEBSF), pepstatin (0.5 mg/ml), and aprotinin (0.5 mg/ml). The homogenate is centrifuged at 70,000g for 30 min at 4°, and the resulting supernatant is adjusted to pH 5.0 with 1 M acetic acid and then centrifuged again at 70,000g for 30 min at 4°. The resulting pellet and supernatant are adjusted to conditions suitable for immunoblot analysis with antibodies to TrxR1 and TrxR2. TrxR1 is detected only in the supernatant, whereas TrxR2 was present mostly in the pellet. Thus, the supernatant and pellet serve as sources for purification of TrxR1 and TrxR2, respectively.

For purification of TrxR1, the supernatant (40 g of protein) from the pH 5 precipitation step is adjusted to pH 7.8 with 1 M ammonium hydroxide and then applied to a DEAE-Sephacel (Pharmacia, Piscataway, NJ) column (10 × 16 cm) that has been equilibrated with 20 mM Tris-HCl (pH 7.8) containing 1 mM EDTA, 1 mM dithiothreitol, and 0.01 mM AEBSF. The column is washed consecutively with 2.5 liters of equilibrium buffer and 2.5 liters of equilibration buffer containing 100 mM NaCl. Proteins are eluted from the column with a linear gradient of 100 to 400 mM NaCl in 5 liters of equilibration buffer, and fractions (25 ml) are collected and TrxR1 is detected by immunoblot analysis. The peak fractions (10.4 g of protein), corresponding to 300 to 380 mM NaCl on the gradient, are pooled, dialyzed overnight against 20 mM Tris-HCl (pH 7.5) containing 1 mM EDTA, 1 mM DTT, and 0.01 mM AEBSF, and then applied to a 2′,5′-ADP-agarose column (2 × 7 cm) that has been equilibrated with 20 mM Tris-HCl (pH 7.5) containing 1 mM EDTA. The column is washed with 100 ml of equilibration buffer, and proteins are then eluted stepwise with 100 ml each of equilibration buffer containing 200 mM KCl, equilibration buffer containing 200 mM sodium phosphate and 200 mM KCl, and equilibration buffer containing 1 M NaCl and 200 mM KCl. TrxR1 is present almost exclusively in the fractions eluted by the buffer containing 200 mM KCl as revealed by SDS–PAGE and Coomassie blue staining and by immunoblot analysis. Peak fractions (19.8 mg of protein) are pooled and then adjusted to 1.2 M ammonium sulfate by addition of 4 M ammonium sulfate. After removal of the resulting precipitate by centrifugation, the supernatant is applied to a Phenyl-5PW high-performance liquid chromatography (HPLC) column (0.75 × 7.5 cm) that has been equilibrated with 20 mM HEPES–NaOH (pH 7.5) containing 1 mM DTT, 1 mM EDTA, and 1.2 M ammonium sulfate. The column is washed with 60 ml of equilibration buffer, and proteins are then eluted with a decreasing linear gradient of 1.2 to 0 M ammonium sulfate in 120 ml of 20 mM HEPES–NaOH (pH 7.5) containing 1 mM DTT and 1 mM EDTA. Peak fractions, corresponding to 0.8 to 0.64 M ammonium sulfate on the gradient, are pooled, concentrated, dialyzed against 20 mM HEPES–NaOH (pH 7.5) containing 1 mM DTT and 1 mM EDTA, divided into portions, and stored at −70°.

Selective Alkylation of Selenocysteine Residue 498 of Thioredoxin Reductase 1

The essential role of selenocysteine residue 498 (Secys-498) in catalysis has been demonstrated by concomitant reduction of enzyme activity to 1% or less with selective alkylation of Secys-498.[24] In the experiment, reaction of native NADPH-reduced enzyme with bromo[1-^{14}C]acetate not only inhibited enzyme activity by 99% but also resulted in incorporation of 1.1 equivalents of alkyl group per subunit, of which >90% was present in the carboxymethyl (CM) derivative of Secys-498 and about 5% was present in the CM derivative of Cys-497. Such a highly selective alkylation can be carried out at pH 6.5 when bromoacetate instead

---

[24] S. N. Gorlotov and T. C. Stadtman, *Proc. Natl. Acad. Sci. U.S.A.* **95**, 8525 (1998).

of iodoacetate is used as alkylating agent, making the reaction more selective for the fully ionized selenol group of selenocysteine. However, if the pK_a of the thiol group of the adjacent cysteine residue is abnormally low, it also might be alkylated under these conditions. Amino acid analysis of the alkylated enzyme, after acid hydrolysis, showed that labeled CM-Secys accounted for at least 80% of the recovered alkyl group and that 20% or less was in CM-Cys. The amount of CM-Secys, when corrected for losses caused by the marked oxygen lability of the selenoether, corresponded to complete derivatization of one Secys-498 per subunit. In this experiment ^{75}Se-labeled HeLa cell TrxR (11.0 nmol; 5.8×10^5 cpm) in 200 μl of 20 mM potassium phosphate (pH 7.0), 1 mM EDTA, and 10% (v/v) glycerol was reduced with 240 nmol of NADPH under argon for 20 min and then reacted with bromo[1-^{14}C]acetic acid (175 nmol; 10 μCi) for 60 min in the dark under argon. The reaction was quenched by the addition of DTT (2 μmol) and the pH was adjusted to 8.0. Guanidine hydrochloride was added to 6 M and after 5 min the enzyme was dialyzed against 1.5 liters of 20 mM Tris-HCl buffer (pH 8.0) and 1 mM EDTA under argon for 2.5 hr. The dialyzed protein was digested with trypsin N-1-tosylamido-2-phenylethyl chloromethyl ketone, 56 μg, for 4 hr under argon and then was adjusted to pH 2.0 with HCOOH and loaded on a C_{18} HPLC column. Peptides were eluted with a 0–50% (v/v) linear gradient of acetonitrile in 0.05% (w/v) trifluoroacetic acid. The ^{75}Se radioactivity of the collected fraction was detected by γ counting and a 10–20% aliquot of each fraction was analyzed for total radioactivity by liquid scintillation counting. The amount of ^{14}C radioactivity was calculated by difference. Alkylation experiments performed at pH 8 use a similar procedure except that the initial enzyme solution is adjusted to pH 8 with potassium phosphate and the incubation time with bromoacetate is 30 min.

Catalytic Role of Selenocysteine Residue

Mammalian TrxR1 has two sets of redox centers, one consisting of Cys-59/Cys-64 adjacent to the flavin ring of FAD and another center consisting of Cys-497/Secys-498 near the C terminus. By selective alkylation of Secys-498 it has been demonstrated that the thioredoxin-induced oxidation of Cys-59-SH/Cys-64-SH is completely blocked, and that the alkylated enzyme shows negligible NADPH-disulfide oxidoreductase activity.[25] Mammalian TrxR might need the redox-active C-terminal sequence for transferring the reducing equivalents from the internal dithiol group to the outer substrate, the oxidized form of thioredoxin, or some other low molecular weight compounds. Speculation as to the ternary structure of mammalian TrxR, made on the basis of mammalian glutathione reductase structure, has led to the conclusion that the redox-active C-terminal sequence

[25] S. R. Lee, S. Bar-Noy, J. Kwon, R. L. Levine, T. C. Stadtman, and S. G. Rhee, *Proc. Natl. Acad. Sci. U.S.A.* **97**, 2521 (2000).

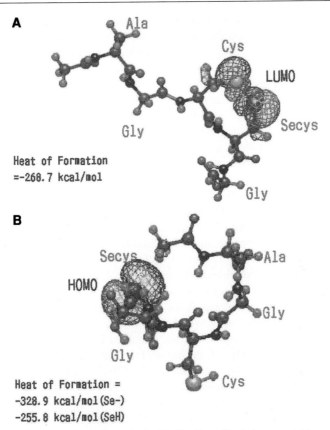

FIG. 3. (A) Oxidized form of *N*-acetyl-Ala-Gly-Cys-Secys-Gly in its most stable conformation.
(B) The reduced form of the peptide in its most stable conformation. The HOMO and LUMO are
designated by the red–blue cages.

appears still too far from the internal redox-active cysteines of the same subunit,
but close to the active site of the other subunit.[26] In a head-to-tail arrangement the
C-terminal Cys-497/Secys-498 residues of one monomer might be located adjacent to the Cys-59 and Cys-64 residues of the second monomer. Accordingly, if
the reductive half-reaction of TrxR is similar to that of glutathione reductase the
charge transfer complex formation may be followed by exchange of the nascent
Cys-59 and Cys-64 dithiol to the selenenylsulfide of the other subunit to generate
the active-site selenolthiol. X-ray crystallography is in progress for elucidating

---

[26] L. Zong, E. S. Arner, and A. Holmgren, *Proc. Natl. Acad. Sci. U.S.A.* **97,** 5854 (2000).

the structure of mammalian thioredoxin reductase,[27] and it should aid in deducing the reaction mechanism and the catalytic roles of Secys. Static protein structure would still require biochemical evidence to deduce the dynamic movement of the Secys-containing C-terminal tail during the course of catalytic turnover.

Computation of frontier molecular orbitals has become one of the routine approaches used by organic chemists to speculate on the chemistry of novel compounds. This could also be useful in speculating on the dynamic biochemical properties of macromolecules. However, it is not realistically possible to compute the highest occupied molecular orbital (HOMO) and lowest unoccupied molecular orbital (LUMO) on the basis of whole protein structure. Instead of the entire protein molecule, we can compute molecular orbitals of the C-terminal sequence N-acetyl-Ala-Gly-Cys-Secys-Gly by the use of PC-based software available on the web.[28] Figure 3A shows the oxidized form of the C-terminal sequence; a thioselenide bridge is protruding from the main chain, which takes a rather straight line at its most stable conformation. The LUMO, localized on the thioselenide bond, indicates the portion where electrons from the reductant are most likely to be accepted. Figure 3B represents the most stable conformation of the reduced form; the selenol group of Secys and thiol group of Cys are oriented in opposite directions. The HOMO, localized on the selenol group of the sequence, suggests that the selenol group of Secys would serve as the nucleophile in reducing the substrate. An interesting implication was obtained from the energy calculation of the two forms of the peptide. Calculation of the heat of formation indicated that the reduced form is more stable than the oxidized thioselenide form provided that the selenol group is deprotonated; when the selenol is protonated the C-terminal sequence is energized as highly as the oxidized form, which has the thioselenide bond between the vicinal Cys and Secys. Further biochemical evidence would be necessary to verify this computer calculation and to elucidate in detail the catalytic role of the penultimate selenocysteine residue.

---

[27] L. Zong, K. Persson, T. Sandalova, G. Schneider, and A. Holmgren, *Acta Crystallogr. D Biol. Crystallogr.* **56,** 1191 (2000).

[28] http://www.fujitsu.co.jp/jp/soft/wimmopac/home-e.html

[28] Mitochondrial Thioredoxin Reductase and Thiol Status

By ALBERTO BINDOLI and MARIA PIA RIGOBELLO

Introduction

Mitoch ... ial thiol groups are involved in several different, strictly integrated functions such as electron transport, oxidative phosphorylation, swelling/contraction, and membrane transport.[1] Mitochondrial thiol groups are also found to be involved in the inner membrane permeability transition[2,3] and cell apoptosis.[4] Mitochondria are able to generate reactive oxygen[5] and nitrogen[6] species, thereby favoring conditions of potential oxidative stress that can alter their thiol status.[7,8] However, the latter is controlled by several enzymes. In particular, the mitochondrial thioredoxin system (comprising NADPH, thioredoxin,[9] and thioredoxin reductase[10,11]) and thioredoxin peroxidase[12,13] appear to play a fundamental role in preventing oxidative stress[11,14,15] and in controlling membrane permeability functions.[10,16,17]

[1] P. C. Jocelyn, "Biochemistry of the SH Group." Academic Press, London, 1972.

[2] A. Bindoli, M. T. Callegaro, E. Barzon, M. Benetti, and M. P. Rigobello, *Arch. Biochem. Biophys.* **342**, 22 (1997).

[3] A. J. Kowaltowski, A. E. Vercesi, and R. F. Castilho, *Biochim. Biophys. Acta* **1318**, 395 (1997).

[4] P. Marchetti, D. Decaudin, A. Macho, N. Zamzami, T. Hirsch, S. A. Susin, and G. Kroemer, *Eur. J. Immunol.* **27**, 289 (1997).

[5] B. Halliwell and J. M. C. Gutteridge, "Free Radicals in Biology and Medicine." Oxford University Press, Oxford, 1999.

[6] C. Richter, M. Schweizer, and P. Ghafourifar, *Methods Enzymol.* **301**, 381 (1999).

[7] N. S. Kosower and E. M. Kosower, *Methods Enzymol.* **251**, 123 (1995).

[8] H. Sies, *Free Radic. Biol. Med.* **27**, 916 (1999).

[9] G. Spyrou, E. Enmark, A. Miranda-Vizuete, and J.-Å. Gustafsson, *J. Biol. Chem.* **272**, 2936 (1997).

[10] M. P. Rigobello, M. T. Callegaro, E. Barzon, M. Benetti, and A. Bindoli, *Free Radic. Biol. Med.* **24**, 370 (1998).

[11] S.-R. Lee, J.-R. Kim, K.-S. Kwon, H. W. Yoon, R. L. Levine, A. Ginsburg, and S. G. Rhee, *J. Biol. Chem.* **274**, 4722 (1999).

[12] S. Watabe, T. Hiroi, Y. Yamamoto, Y. Fujioka, H. Hasegawa, N. Yago, and S. Y. Takahashi, *Eur. J. Biochem.* **249**, 52 (1997).

[13] S. W. Kang, H. Z. Chae, M. S. Seo, K. Kim, I. C. Baines, and S. G. Rhee, *J. Biol. Chem.* **273**, 6297 (1998).

[14] E. S. J. Arnér and A. Holmgren, *Eur. J. Biochem.* **267**, 6102 (2000).

[15] J. R. Pedrajas, E. Kosmidou, A. Miranda-Vizuete, J.-Å. Gustafsson, A. P. H. Wright, and G. Spyrou, *J. Biol. Chem.* **274**, 6366 (1999).

[16] J. Wudarczyk, G. Debska, and E. Lenartowicz, *Arch. Biochem. Biophys.* **327**, 215 (1996).

[17] M. P. Rigobello, F. Turcato, and A. Bindoli, *Arch. Biochem. Biophys.* **319**, 225 (1995).

This article describes the preparation of mitochondrial thioredoxin reductase and the assessment of its activity. In addition a general, rapid, and simple methodology for the estimation of mitochondrial thiols is given. Most of the described methods are based on the reduction of the aromatic disulfide 5,5′-dithiobis (2-nitrobenzoic acid) (DTNB)[18] to form the 2-nitro-5-thiobenzoate anion (CNTP), which has an intense yellow color and is stable at pH 7–9. Thiol content, measured at 412 nm, is calculated employing an ε_M of 13,600 M^{-1} cm^{-1}. DTNB appears to be the disulfide of choice and is probably the most extensively used for thiol group estimation. The reader is also referred to previous articles in this series devoted to the methodologies for assay of protein and nonprotein thiols.[19–22]

Determination of Total Mitochondrial Thiols

Reagents

DTNB (0.1 M) can be dissolved in buffer or directly suspended in distilled water (without buffer) and slowly titrated with 1 M Tris base to pH 7.5, avoiding any local rise above pH 9 to prevent hydrolytic cleavage of the disulfide. Solutions can be stored frozen for several weeks.

Procedure

Mitochondrial proteins (about 0.1–0.4 mg/ml) are added to 0.2 M Tris-HCl buffer (pH 8.1) containing 10 mM EDTA and 1% (w/v) sodium dodecyl sulfate (SDS). Mitochondria undergo a rapid solubilization and the absorption is recorded (this value must be subtracted from the final reading). DTNB (1 mM) is added to both sample and reference. The color develops at 412 nm in about 2–3 min, reaching a plateau that fades slowly.

The determination of thiol groups of whole mitochondria requires treatment with a suitable denaturing agent able to expose all thiol groups to the titrating reagent. SDS is a good detergent and, in our conditions, shows linearity in the range of 0.5–5% (w/v). Total mitochondrial thiol groups measured by this method are about 95 nmol/mg protein (range, 90–100 nmol/mg). These figures are consistent with previous reports obtained by amperometric titration.[23] Lower values indicate incomplete solubilization of mitochondria by the detergent or incomplete titration with the reagent. Instead of SDS, 6–8 M guanidine hydrochloride is also often

[18] G. L. Ellman, *Arch. Biochem. Biophys.* **82**, 70 (1959).
[19] A. F. S. A. Habeeb, *Methods Enzymol.* **25**, 457 (1972).
[20] P. C. Jocelyn, *Methods Enzymol.* **143**, 44 (1987).
[21] L. Packer (ed.), *Methods Enzymol.* **251** (1995).
[22] L. Packer (ed.), *Methods Enzymol.* **252** (1995).
[23] M. V. Riley and A. L. Lehninger, *J. Biol. Chem.* **239**, 2083 (1964).

used, giving substantially similar values of measured thiols. With other detergents, such as Triton X-100 or Igepal CA-360, lower rates and extents of thiol titration are observed. EDTA is necessary to delay the reoxidation of CNTP anion and autoxidation of protein thiols.

To measure small alterations in mitochondrial total thiol status, a procedure in which simultaneous denaturation and thiol assay by the inclusion of DTNB in the reaction mixture might be preferable.[2] This ensures an immediate titration of thiol groups, avoiding nonspecific oxidation after unfolding with SDS. To minimize differences due to the sampling of mitochondria, a preincubation batch of mitochondria (about 4 mg/ml) in the appropriate medium is prepared. Aliquots are transferred to test tubes containing the desired reagents. At the end of the incubation, aliquots of 0.2 ml (containing 0.2–0.8 mg of protein) are transferred to cuvettes containing 2.3 ml of a mixture formed by 0.2 M Tris-HCl (pH 8.1), 1% (w/v) SDS, 10 mM EDTA, and 1 mM DTNB. Readings are taken at 412 nm against a blank containing the complete medium, including DTNB, until stabilization of the reaction is achieved. The apparent absorbance of a blank containing protein solution, in the absence of DTNB, is subtracted from the measured values. Values are reported as nanomoles of thiol per milligram of protein.

If substances interfering with the assay (usually thiol reagents or oxidizing agents) are present during incubation, it is mandatory to stop the reaction with a large volume of cold medium (0–4°), centrifuge, wash, and resuspend the mitochondria before performing the assay. However, precautions must be taken if mitochondrial swelling occurs, because part of the matrix content might be lost during washing procedures.

Determination of Total Protein Thiols, Membrane Thiols, and Acid-Soluble Thiols

The approach is particularly useful when substances interfering with the assay are present in the incubation medium. With the acid precipitation procedure all protein thiols including membrane thiols are measured. Moreover, after acid precipitation, soluble thiols (essentially glutathione) are recovered in the supernatant and can be estimated separately.

Procedure

Mitochondrial proteins (0.5 ml, containing 0.3–0.5 mg of protein) are precipitated with 0.5 ml of 6% (w/v) metaphosphoric acid or 5% (w/v) sulfosalicylic acid. After 5 min of incubation the samples are centrifuged at 15,000g for 10 min at 4° and the pellet is washed twice by resuspension with 0.5 ml of 3% (w/v) metaphosphoric acid or 2.5% (w/v) sulfosalicylic acid. After the final centrifugation, the pellet is dried and treated with 0.5 ml of 0.2 M Tris-HCl (pH 8.1), 1% (w/v) SDS, 10 mM EDTA. For protein estimation, an aliquot is removed or a parallel sample is prepared. Solubilized proteins are transferred to a cuvette and an additional 0.5 ml of 0.2 M Tris-HCl (pH 8.1), 1% (w/v) SDS, and 10 mM EDTA is added. The sample, after measurement of the protein absorption, is reacted with 1 mM DTNB. The reaction is monitored at 412 nm until stabilization.

The supernatant of acid precipitation, after neutralization, can be used for the estimation of acid-soluble thiols, by addition of 1 mM DTNB. Also, total and oxidized glutathione can be estimated by the method of Tietze,[24] modified according to Anderson.[25]

For the quantification of membrane thiols (therefore excluding matrix and intermembrane thiols) the mitochondrial suspension, in hypo-osmotic medium, is subjected to three cycles of freezing and thawing[3] followed by centrifugation in a Centrifuge 5415 B (Eppendorf, Hamburg, Germany), at maximum speed, or in an Airfuge (Beckman, Fullerton, CA). The pellet is resuspended in 0.2 M Tris-HCl (pH 8.1), 1% (w/v) SDS, and 10 mM EDTA. Aliquots are utilized for protein estimation and thiols measured as above.

Determination of Available Protein Thiols

Mitochondrial thiol groups show different rates of reaction and accessibility to DTNB (or other reagents) in response to different functional states,[26] to condensed-orthodox configuration transition,[27] or after addition of specific reagents.[28–30]

Reaction with DTNB is performed in the absence of denaturing agents and only nonhindered thiols are detected. After incubating mitochondria in the presence of 0.1–1 mM DTNB for a defined period of time, they are rapidly centrifuged and CNTP, measured in the supernatant, is an index of the available thiols. Alternatively, the reaction can be monitored with a dual-wavelength spectrophotometer at wavelengths of 412 and 520 nm.[28] Mitochondria (0.3–0.5 mg of protein per milliliter) in buffered (pH 7.4) KCl or sucrose–mannitol medium containing 1 mM EDTA are treated with 1 mM DTNB and the reaction is monitored at 412 nm for 20–30 min. Similar to pure proteins, membranes also show a differential rate and extent of reactivity toward DTNB and fast, slow, and unreactive thiols might be detected.[31,32] The reaction of mitochondria with excess DTNB shows a time course characterized by a rapid increase in absorbance in the first few minutes followed by a slow rise extended over longer periods of time. This indicates the presence

[24] F. Tietze, *Anal. Biochem.* **27,** 502 (1969).
[25] M. E. Anderson, *Methods Enzymol.* **113,** 548 (1985).
[26] N. Sabadie-Pialoux and D. Gautheron, *Biochim. Biophys. Acta* **234,** 9 (1971).
[27] O. Hatase, K. Tsutsui, and T. Oda, *J. Biochem. (Tokyo)* **82,** 359 (1977).
[28] E. J. Harris and H. Baum, *Biochim. J.* **186,** 725 (1980).
[29] P. M. Sokolove, *FEBS Lett.* **234,** 199 (1988).
[30] H.-J. Freisleben, J. Fuchs, L. Mainka, and G. Zimmer, *Arch. Biochem. Biophys.* **266,** 89 (1988).
[31] Y. Ando and M. Steiner, *Biochim. Biophys. Acta* **311,** 26 (1973).
[32] A. Bindoli and S. Fleischer, *Arch. Biochem. Biophys.* **221,** 458 (1983).

of different classes of reacting thiols and a kinetic analysis can be performed. The velocity constants for the various populations of thiol groups reacting with DTNB can be obtained from a semilogarithmic plot of the difference between the final estimated concentration of thiols (C_∞) and the concentration at various times (C_t) expressed as a percentage of the total available thiols (C_∞) versus time, that is, $[(C_\infty - C_t)/C_\infty] \times 100$ versus time.[31] The plot shows a linear portion at longer times, whereas, at shorter times, a positive deviation, inversely related to the incubation time, is apparent. This behavior suggests that at least two simultaneous reactions with different rate constants occur. By drawing a straight line corresponding to the slow reaction, in order to intercept the ordinate axis, and subtracting the extrapolated values from the measured values, a second straight line is obtained. The relative values of the intercepts of the two straight lines indicate the percentages of the two populations of thiols. The composite semilogarithmic plot allows the experimental determination of the $t_{0.5}$ values and the pseudo first-order rate constants, using the relationship $k_{obs} = 0.693/t_{0.5}$ ($\sec^{-1}$). By dividing k_{obs} by DTNB concentration the second-order rate constants ($M^{-1} \sec^{-1}$) can also be obtained.

Mitochondrial 5,5'-Dithiobis(2-Nitrobenzoic Acid) Reductase Activity

Mitochondria have long been known to reduce low molecular weight disulfides. Different reduction mechanisms involving thiol–disulfide exchange with either dihydrolipoate[33] or reduced glutathione[34] have been described. However, the NADPH-dependent DTNB reductase activity of the crude mitochondrial fraction or isolated mitochondrial matrix was assumed to be essentially due to thioredoxin reductase,[17,35,36] although the presence of glutathione and glutathione reductase can lead to an overestimation dependent on the catalytic action of glutathione in the reduction of DTNB.[37] Moreover, in preparations of mitochondrial matrix or of mitochondrial lysates, the concentration of glutathione is diluted and, therefore, its contribution to the overall reduction of DTNB is limited.[36] Besides, contaminant glutathione can be removed by dialysis. The measurement of DTNB reductase activity constitutes a rapid method for thioredoxin reductase evaluation in homogenates, isolated whole mitochondria, and mitochondrial subfractions.

Procedure

Mitochondria (60 mg/ml) are diluted (1 : 9) with distilled water, subjected to three cycles of freezing and thawing, sonicated twice for 30 sec each, and

[33] L. Eldjarn and J. Bremer, *Acta Chem. Scand.* **17**(Suppl. 1), 59 (1963).
[34] P. C. Jocelyn and A. D. Cronshaw, *Biochim. Biophys. Acta* **797**, 203 (1984).
[35] E. Lenartowicz, *Biochem. Biophys. Res. Commun.* **184**, 1088 (1992).
[36] E. Lenartowicz and J. Wudarczyk, *Int. J. Biochem. Cell. Biol.* **27**, 831 (1995).
[37] C. W. I. Owens and R. V. Belcher, *Biochem. J.* **94**, 705 (1965).

centrifuged at 100,000g for 60 min at 4°. The pellet is discarded and the supernatant fraction is (optionally) dialyzed overnight against 20 mM Tris-HCl (pH 7.5) to remove glutathione and finally concentrated in a pressure dialysis system using an Amicon (Danvers, MA) YM10 membrane. Measurement of DTNB reductase activity is performed at 25° in freshly prepared 0.2 M phosphate buffer (pH 7.5), containing 1 mM EDTA. The mitochondrial supernatant fraction (0.3–0.5 mg/ml) and 3 mM DTNB are added to both sample and reference cuvettes and preincubated for 2–3 min. The reaction is initiated by adding 0.2 mM NADPH to the sample cuvette, while the reference cuvette is balanced with an equal volume of medium. Absorbance is read at 412 nm and the activity is expressed as (nanomoles per minute)/milligram of protein.

A modified procedure is based on the strong inhibitory properties of relatively low concentrations of arsenite on thioredoxin reductase,[38] compared with the higher concentrations required to inhibit glutathione reductase. The assay is performed as described above but sodium arsenite (0.1 mM) is included in the reference cuvette and 0.2 mM NADPH is added to both sample and reference cuvettes. This method is particularly useful for samples nondialyzed or obtained from mitochondria lysed with Triton X-100. Because gold compounds are strong inhibitors of thioredoxin reductase,[39,40] 0.5 μM auranofin can be used instead of arsenite.

Purification of Mitochondrial Thioredoxin Reductase

Mitochondrial thioredoxin reductase, designed as TrxR2, is a selenium-containing enzyme,[11,41,42] similar to its cytosolic counterpart,[43,44] indicated as TrxR1. It has been identified in mitochondria of rat, mouse, bovine, and human tissues,[10,11,45–50] and also in yeast[15] and plant[51] mitochondria. On SDS–polyacrylamide gel electrophoresis cytosolic thioredoxin reductase from rat liver

[38] M. Luthman and A. Holmgren, *Biochemistry* **21**, 6628 (1982).

[39] K. E. Hill, G. W. McCollum, and R. F. Burk, *Anal. Biochem.* **253**, 123 (1997).

[40] S. Gromer, L. D. Arscott, C. H. Williams, Jr., R. H. Schirmer, and K. Becker, *J. Biol. Chem.* **273**, 20096 (1998).

[41] Q.-A. Sun, Y. Wu, F. Zappacosta, K.-T. Jeang, B. J. Lee, D. L. Hatfield, and V. N. Gladyshev, *J. Biol. Chem.* **274**, 24522 (1999).

[42] A. Lescure, D. Gautheret, P. Carbon, and A. Krol, *J. Biol. Chem.* **274**, 38147 (1999).

[43] V. N. Gladyshev, K. T. Jeang, and T. C. Stadtman, *Proc. Natl. Acad. Sci. U.S.A.* **93**, 6146 (1996).

[44] T. Tamura and T. C. Stadtman, *Proc. Natl. Acad. Sci. U.S.A.* **93**, 1006 (1996).

[45] S. Watabe, Y. Makino, K. Ogawa, T. Hiroi, Y. Yamamoto, and S. Y. Takahashi, *Eur. J. Biochem.* **264**, 74 (1999).

[46] K. J. Kim, Y. Y. Jang, E. S. Han, and C. S. Lee, *Mol. Cell. Biochem.* **201**, 89 (1999).

[47] A. Miranda-Vizuete, A. E. Damdimopoulos, and G. Spyrou, *Biochim. Biophys. Acta* **1447**, 113 (1999).

[48] H. Kawai, T. Ota, F. Suzuki, and M. Tatsuka, *Gene* **242**, 321 (2000).

[49] P. Y. Gadaska, M. M. Berggren, M. J. Berry, and G. Powis, *FEBS Lett.* **442**, 105 (1999).

[50] A. Miranda-Vizuete, A. E. Damdimopoulos, J. R. Pedrajas, J.-Å Gustafsson, and G. Spyrou, *Eur. J. Biochem.* **261**, 405 (1999).

shows subunits of 57 kDa,[52] whereas mitochondrial thioredoxin reductase, from the same organ, has subunits of about 54 kDa.[10] Cloning and sequencing of mitochondrial thioredoxin reductase has been reported.[11,15,41,42,45,47–50] Mitochondrial thioredoxin reductase is able to reduce chemically unrelated compounds such as DTNB, selenite, and alloxan, in addition to its natural substrate thioredoxin.[10] It is also active in reducing protein disulfide-isomerase of both microsomal and mitochondrial origin[53] that belongs to the thioredoxin family.

Preparation and Processing of Mitochondria

Thioredoxin reductase is purified from rat liver mitochondria isolated by differential centrifugation according to conventional procedures. The preparations of mitochondria can be stored frozen at $-20°$ and combined to obtain about 6 g of protein as measured by the biuret test. Mitochondria are subjected to three cycles of freezing and thawing and then to sonic irradiation (three times for 20 sec each). The enzyme is purified by a modification of the methods of Luthman and Holmgren and[38] and Williams *et al.*[54] used for preparing rat liver cytosolic and *Escherichia coli* thioredoxin reductases, respectively. In particular, the acid precipitation step is omitted. Broken mitochondria are centrifuged at 105,000g for 60 min at 4° and the clear supernatant (mitochondrial matrix and intermembrane space content) is fractionated with ammonium sulfate in three stages at 30, 50, and 85% (w/v) saturation, respectively. Thioredoxin reductase activity is concentrated in the precipitate obtained between 30 and 50% (w/v) saturation, which exhibits about 70% of the activity found in the clear supernatant solution before fractionation. In addition, it has a specific activity, on a protein basis, of about 9-fold that of the original supernatant. The obtained pellet is rapidly dissolved in 10 mM Tris-HCl (pH 7.5) containing 1 mM EDTA and dialyzed overnight against 2 liters of the same buffer. Afterward it is heated at 60° for 3 min and rapidly cooled at 4° in an ice bath. The precipitated proteins are removed by centrifugation at 105,000g for 3 hr at 4°. The resulting supernatant, assayed for thioredoxin reductase by the DTNB reduction procedure (see below), exhibits a specific activity of about 17 times that of the starting supernatant. It is recommended to perform the ammonium sulfate precipitation before heat treatment to obtain a better yield of activity.[54] However, the heat treatment can be omitted,[55] particularly when working with small quantities of mitochondria.

[51] M. Banze and H. Follmann, *J. Plant Physiol.* **156**, 126 (2000).

[52] L. Zhong, E. S. J. Arnér, J. Ljung, F. Åslund, and A. Holmgren, *J. Biol. Chem.* **273**, 8581 (1998).

[53] M. P. Rigobello, A. Donella-Deana, L. Cesaro, and A. Bindoli, *Free Radic. Biol. Med.* **28**, 266 (2000).

[54] C. H. Williams, Jr., G. Zanetti, L. D. Arscott, and J. K. McAllister, *J. Biol. Chem.* **242**, 5226 (1967).

[55] E. S. J. Arnér, L. Zhong, and A. Holmgren, *Methods Enzymol.* **300**, 226 (1999).

DEAE-Sephacel Chromatography

The enzyme solution is transferred to a 2.5 × 15 cm DEAE-Sephacel column equilibrated with 10 mM Tris-HCl (pH 7.5) containing 1 mM EDTA and eluted with 400 ml of NaCl gradient (0 to 0.3 M with a volume of 200 ml for each reservoir). Fractions of 4.5 ml are collected at 3-min intervals and aliquots of 0.1 ml are assayed for enzyme activity by the DTNB reduction method (see below). The protein content of the samples is estimated spectrophotometrically at 280 nm. At variance with the cytosolic isoform, mitochondrial thioredoxin reductase is eluted at 0.13–0.15 M NaCl and the protein fractions obtained in this interval are pooled, concentrated in a pressure dialysis system using an Amicon YM10 membrane, and finally dialyzed against 50 mM Tris-HCl (pH 7.5) containing 0.1 mM EDTA.

2′,5′-ADP-Sepharose 4B Affinity Chromatography

The resulting enzyme solution is applied to a 0.8 × 10 cm 2′,5′-ADP-Sepharose 4B column equilibrated with 50 mM Tris-HCl (pH 7.5), 0.1 mM EDTA and eluted with a step gradient of Na,K-phosphate (0.25 and 0.5 M, respectively) and NaCl (0.8 and 1 M, respectively). Fractions of about 2 ml are collected. Mitochondrial thioredoxin reductase-active fractions are eluted at 0.8 M NaCl, whereas the cytosolic enzyme is eluted at 0.25 M Na,K-phosphate (Fig. 1). The active fractions

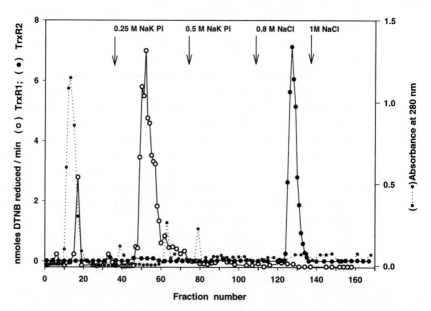

FIG. 1. Chromatographic purification by 2′,5′-ADP Sepharose 4B affinity column chromatography of thioredoxin reductase obtained from rat liver cytosol (TrxR1) and mitochondria (TrxR2). NaK Pi, Na,K-phosphate.

are concentrated in a pressure dialysis system using an Amicon YM10 membrane in the presence of 0.2% (w/v) octylglucoside to limit the loss of enzyme activity. Afterward the fraction is subjected to dialysis against 20 mM Tris-HCl (pH 7.5), 150 mM NaCl, 10% (v/v) glycerol, and 10 mM mercaptoethanol.

Superdex-200 Fast Protein Liquid Chromatography

The dialyzed preparation is applied to a Superdex-200 column equilibrated with 20 mM Tris-HCl (pH 7.5), 150 mM NaCl, 10% (v/v) glycerol, 10 mM mercaptoethanol, and 50 μM phenylmethylsulfonyl fluoride (PMSF) in a fast protein liquid chromatography (FPLC) system from Pharmacia (Piscataway, NJ). Fractions of 0.2 ml are collected at a flow rate of 0.4 ml/min. The obtained enzyme is pure on SDS-polyacrylamide gels stained with Coomassie Brilliant Blue and gives a molecular mass of 54 kDa. The enzyme preparations can be kept in the frozen state for several months without any appreciable loss of activity.

Instead of the freeze–thawing/sonication procedure, Triton X-100 can also be used to disrupt mitochondria, followed by centrifugation and DEAE chromatography.[45]

Mitochondrial thioredoxin reductase can be purified to homogeneity without the preliminary isolation of mitochondria.[11] Tissues from various organs (liver, kidney, heart, and brain) can be stored frozen, thawed, and minced in 20 mM Tris-HCl (pH 7.5) containing 1 mM EDTA and a protease inhibitor cocktail ("Complete"; Roche, Branchburg, NJ) and homogenized with an Ultra Turrax (IKA-Werke, Staufen, Germany) (three times for 15 sec each at 12,000 rpm). The homogenate is centrifuged at 70,000g for 30 min at 4°, the pellet is discarded, and the supernatant is brought to pH 5.0 with 1 M acetic acid and centrifuged again at 70,000g for 30 min at 4°. The pellet, containing most of the TrxR2, is dissolved in 20 mM Tris-HCl buffer (pH 7.5) containing 1 mM EDTA and adjusted to pH 7.5 with 1 M ammonium hydroxide. The enzyme is extensively dialyzed overnight against 10 mM Tris-HCl buffer containing 1 mM EDTA. Afterward, the above-reported procedure is followed.

Assays

Methods used for cytosolic thioredoxin reductase[38,55,56] also apply to mitochondrial thioredoxin reductase. They are essentially based on the direct reduction of DTNB by thioredoxin reductase or by reduction of insulin mediated by thioredoxin.

5,5'-Dithiobis(2-Nitrobenzoic Acid) Reduction Assay

The ability of thioredoxin reductase, in the presence of NADPH, to reduce DTNB, also without thioredoxin,[38] constitutes a simple way to test enzyme activity.

---

[56] A. Holmgren and M. Björnstedt, *Methods Enzymol.* **252**, 199 (1995).

The assay mixture contains 0.2 M Tris-HCl buffer (pH 8.1), 1 mM EDTA, and 0.25 mM NADPH. To the sample cuvette about 0.5–2 μg of TrxR2 protein is added. The reaction is started by the addition of 3 mM DTNB to both sample and reference and the increase in absorbance is monitored at 412 nm over 5 min at 25°. In calculating the enzyme activity it must be considered that 1 mol of NADPH yields 2 mol of CNTP anion.[38]

The other methods are based on a coupled system whereby thioredoxin, reduced by thioredoxin reductase in the presence of NADPH, in turn, reduces insulin.The insulin solution is prepared as described by Arnér et al.[55]

NADPH Oxidation Assay

The assay mixture contains 0.2 M phosphate buffer (pH 7.0), 20 mM EDTA, and 0.160 mM insulin. To the sample cuvette 0.25 mM NADPH and 1–2 μg of TrxR2 protein are also added. After 2 min of incubation at 30°, the reaction is initiated by the addition, to the sample cuvette, of 30 μM E. coli thioredoxin and the decrease of absorbance is monitored at 340 nm for about 5 min. At longer times the solution becomes turbid because of the precipitation of reduced insulin. Activity is calculated as nanomoles of NADPH oxidized per minute, using an ε_M of 6200 M^{-1} cm^{-1} at 340 nm.

Both cytosolic[57] and mitochondrial[10] thioredoxin reductases are able to reduce sodium selenite. The latter, acting as electron acceptor, is therefore a useful tool to test the activity of the enzyme.[55] As noted above, the decrease in absorbance of NADPH is monitored after addition of 0.1 mM sodium selenite, instead of thioredoxin/insulin.

Insulin Reduction Assay

In the insulin reduction assay reduced insulin is titrated with DTNB.[55] Thioredoxin reductase (about 1–2 μg of protein) is incubated in 0.12 M HEPES–Tris buffer (pH 7.0), 3.4 mM EDTA, 0.85 mM NADPH, 30 μM thioredoxin, and 0.35 mM insulin (final volume, 120 μl). Separated samples are incubated at 37° for 5 to 30 min. Reactions are quenched at 5-min intervals by the addition of 0.5 ml of 6 M guanidine hydrochloride in Tris-HCl (0.2 M, pH 8.1) containing 10 mM EDTA and 1 mM DTNB. Reduced insulin content is estimated by the absorbance at 412 nm and the obtained values are plotted versus time. A blank without thioredoxin is subtracted.

Acknowledgment

The authors thank Prof. Fulvio Ursini (University of Padova) for encouragement to write this article.

---

[57] M. Björnstedt, S. Kumar, and A. Holmgren, Methods Enzymol. **252,** 209 (1995).

[29] Protein Electrophoretic Mobility Shift Assay to Monitor Redox State of Thioredoxin in Cells

By Neil A. Bersani, Jason R. Merwin, Nathan I. Lopez, George D. Pearson, and Gary F. Merrill

Introduction

Despite its role as an electron donor in several biochemical reactions involving protein disulfide reduction[1,2] and its potential role as a regulatory molecule affecting transcription factor activity,[3–6] the reduction–oxidation (redox) state of thioredoxin *in vivo* has been investigated only rarely.[7] Past efforts to assess the redox state of proteins *in vivo* have generally relied on producing thiol adducts that cause an increase in protein mass and thus a reduction in electrophoretic mobility during sodium dodecyl sulfate–polyacrylamide gel electrophoresis (SDS–PAGE).[8,9] However, working with purified proteins, Takahashi and Hirose[10] showed that a much stronger shift in electrophoretic mobility is achieved if charged thiol adducts are introduced and proteins are analyzed by urea–PAGE. By including an immunoblotting step, we have adapted their assay to permit analysis of the redox state of thioredoxin *in vivo*. We use the protein electrophoretic mobility shift assay (PEMSA) to determine the effect of the oxidant diamide on the redox state of thioredoxin in human MCF-7 cells, and the effect of deleting the thioredoxin reductase gene (*TRR1*) on the redox state of thioredoxin in yeast.

Methods

Preparation of Samples and Alkylation with Iodoacetamide

Human MCF-7 cells are maintained in a 1 : 1 mixture of Ham's F12 and Dulbecco's modified Eagle's medium (DMEM) containing 10% (v/v) fetal calf

[1] E. C. Moore, P. Reichard, and L. Thelander, *J. Biol. Chem.* **239**, 3445 (1964).

[2] P. G. Porque, A. Baldesten, and P. Reichard, *J. Biol. Chem.* **245**, 2371 (1970).

[3] K. Hirota, M. Matsui, S. Iwata, A. Nishiyama, K. Mori, and J. Yodoi, *Proc. Natl. Acad. Sci. U.S.A.* **94**, 3633 (1997).

[4] J. R. Matthews, N. Wakasugi, J.-L. V. Virelizier, J. Yodoi, and R. T. Hay, *Nucleic Acids Res.* **20**, 3821 (1992).

[5] D. Casso and D. Beach, *Mol. Gen. Genet.* **252**, 518 (1996).

[6] G. D. Pearson and G. F. Merrill, *J. Biol. Chem.* **273**, 5431 (1998).

[7] A. Holmgren and M. Fagerstedt, *J. Biol. Chem.* **257**, 6926 (1982).

[8] T. Kobayashi, S. Kishigami, M. Sone, H. Inokuchi, T. Mogi, and K. Ito, *Proc. Natl. Acad. Sci. U.S.A.* **94**, 11857 (1997).

[9] U. Jakob, W. Muse, M. Eser, and J. C. Bardwell, *Cell* **96**, 341 (1999).

[10] N. Takahashi and M. Hirose, *Anal. Biochem.* **188**, 359 (1990).

serum. About 10^6 cells are plated to 10-cm dishes containing 10 ml of medium and incubated for 3 days at 37°. For experiments involving exposure to diamide, the oxidant is added 30 min before harvesting. Before harvesting, cultures are rinsed twice in phosphate-buffered saline (PBS), pH 7.4.

To simultaneously lyse cells and initiate carboxymethylation of protein sulfhy-dryls, 200 μl of urea buffer [8 M urea, 100 mM Tris (pH 8.2), 1 mM EDTA] adjusted to 30 mM iodoacetic acid (IAA), using a 600 mM stock solution, freshly prepared in 1 M Tris, pH 8.2, is applied to cell monolayers, and lysates are rapidly triturated and transferred to 1.5-ml microcentrifuge tubes. After 15 min at 37°, samples are clarified by microcentrifugation (30 sec), and supernatants are transferred to fresh tubes. At this point, samples can be stored frozen for later processing. We have had success in assaying samples that were IAA-treated in other laboratories and shipped to us on dry ice.

Yeast are grown at 25° in yeast extract–peptone–dextrose (YEPD) medium to a density of 10^7 cells/ml (OD_{600} of 0.4). A 50-ml culture yields sufficient material for several replicate assays. To achieve consistent aeration, a shaker speed of 100 rpm and a 600-ml flask are used. Yeast are harvested at 25° by centrifugation at 700g for 5 min, resuspended in TE buffer [10 mM Tris (pH 7), 1 mM EDTA], centrifuged as described above, and flash frozen in liquid nitrogen. Using a hammer, the frozen pellet is knocked into a chilled mortar, and ground under liquid nitrogen until microscopic examination reveals >90% cell breakage. The frozen powder of broken cells is scraped into chilled 1.5-ml tubes, using a chilled spatula, and is stored frozen only as long as it takes to collect a set of samples (usually less than 2 hr). Alkylation is initiated by thawing the frozen powder in 200 μl of urea buffer containing 30 mM IAA, triturating the sample vigorously, and incubating the lysate at 37°. After 15 min, the IAA-treated lysate is clarified by microcentrifugation (30 sec), and the supernatant is transferred to fresh tubes for either immediate processing or storage at −80°.

For both mammalian cells and yeast, the initial alkylation reaction (with IAA) is done at the time of harvest to avoid *in vitro* thiol oxidation during storage. In control experiments, in which we store frozen yeast powder for several days before alkylating samples, a gradual increase in the amount of oxidized thioredoxin is observed.

Reduction with Dithiothreitol and Alkylation with Iodoacetamide

Unreacted IAA is removed by precipitating proteins with 10 volumes of cold acetone–1 N HCl (98 : 2, v/v), followed by microcentrifugation for 5 min at 11,000g at 4°. The pellet is washed three times by resuspension in cold acetone–1 N HCl–H_2O (98 : 2 : 10, v/v/v) followed by microcentrifugation. The washing steps prevent carboxymethylation of protein by residual IAA during the subsequent reduction reaction. Reduction of protein disulfides is achieved by resuspending the final acetone precipitate in 95 μl of urea buffer containing 3.5 mM dithiothreitol (DTT) and incubating samples for 30 min at 37°.

After reduction, newly generated thiols are amidomethylated by adjusting the sample to 10 mM iodoacetamide (IAM), using a 200 mM stock solution, prepared fresh in 1 M Tris, pH 8.2. After incubation for 15 min at 37°, samples are clarified by microcentrifugation (30 sec), and the protein concentration of the supernatant is determined[11] with bovine serum albumin (BSA) as a standard.

Electrophoretic migration markers for the charged isoforms of thioredoxin are prepared by harvesting cells in urea buffer containing 3.5 mM DTT and incubating lysates for 30 min at 37° to reduce all cysteine residues to the sulfhydryl form. Reduced lysates are adjusted to either 30 mM IAA, 10 mM IAM, or 15 mM IAA–15 mM IAM and alkylated at 37° for 15 min.

Urea–Polyacrylamide Gel Electrophoresis and Immunoblotting

After the sulfhydryl alkylation reactions, proteins are analyzed on a discontinuous polyacrylamide slab gel (6 × 8 × 0.7 cm). Both the stacking gel [2.5% (w/v) acrylamide, 0.075% (w/v) bisacrylamide, 0.12 M Tris-HCl (pH 6.8)] and running gel [9% (w/v) acrylamide, 0.27% (w/v) bisacrylamide, 0.037 M Tris-HCl (pH 8.8)] contain 8 M urea. The chamber buffer is 0.025 M Tris, 0.192 M glycine (pH 8.3). Samples for electrophoresis contain 15 μg of alkylated protein in a final volume of 20 μl of urea buffer containing 3.5 mM DTT, 8% (v/v) glycerol, and 0.1% (w/v) bromphenol blue. Gels are run at a constant current of 5 mA for 2.5 hr. After electrophoresis, gels are given three 5-min rinses with 50 mM Tris, pH 8.3, and are equilibrated with Towbin buffer [25 mM Tris base, 192 mM glycine, 20% (v/v) methanol, 0.1% (w/v) SDS].[12] Proteins are transferred to Hybond enhanced chemiluminescence (ECL) membrane (Amersham, Arlington Heights, IL) for 0.8 hr at a constant current of 60 mA, using a semidry transfer apparatus (Bio-Rad, Hercules, CA). After transfer, membranes are blocked by incubation overnight at 4° in 40 ml of TBS-T [15 mM Tris (pH 7.6), 140 mM NaCl, 0.1% (v/v) Tween]. All subsequent incubations are done at room temperature. To probe for human or yeast thioredoxin, blocked blots are incubated for 1 hr in 40 ml of TBS-T containing a 1 : 10,000 dilution of either polycolonal goat anti-human thioredoxin antibody (American Diagnostica, Greenwich, CT) or polyclonal rabbit anti-yeast thioredoxin antibody[13] (provided by E. Muller, University of Washington, Seattle, WA). After incubation with primary antibody, blots are rinsed twice for 10 min in 40 ml of TBS-T, and incubated for 1 hr in 40 ml of TBS-T containing a 1 : 10,000 dilution of horseradish peroxidase-conjugated rabbit anti-goat IgG antibody (Accurate, Westbury, NY) or donkey anti-rabbit IgG antibody (Amersham). Both primary and secondary antibody solutions are stored at 4° and reused several times without loss of sensitivity. After incubation with secondary antibody, blots are washed twice for 10 min each

[11] M. Bradford, *Anal. Biochem.* **72,** 248 (1976).

[12] H. Towbin, T. Staehelin, and J. Gordon, *Proc. Natl. Acad. Sci. U.S.A.* **76,** 4350 (1979).

[13] E. G. D. Muller, *Arch. Biochem. Biophys.* **318,** 356 (1994).

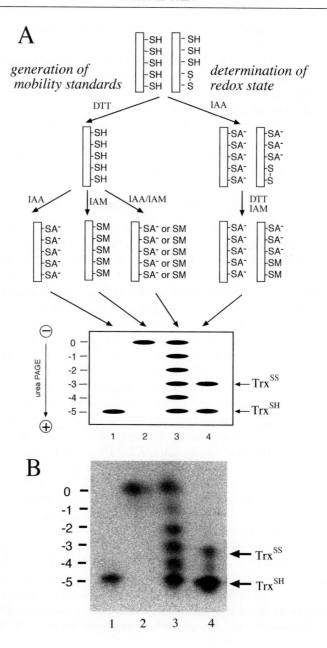

in 40 ml of TBS-T. Antibody binding is visualized by chemiluminescence with ECL reagents (Amersham) and BioMax-ML X-ray film (Kodak, Rochester, NY). Blots can be stripped by incubation for 30 min at 50° in stripping buffer [100 mM 2-mercaptoethanol, 2% (w/v) SDS, 62.5 mM Tris (pH 6.7)], rinsing twice in TBS-T, and blocking and reprobing as described above. Stripping results in about a 4-fold reduction in sensitivity.

Densitometry data are obtained with a Molecular Dynamics (Sunnyvale, CA) densitometer and analyzed using ImageQuant software. Equal area rectangles are used to quantitate pixel density of bands within each lane, and to correct for background pixel density. When thioredoxin migrates as a doublet, the signals of both bands are summed.

In side-by-side comparisons, we have found that protein bands obtained after urea–PAGE are not as sharp as those obtained after SDS–PAGE. Also, during immunostaining, a higher background is observed when blotting from urea gels rather than SDS gels. For both anti-human and anti-yeast thioredoxin antibodies, background staining is reduced if blocking reagents such as dried milk or gelatin are omitted.

Results

Protein Electrophoretic Mobility Shift Assay of Thioredoxin Redox State in Human MCF-7 Cells

The strategy used to determine the *in vivo* redox state of thioredoxin is outlined in Fig. 1A. Human thioredoxin contains five cysteines. Cysteines at positions 61, 68, and 72 are not present in the thioredoxins of prokaryotes and lower eukaryotes

---

FIG. 1. Protein electrophoretic mobility shift assay (PEMSA) of thioredoxin redox state. (A) Assay scheme. Mobility standards, corresponding to fully reduced thioredoxin (lane 1), fully oxidized thioredoxin (lane 2), or all six oxidation states of thioredoxin (lane 3), are prepared by treating lysates with dithiothreitol (DTT) and either iodoacetate (IAA), iodoacetamide (IAM), or a mixture of IAA and IAM, respectively. IAA adds acidic carboxymethyl thiol adducts (–SA⁻) and IAM adds neutral amidomethyl thiol adducts (–SM). During urea–PAGE at pH 8.5, the ionized carboxymethyl adducts result in faster protein migration toward the anode. To determine the redox state of thioredoxin *in vivo* (lane 4), lysate is treated immediately with IAA to carboxymethylate thiols, and then treated with DTT and IAM, to reduce disulfides and amidomethylate newly generated thiols. The IAM step, although not obligatory, is nonetheless included as a precaution against further thiol chemistry having unanticipated effects on protein mobility. Urea is present during all steps of the procedure to ensure that all thiols are accessible to alkylating and reducing agents. To visualize thioredoxin in the lysate, the electrophoretically separated proteins are transferred to nitrocellulose and immunostained. (B) Thioredoxin redox state in MCF-7 cells. Mobility standards (lanes 1–3) were prepared as described in (A). To determine the *in vivo* redox state of thioredoxin, lysate was treated sequentially with IAA, DTT, and IAM (lane 4). Bars at left represent the expected mobilities for the six potential charge isomers of thioredoxin. Trx^{SS} and Trx^{SH} represent the mobilities expected for the disulfide and fully reduced form, respectively.

and have been referred to as structural cysteines.[14] Cysteines at positions 32 and 35 are conserved in all thioredoxins and are the active site cysteines that interconvert between the dithiol and disulfide forms (Trx^{SH} and Trx^{SS}, respectively) during oxidoreductase reactions. In Fig. 1, thioredoxin in cells is depicted as being a mixture of the fully reduced and disulfide forms. If cells are lysed in the presence of urea and iodoacetate (IAA), all free thiols acquire acidic carboxymethyl adducts. The presence of urea ensures that all protein thiols are accessible for reaction with the alkylating agent. If lysates are subsequently treated with dithiothreitol (DTT) and iodoacetamide (IAM), protein disulfides are reduced and the newly formed thiols acquire neutral amidomethyl adducts. When lysate proteins are subsequently analyzed by urea–PAGE at pH 8.3, the fully reduced form of thioredoxin, which has five negatively charged carboxymethyl adducts, migrates more rapidly toward the cathode than the disulfide form, which has only three carboxymethyl adducts.

Figure 1A also shows how migration standards for fully reduced thioredoxin (lane 1), fully oxidized thioredoxin (lane 2), and intermediate oxidation states of thioredoxin (lane 3) are prepared.

Use of the assay to determine the redox state of thioredoxin in human MCF-7 cells is shown in Fig. 1B. Lane 4, which contains samples treated sequentially with IAM, DTT, and IAM, shows that 90% of total thioredoxin migrated with the −5 mobility expected for the fully reduced protein. About 6% migrated with the −3 mobility expected for the disulfide form of the protein and 4% migrated with a −4 mobility, which matches that expected for a mixed disulfide intermediate in which one active site cysteine is a free thiol and the other is engaged in a disulfide bond to another polypeptide or compound. Alternatively, the −4 species may be due to a thioredoxin in which both active site cysteines are reduced but one of the structural cysteines is oxidized.

Figure 1B also shows a number of control samples. Lane 1, which contains lysates treated with DTT and IAA, shows that all five thioredoxin cysteines were accessible to the reducing and alkylating agents, and that carboxymethylation did not interfere with immunoreactivity. Lane 2, which contains samples treated with DTT and IAM, shows that amidomethylation did not interfere with immunoreactivity. Lane 3, which contains samples treated with DTT and a mixture of IAA and IAM, shows the migration rates of the six expected charge isoforms of thioredoxin.

One potential use of the PEMSA procedure is to determine the effects of various conditions on the redox state of thioredoxin. The effect of acute exposure to the thiol-specific oxidant diamide is shown in Fig. 2. A 30-min treatment with 1 mM diamide (lane 3) decreased the −5 signal (fully reduced form) and increased the −3 signal (disulfide form) relative to nontreated cells (lane 2), suggesting that the active site cysteines of thioredoxin were either directly reducing diamide or participated in reducing disulfides generated elsewhere in the cell in response to the

[14] X. Ren, M. Bjornstedt, B. Shen, M. L. Ericson, and A. Holmgren, *Biochemistry* **32,** 9701 (1993).

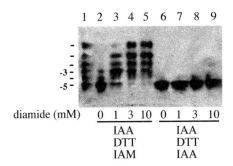

FIG. 2. Effect of diamide on thioredoxin redox state in MCF-7 cells. Lysates from MCF-7 cells treated for 30 min with diamide at the indicated concentrations were alkylated with IAA and reduced with DTT. Reduced lysates were given a secondary alkylation with either IAM (lanes 2–5) or IAA (lanes 6–9). The absence of slower migrating electrophoretic species when IAA was used as the secondary alkylation reagent indicates that the slower migrating species in lanes 2–5 were due to disulfide bond protection of cysteines during the first alkylation reaction.

diamide challenge. In cells treated with 3 or 10 mM diamide (lanes 4 and 5, more highly oxidized thioredoxin species were observed. The appearance of thioredoxin charge isoforms with more than two oxidized cysteines in diamide-treated cells indicated that cysteines other than the active site cysteines were becoming oxidized. Perhaps as cysteines at the active site participated in disulfide reduction reactions, sulfhydryl transfer reactions resulted in the oxidation of vicinal cysteines on the dimeric thioredoxin protein. Alternatively, oxidation of structural cysteines may be due to general oxidation of all protein thiols at high diamide concentrations.

To confirm that the slower migrating species observed in diamide-treated cells were due to the presence of S–S bonds during the initial alkylation step with IAA, we tested whether the slower migrating species were converted to the fast-migrating −5 form if, after reduction with DDT, samples were treated with IAA, rather than IAM, during the second alkylation step. As shown in Fig. 2, lanes 6 through 9, if lysates from diamide-challenged MCF-7 cells were treated with IAA during the second alkylation step, only the fast-migrating −5 form of thioredoxin was observed. Elimination of the slower migrating species by performing the second alkylation step with IAA rather than IAM established that all of the slower migrating bands in cells analyzed by the standard method (lanes 2–4) were due to cysteine sequestration in disulfide bonds.

To assess the variability between experiments, the redox state of thioredoxin in diamide-challenged cells was determined three times. The results, summarized in Table I, show that in nonchallenged cells, between 75 and 94% of total thioredoxin migrated at the −5 mobility expected for the fully reduced protein. In contrast, in cells exposed to 1 mM diamide, only between 14 and 20% of total thioredoxin migrated as the fully reduced protein and between 30 and

TABLE I
DENSITOMETRIC DATA ANALYSIS OF THIOREDOXIN REDOX STATE IN
DIAMIDE-CHALLENGED MCF-7 CELLS[a]

Diamide (mM)	Charge isomer	Fraction of total thioredoxin in each isomer			$\dfrac{\text{Oxidized Cys in isomer}}{\text{Total Cys in isomer}}$	$\dfrac{\text{Oxidized Cys in isomer(s)}}{\text{Total Cys in sample}}$		
		i	ii	iii		i	ii	iii
0	−5	0.75	0.94	0.90	0.0	0.00	0.00	0.00
	−4	0.17	0.06	0.09	0.2	0.03	0.01	0.02
	−3	0.06	0.00	0.01	0.4	0.02	0.00	0.00
	−2	0.01	0.00	0.00	0.6	0.01	0.00	0.00
	−1	0.00	0.00	0.00	0.8	0.00	0.00	0.00
	0	0.00	0.00	0.00	1.0	0.00	0.00	0.00
	All	1.00	1.00	1.00	NA	0.06	0.01	0.02
1	−5	0.15	0.20	0.14	0.0	0.00	0.00	0.00
	−4	0.13	0.16	0.12	0.2	0.03	0.03	0.02
	−3	0.43	0.34	0.30	0.4	0.17	0.14	0.12
	−2	0.15	0.12	0.12	0.6	0.09	0.07	0.07
	−1	0.13	0.15	0.22	0.8	0.10	0.12	0.18
	0	0.01	0.03	0.09	1.0	0.01	0.03	0.09
	All	1.00	1.00	1.00	NA	0.40	0.39	0.48
3	−5	0.01	0.00	0.00	0.0	0.00	0.00	0.00
	−4	0.08	0.00	0.00	0.2	0.02	0.00	0.00
	−3	0.23	0.02	0.05	0.4	0.09	0.01	0.02
	−2	0.20	0.15	0.16	0.6	0.12	0.09	0.10
	−1	0.29	0.35	0.34	0.8	0.23	0.28	0.27
	0	0.19	0.48	0.45	1.0	0.19	0.48	0.45
	All	1.00	1.00	1.00	NA	0.65	0.86	0.84

[a] Data sets for three experiments (*i, ii,* and *iii*) are shown. In each case, cells were incubated with diamide for 30 min, and the redox state of thioredoxin was determined by PEMSA as shown in Fig. 1. Data set *i* was obtained from the gel shown in Fig. 2. For each lane, the density of each band was normalized to the summed density of all bands to obtain the fraction of total thioredoxin in each isomer. This fraction was then multiplied by the fraction of oxidized cysteines in each isomer, to determine the amount of oxidized cysteine in the isomer(s) as a fraction of the total cysteine in the sample. NA, Not applicable.

43% migrated at the −3 mobility expected for the disulfide form. In cells exposed to 3 mM diamide, essentially no fully reduced thioredoxin was detected and between 68 and 98% migrated at mobilities more oxidized than the disulfide form.

An alternative way of summarizing and comparing the redox state of thioredoxin under various conditions is to calculate the percentage of total thioredoxin cysteines that are oxidized. As shown in the right-hand set of columns in Table I,

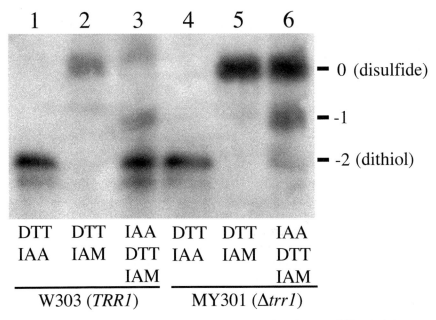

FIG. 3. Effect of deleting the thioredoxin reductase gene on the redox state of thioredoxin in yeast. A yeast strain (MY301) carrying a disrupted thioredoxin reductase gene allele ($\Delta trr1$), or an isogenic strain (W301) carrying a wild-type allele (*TRR1*) were analyzed. For both strains, mobility standards for fully reduced (lanes 1 and 4) and fully oxidized (lanes 2 and 5) thioredoxin were determined. PEMSA analysis showed that the majority of thioredoxin was in the dithiol form in wild-type cells (lane 3) and in the disulfide form in $\Delta trr1$ cells (lane 6).

in cells that were not treated with diamide, only between 1 and 6% of the total cysteines in thioredoxin were oxidized. In contrast, in cells treated with 1 mM diamide, between 37 and 48% of total cysteines in thioredoxin were oxidized, and in cells treated with 3 mM diamide, between 65 and 86% of the cysteines in thioredoxin were oxidized.

Determination of Thioredoxin Redox State in Wild-Type and Thioredoxin Reductase Null Yeast

In the budding yeast *Saccharomyces cerevisiae*, all the genes in the genome have been identified, and the effects of eliminating specific genes on cell phenotype can be investigated. Yeast lacking the *TRR1* gene encoding thioredoxin reductase are viable.[15] To test the relationship between thioredoxin reductase

[15] A. K. Machado, B. A. Morgan, and G. M. Merrill, *J. Biol. Chem.* **272**, 17045 (1997).

activity and thioredoxin redox state, the redox state of thioredoxin in wild-type and thioredoxin reductase null ($\Delta trr1$) yeast was determined. In contrast to mammalian thioredoxin, yeast thioredoxin contains only the two cysteines present at the active site (Cys-32 and Cys-35). Thus, when assayed by the PEMSA method, only three charge isoforms (0, -1, and -2) are expected.

Densitometric analysis of the data in Fig. 3 shows that 58% of total thioredoxin was in the reduced form in wild-type yeast (lane 3), but only 14% was in the reduced form in $\Delta trr1$ yeast (lane 6). In both yeasts, if lysates were treated first with DTT and then with IAM, all thioredoxin migrated at the slow (zero charge) mobility (lanes 2 and 5). Conversely, if lysates were treated first with DTT and then IAA, all thioredoxin migrated at the fast (-2 charge) mobility (lanes 1 and 4), confirming that the slower migrating bands in lanes 3 and 6 were due to disulfide bond protection of cysteines during the carboxymethylation reaction. Full conversion to the fast-migrating isoform in lanes 1 and 4 also confirms that all free thiols are accessible to IAA during the alkylation reaction and that carboxymethylation does not interfere with the immunoreactivity of the protein.

The results shown in Fig. 3 provide *in vivo* evidence of the link between thioredoxin reductase activity and thioredoxin redox state. In addition, the results provide genetic validation of the PEMSA method as a procedure for determining the *in vivo* redox state of thioredoxin.

Acknowledgments

We thank Dean Malencik for suggesting urea–PAGE as a means of separating thioredoxin charge isomers, and Eric Muller for supplying thioredoxin antibody. This work was supported by NSF Grant 9728782-MCB and NIH Grant CA82633 to G.F.M.

[30] Recycling of Vitamin C by Mammalian Thioredoxin Reductase

By JAMES M. MAY

Introduction

Mammalian thioredoxin reductase (TrxR) (EC 1.6.4.5) functions optimally to reduce disulfide bonds in conjunction with thioredoxin (Trx).[1] However, the enzyme alone has a broad substrate specificity, and can carry out NADPH-dependent reduction of a variety of redox-active molecules, including lipoic acid, vitamin K_3, 5,5′-dithiobis(2-nitrobenzoic acid) (DTNB), alloxan, and hydroperoxides.[1] This diversity may reflect the ability of the penultimate selenocysteine in the C-terminal portion of TrxR to function as a "super" thiol. Because thiols such as glutathione (GSH)[2] and lipoic acid[3] effectively reduce dehydroascorbic acid (DHA) to ascorbate, it is not surprising that TrxR can do the same. We found that purified rat liver TrxR reduces DHA to ascorbate, and that this reduction is enhanced by Trx.[4] Activity measured as either the disappearance of NADPH or appearance of ascorbate is stoichiometric. The reaction of DHA with TrxR in the presence of Trx has an apparent K_m of 0.7 mM and a k_{cat} of 71 min^{-1}. Although the latter is only about 2% of the rate at which the thioredoxin system reacts with disulfide substrates,[5] hydrophobic derivatives of DHA show up to 10-fold higher k_{cat} values.[4] The most likely mechanism for the reaction of TrxR with DHA is a two-step nucleophilic substitution as predicted for GSH reduction of DHA, especially because the ascorbate free radical (AFR) is not generated during the latter reaction.[6]

More surprising from a mechanistic standpoint was the finding that the thioredoxin system (TrxR plus Trx) can also reduce the AFR to ascorbate.[7] Using purified rat liver TrxR and bacterial Trx, this activity has an apparent K_m of 3 μM for the AFR and a k_{cat} of 135 min^{-1}. When measured either as loss of NADPH or as a decrease in the steady state AFR concentration determined by electron paramagnetic resonance (EPR) spectroscopy, the reaction is inhibited by aurothioglucose (ATG) and stimulated by selenocystine. The latter has been shown to enhance the

[1] E. S. J. Arnér and A. Holmgren, *Eur. J. Biochem.* **267,** 6102 (2000).

[2] B. S. Winkler, *Biochim. Biophys. Acta* **1117,** 287 (1992).

[3] B. C. Scott, O. I. Aruoma, P. J. Evans, C. O'Neill, A. Van der Vliet, C. E. Cross, H. Tritschler, and B. Halliwell, *Free Radic. Res.* **20,** 119 (1994).

[4] J. M. May, S. Mendiratta, K. E. Hill, and R. F. Burk, *J. Biol. Chem.* **272,** 22607 (1997).

[5] M. Luthman and A. Holmgren, *Biochemistry* **21,** 6628 (1982).

[6] J. M. May, Z. C. Qu, R. R. Whitesell, and C. E. Cobb, *Free Radic. Biol. Med.* **20,** 543 (1996).

[7] J. M. May, C. E. Cobb, S. Mendiratta, K. E. Hill, and R. F. Burk, *J. Biol. Chem.* **273,** 23039 (1998).

activity of TrxR in response to disulfide substrates.[8] The molecular mechanism by which the TrxR system can accomplish a one-electron reduction of the AFR has not been established, although there are several possible mechanisms.[7]

The question arises as to whether reduction of either DHA or the AFR by TrxR or the thioredoxin system has physiologic relevance. About 80% of the NADPH-dependent reduction of DHA to ascorbate in overnight-dialyzed rat liver cytosolic extracts can be attributed to TrxR, on the basis of inhibition by 10 μM ATG.[4] At this concentration ATG is specific for TrxR over glutathione peroxidase.[9] The remaining activity may be due to a 3α-hydroxysteroid dehydrogenase in rat liver, which has a K_m for DHA of 4 mM and a k_{cat} of 59 min^{-1}.[10] Further, dialysates from livers of selenium-deficient rats have only about 12% of the ATG-inhibitable DHA reductase activity of liver dialysates from control animals.[4] NADPH-dependent and ATG-sensitive DHA reductase activity is also present in dialyzed hemolysates prepared from human erythrocytes.[11] In liver dialysates, the rate of TrxR-mediated DHA reduction is about 25–30% that of GSH-dependent reduction, whereas in erythrocytes, it is only about 10%.

NADPH-dependent AFR reductase activity is also present in overnight-dialyzed rat liver cytosolic extracts.[7] Several features suggest that this activity is due to TrxR. It is (1) markedly decreased in dialysates from selenium-deficient rats, (2) inhibited by concentrations of ATG known to be selective for TrxR, and (3) stimulated by selenocystine. However, AFR reduction due to TrxR in dialysates is less than 2% of that of NADH-dependent AFR reduction in microsomes.

The TrxR system can reduce both the AFR and DHA to ascorbate, and this activity is readily apparent in dialyzed cell extracts. Because the techniques developed in the studies noted above might be useful in comparing rates of TrxR-dependent DHA and AFR reduction with rates of ascorbate recycling by other routes in different cell types or tissues, the following outlines the methodological approaches that we have found most useful.

Methods

Tissue Preparation and Removal of Low Molecular Weight Cofactors

TrxR-dependent reduction of either DHA or the AFR can be measured using crude extracts from various tissues, so long as low molecular weight cofactors for DHA reduction such as NADPH and GSH are removed. This is probably

---

[8] M. Björnstedt, M. Hamberg, S. Kumar, J. Xue, and A. Holmgren, *J. Biol. Chem.* **270,** 11761 (1995).

[9] K. E. Hill, G. W. McCollum, M. E. Boeglin, and R. F. Burk, *Biochem. Biophys. Res. Commun.* **234,** 293 (1997).

[10] B. Del Bello, E. Maellaro, L. Sugherini, A. Santucci, M. Comporti, and A. F. Casini, *Biochem. J.* **304,** 385 (1994).

[11] S. Mendiratta, Z.-C. Qu, and J. M. May, *Free Radic. Biol. Med.* **25,** 221 (1998).

most easily accomplished by dialyzing cytosolic extracts of tissue or cells. For tissue, homogenize 100–400 mg in 1.0–1.5 ml of an ice-cold buffer consisting of 50 mM Tris-HCl and 1 mM EDTA, pH 7.4 (Tris–EDTA). The type and duration of homogenization depend on the tissue, but for liver homogenization using a conical polytetrafluoroethylene (Teflon) pestle in microcentrifuge tubes is sufficient. The homogenate is centrifuged or microcentrifuged at 4000–6000g at 3° for 15 min, and the resulting supernatant is centrifuged at 100,000g for 1 hr at 3°. The clear supernatant is then dialyzed overnight against three changes of 250 ml of the Tris–EDTA buffer at 3°. The dialyzed sample (typically 5–25 mg of protein per milliliter) is saved on ice for assay and for protein determination.

Assay of Dehydroascorbic Acid Reduction by Thioredoxin Reductase

The activity can be measured either by monitoring the appearance of ascorbate or the disappearance of NADPH. There is a one-to-one concordance between the two assays, using purified rat liver TrxR.[4]

Measurement of Rate of NADPH Loss. The assay is more rapid than measuring the rate of ascorbate generation, and it confirms directly that the reaction is linear under the selected conditions. However, it is not specific and will detect NADPH oxidase activity. This can be controlled for by including a blank with NADPH, but no DHA. The reaction is carried out in a UV spectrophotometric semimicrocuvette at room temperature. Add 0.35 ml of Tris–EDTA buffer, 0.05 ml of enzyme, and 0.05 ml of 4 mM NADPH in the same buffer, and start the assay by adding 0.05 ml of freshly prepared 10 mM DHA. After mixing, record the change in absorption at 340 nm for 1–3 min in a recording spectrophotometer. From the linear slope of this line calculate the amount of NADPH consumed, using an extinction coefficient of 6.22×10^3 M^{-1} cm^{-1} for NADPH. The rate of loss of NADPH equals the rate of ascorbate generated, corrected for the background rate of NADPH oxidation (the blank reaction). Commercial DHA (Sigma-Aldrich, St. Louis, MO) is not pure, but can be used if dissolved in the Tris–EDTA buffer just before addition to the assay (it is not stable at physiologic pH). The amount of DHA added in this assay is not saturating for TrxR, and therefore a decrease in the added DHA concentration during multiple assays over time may result. It is preferable to use DHA prepared by bromine oxidation of 10 mM ascorbic acid in water for most studies.[12] The resulting DHA is stabilized by the low pH of the reaction medium, so that multiple samples can be analyzed without significant loss of DHA. It is necessary to confirm that the pH of the reaction mixture is not affected by the addition.

Measurement of Dehydroascorbic Acid Reduction by Appearance of Ascorbate. For this assay, the reaction volume is scaled down to 100 μl and the reaction is carried out in 1.5-ml microcentrifuge tubes. Otherwise, the agent concentrations

[12] P. W. Washko, Y. Wang, and M. Levine, *J. Biol. Chem.* **268,** 15531 (1993).

and conditions are the same as those described above. After adding DHA, incubate the reaction mixture for 10 min at room temperature. Stop the reaction by adding 10 volumes of ice-cold 80% (v/v) methanol containing 1 mM EDTA. After 5 min on ice, microcentrifuge the sample at 6000g at 3° for 15 min and take aliquots of the supernatant for assay of ascorbate by high-performance liquid chromatography (HPLC). We use HPLC with electrochemical detection of ascorbate,[4] but UV detection should also have adequate sensitivity and low background in dialyzed samples. A blank sample with no enzyme should be run in assays using commercial DHA to correct for ascorbate contamination, which can vary depending on the lot of DHA. In liver dialysates, increases in ascorbate are linear over about 20 min, but this should be confirmed in preliminary studies.

Assay of Thioredoxin Reductase Reduction of Ascorbate Free Radical

It is important in this assay to generate stable steady state concentrations of the AFR in the reaction mixture. Because TrxR reduces DHA to ascorbate, it is not possible to use the reverse dismutation reaction of a mixture of DHA and ascorbate to generate the AFR. Stable concentrations of the AFR are generated for several minutes by using ascorbate oxidase (0.05–0.4 U/ml, from *Cucurbita* species; Sigma-Aldrich) to oxidize 1–10 mM ascorbate. This reaction generates only the AFR, which then will dismutate to ascorbate and DHA. Although it is possible to monitor the disappearance of the AFR signal by EPR spectroscopy, the traditional indirect assay of NAD(P)H disappearance is much easier, can be quantified, and does not require an EPR spectrometer.

In a UV semimicrocuvette containing 0.3 ml of Tris–EDTA buffer, add 0.05 ml of 2 mM NADPH, 0.05 ml of enzyme, 0.02 ml of ascorbate oxidase (1 U/ml), and, finally, to start the assay, 0.05 ml of freshly prepared 10 mM ascorbic acid in Tris–EDTA buffer. Mix and monitor the decrease in absorbance at 340 nm at room temperature for up to 3 min. A blank sample without enzyme is used to correct for NADPH oxidation in the buffer, and the rate of AFR reduction is calculated as described above, with the caveat that 1 mol of NADPH will reduce 2 mol of the AFR. Activity then can be normalized to protein or to the amount of purified enzyme present, if known.

Because DHA will be generated in this reaction mixture by dismutation of two molecules of the AFR to one molecule each of DHA and ascorbate, it is necessary to establish that this reaction is a small component of the AFR reduction for each system, and to correct for it. This is done by incubating ascorbate and the oxidase under the conditions of the assay in a UV cuvette and monitoring the loss of ascorbate over the time of the assay by the decrease in absorbance at 265 nm. On the basis of the dismutation reaction of the AFR, the amount of DHA generated will be half the amount of ascorbate lost over the short course of the assay. By including a sample containing that concentration of DHA as a control, the maximal component of DHA-dependent NADPH loss can be determined. In our previous studies, at 1 mM ascorbate and ascorbate oxidase at 0.05 U/ml, this rate was found to be less than 5% of the AFR-dependent rate.[7]

Establishing Specificity for Thioredoxin Reductase in Crude Tissue Dialysates

Because there may be other NADPH-dependent DHA or AFR reductases in crude tissue extracts, it is necessary to determine the fraction of measured activity that is specific for TrxR. This can be done by including an incubation containing 10 μM ATG. Activity dependent on TrxR is then calculated by subtracting the rate in the presence of ATG from that found in its absence. Other maneuvers to show specificity for TrxR are to include either 5 μM commercial Trx or 50 μM selenocystine. Both will increase the observed rates of DHA or AFR reduction. The effect of mammalian Trx is to increase activity by up to 2-fold, whereas selenocystine enhances ATG-sensitive activity by 3- to 5-fold.

Conclusion

Although TrxR-dependent reduction of both DHA and the AFR can be documented in tissue extracts, whether this contributes to ascorbate recycling within cells is yet to be determined. Optimal DHA reduction requires GSH in erythrocytes,[6] lens epithelium,[13] and in several types of cultured cells. The latter include HepG2 cells, H4IIE cells, and primary cultures of bovine artery endothelial cells.[13a] However, neither HL60 cells[14] nor skin keratinocytes[15] appear to require GSH for DHA reduction. DHA reduction by the TrxR system could well play an important role in such cells.

There is also a question of whether ascorbate is regenerated from DHA or the AFR in cells. DHA reduction is certainly important when ascorbate is oxidized outside cells to DHA, which enters cells by facilitated diffusion on GLUT-type glucose transporters and is rapidly reduced to ascorbate. On the other hand, it may be more efficient to recycle from the AFR stage when ascorbate is oxidized within cells by reactive oxygen species generated in mitochondria or by cellular metabolism. The AFR is relatively long-lived, and its reduction will prevent possible loss at the DHA stage due to irreversible ring opening. Coassin *et al.*[16] found greater AFR reductase activity than GSH-dependent DHA reductase activity in various pig tissues. Although they did not measure NADPH-dependent DHA reductase activity,

[13] F. J. Giblin, B. S. Winkler, H. Sasaki, B. Chakrapani, and V. Leverenz, *Invest. Ophthalmol. Vis. Sci.* **34,** 1298 (1993).

[13a] J. M. May, Z.-C. Qu, and X. Li, *Biochem. Pharmacol.* **63,** 873 (2001).

[14] V. H. Guaiquil, C. M. Farber, D. W. Golde, and J. C. Vera, *J. Biol. Chem.* **272,** 9915 (1997).

[15] I. Savini, S. Duflot, and L. Avigliano, *Biochem. J.* **345,** 665 (2000).

[16] M. Coassin, A. Tomasi, V. Vannini, and F. Ursini, *Arch. Biochem. Biophys.* **290,** 458 (1991).

there was little evidence of NADH-dependent DHA reduction at DHA concentrations under 100 μM. Further, the measured affinities for DHA of glutaredoxin (0.26 mM),[17] and of the thioredoxin system (0.7 mM)[4] are not favorable for reducing the low concentrations of DHA expected to be generated within cells. On the other hand, the apparent K_m of TrxR for the AFR (3 μM)[7] is in the range expected for the AFR in oxidatively stressed cells.[18,19] Although NADH-dependent mitochondrial and microsomal AFR reduction shows a low apparent K_m and much greater activity when normalized to protein than does TrxR-mediated activity,[7] sequestration of the NADH-dependent enzymes because of their membrane-bound nature may limit access to the AFR. The cytoplasmic location of the thioredoxin system in most cells might allow it to play a significant role in reducing the AFR, but this is yet to be demonstrated.

[17] W. W. Wells, D. P. Xu, and M. P. Washburn, *Methods Enzymol.* **252**, 30 (1995).
[18] S. Pietri, M. Culcasi, L. Stella, and P. J. Cozzone, *Eur. J. Biochem.* **193**, 845 (1990).
[19] G. R. Buettner and B. A. Jurkiewicz, *Free Radic. Biol. Med.* **14**, 49 (1993).

[31] Thioredoxin Cytokine Action

By Yumiko Nishinaka, Hajime Nakamura, and Junji Yodoi

Introduction

Human thioredoxin (TRX) is secreted from various types of cells including activated normal B lymphocytes and virus-transformed cells, and the secreted TRX exhibits multiple cytokine-like activities.[1,2] In 1985, human TRX was first purified from the culture supernatant of an adult T cell leukemia (ATL) cell line, ATL-2, transformed by human T cell leukemia virus I (HTLV-I), as ATL-derived factor (ADF).[3] It induces interleukin 2 (IL-2) receptor α-chain (IL-2Rα) expression on a human large granular lymphocyte (LGL) cell line YT and it was thus considered a novel cytokine with IL-2Rα-inducing activity.[4] In 1989, cDNA sequencing revealed that ADF is a human homolog of TRX.[5] Expression and production of

[1] J. Yodoi and T. Uchiyama, *Immunol. Today* **13**, 405 (1992).
[2] H. Nakamura, K. Nakamura, and J. Yodoi, *Annu. Rev. Immunol.* **15**, 351 (1997).
[3] K. Teshigawara, M. Maeda, K. Nishino, T. Nikaido, T. Uchiyama, M. Tsudo, Y. Wano, and J. Yodoi, *J. Mol. Cell. Immunol.* **2**, 17 (1985).
[4] J. Yodoi, K. Teshigawara, T. Nikaido, K. Fukui, T. Noma, T. Honjo, M. Takigawa, M. Sasaki, N. Minato, M. Tsudo, T. Uchiyama, and M. Moeda, *J. Immunol.* **134**, 1623 (1985).
[5] Y. Tagaya, Y. Maeda, A. Mitsui, N. Kondo, H. Matsui, J. Hamuro, N. Brown, K. Arai, T. Yokota, H. Wakasugi, and J. Yodoi, *EMBO J.* **8**, 757 (1989).

TRX are markedly enhanced in Epstein–Barr virus (EBV)-infected lymphoblastoid B cell lines as well as in HTLV-I-infected T cell lines. During the study of B cell transformation by EBV, Wakasugi and co-workers independently reported an IL-1-like soluble factor produced by EBV-transformed B lymphoid cells named 3B6-IL-1, which was later found to be identical to ADF/TRX.[6–8] This factor has comitogenic activity in thymocytes and the human HSB-2 cell line, despite completely lacking pyrogenic activity, which is a characteristic of macrophage-derived IL-1s. The producer cell line 3B6 uses this factor as an autocrine growth factor. Later, several other cytokine-like factors previously reported were shown to be identical or closely related to TRX. These include MP6-BCGF,[9] B cell growth factor derived from the T cell hybridoma MP6, which has comitogenecity with IL-4 and other cytokines; eosinophil cytotoxicity-enhancing factor (ECEF),[10] produced by activated U937 cells; one component of early pregnancy factor (EPF),[11] originally defined as an immunosuppressive serum factor; and surface-associated sulfhydryl protein (SASP).[12]

TRX is a small multifunctional and ubiquitous protein having a redox-active disulfide/dithiol within the conserved active site sequence: -Cys-Gly-Pro-Cys-.[13] Although TRX is predominantly localized in cytosol, playing pivotal roles in the maintenance of the redox status in cells, it is secreted by lymphocytes, hepatocytes, fibroblasts, virus-infected cells, and cancer cells in response to a variety of stimuli, and acts as a cytokine.[3,7,14,15] Because the protein has no classic secretory signal sequence, the mechanism by which secretion occurs is not known. The secretion of TRX is enhanced by drugs that block transport along the classic secretory pathway as well as IL-1β,[15] suggesting that TRX and IL-1β use a novel leaderless pathway for secretion. Although the mechanism of TRX action is not fully understood, the enzymatic activity of TRX seems to be crucial for most of its biological functions. Indeed, a catalytic site mutant TRX (C32S/C35S), in which the two active cysteines are substituted with serine residues, has no

---

[6] H. Wakasugi, L. Rimsky, Y. Mahe, A. M. Kamel, D. Fradelizi, T. Tursz, and J. Bertoglio, *Proc. Natl. Acad. Sci. U.S.A.* **84**, 804 (1987).

[7] N. Wakasugi, Y. Tagaya, H. Wakasugi, A. Mitsui, M. Maeda, J. Yodoi, and T. Tursz, *Proc. Natl. Acad. Sci. U.S.A.* **87**, 8282 (1990).

[8] E. E. Wollman, L. d'Auriol, L. Rimsky, A. Shaw, J. P. Jacquot, P. Wingfield, P. Graber, F. Dessarps, P. Robin, F. Galibert, J. Bertoglio, and D. Fradelizi, *J. Biol. Chem.* **263**, 15506 (1988).

[9] M. Carlsson, C. Sundstrom, M. Bengtsson, T. H. Totterman, A. Rosen, and K. Nilsson, *Eur. J. Immunol.* **19**, 913 (1989).

[10] D. S. Silberstein, M. H. Ali, S. L. Baker, and J. R. David, *J. Immunol.* **143**, 979 (1989).

[11] K. F. Tonissen and J. R. Wells, *Gene* **102**, 221 (1991).

[12] H. Martin and M. Dean, *Biochem. Biophys. Res. Commun.* **175**, 123 (1991).

[13] A. Holmgren, *J. Biol. Chem.* **264**, 13963 (1989).

[14] M. L. Ericson, J. Horling, V. Wendel-Hansen, A. Holmgren, and A. Rosen, *Lymphokine Cytokine Res.* **11**, 201 (1992).

[15] A. Rubartelli and R. Sitia, *Biochem. Soc. Trans.* **19**, 255 (1991).

growth-promoting activity.[16] The requirement of the catalytic site has also been shown for other biological functions including chemokine-like effects (see below) and the regulation of transcription factors. In the interaction with transcription factors such as Ref-1[17] and NF-κB,[18,19] TRX has been demonstrated to be directly associated with target proteins and activate those proteins by dithiol-dependent reduction. The importance of the TRX catalytic site has also been shown in the interaction between TRX and TRX-binding proteins such as TRX-binding protein 2/vitamin D_3-upregulated protein 1 (TBP-2/VDUP1)[20] and apotosis signal-regulating kinase 1 (ASK-1).[21] The biological functions of TRX may be strictly regulated by its enzymatic reaction as well as structure-dependent targeting.[19]

Truncated Form of Thioredoxin

A truncated form of TRX was first described as eosinophil cytotoxicity-enhancing factor (ECEF) with a 10-kDa molecular mass,[10] which lacks the C-terminal 16 or 24 amino acids of TRX. It shows significant cytotoxicity-enhancing activity in eosinophils and U937 cells at concentrations as low as 10 pM, whereas full-length TRX shows no such activity. Although the truncated TRX retains the conserved active site, it has no dithiol reductase activity. These findings indicate that the enzymatic activity does not correlate with the ECEF activity. The cleavage of exogenous recombinant TRX by uninfected and human immunodeficiency virus (HIV)-infected macrophages to a truncated form was described in an earlier report.[22] However, the mechanism of proteolytic cleavage has not been clarified. More recently, it was shown that endogenous truncated TRX is produced and released from normal monocytes and other cell lines such as MP6, 3B6, and U937 by physiological stimuli and oxidative stress.[23,24] It should be noted that all these cells produce and release full-length TRX as well. Previous studies of the MP6-derived TRX, which is a mixture of TRX and truncated TRX, showed that its B cell stimulatory activity was >500-fold higher than that of purified placental

[16] G. Powis, D. L. Kirkpatrick, M. Angulo, and A. Baker, *Chem. Biol. Interact.* **111–112,** 23 (1998).

[17] K. Hirota, M. Matsui, S. Iwata, A. Nishiyama, K. Moroi, and J. Yodoi, *Proc. Natl. Acad. Sci. U.S.A.* **94,** 3633 (1997).

[18] J. R. Matthews, N. Wakasugi, J. L. Virelizier, J. Yodoi, and R. T. Hay, *Nucleic Acids Res.* **20,** 3821 (1992).

[19] J. Qin, G. M. Clore, W. M. Kennedy, J. R. Huth, and A. M. Gronenborn, *Structure* **3,** 289 (1995).

[20] A. Nishiyama, M. Matsui, S. Iwata, K. Hirota, H. Masutani, H. Nakamura, Y. Takagi, H. Sono, Y. Gon, and J. Yodoi, *J. Biol. Chem.* **274,** 21645 (1999).

[21] M. Saitoh, H. Nishitoh, M. Fujii, K. Takeda, K. Tobiume, Y. Sawada, M. Kawabata, K. Miyazono, and H. Ichijo, *EMBO J.* **17,** 2596 (1998).

[22] G. W. Newman, M. K. Balcewicz-Sablinska, J. R. Guarnaccia, H. G. Remold, and D. S. Silberstein, *J. Exp. Med.* **180,** 359 (1994).

[23] B. Sahaf, A. Soderberg, G. Spyrou, A. M. Barral, K. Pekkari, A. Holmgren, and A. Rosen, *Exp. Cell. Res.* **236,** 181 (1997).

[24] B. Sahaf and A. Rosen, *Antioxid. Redox. Signal.* **2,** 717 (2000).

TRX,[25] suggesting that TRX and truncated TRX are synergistic for the activity. In contrast, opposing effects of TRX and truncated TRX have been shown in the development of HIV-1 infection. Exogenous recombinant TRX inhibits HIV expression, whereas truncated TRX enhances it. A report by Pekkari et al. showed that a recombinant truncated form of TRX (amino acids 1–80) is, by itself (it is a dimer in solution), a potent mitogenic cytokine stimulating growth of resting human peripheral blood mononuclear cells (PBMCs), whereas full-length TRX is not so potent at the same dose.[26] The cytokine-like effects of the two proteins may be mediated via different mechanisms.

Thioredoxin as Costimulatory Molecule of Cytokine Action

TRX is also a potent costimulus in the expression and release of cytokines.[27] The expression of cytokines is redox-regulated, and TRX can augment the expression and release of several cytokines including IL-1, IL-2, IL-6, IL-8, and tumor necrosis factor α (TNF-α).[27] TRX exhibits growth-promoting activity in combination with IL-2. In addition, TRX augments the growth-promoting effects of other cytokines or growth factors on lymphocytes as well as nonlymphoid cells.[7,28] The cytoprotective activity of TRX is mainly explained by its scavenging activity for reactive oxygen species (ROS).[2] As ROS are considered to be signal messengers, the scavenging effect of TRX may contribute to the modulation of immune responses by cytokines and growth factors through ROS-mediated stress signals.

Chemokine-Like Activity of Thioredoxin

In 1993, the first report about the chemokine-like activity of TRX provided evidence that it can induce migration of eosinophils from patients with hypereosinophilia, although TRX exhibited little activity on eosinophils from healthy donors.[29] TRX also shows enhancing effects on both the chemotactic and chemokinetic activity of the complement anaphylatoxin peptide C5a in eosinophil migration. In contrast, TRX shows no effect on modulation of migratory behavior of human eosinophils by IL-3, IL-5, or granulocyte–macrophage colony-stimulating factor. A catalytic site mutant TRX (C32S/C35S) shows neither migration activity nor an enhancing effect. More recently, the chemokine activity of TRX toward

[25] A. Rosen, P. Lundman, M. Carlsson, K. Bhavani, B. R. Srinivasa, G. Kjellstrom, K. Nilsson, and A. Holmgren, Int. Immunol. 7, 625 (1995).

[26] K. Pekkari, R. Gurunath, E. S. Arner, and A. Holmgren, J. Biol. Chem. 275, 37474 (2000).

[27] H. Schenk, M. Vogt, W. Droge, and K. Schulze-Osthoff, J. Immunol. 156, 765 (1996).

[28] G. Powis, J. R. Gasdaska, P. Y. Gasdaska, M. Berggren, D. L. Kirkpatrick, L. Engman, I. A. Cotgreave, M. Angulo, and A. Baker, Oncol. Res. 9, 303 (1997).

[29] K. Hori, M. Hirashima, M. Ueno, M. Matsuda, S. Waga, S. Tsurufuji, and J. Yodoi, J. Immunol. 151, 5624 (1993).

polymorphonuclear leukocytes (PMNs), monocytes, and T lymphocytes from normal individuals was clearly demonstrated by Bertini *et al.* in a standard *in vitro* chemotaxis assay with micro-Boyden chambers.[30] The potency of the chemotactic action of TRX was comparable to that of known chemokines such as IL-8 for PMNs, monocyte chemoattractant protein 1 (MCP-1) for monocytes, and RANTES for T cells, at optimal concentrations (0.1–2.5 nM; 1–30 ng/ml). The chemokine activity of TRX was also demonstrated *in vivo* using a murine air pouch model. As mutant TRX (C32S/C35S) is not chemotactic, the chemokine activity is related to its enzymatic action, as are most of the biological effects of TRX. The entire family of chemokines is known to be G-protein dependent, whereas TRX is G-protein independent.[30] This finding suggests that the chemokine activity of TRX is not mediated via any known chemokine receptors. TRX may modulate thiols of membrane target proteins (receptors?) to transduce its signal. Neither specific proteins nor reservoirs on cell membrane have been discovered.

Circulating Thioredoxin Levels in Human Plasma

Circulating TRX can be detected in the plasma of healthy donors. The plasma levels of full-length TRX and truncated TRX were determined to be 16–55 ng/ml (median, 29 ng/ml) and 2–175 ng/ml (median, 20 ng/ml), respectively, by sandwich enzyme-linked immunosorbent assay (ELISA).[26,31,32] Both levels are within the physiological range in terms of cytokine- or chemokine-like activities. These data suggest that plasma TRX levels reflect a variety of states of immune responses against extracellular stimuli such as infection or inflammation. In fact, TRX levels are increased in plasma of HIV- or hepatitis C virus (HCV)-infected individuals.[31,33] In addition, an association between elevated plasma TRX and the decreased survival of HIV-infected patients has been reported.[34] An increase in plasma or serum TRX has also been described in other diseases such as hepatocellular carcinoma (HCC)[35] rheumatoid arthritis (RA).[36] Considering that the two forms of TRX (full length and truncated) have different biological functions as

[30] R. Bertini, O. M. Howard, H. F. Dong, J. J. Oppenheim, C. Bizzarri, R. Sergi, G. Caselli, S. Pagliei, B. Romines, J. A. Wilshire, M. Mengozzi, H. Nakamura, J. Yodoi, K. Pekkari, R. Gurunath, A. Holmgren, L. A. Herzenberg, and P. Ghezzi, *J. Exp. Med.* **189,** 1783 (1999).

[31] H. Nakamura, S. DeRosa, M. Roederer, M. T. Anderson, J. G. Dubs, J. Yodoi, A. Holmgren, L. A. Herzenberg, and L. A. Herzenberg, *Int. Immunol.* **8,** 603 (1996).

[32] H. Kogaki, Y. Fujiwara, A. Yoshiki, S. Kitajima, T. Tanimoto, A. Mitsui, T. Shimamura, J. Hamuro, and Y. Ashihara, *J. Clin. Lab. Anal.* **10,** 257 (1996).

[33] Y. Sumida, T. Nakashima, T. Yoh, Y. Nakajima, H. Ishikawa, H. Mitsuyoshi, Y. Sakamoto, T. Okanoue, K. Kashima, H. Nakamura, and J. Yodoi, *J. Hepatol.* **33,** 616 (2000).

[34] H. Nakamura, S. C. De Rosa, J. Yodoi, A. Holmgren, P. Ghezzi, and L. A. Herzenberg, *Proc. Natl. Acad. Sci. U.S.A.* **98,** 2688 (2001).

[35] M. Miyazaki, N. Noda, M. Terada, and H. Wakasugi, *in* "Redox Regulation of Cell Signaling and Its Clinical Application" (L. Packer and J. Yodoi, eds.), p. 235. Marcel Dekker, New York, 1999.

[36] S. Yoshida, T. Katoh, T. Tetsuka, K. Uno, N. Matsui, and T. Okamoto, *J. Immunol.* **163,** 351 (1999).

described above, levels of these proteins in plasma samples should preferably be analyzed separately. Here we describe a sensitive ELISA for human TRX. This method is suitable for analyzing TRX levels in any fluid samples including blood samples and cell lysates.

Sandwich Enzyme-Linked Immunosorbent Assay for Human Thioredoxin

See Nakamura et al.[31] and Kogaki et al.[32] for additional details.

Reagents

Citrate buffer (pH 3.5), 0.1 M
Blocking buffer: 50 mM phosphate buffer, 1% (w/v) bovine serum albumin (BSA), 0.05% (v/v) Tween 20 (pH 6.0)
Dilution buffer: 50 mM phosphate buffer, 1% (w/v) BSA, 0.05% (v/v) Tween 20, pH 8.0
Recombinant human TRX, provided by Ajinomoto (Kawasaki, Japan)
Specific murine monoclonal antibodies to human TRX (ADF-11 and ADF-21), provided by Fuji Rebio (Tokyo, Japan)
Horseradish peroxidase-labeled ADF-11
Tween 20-supplemented phosphate-buffered saline (TPBS)
Substrate solution: 2,2'-azinobis(3-ethylbenzothiazoline 6-sulfuric acid), 0.5 mg/ml, dissolved in 0.1 M triethanolamine–succinate buffer containing 0.03% (v/v) hydrogen peroxide, pH 4.2
Stopping solution: 1% (w/v) oxalic acid, pH 1.9

Procedure

Two specific murine monoclonal antibodies to nonoverlapping epitopes of human TRX (ADF-11 and ADF-21) are used in the sandwich ELISA. Recombinant TRX is used as a standard, at a 2-fold dilution from 320 to 5 ng/ml. Standard samples of TRX should be stored in small aliquots at −70°. Each aliquot should be discarded after use once being thawed. Ninety-six-well microplates (Nunc, Roskilde, Denmark) are precoated with 100 μl of ADF-21 (15 μg/ml) in 0.1 M citrate buffer (pH 3.5). Then, 200 μl of blocking buffer [50 mM phosphate buffer, 1% (w/v) BSA, 0.05% (v/v) Tween 20 (pH 6.0)] and 20 μl of TRX standards or samples are added to the wells in duplicate and incubated at room temperature for 2 hr. After washing three times with TPBS, 200 μl of horseradish peroxidase-labeled ADF-11 (0.1 μg/ml) diluted in dilution buffer [50 mM phosphate buffer, 1% (w/v) BSA, 0.05% (v/v) Tween 20 (pH 8.0)] is added and incubated at room temperature for 2 hr. After washing three times with TPBS, 100 μl of substrate solution is added and incubated at room temperature for 1 hr. The reaction is terminated by the addition of 100 μl of stopping solution [1% (w/v) oxalic acid, pH 1.9], and the absorption at 405 nm is measured by ELISA reader (e.g., Vmax kinetic microplate reader; Molecular Devices, Menlo Park, CA).

Analysis of Results

Data are calculated by Softmax (version 2.01; Molecular Devices). The TRX concentration of plasma samples is calculated by fitting the standard curve values for recombinant TRX to a four-parameter logit–log curve model shown as follows: $Y = (A - D)/[1 + (X/C)^B] + D$, where A is the maximal absorption, B is the reaction order, C is the background absorption, and D is the background absorption. If blood samples such as plasma are measured, erythrocyte lysis should be avoided during the collection, as erythrocytes contain high levels of TRX. In case it is inevitable, correction can be made by estimating the level of lysis based on plasma hemoglobin (Hb) contents, and subtracting the corresponding amount of erythrocyte-derived TRX from the overall TRX. Better results can be obtained when hemolyzed plasma with plasma Hb < 15 mg/dl is used. To establish correction values, the correspondence between the amount of TRX and the amount of Hb in fully hemolyzed samples is determined. For example, the average-hemolyzed TRX : Hb ratio in red blood cells (RBCs) from 77 HIV-infected and 18 control individuals was determined as 0.937 (ng/ml)/(mg/dl) in our previous study.[31] Plasma TRX was corrected by the following formula: corrected plasma TRX (ng/ml) = TRX measured in plasma (ng/ml) − [(TRX : Hb ratio) × (plasma Hb)], where the TRX : Hb ratio is the average ratio (0.937), and plasma Hb is the Hb measured for each plasma sample.

Concluding Remarks

TRX, a stress-inducible antioxidant protein, has been shown to play crucial roles in cytoprotection against oxidative stress of various sorts. The cytoprotective effects of TRX are attributed mostly to its scavenging activity against ROS. However, as described above, secreted TRX shows cytokine- or chemokine-like activities and the levels of plasma TRX are within the physiological range. These facts strongly suggest that TRX secreted in response to oxidative stress or other stimuli is a stress sensor to mediate signals from the outer membrane. Of special interest is the finding of target proteins on the cell membrane to mediate signals from secreted or exogenously added TRX. The association of plasma TRX levels with oxidative stress-related diseases is also interesting, and further studies may provide a new therapeutic approach to those diseases.

Acknowledgment

This work was supported by a grant-in-aid of Research for the Future from the Japan Society for the Promotion of Science.

[32] Identification of Thioredoxin-Linked Proteins by Fluorescence Labeling Combined with Isoelectric Focusing/Sodium Dodecyl Sulfate–Polyacrylamide Gel Electrophoresis

By JOSHUA H. WONG, HIROYUKI YANO, YOUNG-MOO LEE, MYEONG-JE CHO, and BOB B. BUCHANAN

Introduction

Thioredoxins are a class of small (12-kDa) multifunctional regulatory proteins occurring in all types of cells.[1–4] Plants contain three types of thioredoxin—two in chloroplasts (f and m) and one (h) in the cytosol.

Thioredoxins f and m are reduced in oxygenic photosynthesis by photoreduced ferredoxin (Fdx) via the iron–sulfur enzyme, ferredoxin–thioredoxin reductase (FTR) [Eq. (1)]:

$$\text{Thioredoxin } f \text{ or } m_{\text{ox}} + 2\text{Fdx}_{\text{red}} + 2\text{H}^+ \xrightarrow{\text{FTR}} \text{thioredoxin } f \text{ or } m_{\text{red}} + 2\text{Fdx}_{\text{ox}}$$
$$(\text{–S–S–}) \qquad\qquad\qquad\qquad\qquad\qquad (\text{–SH HS–})$$
$$(1)$$

By contrast, thioredoxin h, a member of a multigene family, is reduced via NADP-thioredoxin reductase (NTR) by NADPH generated metabolically [Eq. (2)]. The reduced thioredoxin, in turn, reduces disulfide bonds of a broad spectrum of target proteins [Eq. (3)] including photosynthetic enzymes in the case of thioredoxin f or m and a variety of seed proteins with thioredoxin h—all with a redox-active disulfide site.

$$\text{Thioredoxin } h_{\text{ox}} + \text{NADPH} + \text{H}^+ \xrightarrow{\text{NTR}} \text{thioredoxin } h_{\text{red}} + \text{NADP}^+ \qquad (2)$$
$$(\text{–S–S–}) \qquad\qquad\qquad\qquad\qquad (\text{–SH HS–})$$

$$\text{Thioredoxin } h_{\text{red}} + \text{target protein}_{\text{ox}} \longrightarrow \text{thioredoxin } h_{\text{ox}} + \text{target protein}_{\text{red}} \qquad (3)$$
$$(\text{–SH HS–}) \qquad (\text{–S–S–}) \qquad\qquad\qquad (\text{–S–S–}) \qquad\qquad (\text{–SH HS–})$$

In reducing specific target enzymes, thioredoxins f and m enable chloroplasts to distinguish light from dark and thereby regulate photosynthetic and dissimilatory processes. Thioredoxin h has a number of regulatory roles, including a function in germinating seeds. Its ability to reduce seed proteins (storage proteins, proteases,

[1] I. Besse and B. B. Buchanan, *Bot. Bull. Acad. Sin. (Taipei)* **38**, 1 (1997).
[2] J. P. Jacqout, J.-M. Lancelin, and Y. Meyer, *New Phytol.* **136**, 543 (1999).
[3] E. Ruelland and M. Miginiac-Maslow, *Trends Plant Sci.* **4**, 136 (1999).
[4] P. Shürmann and J. P. Jacqot, *Annu. Rev. Plant Physiol. Plant Mol. Biol.* **51**, 371 (2000).

protease inhibitors, and inhibitors of starch-degrading enzymes) enables thiore-doxin *h* to facilitate the mobilization of stored nitrogen and carbon reserves in the endosperm, thereby regulating germination and seedling development in cereals.

A better understanding of the role of thioredoxin could be achieved if more were known about the target proteins, especially for the thioredoxins occurring outside plastids. A strategy has been developed to identify proteins that are specif-ically reduced by thioredoxin.[5] An uncharged probe, monobromobimane (mBBr), which covalently binds the sulfhydryl groups newly generated by thioredoxin, renders the protein–mBBr derivative fluorescent. The fluorescent mBBr–protein adducts so formed can be separated by a combination of isoelectric focusing (IEF)/reducing sodium dodecyl sulfate–polyacrylamide gel electrophoresis (SDS–PAGE) and nonreducing/reducing two-dimensional (2D) PAGE, two well-known gel-based electrophoresis methods. The thioredoxin targets can be distinguished from other proteins by comparing the fluorescent and protein-stained gels. Those proteins emerging as thioredoxin targets are excised and identified by amino acid sequencing.

The strategy can be applied to identify thioredoxin target proteins *in vitro* (e.g., dry seeds)[5] or to assess thiol status *in vivo* (e.g., germinating seeds).[6] This approach should be effective for studying the redox properties of a variety of systems, including cells and organelles such as nuclei and mitochondria.

Sources of Materials and Chemicals

The protocol below has been devised for peanut seeds.

Materials

Peanut (*Arachis hypogaea* L.) seeds are obtained from a local market. Plant *h*-type thioredoxins and NTR are preferred although the commercially available *Escherichia coli* counterparts are effective with the seed proteins tested. In our case, we used *Chlamydomonas reinhardtii* thioredoxin *h* and *Arabidopsis thaliana* NTR.

Chemicals

Reagents for IEF and SDS–polyacrylamide gel electrophoresis are purchased from Bio-Rad Laboratories (Hercules, CA). mBBr (trade name Thiolyte) is ob-tained from Calbiochem (San Diego, CA). Other chemicals and biochemicals are purchased from commercial sources and are of the highest quality available.

---

[5] H. Yano, J. H. Wong, Y. M. Lee, M.-J. Cho, and B. B. Buchanan, *Proc. Natl. Acad. Sci. U.S.A.* **98**, 4794 (2001).

[6] H. Yano, J. H. Wong, M.-J. Cho, and B. B. Buchanan, *Plant Cell Physiol.* **42**, 879 (2001).

Methods

Isolation of Protein Fractions

Chemicals and Materials

Peanut seeds, 12 for each preparation
Ethyl ether
Tris-HCl buffer (pH 7.5), 0.1 M, containing 1.5 M NaCl, 2 mM phenylmethyl-
sulfonyl fluoride (PMSF), and
NaN$_3$, 0.2% (w/v)

Procedure. Protein is extracted according to a modification of the protocol of Shokraii and Esen[7] as follows. The peanut meal is prepared by grinding 12 randomly selected deskinned peanut seeds to a fine powder in a prechilled mortar with a pestle. The meal is defatted three times with ethyl ether, using a solvent-to-meal ratio of 10 : 1 (v/w). The defatted meal is air dried and then extracted for 2 hr by shaking in 0.1 M Tris-HCl buffer (pH 7.5) containing 1.5 M NaCl, 2 mM PMSF, and 0.2% (w/v) NaN$_3$. The meal-to-buffer ratio is 1 : 20 (w/v). The slurry is clarified by centrifugation (14,000g, 5 min, 4°) and the supernatant fraction is saved for analysis after protein determination with Coomassie blue.[8]

Reduction and in Vitro Labeling of Proteins

Reagents

Tris-HCl buffer (pH 7.9), 50 mM
Thioredoxin h, *C. reinhardtii*, 0.3 mg/ml in 30 mM Tris-HCl, pH 7.9
NTR, *A. thaliana*, 0.3 mg/ml in 30 mM Tris-HCl, pH 7.9
NADPH, 20 mM in 30 mM Tris-HCl, pH 7.9
Dithiothreitol (DTT), 50 mM
mBBr, 20 mM, dissolved in acetonitrile and stored protected from light
SDS, 10% (w/v)
2-Mercaptoethanol (2-ME), 100 mM
Target proteins, 30–50 μg, in extraction solvent

Procedure. Reduction of the disulfide bonds of peanut proteins is achieved with the NADP/thioredoxin system incubated for 3 or 5 hr at 37°. Reduction is carried out in a 1.5-ml microcentrifuge tube containing a solution consisting of 50 μl of Tris-HCl buffer (50 mM, pH 7.9), 0.8 μl (0.125 μmol) of NADPH, 8 μl (2.4 μg) of *C. reinhardtii* thioredoxin h, 7 μl (2.1 μg) of *A. thaliana* NTR, and 20 μl (50 μg) of peanut extract. The final volume is 100 μl. For complete reduction,

---

[7] E. H. Shokraii and A. Esen, *J. Agric. Food Chem.* **40**, 1491 (1992).
[8] M. Bradford, *Anal. Biochem.* **72**, 248 (1976).

samples are boiled with 5 mM DTT for 5 min. After the incubation, 20 μl of mBBr (0.2 μmol) is added and the incubation is continued for another 20 min at 25°. To stop the reaction and derivatize excess mBBr, 10 μl of 10% (w/v) SDS and 10 μl of 100 mM 2-ME are added and the samples are ready for processing before application to the gels. *Note:* Peanut proteins are especially difficult to reduce. Proteins from other sources typically require shorter reaction times for reduction by thioredoxin.

Processing of Labeled Protein Samples

Removal of Fluorescent Probe. The unbound mBBr can be removed from the sample by centrifugation in an ultrafree centrifugal filter device with a 5000 molecular weight cutoff membrane (Millipore, Bedford, MA), using a Beckman (Fullerton, CA) model TJ-6 centrifuge (5000 rpm). Alternatively, for nonreducing/reducing 2D PAGE, the reduced and labeled samples can be precipitated with trichloroacetic acid (TCA) and the precipitate analyzed by electrophoresis as described below. TCA precipitation is not appropriate for samples to be analyzed by IEF in the first dimension.

Precipitation of Labeled Proteins with Trichloroacetate. An equal volume of 20% (w/v) TCA is added to the sample, which is incubated on ice for 20 min. The precipitated protein is collected by centrifugation at 14,000 rpm for 15 min at 4° in an Eppendorf (Hamburg, Germany) microcentrifuge. After removal of the supernatant fraction, the pellet is washed with 1.0 ml of acetone, collected by centrifugation, and air dried. The dried sample is prepared for electrophoresis in the first dimension by dissolving in 10–20 μl of Laemmli sample buffer without reductant.[9]

Sample Preparation for Isoelectric Focusing/Sodium Dodecyl Sulfate–Polyacrylamide Gel Electrophoresis. Samples are concentrated to yield 10 to 20 mg of protein per milliliter by centrifugation as described above by using the appropriate molecular weight cutoff filter, in our case 5 kDa.

Analysis of Labeled Protein Fractions

NONREDUCING/SODIUM DODECYL SULFATE-REDUCING TWO-DIMENSIONAL POLYACRYLAMIDE GEL ELECTROPHORESIS

Reagents

Laemmli sample buffer:
Tris-HCl (pH 6.8), 62.5 mM
SDS, 2% (w/v)
Glycerol, 25% (v/v)
Bromphenol blue 0.01% (w/v)
2-ME, 5% (v/v)

---

[9] U. K. Laemmli, *Nature* (*London*) **227,** 680 (1970).

Tris–glycine–SDS electrophoresis buffer:
Tris, 25 mM
Glycine, 192 mM
SDS (pH 8.3), 0.1% (w/v)

Procedure. Thioredoxin-reduced and control mBBr-labeled protein samples are dissolved in Laemmli sample buffer free of 2-ME or other reducing agents.[9] Gels [10–20% (w/v) polyacrylamide gradient, 1.0-mm thickness] are prepared as described by Laemmli and subjected to electrophoresis in the first dimension for 16 hr at a constant current of 7 mA. After electrophoresis, the narrow gel lane containing the separated proteins is excised from the gel and immersed in complete Laemmli sample buffer (with 2-ME) for 20 min at room temperature. The gel strip is applied horizontally to a second gel [10–20% (w/v) polyacrylamide gradient, 1.5-mm thickness], and electrophoresis is carried out in the second dimension as described above.

ISOELECTRIC FOCUSING/SODIUM DODECYL SULFATE-REDUCING TWO-DIMENSIONAL POLYACRYLAMIDE GEL ELECTROPHORESIS. Isoelectric focusing and the subsequent SDS–PAGE are performed with a PROTEAN IEF cell and Criterion precast gel system (Bio-Rad) according to the manufacturer instructions. Similar equipment from other manufacturers can also be used.

Reagents and Equipment

IPG strips (pH 3–10), 11 cm
Rehydration buffer:
3-[(3-Cholamidopropyl)dimethylammonio]-1-propanesulfonate (CHAPS), 0.5% (v/v)
Urea, 8 M
DTT, 10 mM
Bio-Lytes, 0.1% (w/v), pH 3–10
Bromphenol blue, 0.001% (w/v)
Equilibration buffer:
Tris-HCl (pH 8.8), 0.375 M
SDS, 2% (w/v)
Glycerol, 20% (v/v)
Bromphenol blue, 0.01% (w/v)
Urea, 6 M
DTT, 130 mM
Reduced and mBBr-labeled protein sample, 100–500 μg
Polyacrylamide gradient Criterion precast gel with IPG sample well, 10–20% (w/v)
PROTEAN IEF cell
Criterion precast gel system

Procedure. IPG strips (pH 3–10) are rehydrated in rehydration buffer for 12 hr at 20° in the standard tray on the PROTEAN IEF cell. Thioredoxin-reduced and control mBBr-labeled protein samples (after dialysis to remove salt and concentration by an ultrafree centrifugal filter device) are dissolved in 30 μl of rehydration buffer and applied to the rehydrated IPG strip in the tray. Isoelectric focusing is performed in a PROTEAN IEF cell using a preset program with 35,000 V·hr in total and an upper voltage limit of 8000 V. After the termination of isoelectric focusing, the IPG strip is removed and dipped in equilibration buffer for 20 min. The strip is then applied horizontally to a 10–20% (w/v) polyacrylamide gradient Criterion gel with IPG sample well and electrophoresis is performed in the second dimension at constant 125 V at 25° for 1.5 hr, using a Criterion precast gel system.

Processing of Fluorescent/Protein Gels

FLUORESCENCE PHOTOGRAPHY

Reagents and Equipment

Methanol (40%, v/v)–5% (v/v) acetic acid
Spectroline (Spectronic, Westbury, NY) transilluminator, model TS-365, 365-nm ultraviolet
Gel Doc 1000 (Bio-Rad)

Procedure. After electrophoresis, gels are placed in 200 ml of a solution containing 40% (v/v) methanol–5% (v/v) acetic acid and soaked for 30 min to fix the proteins. The gel is first examined with a 365-nm UV lightbox (Spectroline; Spectronic) to observe fluorescence of the mBBr-labeled proteins. The fluorescence of mBBr-labeled protein bands is then visualized by placing the gels on a light box fitted with an ultraviolet light source (365 nm) inside a Gel Doc 1000 (Bio-Rad) and the image is captured with the camera on the Gel Doc and analyzed by the Quantity One software program.

PROTEIN STAINING/DESTAINING/PHOTOGRAPHY

Reagents

Coomassie Brilliant Blue R-250, 0.25% (w/v)
Methanol (40%, v/v)–5% (v/v) acetic acid

Procedure. Nonreducing/reducing SDS gels and IEF/SDS-reducing 2D gels are stained with 100 ml of 0.25% (w/v) Coomassie Brilliant Blue R-250 in 40% (v/v) methanol–5% (v/v) acetic acid for 1 to 2 hr. Proteins are destained overnight with several changes (200 ml) of the above-described solvent without the dye. Images of destained gels are captured with a Gel Doc on a white light box as described above.

Identification of Protein Targets

In-Gel Digestion and Peptide Fractionation. Reduction/alkylation and trypsin in-gel digestion of mBBr-labeled proteins are carried out essentially by the procedure described by Shevchenko *et al.*[10] Extracted trypsin-digested peptides from gels are separated by microbore C_{18} reversed-phase column (1 mm × 25 cm; Vydac, Hesperia, CA) on an ABI 172 HPLC system (Applied Biosystems, Foster City, CA). After injection of the sample, the column is washed for 5 min under equilibration conditions: 95% (v/v) solvent A [0.1% (w/v) trifluoroacetic acid in water], 5% (v/v) solvent B [0.075% (w/v) trifluoroacetic acid in 70% (v/v) acetonitrile], each for 5 min for column equilibration. The column is eluted first with a gradient from 5 to 10% (v/v) solvent B for 10 min, second with a linear gradient from 10 to 70% (v/v) solvent B for 70 min that is then increased to 90% (v/v) solvent B over 15 min.

Amino Acid Sequence Analysis of Peptides. Sequence analysis of C_{18}-purified peptides is performed by automated Edman degradation on an ABI model 494 Procise sequencer (Applied Biosystems). Nontarget proteins are also analyzed by nanoelectrospray ionization tandem mass spectrometry (nano-ESI/MS/MS), using a hybrid mass spectrometer (QSTAR; Applied Biosystems, Foster City, CA). Nanospray capillaries are obtained from Protana (Odense, Denmark). For nano-ESI/MS/MS, in-gel-digested peptide mixture is analyzed directly without any C_{18} column fractionation.

Applications

Reduction of Proteins by Thioredoxin

As seen previously with proteins from other seeds, thioredoxin *h,* reduced with NADPH via NTR, was effective in the reduction of peanut proteins. On the basis of results with one-dimensional electrophoresis gels, thioredoxin *h* appeared to show a preference for intramolecular disulfide bonds (data not shown).

To isolate and characterize the proteins targeted by thioredoxin and confirm the specificity of their disulfide bonds, we subjected the preparation to more complete two-dimensional separation procedures. We first applied nonreducing/reducing SDS–PAGE to identify the thioredoxin-linked proteins and determine the nature of their disulfide bonds. We then analyzed the samples by IEF/reducing SDS–gel electrophoresis for better resolution and a more complete analysis of the target proteins. In accord with longstanding findings from this laboratory, reduced glutathione was consistently without effect (data not shown).

---

[10] A. Shevchenko, M. Wilm, O. Vorm, and M. Mann, *Anal. Chem.* **68,** 850 (1996).

Isolation of Thioredoxin Target Proteins: Nonreducing/Reducing Two-Dimensional Sodium Dodecyl Sulfate–Gel Electrophoresis

As shown in Fig. 1,[5] the nonreducing/reducing SDS–gel electrophoresis system provides a direct means both to identify disulfide proteins, and to determine the nature of their disulfide bonds. Proteins with either inter- or intramolecular disulfide bonds (potential targets I and II, respectively) that have undergone reduction before application to the gel (Fig. 1A) resemble proteins without disulfide bonds and are recovered on the diagonal line (line of protein monomers) (Fig. 1B, right-hand side). By contrast, if the proteins are not reduced before electrophoresis in the first dimension, those having intermolecular disulfide bonds are recovered below the diagonal line owing to dissociation and the attendant decrease in molecular mass after reduction in the second dimension (Fig. 1B, left-hand side). Counterparts with intramolecular disulfide bonds, on the other hand, are recovered above the diagonal line as a result of the change in the apparent migration in reducing SDS–gel electrophoresis after reduction. An analysis of protein-stained gels can

A. Reduction of Possible Target Proteins by Trx and Labeling with mBBr

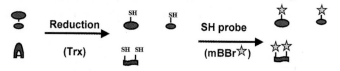

B. Separation of Proteins by Electrophoresis

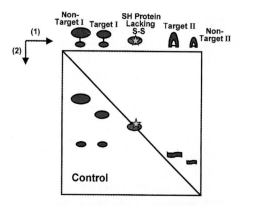
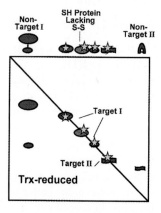

(1) = Non-reducing Direction (2) = Reducing Direction

FIG. 1. Distinction between proteins with intra- vs. intermolecular disulfide bonds after labeling with mBBr (A) and nonreducing/reducing two-dimensional SDS–gel electrophoresis (B). Trx, Thioredoxin. [Reproduced with permission from H. Yano, J. H. Wong, Y. M. Lee, M.-J. Cho, and B. B. Buchanan, *Proc. Natl. Acad. Sci. U.S.A.* **98,** 4794 (2001).]

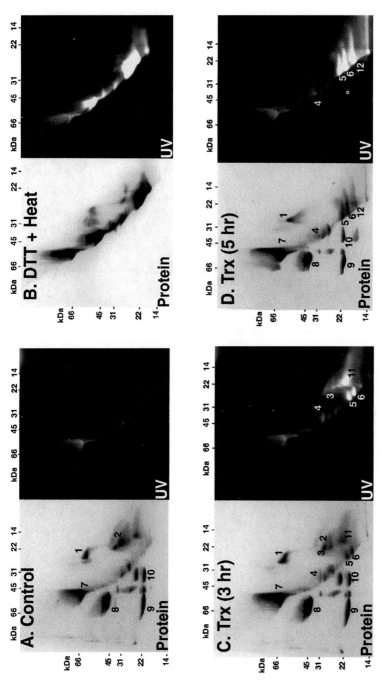

Fig. 2. Analysis of peanut seed proteins after reduction by the NADP/thioredoxin system or DTT and labeling with mBBr, using nonreducing/reducing two-dimensional SDS–gel electrophoresis. Fifty micrograms of peanut protein (20–μl extract) were incubated in 50 mM Tris-HCl buffer, pH 7.9, in a final volume of 100 μl. (A) Control: no addition; (B) DTT plus heat: the sample was heated in boiling water after addition of 5 mM DTT; (C) Trx (3 hr): incubation for 3 hr at 37° in the presence of 0.125 μmol of NADPH, 2.4 μg of NADPH, 2.4 μg of C. reinhardtii thioredoxin h, and 2.1 μg of A. thaliana NTR; (D) Trx (5 hr): Incubation under the same conditions as (C) except that it was for 5 hr. Trx, Thioredoxin. [Reproduced with permission from H. Yano, J. H. Wong, Y. M. Lee, M.-J. Cho, and B. B. Buchanan, Proc. Natl. Acad. Sci. U.S.A. 98, 4794 (2001).]

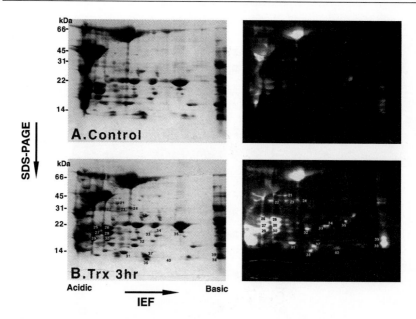

FIG. 3. Analysis of peanut seed proteins after reduction by the NADP/thioredoxin system and labeling with mBBr, using IEF/reducing SDS–PAGE/mBBr. One hundred and fifty micrograms of peanut protein (60-μl extract) was incubated in 50 mM Tris-HCl buffer, pH 7.9, in a final volume of 300 μl. (A) Control: no addition; (B) Trx (3 hr): Incubation for 3 hr in the presence of 0.375 μmol of NADPH, 7.2 μg of *C. reinhardtii* thioredoxin *h*, 6.3 μg of *A. thaliana* NTR. Trx, Thioredoxin. [Reproduced with permission from H. Yano, J. H. Wong, Y. M. Lee, M.-J. Cho, and B. B. Buchanan, *Proc. Natl. Acad. Sci. U.S.A.* **98,** 4794 (2001).]

thus serve to identify not only proteins with disulfide bonds, but also the nature of these bonds. The nonreducing/reducing two-dimensional gel system has been widely used for this purpose.[11]

In the present study, the primary objective was somewhat different. The gel system was used to identify not all disulfide proteins, but only those targeted by a specific reductant, thioredoxin *h*, using mBBr as a probe. The bond that had been reduced in these target proteins could then be shown to be either inter- or intramolecular.

Nonreducing Sodium Dodecyl Sulfate-Reducing Two-Dimensional Polyacrylamide Gel Electrophoresis

Application of the nonreducing/SDS-reducing 2D PAGE system to peanut extract reveals the specificity of thioredoxin in the reduction of intramolecular

---

[11] A. T. Endler and R. P. Tracy, *Cancer Invest.* **5,** 127 (1987).

versus intermolecular disulfide bonds. Thus, Fig. 2[5] shows a preponderance of proteins with intermolecular disulfide bonds (below the line) and a significant number of proteins with intramolecular disulfide bonds (above the line). Few of the proteins had free sulfhydryl groups (on the diagonal line).

Isoelectric Focusing/Sodium Dodecyl Sulfate-Reducing Two-Dimensional Polyacrylamide Gel Electrophoresis

Application of the IEF/SDS-reducing 2D PAGE system to peanut extract improved the separation of the labeled proteins (Fig. 3).[5] By application of both methods to peanut extracts, we isolated at least 20 thioredoxin targets and identified 5, including 3 allergens (*Ara h 2, Ara h 3*, and *Ara h 6*) and 2 proteins not known to occur in peanut (desiccation-related and seed maturation protein).[5] These techniques open the door to the identification of thioredoxin-linked protein targets in a wide range of systems using either single or multiple time points.

[33] Thioredoxin and Mechanism of Inflammatory Response

By Takashi Okamoto, Kaori Asamitsu, and Toshifumi Tetsuka

Introduction

Accumulating evidence has implicated various cytokines and cell adhesion molecules in inflammatory responses. For example, in the pathophysiology of rheumatoid arthritis (RA) cytokines including tumor necrosis factor (TNF-α), interleukin 1 (IL-1), IL-6, IL-8, and granulocyte–macrophage colony-stimulating factor (GM-CSF) are abundantly produced in the inflamed rheumatoid synovial fluid.[1–6] Among these cytokines, TNF-α and IL-1 are considered to play crucial roles because they are known to induce IL-6, IL-8, GM-CSF, and themselves.[5–8]

[1] R. Badolato and J. J. Oppenheim, *Semin. Arthritis Rheum.* **26,** 526 (1996).

[2] M. Feldmann, F. M. Brennan, and R. N. Maini, *Ann. Rheum. Dis.* **14,** 397 (1996).

[3] D. G. Remick, L. E. DeForge, J. F. Sullivan, and H. J. Showell, *Immun. Invest.* **21,** 321 (1992).

[4] S. Sakurada, T. Kato, and T. Okamoto, *Int. Immunol.* **8,** 1483 (1996).

[5] S. Yoshida, T. Kato, S. Sakurada, C. Kurono, Y. J. Yang, N. Matsui, T. Soji, and T. Okamoto, *Int. Immunol.* **11,** 151 (1999).

[6] S. Yoshida, T. Katoh, T. Tetsuka, K. Uno, N. Matsui, and T. Okamoto, *J. Immunol.* **163,** 351 (1999).

[7] J. M. Alvaro-Gracia, N. J. Zvaifler, C. B. Brown, K. Kaushansky, and G. S. Firestein, *J. Immunol.* **146,** 3365 (1991).

[8] W. P. Arend and J. M. Dayer, *Arthritis Rheum.* **38,** 151 (1995).

The local production of these inflammatory cytokines also accounts for the systemic manifestations. In addition, expression of certain cellular adhesion molecules including intercellular adhesion molecule 1 (ICAM-1), vascular cell adhesion molecule 1 (VCAM-1), and E-selectin that are induced by TNF-α and IL-1,[4,9] are responsible for accumulation of the inflammatory cells in the inflamed tissue. Thus, the persistent production of proinflammatory cytokines TNF-α and IL-1 has been ascribed to the maintenance and expansion of inflammatory responses both intra- and interarticularly, which eventually lead to synovial proliferation and progressive joint destruction.[1,2]

Interestingly, gene exression of these cytokines and cell adhesion molecules are under the control of a common transcription factor, nuclear factor κB (NF-κB), and TNF-α and IL-1 are known to stimulate inducible expression of these genes through a signal transduction pathway leading to activation of NF-κB.[4–6,10,11] In concordance with these observations, Handel *et al.*[12] have demonstrated the nuclear localization of NF-κB in the synovial lining cells of freshly isolated tissues from RA patients, indicating the activation of NF-κB *in situ*. It was also noted that some of the drugs effective in the treatment of rheumatoid arthritis have been demonstrated to have inhibitory actions on NF-κB activity or its activation cascade.[5,13–17]

NF-κB is an inducible transcription factor present in cells of the primordial mesenchymal cell lineage, including lymphocytes, macrophages, and fibroblasts.[10,17] NF-κB is composed of two subunits, p50 and p65, and exists as an inactive form in the cytoplasm in association with the inhibitory molecule I-κB. On stimulation by proinflammatory cytokines, NF-κB is dissociated from I-κB in conjunction with its phosphorylation followed by proteolytic degradation, and then NF-κB moves into the nucleus, where it activates target genes.[10,17] Although NF-κB is by no means the sole determinant for inducible expression of these genes, molecular genetic studies, such as elucidation of the responsible *cis*-regulatory elements in promoter regions of inducible expression of these genes, have indicated that NF-κB plays an indispensable role.[17]

It has long been assumed that oxidative stress is involved in the pathophisiology of RA. Activated macrophages in the inflamed RA joint were considered to be responsible for the production of radical oxygen intermediates (ROIs).[18] There is accumulating evidence indicating that ROIs are involved in the initial stage of the

[9] L. S. Davis, A. F. Kavanaugh, L. A. Nichols, and P. E. Lipsky, *J. Immunol.* **154**, 3525 (1995).

[10] P. A. Baeuerle, *Biochim. Biophys. Acta* **1072**, 63 (1991).

[11] H. C. Ledebur and T. P. Parks, *J. Biol. Chem.* **270**, 933 (1995).

[12] M. L. Handel, L. B. McMorrow, and E. M. Gravallese, *Arthritis Rheum.* **38**, 1762 (1995).

[13] E. Kopp and S. Ghosh, *Science* **265**, 956 (1994).

[14] N. Auphan, J. A. DiDonato, C. Rosette, A. Helmberg, and M. Karin, *Science* **270**, 286 (1995).

[15] J. P. Yang, J. P. Merin, T. Nakano, T. Kato, Y. Kitade, and T. Okamoto, *FEBS Lett.* **361**, 89 (1995).

[16] M. J. Yin, Y. Yamamoto, and R. B. Gaynor, *Nature (London)* **396**, 77 (1998).

[17] T. Okamoto, S. Sakurada, J. P. Yang, and J. P. Merin, *Curr. Top. Cell Regul.* **35**, 149 (1997).

[18] B. Halliwell, J. R. Hoult, and D. R. Blake, *FASEB J.* **2**, 2867 (1988).

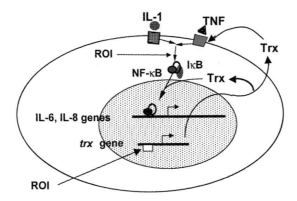

FIG. 1. Involvement of Trx in the signal transduction pathway for NF-κB activation. The first step presumably involves receptor-mediated signal initiation. Although the precise mechanism is not known, exogenous Trx facilitates I-κB phosphorylation and NF-κB nuclear translocation [S. Yoshida, T. Katoh, T. Tetsuka, K. Uno, N. Matsui, and T. Okamoto, *J. Immunol.* **163**, 351 (1999)]. The second step involves the DNA-binding of NF-κB [T. Okamoto, H. Ogiwara, T. Hayashi, A. Mitsui, T. Kawabe, and J. Yodoi, *Int. Immunol.* **4**, 811 (1992); T. Hayashi, Y. Ueno, and T. Okamoto, *J. Biol. Chem.* **268**, 11380 (1993)]. It was shown that Trx comigrates with NF-κB to the nucleus on signaling [S. Sakurada, T. Kato, and T. Okamoto, *Int. Immunol.* **8**, 1483 (1996)]. It is considered, although not yet proven, that radical oxygen species (ROS) are involved as well because the signaling process is effectively blocked by antioxidants. It is thus presumed that intracellular locations for the actions of Trx and ROS are distinct. T. Okamoto, S. Sakurada, J. P. Yang, and J. P. Merin, *Curr. Top. Cell Regul.* **35**, 149 (1997).

NF-κB activation cascade.[4,17,19,20] The increase in intraarticular pressure[21,22] and the subsequent hypoxia–reperfusion injury in synovial tissue was considered to generate ROIs owing to the uncoupling of intracellular redox systems.[21–22] These were evidenced by demonstration of the increase in lipid peroxidation products,[22] the presence of ROI-mediated degradation of hyaluronic acid, and depletion of ascorbate in synovial fluids of patients with RA (reviewed in Mapp *et al.*[22]). More importantly, it has been demonstrated that thioredoxin (Trx) is abundantly produced in the inflamed joint fluid of RA patients.[6]

Trx production is known to be induced by ROIs[23,24] and is considered to be a surrogate marker for the oxidative stress response. Trx appears to stimulate the signal transduction pathway leading to activation of NF-κB at least at two independent steps (Fig. 1).[4,6,17,28,29] First, Trx acts as a costimulatory factor for

[19] K. Tozawa, S. Sakurada, K. Kohri, and T. Okamoto, *Cancer Res.* **55**, 4162 (1995).

[20] R. Schreck, P. Rieber, and P. A. Baeuerle, *EMBO J.* **10**, 2247 (1991).

[21] D. R. Blake, P. Merry, J. Unsworth, B. L. Kidd, J. M. Outhwaite, R. Ballard, C. J. Morris, L. Gray, and J. Lunec, *Lancet* **1**, 289 (1989).

[22] P. I. Mapp, M. C. Grootveld, and D. R. Blake, *Br. Med. Bull.* **51**, 419 (1995).

[23] Y. Sachi, K. Hirota, H. Masutani, K. Toda, T. Okamoto, M. Takigawa, and J. Yodoi, *Immunol. Lett.* **44**, 189 (1995).

[24] Y. Taniguchi, Y. Taniguchi-Ueda, K. Mori, and J. Yodoi, *Nucleic Acids Res.* **24**, 2746 (1996).

the inflammatory response elicited by proinflammatory cytokines when added extracellularly.[6,25–27] For example, Yoshida et al.[6] found that Trx is present in high concentrations in the joint fluid of RA patients and demonstrated that Trx could accelerate the nuclear translocation of NF-κB and accordingly the extents of IL-6 and IL-8 production in response to TNF-α were greatly augmented. In fact, the I-κBα phosphorylation at Ser-32 and its subsequent degradation in response to TNF-α was facilitated by Trx. These findings indicate that the elevated Trx level in the synovial fluid (SF) of RA patients might be involved in the aggravation of rheumatoid inflammation by augmenting the NF-κB activation pathway. Second, Trx augments the DNA-binding activity of NF-κB at least *in vitro*.[17,28–30] In addition, Sakurada et al.[4] demonstrated that Trx transiently comigrated to the nucleus together with NF-κB. Qin et al.[30] revealed by nuclear magnetic resonance (NMR) the formation of a molecular complex between Trx and the DNA-binding loop of NF-κB p50 subunit with two redox-sensitive cysteines on each molecule in close proximity. These observations collectively indicate that Trx acts as a signaling effector for the activation of NF-κB[17,28,29] and that inflammatory responses are controlled by the redox mechanism with NF-κB as a major and direct target.

Here we describe experimental procedures to elucidate the mechanism by which Trx activates NF-κB: immunostaining and electrophoretic mobility shift assay (EMSA).

NF-κB Immunostaining with Synovial Fibroblast Cultures

Preparation of Synovial Cells

Rheumatoid synovial fibroblasts (RSFs) were isolated from fresh synovial tissue biopsy samples from patients with RA.[4–6,31] Tissue samples were minced into small pieces and treated with collagenase/dispase (1 mg/ml; Boehringer, Mannheim, Germany) for 10–20 min at 37°. The cells obtained are cultured in Ham's F12 (GIBCO-BRL, Grand Island, NY) supplemented with 10% (v/v) fetal calf serum (FCS) (Irvine Scientific, Santa Ana, CA), penicillin (100 U/ml),

[25] H. Wakasugi, L. Rimsky, Y. Mahe, A. M. Kamel, D. Fradelizi, T. Tursz, and J. Bertoglio, *Proc. Natl. Acad. Sci. U.S.A.* **84**, 804 (1987).

[26] Y. Tagaya, Y. Maeda, A. Mitsui, N. Kondo, H. Matsui, J. Hamuro, N. Brown, K. Arai, T. Yokota, H. Wakasugi, and J. Yodoi, *EMBO J.* **8**, 757 (1989).

[27] H. Schenk, M. Vogt, W. Droge, and K. Schulze-Osthoff, *J. Immunol.* **156**, 765 (1996).

[28] T. Okamoto, H. Ogiwara, T. Hayashi, A. Mitsui, T. Kawabe, and J. Yodoi, *Int. Immunol.* **4**, 811 (1992).

[29] T. Hayashi, Y. Ueno, and T. Okamoto, *J. Biol. Chem.* **268**, 11380 (1993).

[30] J. Qin, G. M. Clore, W. M. P. Kennedy, J. R. Huth, and A. M. Gronenborn, *Structure* **3**, 289 (1995).

[31] K. Asamitsu, S. Sakurada, T. Kawabe, K. Mashiba, K. Nakagawa, K. Torikai, K. Onozaki, and T. Okamoto, *Arch. Virol.* **143**, 1 (1998).

streptomycin (100 μg/ml), and 0.5 mM mercaptoethanol. The culture medium is changed every 3–5 days and nonadherent lymphoid cells are removed. Adherent cell subcultures are maintained in the same medium and harvested by trypsinization every 7–10 days, before they reach confluency. Usually stable RSF cultures are obtained during the third to thirteenth passage. Whenever we have tested for cytokine induction and the sensitivity for NF-κB nuclear translocation, no significant difference has been noticed among different RSF passages. To characterize the phenotype of adherent cells, the cells are stained with mouse monoclonal antibodies against human HLA-DR, von Willebrand factor, desmin, smooth muscle α-actin, CD1α, CD68, and 5B5. Only 5B5 is positive for RSF, suggesting its fibroblast-like phenotype.

Immunofluorescence

Approximately 1×10^4 RSFs are seeded and cultured in four-well LabTek chamber slides (Nunc, Naperville, IL) and allowed to adhere for 72 hr. To examine the costimulatory effect of Trx, the cells are first incubated with or without Trx (100 ng/ml), a physiological concentration observed in the joint fluids of RA patients, and then stimulated with various concentrations (0, 10, 100, or 250 pg/ml) of TNF-α for 30 min.

Indirect immunofluorescence using the specific anti-p65 antibody is performed basically as follows.

1. Spread 1–5 $\times$ 10^4 RSF cells per well in a 4-well LabTek chamber slide and let them grow for 1–3 days, until they reach 50–80% confluency. Treat the cells with various reagents such as TNF-α and Trx during this period.

2. Wash gently three times with phosphate-buffered saline (PBS) for 3 min each. Try not to dry the cells because inappropriately dried cells often cause high background staining.

3. Remove only the chambers from the slides and leave the bonding material.

4. Fix the RSFs in PBS containing 4% (w/v) paraformaldehyde in PBS for 10 min at room temperature.

5. Wash once with PBS for 3 min.

6. Permealize the cells with 0.5% (w/v) Triton X-100 in PBS for 20 min at room temperature.

7. Incubate the RSFs with the primary antibody, rabbit polyclonal antibody against p65 subunit (e.g., anti-p65 C-20 rabbit antibody; Santa Cruz Biotechnology, Santa Cruz, CA) at a 1 : 100 dilution (depending on the antibody titer), in PBS containing 1% (w/v) bovine serum albumin (BSA) and 0.1% (w/v) NaN$_3$ in a humidified box at 37° for 1 hr. The antibody incubation conditions (antibody dilution, temperature, and duration) are subject to modification depending on the antibody used. When using the 4-well chamber slide, 80–100 μl of the diluted primary antibody per well is sufficient to cover the cells.

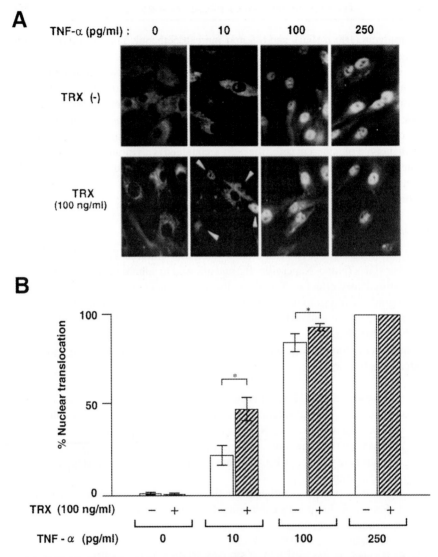

FIG. 2. Trx facilitated the nuclear translocation of NF-κB on stimulation with TNF-α. (A) Nuclear translocation of NF-κB in the cultured RSF. RSF cultures were stimulated with various concentration of TNF-α for 1 hr in the presence or the absence of Trx (100 ng/ml). The cells were immunostained with rabbit polyclonal antibody against p65 subunit of NF-κB and subsequently stained with FITC-conjugated goat anti-rabbit IgG. NF-κB (p65) was localized in the cytoplasm of unstimulated RSF (TNF-α,0), and was translocated to the nucleus by stimulation of TNF-α (10, 100, or 250 pg/ml). Note that addition of TRX accelerated the nuclear translocation of NF-κB when cells were stimulated with lower concentrations of TNF-α ($<$100 pg/ml). (B) Quantitation of the nuclear translocation of NF-κB (p65). Thirty cells in each experimental setting of the results in Fig. 2B were counted from 10 different areas (a total of 300 cells was counted). Significant effects of the addition of Trx in promoting the nuclear translocation of NF-κB were observed at TNF-α concentrations between 10 and 100 pg/ml. *$p < 0.05$ and **$p < 0.01$.

8. Incubate in 0.05% (v/v) Triton-X in PBS three times for 3 min at room temperature for 20 min.

9. Incubate with the secondary antibody, fluorescein isothiocyanate (FITC)-conjugated goat anti-rabbit (whole IgG) antibody (Cappel Organon Teknika, Durham, NC), in PBS [or PBS containing 1% (w/v) BSA and 0.1% (w/v) NaN$_3$] at a suitable dilution, (e.g., 1 : 100) for 20 min at 37°, using a humidified box. To minimize photobleaching of the fluorophore, keep the cells in the dark (e.g., cover with foil) after applying the fluorescent conjugated antibody. It is optional to counterstain the nucleus by adding fluorescent dyes such as diamidinophenylindole (DAPI, 1–2 mg/ml) and Hoechst 33342 (1 mM) to the secondary antibody solution.

10. Wash three times with PBS for 3 min each and remove the bonding material by peeling it away with a thumbnail.

11. Mount a coverslip with a drop of mounting medium such as Fluorsave (Calbiochem, La Jolla, CA) and seal the coverslip with nail polish to prevent drying and movement under the microscope. The sample slides can be stored below −20° for at least 1 month.

Analysis of Results

In Fig. 2, indirect immunostaining using antibody to the p65 NF-κB subunit was carried out with one of the RSF cultures that were stimulated with various concentrations of TNF-α.[6] Nuclear translocation of NF-κB was demonstrated after 30 min of the treatment with TNF-α at concentrations greater than 10 pg/ml. However, when Trx at 100 ng/ml (an average Trx concentration in the synovial fluid of RA patients) was added 1hr before TNF addition, the NF-κB nuclear translocation was significantly promoted (Fig. 2A). As shown in Fig. 2B, the percent nuclear translocation of NF-κB was augmented by addition of Trx in the presence of TNF-α at 10 pg/ml (48.7 ± 7.9 vs. $23.3 \pm 6.3\%, p < 0.01$). Similar experiments were performed with RSF cultures obtained from other RA patients and similar results were obtained.

Electrophoretic Mobility Shift Assay

Preparation of Nuclear and Cytosolic Extracts

Crude nuclear and cytosolic extracts are prepared from semiconfluent cell cultures according to the method of Dignam *et al.*[32] The crude nuclear extract is fractionated on heparin-Sepharose and DEAE-Sepharose columns.[33] Briefly, the

[32] J. P. Dignam, R. M. Lebovitz, and R. G. Roeder, *Nucleic Acids Res.* **11,** 1475 (1983).
[33] T. Okamoto and F. Wong-Staal, *Cell* **47,** 29 (1986).

extract is applied to a heparin-Sepharose column and the 0.4 M KCl eluate is subsequently dialyzed and applied to a DEAE-Sepharose column. The NF-κB activity recovered in the 0.225 M KCl eluate (called "DE 0.225") is then dialyzed against buffer containing 20 mM HEPES–KOH (pH 7.9), 0.1 M KCl, 0.2 mM EDTA, 0.5 mM phenylmethylsulfonyl fluoride (PMSF), and 20% (v/v) glycerol and stored in liquid nitrogen until use. The crude cytosolic extract S100 obtained by the method described above[32] is fractionated to obtain NF-κB/I-κB complex.[29] The S100 fraction is first passed through a DEAE-Sepharose column in the presence of 0.28 M KCl to remove nucleic acids. The flowthrough fraction is dialyzed and loaded onto another DEAE-Sepharose column and protein is eluted by a continuous KCl gradient. The cytosolic NF-κB/I-κB complex will be detected by electrophoretic mobility shift assay (EMSA) in the presence of a detergent mixture containing 1.2% (v/v) Nonidet P-40 (NP-40) and 0.8% (v/v) deoxycholate (DOC). The DEAE fractions containing this complex are further fractionated through phosphocellulose, heparin-Sepharose, and DEAE-Sepharose columns. The 0.2 M KCl eluate from the last DEAE column is concentrated by ultrafiltration (Centriprep; Amicon, MA) and further fractionated by a glycerol gradient centrifugation. To examine the redox status of NF-κB, either in the nucleus or in the cytoplasm, preparation and fractionation of the cytosolic NF-κB is carried out with deaerated and N_2-saturated buffers and no artificial reductant such as dithiothreitol (DTT) or 2-mercaptoethanol (2-ME) is added.

DNA Probe and Competitor Oligonucleotides

Oligodeoxynucleotides for the κB-binding site used for DNA-binding assays as probes or competitors are synthesized.[29] The sequences of the double-stranded oligonucleotides encompass the κB motif from the human immunodeficiency virus type 1 (HIV-1) or interleukin 2 receptor α chain (IL-2Rα) enhancer. DNA sequences of these synthetic oligonucleotides for κB-binding sites and of the mutant are as follows (the underlined sequence represents the NF-κB-binding site).

HIV-1 κB wild type: GATCTA<u>GGGACTTTCCGCTGGGGACTTTCC</u>AG

HIV-1 κB mutant: GATCTActcACTTTCCGCTGctcACTTTCCAG

IL-2Rα κB: GATCTCAG<u>GGGAATCTCCC</u>TCTCCTTTTATGGGCGTAGCG

The double-stranded κB oligonucleotide with overhangs at both ends can be filled in with labeled nucleotide (in the case of HIV-1 κB wild type with [α-^{32}P]dATP) by the Klenow fragment of *Escherichia coli* DNA polymerase I. The radiolabeled deoxynucleotide is subject to change according to the composition of the protruding end sequence of the double-stranded oligonucleotide. After isolation of the probe, typically by passing it through a gel-filtration spin column packed with Sephardex G-25, its specific activity (cpm/μl) is determined by measuring 1 μl by Cerenkov counting in a scintillation counter. A typical binding reaction will contain about 5000 to 20,000 cpm and about 10 to 100 fmol of probe DNA (10 fmol of DNA in a final reaction volume of 10 μl gives a 1 nM DNA concentration).

Reduction–Oxidation (Redox) Reagents and Reactions

Recombinant human Trx or its mutant with a single amino acid replacement of Cys-31 by serine,[28,29] and thioredoxin reductase (TrxR), are purified from rat liver as described.[29,34] Trx and its mutant are purified to homogeneity, incubated with 1 mM DTT for 16 hr at 4°, dialyzed against a buffer containing 20 mM HEPES–KOH (pH 7.9), 60 mM KCl, 0.1 mM EDTA, and 5% (v/v) glycerol, and stored in liquid nitrogen ("fully reduced form"). Some is stored at $-20°$ to be naturally oxidized.

Chemical oxidation of the thiols on NF-κB is performed by treating the nuclear and the cytosolic fractions containing NF-κB with diamide, an inorganic catalyst of oxidation of dithiols [(SH)$_2$] to generate disulfides (–S–S–), on ice for 5 min in the EMSA buffer. Reversion of the disulfides to dithiols is carried out by treatment with various reducing reagents. Reactions with the Trx system, containing various concentrations of human Trx (naturally oxidized form), TrxR, and 1 mM NADPH are performed at 25° for 5 min. These reaction conditions are sufficient to fully reduce the naturally oxidized Trx as examined by insulin reduction assay.[35]

Electrophoretic Mobility Shift Assay

Binding reactions of the DNA probe with protein are performed at 30° for 10 min in a total volume of 10 μl op buffer containing 22 mM HEPES–KOH (pH 7.9), 60 mM KCl, 1 mM MgCl$_2$, 0.02 mM EDTA, 5% (v/v) glycerol, 0.1% (v/v) NP-40, 1.0 μg of poly(dI-dC), and 10–100 fmol (5000–20,000 cpm) of the labeled probe (0.1 to 0.5 ng).[28,29] The DNA–protein complexes are resolved on nondenaturing 6% (w/v) polyacrylamide gels. Electrophoresis is performed with 0.5× TBE buffer (4.5 mM Tris, 4.5 mM boric acid, 0.1 mM EDTA, pH 8.0) at 4° at approximately 30–35 mA for the time required to give good separation of free probe and the protein–DNA complexes (2–2.5 hr for a 20-cm-long gel). On EMSA, NF-κB usually gives two retarded bands.[28,29] Because the lower band is first seen with less amount of the NF-κB fractions (DE 0.225), the upper and lower bands might represent tetramer and dimer complexes, respectively. The intensity of the upper

[34] M. Luthman and A. Holmgren, *Biochemistry* **21**, 6628 (1982).

[35] A. Mitsui, T. Hirakawa, and J. Yodoi, *Biochem. Biophys. Res. Commun.* **186**, 1220 (1992).

band relative to the lower band varies among the different preparations and may reflect variations during the protein extraction procedures.

For the competition experiments, a 50-fold molar excess of unlabeled HIV-1 κB wild type (wt), HIV-1 κB mutant (mut), or IL-2Rα oligonucleotide is preincubated with the protein on ice for 5 min before adding the radioactive probe. Another useful variation to identify proteins present in the protein–DNA complex is to use the antibody. If the protein that forms the complex is recognized by the antibody, addition of the antibody can either block complex formation, or it can form an antibody–protein–DNA ternary complex and thereby specifically result in a further reduction in the mobility of the complex (supershift). Results may be different depending on the antibody epitopes on the DNA-binding protein or on whether the antibody is added before or after the protein binds DNA.

Analysis of the Results

In Fig. 3, the DNA-binding activity of the nuclear and the cytosolic NF-κB was examined on oxidoreductive modifications of sulfhydryls on the cysteine residues by various oxidizing or reducing reagents. In the experiment described in Fig. 3A (left), results with the nuclear NF-κB are demonstrated. When the DE 0.225 fraction containing the nuclear NF-κB was treated with diamide, the DNA-binding activity was totally abolished (compare lanes 1 and 2 in Fig. 3B, left). However, the subsequent addition of 10 mM DTT (lane 3 in Fig. 3B, left) or 2-ME (not shown) fully restored the DNA-binding activity of NF-κB. The effect of reduction by Trx (20 mM, reduced form) was demonstrated with much greater efficiency (lane 5 Fig. 3B, left) but not by the naturally oxidized form of Trx (lane 6, Fig. 3B, left).

We then examined whether TRX requires H^+ proton recruitment to reactivate the oxidized NF-κB (see Fig. 3A, and Fig. 3B, left, lanes 8–15). The reducing activity of Trx appeared to require proton recruitment from NADPH through Trx reductase (TrxR), because omission of one component from the Trx system, either NADPH, TrxR, or Trx itself, did not support restoration of the NF-κB DNA-binding activity whereas a complete Trx system efficiently restored the NF-κB activity in a dose-dependent manner for Trx (lanes 10–12 in Fig. 3B, left). On the other hand, a mutant Trx (a single amino acid replacement of the catalytic active center Cys-31 by serine)[28,29,34] failed to recover the NF-κB activity even with the Trx-regenerating system (lanes 13–15 in Fig. 3B, left). These observations indicate that Trx restores the DNA-binding avtivity of the oxidized NF-κB by facilitating proton transfer to the disulfides on the NF-κB molecule through its catalytic action.

Similarly, the redox regulation of the DNA-binding activity was examined with cytosolic NF-κB. In the experiment described in Fig. 3B (right), the NF-κB/I-κB

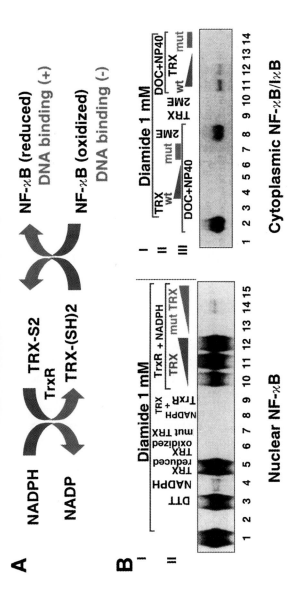

FIG. 3. The Trx system and stimulation of NF-κB DNA binding *in vitro*. (A) The schematic diagram of the actions of Trx and TrxR (Trx reductase) in the presence of NADPH. Reactions of the Trx system, containing human Trx, TrxR, and NADPH are performed at 25° for 5 min. The DNA-binding activity of NF-κB was then evaluated by EMSA *in vitro*. (B) Representative results of the oxidoreductive ("redox") modulation of the DNA-binding activity of NF-κB by Trx. *Left*: Nuclear NF-κB fraction was treated by diamide (I), an oxidizing agent, and activities of various reducing agents were examined for the ability to restore the DNA-binding activity (II). Oxidation was performed with 1 m*M* (final concentration) diamide on ice for 5 min. Treatment with diamide was performed with diamide (1 m*M*) on ice for 5 min. Reduction with various reducing agents including the Trx system was performed at 30° for 5 min. The oxidized NF-κB was treated with 10 m*M* DTT (lane 3), 1 m*M* NADPH (lane 4), 20 µ*M* "fully reduced" Trx (lane 5), 20 µ*M* "naturally oxidized" Trx (lane 6), 20 µ*M* mutant Trx (lane 7), Trx and TrxR (lane 8), or Trx and 1 m*M* NADPH (lane 9). In lanes 10 through 15, various concentrations of Trx (lanes 10–12), TrxR and NADPH (lanes 13–15) were added in addition to the fully competent Trx-reducing system, TrxR and its inactive mutant (mTrx) (lanes 13–15) were added in addition to the fully competent Trx-reducing system, TrxR and NADPH. *Right*: Cytosolic NF-κB/I-κB complex was treated with diamide (I) followed by reducing agents or detergent including 1.2% (v/v) NP-40 and 0.8% (v/v) deoxycholate (DOC) (II or III), and the NF-κB DNA-binding activity was examined by EMSA. Final concentrations of wild-type and mutant Trx, and of 2-ME, were from 1 to 20 µ*M*, 20 µ*M*, and 10 m*M*, respectively.

complex in the cytosolic fraction, as described above, was treated first with detergents (DOC and NP-40) and the effects of oxidoreductive modifications were then examined. Either oxidation or reduction alone could not dissociate the cytosolic NF-κB from the NF-κB/I-κB complex. The cytosolic NF-κB, showing strong binding to the κB DNA probe on detergent treatment (lane 2 in Fig. 3B, right), lost its activity by oxidation with diamide (lane 3 in Fig. 3B, right). Effects of Trx in restoring the DNA-binding activity of the oxidized cytosolic NF-κB were evident only when Trx was added after the treatment with detergent (compare lanes 4–6 and lanes 11–13 in Fig. 3B, right). The full components of the Trx system as well as the reduced form of Trx restored the DNA-binding activity of the oxidized NF-κB. The relative activity of Trx in restoring the DNA-binding activity of NF-κB was about 500-fold greater than 2-ME or DTT because 20 μM Trx had nearly the same activity as 10 mM DTT (Fig. 3B, left, compare lanes 3 and 5) or 10 mM 2-ME (not shown).

[34] Redox State of Cytoplasmic Thioredoxin

By DANIEL RITZ and JON BECKWITH

Introduction

Thioredoxins are ubiquitous small soluble proteins capable of catalyzing thiol–disulfide redox reactions. The two cysteines of the conserved active site of a thioredoxin (W-C-X_1-X_2-C) can be oxidized to form a reversible disulfide bond in the process of reducing exposed disulfides in substrate proteins. The redox potential of thioredoxins is low (-270 mV for *Escherichia coli* thioredoxin 1),[1] allowing them to catalyze thiol–disulfide reductions *in vitro* and *in vivo*.[2] Thioredoxins are kept in the reduced state by the enzyme thioredoxin reductase (TrxB), a dimeric flavoenzyme that specifically catalyzes the NADPH-dependent reduction of thioredoxin.[3]

An important function of thioredoxins is to maintain the activity of a small number of metabolic reductases by reducing their disulfide bonds. These reductases, among them ribonucleotide reductase,[4] become oxidized during their enzymatic action, and must be restored to the active reduced state. Other metabolic enzymes whose active sites become oxidized as part of their reaction cycle and are reduced

[1] G. Krause, J. Lundstrom, J. L. Barea, C. Pueyo de la Cuesta, and A. Holmgren, *J. Biol. Chem.* **266,** 9494 (1991).
[2] A. Holmgren, *Annu. Rev. Biochem.* **54,** 237 (1985).
[3] C. H. Williams, Jr., *FASEB J.* **9,** 1267 (1995).
[4] A. Jordan and P. Reichard, *Annu. Rev. Biochem.* **67,** 71 (1998).

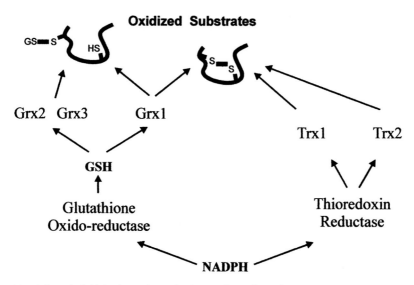

FIG. 1. Branched thiol redox pathways in the cytoplasm of *E. coli*. NADPH provides the reducing equivalents to reduce thioredoxins directly via thioredoxin reductase and glutaredoxins indirectly via glutathione and glutathione reductase.

by thioredoxins include phosphoadenosine-phosphosulfate (PAPS) reductase,[5] and methionine sulfoxide reductase (MSR).[6] In addition to thioredoxins, many cells contain glutaredoxins, which can also carry out the reduction of ribonucleotide reductase and other enzymes. Glutaredoxins are members of a protein class sharing the structural motif of a thioredoxin fold[7,8] but are themselves maintained in the reduced state by glutathione and glutathione reductase (Fig. 1).

In this article we describe how to assess the redox state of thioredoxin *in vivo* directly and indirectly on the basis of its role as a reductase. Our studies are based on work with the thioredoxins of *E. coli,* Trx1 and Trx2. *Escherichia coli* also expresses three glutaredoxins: Grx1, Grx2 and Grx3.

In addition to its function as an important cellular reductant, thioredoxin, in its oxidized state, can play a role as an oxidase. In a thioredoxin reductase mutant (*trxB*), high levels of disulfide bonds accumulate in the cytoplasm in proteins that

[5] C. H. Lillig, A. Prior, J. D. Schwenn, F. Åslund, D. Ritz, A. Vlamis-Gardikas, and A. Holmgren, *J. Biol. Chem.* **274,** 7695 (1999).

[6] W. T. Lowther, N. Brot, H. Weissbach, J. F. Honek, and B. W. Matthews, *Proc. Natl. Acad. Sci. U.S.A.* **97,** 6463 (2000).

[7] H. J. Dyson, A. Holmgren, and P. E. Wright, *Biochemistry* **28,** 7074 (1989).

[8] T. H. Xia, J. H. Bushweller, P. Sodano, M. Billeter, O. Bjørnberg, A. Holmgren, and K. Wuthrich, *Protein Sci.* **1,** 310 (1992).

are usually exported. The formation of these bonds is catalyzed by the oxidized thioredoxins that accumulate.[9] When both of the major reductive pathways are eliminated by combining mutations in the gene for glutathione reductase (*gor*) and the *trxB* gene (Fig. 1), disulfide bonds accumulate to even higher levels.[10]

The redox state of cytoplasmic thioredoxins is important not only because of its role as electron donor to various enzymes but also because the function of a variety of regulatory proteins is controlled by the thiol redox state of the cytoplasm. Oxidation of critical cysteine residues can either activate or inactivate a given protein in physiologically significant reactions. For example, thiol redox reactions are important for oxidative stress responses (reviewed in Carmel-Harel and Storz[11]), the regulation by light of photosynthetic enzymes in the chloroplasts of plants,[12] the regulation of transcription factors such as NF-κB,[13] or in the regulation of apoptosis.[14] In *E. coli,* reduced Trx1, but not the oxidized form, is a subunit of bacteriophage T7 DNA polymerase regulating its activity.[15] The determination of the redox state of thioredoxins can therefore be the first step toward understanding a regulatory redox mechanism or the identification of a new redox pathway.

Determination of Redox State *in Vivo*

To be able to detect the *in vivo* redox state of any protein it is necessary to efficiently block thiol redox exchanges and then separate and detect the oxidized and reduced species. A combination of acid quenching followed by alkylation of free cysteines and detection of reduced (alkylated) and oxidized (nonalkylated) proteins has proven to be a simple and reliable method to determine the redox state of thiol-active proteins *in vitro* and *in vivo*.[16] The method has been used to detect the redox state of disulfide bond catalysts DsbA, DsbB, and DsbC[17,18] and more recently to identify disulfide-bonded intermediates in the thiol redox cascade with

[9] E. J. Stewart, F. Åslund, and J. Beckwith, *EMBO J.* **17,** 5543 (1998).

[10] P. H. Bessette, F. Åslund, J. Beckwith, and G. Georgiou, *Proc. Natl. Acad. Sci. U.S.A.* **96,** 13703 (1999).

[11] O. Carmel-Harel and G. Storz, *Annu. Rev. Microbiol.* **54,** 439 (2000).

[12] B. B. Buchanan, *Arch. Biochem. Biophys.* **288,** 1 (1991).

[13] K. Schulze-Osthoff, H. Schenk, and W. Droge, *Methods Enzymol.* **252,** 253 (1995).

[14] M. Saitoh, H. Nishitoh, M. Fujii, K. Takeda, K. Tobiume, Y. Sawada, M. Kawabata, K. Miyazono, and H. Ichijo, *EMBO J.* **17,** 2596 (1998).

[15] H. E. Huber, S. Tabor, and C. C. Richardson, *J. Biol. Chem.* **262,** 16224 (1987).

[16] T. Zander, N. D. Phadke, and J. C. Bardwell, *Methods Enzymol.* **290,** 59 (1998).

[17] J. C. Joly and J. R. Swartz, *Biochemistry* **36,** 10067 (1997).

[18] T. Kobayashi, S. Kishigami, M. Sone, H. Inokuchi, T. Mogi, and K. Ito, *Proc. Natl. Acad. Sci. U.S.A.* **94,** 11857 (1997).

DsbD at its center.[19] The addition of acid to a growing culture almost instantly stops any thiol–disulfide exchanges by protonating the active thiolate anion and at the same time also denatures cellular proteins. The reactivity of the active site thiols is restored as the denatured and precipitated proteins are solubilized at neutral pH, but all free thiols are immediately alkylated at this step. The technique is powerful and is broadly applicable, but the results can be flawed in the rare instance when thiol groups are poorly accessible to the acid or more frequently the trapping reagent.

The protocol described here has been optimized for the detection of the redox state of *E. coli* thioredoxins in Western blots using anti-thioredoxin antibody (Sigma, St. Louis, MO) after separation of the alkylated and nonalkylated forms by standard sodium dodecyl sulfate–polyacrylamide gel electrophoresis (SDS–PAGE).

Reagents

4-Acetamido-4′-maleimidylstilbene 2,2′-disulfonic acid (AMS; Molecular Probes, Eugene, OR)
Stock solution: 250 mM in dimethyl sulfoxide (DMSO)
Incubation buffer: 200 mM Tris-HCl (pH 7.5), 2% (w/v) SDS
Red loading buffer, 3× (New England BioLabs, Beverly, MA)

Procedure

1. Grow cells to midlog phase (OD$_{600}$ of ~0.5). Prepare microcentrifuge tubes for four samples of each culture: the first sample is dithiothreitol (DTT) reduced only, the second sample is DTT reduced and AMS alkylated, the third sample is not treated, and the fourth is AMS alkylated (the actual experiment).

2. Reduce the cells (DTT-reduced controls only) by adding 50 μl of a 1 M DTT solution and incubate for 20 min on ice.

3. Precipitate 500 μl of cells with 50 μl of 100% (w/v) trichloroacetic acid (TCA) solution.

4. Incubate for 1 hr on ice, and collect precipitated proteins by centrifugation (10 min at 4°).

5. Wash the pellets with 1 ml of acetone. Vortex and incubate for 15 min on ice. Collect by centrifugation as described above.

6. Discard the acetone by inverting the tubes and air dry the pellets for 15 min.

7. Mix incubation buffer with the AMS stock solution to a final concentration of 15 mM AMS. Use just incubation buffer for nontreated controls.

8. Add 20 μl of AMS solution to each sample, resuspend by pipetting, and incubate for at least 60 min at room temperature in the dark.

9. Add 10 μl of 3× red loading buffer and load 5–10 μl on a 15% (w/v) SDS gel for electrophoresis and subsequent detection by Western blot.

---

[19] F. Katzen and J. Beckwith, *Cell* **103,** 769 (2000).

Comments

The use of AMS as alkylating reagent is advantageous because the conjugation of the large negatively charged group of 536 Da with each of the sulfhydryl groups of the active site allows the separation of the reduced form from the oxidized form of thioredoxin on the basis of the difference in molecular mass by SDS–PAGE. In an *E. coli* wild-type strain 70–90% of Trx1 is found in a reduced state, consistent with its *in vivo* role as a disulfide bond reductase (D. Ritz, unpublished results, 2000). The migration of the oxidized protein is identical to that of the nonalkylated protein as Trx1 contains only the two cysteine residues that are part of its active site. Thioredoxin 2, on the other hand, contains an N-terminal extension of 30 amino acids that includes an additional four cysteine residues.[20] In this case, the nonalkylated protein migrates faster than the oxidized protein with an active site disulfide bond. A sample containing oxidized thioreodoxins can be obtained either from a mutant strain lacking thioredoxin reductase activity (*trxB⁻*) or by transiently oxidizing the active site *in vivo* by adding H_2O_2 to a final concentration of 200 μM 30 sec before acid quenching.

Sometimes disulfide bonds are not reducible by adding 100 mM DTT to a culture. Usually this is not the case with disulfide bonds formed involving the active site cysteines, but it has been observed with structural disulfide bonds (e.g., in DsbC). In this case the procedure for the DTT-reduced and AMS-alkylated control might be altered as follows.

1. Precipitate 500 μl of cells with 50 μl of 100% (w/v) TCA solution.
2. Incubate for 1 hr on ice, and collect precipitated proteins by centrifugation (10 min at 4°). Remove carefully the acid solution.
3. Resuspend the pellet in 50 μl of incubation buffer and reduce denatured proteins by adding 5 μl of a 1 M DTT solution. Incubate for 20 min on ice.
4. Precipitate proteins by centrifugation (10 min at 4°). Carry out alkylation as described above.

Consider using other alkylating reagents in case AMS does not seem to react completely with the free thiols in the protein. Biotin-conjugated iodoacetamide (BIAM; Molecular Probes) has been used successfully to label proteins that acquired a disulfide bond after oxidative stress.[21] The molecular mass (454 Da) is comparable to that of AMS, but it does not contain any charges and the reactive group is iodoacetamide instead of maleimide.

---

[20] A. Miranda-Vizuete, A. E. Damdimopoulos, J. Gustafsson, and G. Spyrou, *J. Biol. Chem.* **272,** 30841 (1997).
[21] J. R. Kim, H. W. Yoon, K. S. Kwon, S. R. Lee, and S. G. Rhee, *Anal. Biochem.* **283,** 214 (2000).

Thioredoxin as Disulfide Bond Reductase

Traditionally, assays for thioredoxin activity relied on its ability to reduce disulfide-bonded substrates such as insulin *in vitro*.[22] In these experiments the electron donor is either thioredoxin reductase or a small thiol-reducing agent (e.g., DTT). Even DsbA, which catalyzes protein disulfide formation *in vivo*, is capable of promoting the reduction of the intermolecular disulfide bond in insulin.[23] In the following paragraphs we outline experimental approaches to assay thioredoxin function *in vivo* on the basis of the ability of a recombinant thioredoxin to substitute for the *E. coli* enzyme in mutant strains lacking the gene encoding thioredoxin.

Using different mutant strains the procedure should be useful for studying several different aspects of thiol redox chemistry *in vivo*, including the role of specificity of the interaction between thioredoxin and substrate proteins and the importance of the redox potential for the *in vivo* function of the protein. Moreover, the method can be adapted to look for new proteins with thiol-reducing activity either by screening clone banks for genes that will complement a strain deficient in that activity or by selecting for mutants that acquire a thioredoxin activity. The following paragraphs describe mutant strains that can provide useful phenotypes for genetic selections.

Escherichia coli mutants that lack either Trx1, Trx2, or Grx1 are not impaired in growth. However, a *trxA, grxA* double mutant cannot grow on defined medium and a triple mutant with knockouts in the genes encoding Trx1, Trx2, and Grx1 is not viable (Table I). The growth defect of the latter strain cannot be rescued by the addition of small thiol-reducing molecules, suggesting that the thioredoxins and the glutaredoxin 1 must specifically interact with an essential substrate. Because ribonucleotide reductase is the only known essential substrate of these reductants, it is assumed that the nonviable phenotype of the triple mutant is due to the absence of deoxyribonucleotides. The failure of the *trxA, grxA* double mutant to grow on

TABLE I

REDUNDANCY OF CYTOPLASMIC THIOL REDOX ENZYMES IN *Escherichia coli*[a]

Mutant background	Minimal medium with Met-sulfoxide[b]	Minimal medium	Rich medium
trxA	−	+	+
trxC	+	+	+
grxA	+	+	+
trxA, grxA	−	−	+
trxA, trxC, grxA	−	−	−

[a] The ability of mutant strains lacking one or several genes encoding thiore-doxin/glutaredoxin to grow on the specified medium is shown.

[b] For this experiment a methionine auxotrophic strain (*metE*) was used.

defined medium is due to the inability to reduce PAPS reductase, which is essential for reductive sulfur assimilation.

Finally, methionine sulfoxide reductase, whose function it is to reverse the effects of oxidative damage on free methionine and methionine residues in proteins[24,25] becomes essential in a methionine auxotroph (e.g., *metE*). The mutant cannot grow on defined media with methionine sulfoxide as sole methionine source if the strain has a mutation in *trxA*,[26] suggesting that thioredoxin 1 is the only member of the thiol redox family that can reduce the enzyme.

In summary, these three metabolic enzymes have varying requirements for thiol reductants. These differences may be due either to variations in redox potential or because specific protein–protein interactions are necessary.

The growth defects in the mutant backgrounds can be complemented by introducing the missing gene in *trans* on a plasmid. In addition, when Grx1 is expressed at high levels, it rescues the growth defect of a *trxA, metE* double mutant.[9] Similarly, Trx2, when overexpressed, complements a *trxA, grxA* mutant for growth on minimal medium. Interestingly, a selection for mutants of a *trxA, grxA* strain that restored the ability to grow on minimal medium yielded strains that constitutively induced the OxyR stress response system. In this new mutant strain, Trx2, whose expression is dependent on OxyR, is expressed at high levels.[27]

Surprisingly, when overexpressed, a glutaredoxin-like protein, NrdH, which is homologous to the electron donor for a ribonucleotide reductase in *Lactococcus lactis*,[28,29] complements the growth defect of a triple mutant (*trxA, trxC, grxA*). This result clearly shows that other proteins can serve as electron donors to ribonucleotide reductase *in vivo* even if they cannot perform this function under physiological conditions because of expression levels.[9] These studies confirmed that NrdH, although glutaredoxin-like, is reduced *in vivo* by thioredoxin reductase (confirming the *in vitro* result[30]). In general, a protein with suspected thioredoxin/glutaredoxin activity will complement the growth defect of the triple *trxA, trxC, grxA* mutant if it is expressed at a sufficient level and it can donate the

[22] A. Holmgren, *J. Biol. Chem.* **254,** 9627 (1979).

[23] J. C. Bardwell, K. McGovern, and J. Beckwith, *Cell* **67,** 581 (1991).

[24] W. T. Lowther, N. Brot, H. Weissbach, and B. W. Matthews, *Biochemistry* **39,** 13307 (2000).

[25] J. Moskovitz, M. A. Rahman, J. Strassman, S. O. Yancey, S. R. Kushner, N. Brot, and H. Weissbach, *J. Bacteriol.* **177,** 502 (1995).

[26] M. Russel and P. Model, *Proc. Natl. Acad. Sci. U.S.A.* **82,** 29 (1985).

[27] D. Ritz, H. Patel, B. Doan, M. Zheng, F. Åslund, G. Storz, and J. Beckwith, *J. Biol. Chem.* **275,** 2505 (2000).

[28] A. Jordan, E. Pontis, M. Atta, M. Krook, I. Gibert, J. Barbe, and P. Reichard, *Proc. Natl. Acad. Sci. U.S.A.* **91,** 12892 (1994).

[29] A. Jordan, E. Pontis, F. Åslund, U. Hellman, I. Gibert, and P. Reichard, *J. Biol. Chem.* **271,** 8779 (1996).

[30] A. Jordan, F. Åslund, E. Pontis, P. Reichard, and A. Holmgren, *J. Biol. Chem.* **272,** 18044 (1997).

electrons to ribonucleotide reductase. Under these conditions the unknown thioredoxin/glutaredoxin can be kept in a reduced state by either thioredoxin reductase, glutathione/glutathione reductase, or an externally added small reducing agent.[9]

Trx1, Trx2, and Grx1 exhibit redundant functions as they can all serve as electron donors to ribonucleotide reductase *in vivo*. Even NrdH, a protein that is not expressed at a significant level in *E. coli*, can substitute for the physiological electron donors. Still, Grx3, which reduces ribonucleotide reductase *in vitro*,[31] is not able to function in this role *in vivo*. This suggests that the thioredoxins and glutaredoxin exhibit specificity in their interactions with protein substrates. This specificity becomes even more apparent when considering the specificity of PAPS reductase and MSR (compare in Table I). In *E. coli* thioredoxin 1 is the most active disulfide reductase in the cytoplasm and may provide most of the reductive power for the enzymes we have listed. Trx2 and Grx1 may be used as alternative electron donors under stress conditions. Grx1 may also come into play under conditions in which the thioredocin reductase pathway is blocked, because glutaredoxins use a different regeneration system.

The specificity of the interactions we have described for the thioredoxins and glutaredoxins may be largely due to their differing redox potentials. The redox properties of thioredoxin-like proteins are determined to a significant extent by the nature of the two amino acids between the active site cysteines; thioredoxin variants that have a higher redox potential due to point mutations in that region have been engineered.[32,33] These variants perform less efficiently or not at all *in vivo* as reductants for PAPS reductase and MSR.[34] This finding confirmed that the redox potential is important for the function of thiol redox enzymes, as the more oxidizing variants are less able to reduce substrate proteins in the cytoplasm.

Strains for Detecting Novel or Altered Reductant Activities

The mutant strains that can be used to assay thioredoxin function are shown in Table II.[35,36] FÅ174 grows only if Trx2 is expressed from the plasmid pBAD39-*trxC*. The plasmid must be maintained with ampicillin (100 μg/ml) and 0.5 mM isopropyl-β-D-thiogalactopyranoside (IPTG)[37] and the thioredoxin gene is expressed with 0.2% (w/v) arabinose. In addition to missing the physiological electron donors, this strain is also missing the gene encoding NrdH, because a *trxA*,

[31] F. Åslund, B. Ehn, A. Miranda-Vizuete, C. Pueyo, and A. Holmgren, *Proc. Natl. Acad. Sci. U.S.A.* **91**, 9813 (1994).

[32] M. Huber-Wunderlich and R. Glockshuber, *Fold Des.* **3**, 161 (1998).

[33] E. Mössner, M. Huber-Wunderlich, and R. Glockshuber, *Protein Sci.* **7**, 1233 (1998).

[34] E. Mössner, M. Huber-Wunderlich, A. Rietsch, J. Beckwith, R. Glockshuber, and F. Åslund, *J. Biol. Chem.* **274**, 25254 (1999).

[35] W. A. Prinz, F. Åslund, A. Holmgren, and J. Beckwith, *J. Biol. Chem.* **272**, 15661 (1997).

[36] M. Russel and P. Model, *J. Biol. Chem.* **261**, 14997 (1986).

[37] J. M. Ghigo and J. Beckwith, *J. Bacteriol.* **182**, 116 (2000).

TABLE II
GENOTYPE OF STRAINS USED TO ANALYZE ABILITY OF THIOREDOXIN TO ACT AS REDUCTASE

Assay for:	Strain name	Relevant genotype
Ribonucleotide reductase	FÅ174[a]	$\Delta trxA$ $\Delta trxC$ $grxA$::kan $nrdH$::spc$^+$ pBAD39-$trxC$
PAPS reductase	WP813[b]	$\Delta trxA$ $grxA$::kan
Met-sulfoxide reductase	A313[c]	$trxA$::kan $metE$::Tn10

[a] F. Åslund (unpublished, 1998).
[b] See W. A. Prinz, F. Åslund, A. Holmgren, and J. Beckwith, *J. Biol. Chem.* **272,** 15661 (1997).
[c] See M. Russel and P. Model, *J. Biol. Chem.* **261,** 14997 (1986).

trxC, grxA strain accumulates suppressors that express NrdH at high levels (F. Åslund, personal communication, 1998).

Growth Medium

M63 medium contains the following:

KH$_2$PO$_4$	3 g/liter
K$_2$HPO$_4$	7 g/liter
(NH$_4$)$_2$SO$_4$	2 g/liter
FeSO$_4$	0.5 mg/liter
Mg$_2$SO$_4$	0.12 g/liter
Thiamin	10 mg/liter
Glycerol	2 g/liter
Leucine and isoleucine	50 mg/liter each

Plates contain 1.5% (w/v) noble agar (Sigma) in order to minimize background growth due to contamination of the minimal medium. The plates also contain methionine sulfoxide (50 μg/ml; Sigma) for the MSR assay.

Procedure

The gene to be analyzed should be cloned into a vector in which its expression can be controlled. Suitable vectors contain a p15A origin of replication and have an inducible promoter such as in the arabinose-inducible plasmid pBAD33.[38] The recombinant plasmid is then transformed into the tester strain and plated onto the selective medium in the presence and absence of inducer. For the analysis of

[38] L. M. Guzman, D. Belin, M. J. Carson, and J. Beckwith, *J. Bacteriol.* **177,** 4121 (1995).

the complementation of the growth defect, selection must be carried out for the loss of the original complementing plasmid that allowed the strain to grow. To that end the selective medium contains streptomycin (1.5 mg/ml) but no ampicillin or IPTG. This double selection demands loss of the gene conferring streptomycin sensitivity carried by the pBAD39-*trxC* plasmid and also results in failure of this IPTG-dependent plasmid to replicate. Growth is scored after 48 hr by measuring the size of the colonies.

Thioredoxin as Thiol Oxidant

Whereas, ordinarily, stable disulfide bonds do not form in proteins in the strongly reducing environment of the cytoplasm, mutations that eliminate certain components of the disulfide bond reduction pathways allow these bonds to form. A signal sequenceless variant of the periplasmic alkaline phosphatase, which is inactive in the cytoplasm of wild-type cells, acquires disulfide bonds and becomes enzymatically active in the cytoplasm of a strain lacking thioredoxin reductase. However, the formation of disulfide bonds is dependent on the presence of the oxidized thioredoxins (Trx1 and Trx2) that accumulate in the *trxB* mutant. They actively participate in the formation of disulfides in the cytoplasm and thus function as thiol oxidases.[9]

Strains have been constructed that allow even more efficient disulfide bond formation in the cytoplasm than in the *trxB* strain. A *trxB, gor* double-mutant strain, which lacks the reductive power of both branches of the thiol redox pathways (Fig. 1), grows poorly under aerobic conditions. In contrast to the *trxA, trxC, grxA* mutant, growth can be restored by addition of a reductant such as DTT.[35] Suppressor mutations of the *trxB, gor* mutant that confer growth at normal rates can be obtained after selection in the absence of external reductant. Despite the fact that some source of thiol-reducing power must have been restored to the cytoplasm by these suppressors, the suppressed strains are still able to efficiently catalyze disulfide bond formation in the cytoplasm in proteins such as alkaline phosphatase.[10]

The mutant strains have been successfully used to express foreign proteins containing multiple disulfide bonds such as tissue plasminogen activator (tPA) in the cytoplasm of *E. coli*. The coexpression of a high-copy (pBR322-based) plasmid expressing tPA with a chimeric thioredoxin containing the active site of glutaredoxin 1 or the expression of the signal sequenceless disulfide bond isomerase DsbC increased the yield of enzymatically active protein 15- and 21-fold, respectively.[10] This increase is most likely due to isomerase activity of these proteins enabling incorrectly formed disulfide bonds to be reshuffled and thereby increasing the active portion of the overexpressed protein. As many secreted proteins contain disulfide bonds that are required for proper function, expression in these strain backgrounds obviates the requirement for export to the periplasm, a process that has its own limitations. These approaches may facilitate the production of significant amounts of biotechnologically important proteins.

Acknowledgments

This work was supported by grants from the National Institutes of Health (GM41883 and GM55090). J.B. is an American Cancer Society Research Professor.

[35] Thioredoxin, Thioredoxin Reductase, and Thioredoxin Peroxidase of Malaria Parasite *Plasmodium falciparum*

By Stefan M. Kanzok, Stefan Rahlfs, Katja Becker, and R. Heiner Schirmer

Introduction

The malaria parasite *Plasmodium falciparum* is known to be highly sensitive to reactive oxygen species (ROS)[1,2] although it possesses antioxidant enzymes such as glutathione reductase (GR), superoxide dismutase, and, as described here, the thioredoxin system, which comprises NADPH, thioredoxin reductase (TrxR), and thioredoxin (Trx).[3,4] Substrates of the thioredoxin system studied under the aspect of antioxidative defense in *P. falciparum* are the enzyme thioredoxin peroxidase[5–7] (TPx), and the peptides glutathione disulfide (GSSG) and *S*-nitrosoglutathione (GSNO).[8]

Plasmodium falciparum thioredoxin reductase is a homodimeric (61 kDa per subunit) enzyme with FAD as a prosthetic group, which catalyzes the reaction

---

[1] N. H. Hunt and R. Stocker, *Blood Cells* **16**, 499 (1990).

[2] R. H. Schirmer, T. Schöllhammer, G. Eisenbrand, and R. L. Krauth-Siegel, *Free Radic. Res. Commun.* **3**, 3 (1987).

[3] C. H. Williams, Jr., L. D. Arscott, S. Müller, B. W. Lennon, M. L. Ludwig, P. F. Wang, D. M. Veine, K. Becker, and R. H. Schirmer, *Eur. J. Biochem.* **267**, 6110 (2000).

[4] K. Becker, S. Gromer, R. H. Schirmer, and S. Müller, *Eur. J. Biochem.* **267**, 6118 (2000).

[5] H. Sztajer, B. Gamain, K. D. Aumann, C. Slomianny, K. Becker, R. Brigelius-Flohé, and L. Flohé, *J. Biol. Chem.* **276**, 7397 (2001).

[6] S.-I. Kawazu, N. Tsuji, T. Hatabu, S. Kawai, Y. Matsumoto, and S. Kano, *Mol. Biochem. Parasitol.* **109**, 165 (2000).

[7] S. Rahlfs and K. Becker, *Eur. J. Biochem.* **268**, 1404 (2001).

[8] S. M. Kanzok, R. H. Schirmer, I. Türbachova, R. Iozef, and K. Becker, *J. Biol. Chem.* **275**, 40180 (2000).

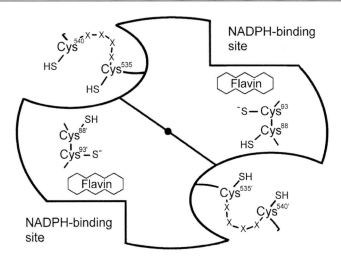

FIG. 1. Dimeric *P. falciparum* thioredoxin reductase viewed along the molecular dyad. The model is based on the structure of human glutathione reductase and physicochemical data of thioredoxin reductases [C. H. Williams, Jr., L. D. Arscott, S. Müller, B. W. Lennon, M. L. Ludwig, P. F. Wang, D. M. Veine, K. Becker, and R. H. Schirmer, *Eur. J. Biochem.* **267**, 6110 (2000)]. The central black circle indicates the molecular 2-fold axis. Looking at the upper right-hand subunit, the electron flow is from the nicotinamide of NADPH via flavin to the cysteine pair C88/C93; hence electrons are transferred to the cysteine pair C535'/C540' of the other subunit while a second NADPH keeps the cysteine-pair C88/C93 reduced. The cysteine pair C535'/540' is now ready for reducing TrxS₂ (not shown).

NADPH + H⁺ + TrxS₂ → NADP⁺ + Trx(SH)₂. TrxR closely resembles glutathione reductase (NADPH + H⁺ + GSSG → NADP⁺ + 2GSH) in sequence and structure. Like dihydrolipoamide dehydrogenase, trypanothione reductase, and mercuric ion reductase,[9] both enzymes belong to a family of flavoenzyme oxidoreductases. *Pf*TrxR contains two catalytic relay units. In each unit electrons are transferred from NADPH via flavin to the cysteine pair C88/C93, subsequently to the C-terminal cysteine pair C535'/C540' of the other subunit, and from there to the substrate thioredoxin disulfide (Fig. 1).[3,10] Unlike mammalian, *Caenorhabditis elegans,* or insect TrxR the C-terminal redox group of *P. falciparum* thioredoxin reductase is a disulfide loop and not a sequential Cys–Cys or Cys–Sec (selenocysteine) pair. (See Refs. 11 and 12, and see [36] in this volume[13].)

[9] L. D. Arscott, S. Gromer, R. H. Schirmer, K. Becker, and C. H. Williams, Jr., *Proc. Natl. Acad. Sci. U.S.A.* **94,** 3621 (1997).
[10] Z. Krnajski, T. W. Gilberger, R. D. Walter, and S. Müller, *J. Biol. Chem.* **275,** 40874 (2000).
[11] S. M. Kanzok, A. Fechner, H. Bauer, J. K. Ulschmid, H.-M. Müller, J. Botella-Munoz, S. Schneuwly, R. H. Schirmer, and K. Becker, *Science* **291,** 643 (2001).
[12] T. Tamura and T. C. Stadtman, *Proc. Natl. Acad. Sci. U.S.A.* **93,** 1006 (1996).
[13] S. Gromer, H. Merkle, R. H. Schirmer, and K. Becker, *Methods Enzymol.* **347,** [36], 2002 (this volume).

The enzyme reduces $TrxS_2$ with high turnover, whereas GSNO is reduced with low efficiency and GSSG is no substrate at all. At high concentrations of $NADP^+$ and $Trx(SH)_2$ the rate of the reverse reaction is significant. In contrast, the closely related enzyme glutathione reductase (which occurs at high concentration in erythrocytic forms of P. falciparum[14]) reduces GSSG with high efficiency,[15] and does not reduce $TrxS_2$, GSNO, and $TPxS_2$.[7,8]

The artificial substrate 5,5′-dithiobis(2-nitrobenzoate) (DTNB, Ellman's reagent) is useful for standard assays of PfTrxR but, even though DTNB is a poor substrate of GR, contaminating GR can interfere with the DTNB reduction assay for TrxR.

The physiologic substrate of Pf TrxR, thioredoxin, is a 12-kDa multifunctional protein of ubiquitous occurrence.[16,17] The active site of P. falciparum Trx is represented by the characteristic -Trp-Cys-Gly-Pro-Cys- sequence, which is conserved in most known thioredoxins. In the oxidized state of the protein the two active site cysteines form a disulfide bridge ($TrxS_2$) that, on reduction by thioredoxin reductase, transforms into a dithiol [$Trx(SH)_2$].

$PfTrx(SH)_2$ significantly reduces GSSG and GSNO at physiologic concentrations. Furthermore, it is an efficient electron donor to thioredoxin peroxidase 1 ($PfTPx-1$).[7] $PfTPx-1$ is a 21.8-kDa per subunit protein that belongs to a widely distributed family of antioxidant enzymes (peroxiredoxins) which function by reducing hydrogen peroxide, alkyl hydroperoxides, and aromatic hydroperoxides to water or the corresponding alcohol [$Trx(SH)_2 + ROOH \rightarrow TrxS_2 + H_2O + ROH$]. As a 2-Cys thioredoxin peroxidase, the enzyme contains a characteristic VCP motif both in its N-terminal and its C-terminal region. Apart from $PfTPx-1$, a putative $PfTPx-2$ has been reported in P. falciparum[7] as well as a 1-Cys peroxiredoxin.[6] A putative glutathione-dependent peroxidase sequence also turned out to encode a thioredoxin-dependent peroxidase.[5] In this article we describe current methods used in studying the P. falciparum thioredoxin system and its substrates (see Scheme 1).

Expression and Purification of Recombinant Thioredoxin Reductase, Thioredoxin 1, and Thioredoxin Peroxidase 1

On the basis of sequences obtained by screening the partially known P. falciparum genome (Sanger and TIGR databases) we have cloned the genes of TPx-1[7]

[14] R. L. Krauth-Siegel, J. G. Müller, F. Lottspeich, and R. H. Schirmer, Eur. J. Biochem. 235, 345 (1996).

[15] C. C. Böhme, L. D. Arscott, K. Becker, R. H. Schirmer, and C. H. Williams, Jr., J. Biol. Chem. 275, 37317 (2000).

[16] A. Holmgren and M. Björnstedt, Methods Enzymol. 252, 199 (1995).

[17] E. S. J. Arnér and A. Holmgren, Eur. J. Biochem. 267, 6102 (2000).

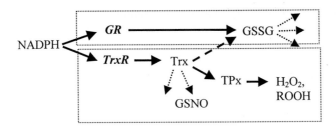

SCHEME 1. Flow of reducing equivalents from NADPH via the glutathione redox system and the thioredoxin system to oxidized substrates.

and Trx-1[8] (Table I). *Plasmodium falciparum* TrxR has been studied as a recombinant protein since 1996.[18] The expression of this protein in *Escherichia coli* was optimized by introducing multiple C/G-enriched silent mutations into the ~70-bp-long 3′ end of the open reading frame.[8] This modified TrxR gene is referred to as *trop*. The protocols for expressing the three genes *trop, trx1,* and *tpx1* are similar. In each case *E. coli* M15 cells are used, the expression vector being pQE30. This vector adds a hexahistidyl (His$_6$) tag to the N-terminus of the recombinant protein, which allows rapid and efficient purification with a Ni-NTA agarose column (Qiagen, Chatsworth, CA).

The isolation procedure results in 10–30 mg of >98% pure protein per liter of culture medium as judged from silver-stained sodium dodecyl sulfate (SDS)–polyacrylamide gels. The gels representing Trx-1, TPx-1, and TrxR show a protein band of 13, 23, and 62 kDa, respectively. It should be noted that, because of the His$_6$ tag, the relative molecular weight of gene products is increased by approximately 1100 Da.

TABLE I
DATABASE ACCESSION NUMBERS OF GENES AND PROTEINS

Protein	Accession no. for:	
	Nucleotide sequence	Amino acid sequence
Pf Trx-1	AF202664	AAF34541
Pf TrxR	X87095	CAA60574
Pf TPx-1	AF225977	AAF67110
Putative *Pf* TPx-2	AF225978	AAK20024
Pf GR	X93462	CAA63747
Human GR	X15722	P00390
Drosophila melanogaster Trx-1	L27072	AAA28937
Drosophila melanogaster TrxR-1	AF301144	AAG25639
Escherichia coli Trx	M54881	AAA24696

All preparations are practically free of disturbing contaminations; specifically, all preparations contained *E. coli* GR activities below the detection limit. In contrast, the untagged proteins are difficult to purify without contaminating disulfide reductases and thioredoxins from *E. coli*. In spite of these obvious advantages it is necessary to keep in mind that the N-terminal tag might influence kinetic and cell biological properties of a recombinant protein. Comparisons with unmodified protein are therefore recommended.

Assays for Reactions and Compounds of Thioredoxin System

Enzymes, Reagents, and Buffers

For proteins, the molar concentrations refer to the subunit. The pH values of buffers were adjusted at 25°.

Buffer T: 100 mM Potassium phosphate, 2 mM EDTA; pH 7.4, 25°

DTNB: 100 mM DTNB in dimethyl sulfoxide (DMSO) (store frozen, keep dark)

NADPH: 4 mM NADPH in buffer (prepare freshly every day, keep on ice)

GSSG: 20 mM in buffer T (store frozen)

GSNO: 10 mM in buffer T (prepare freshly every day, keep on ice)

GSH: 10 mM in buffer T [store at $-20°$; concentrations of GSH and GSSG should be determined before experiments; commercial GSH preparations contain up to 1% (w/w) GSSG]

Dithioerythritol (DTE): 200 mM in water (store frozen)

*Pf*GR: 200 μM in 20 mM potassium phosphate, (pH 7.5), 1 mM EDTA, 2 mM DTE, $\sim$5% (w/v) ammonium sulfate (store at 4°); dilutions are made in buffer T before the experiments

*Pf*TrxR: 200 μM in 20 mM Tris-HCl (pH 7.4), 200 mM KCl (store at 4°)

*Pf*Trx: 200 μM in 20 mM Tris-HCl, pH 7.4, or in buffer T (store at 4°)

*Pf*TPx-1: 200 μM in 50 mM HEPES (pH 7.2), 1 mM DTE (stability crucially depends on the presence of reducing agents)

Unless otherwise stated the assays are carried out in buffer T at 25° in a final volume of 1 ml. Most assays are based on the oxidation of NADPH by thioredoxin reductase, the absorbance coefficient of NADPH at 340 nm being 6.22 mM^{-1} cm^{-1}. The following precautions should be taken.

1. Dithioerythritol (DTE) must not be inadvertantly carried over to the assay mixture because DTE reduces thioredoxin. This reaction prevents NADPH-dependent TrxS$_2$ reduction and reactions coupled to NADPH-dependent thioredoxin reduction.

---

[18] S. Müller, T. W. Gilberger, P. M. Färber, K. Becker, R. H. Schirmer, and R. D. Walter, *Mol. Biochem. Parasitol.* **80**, 215 (1996).

2. Dilute solutions of TrxR ($<1 \; \mu M$) should not be incubated with NADPH in the absence of a disulfide substrate because NADPH destabilizes and inactivates the enzyme. At 3.2 nM TrxR, for instance, 100 μM NADPH leads to 90% inactivation within 15 min at 37° and within 30 min at 25°. At 322 nM (1 enzyme unit/ml), 10% activity is lost under these conditions. The general suggestion for NADPH-dependent enzymes, that each incubation should be started by adding NADPH to the enzyme and determining the NADPH oxidase activity, does not apply for *P. falciparum* thioredoxin reductase and other NADPH-sensitive disulfide reductases. Under most conditions the NADPH oxidase activity is negligible.

Determination of Thioredoxin Reductase Activity

Pf TrxR has a broad substrate specificity. It reduces, for instance, thioredoxins from other species as well as low molecular weight compounds such as DTNB.[3,8] For determining TrxR activity two different assay systems were employed, using Trx and DTNB as disulfide substrates, respectively. In a third assay system thioredoxin reduction by TrxR was coupled to GSSG reduction. This GSSG reduction assay for TrxR and/or Trx (GHOST assay) is also described here.

Thioredoxin Reduction Assay

One milliliter of buffer T at 25° contains 100 μM NADPH and 20.8 μM ($= 2 \times K_m$) TrxS$_2$. The reaction is started with TrxR (0.32–3.2 nM final concentration) and the decrease in absorbance at 340 nm is monitored. One enzyme unit is defined as the consumption of 1 μmol of NADPH per minute. The Michaelis–Menten equation with a K_m of 10.4 μM for TrxS$_2$ is applied for obtaining the turnover number (3100 μmol of TrxS$_2$ reduction per micromole of subunit per minute). Under substrate saturation conditions, 322 nM TrxR supports a TrxS$_2$ turnover (or flux) of 1000 μM min^{-1} (Table II), that is, 322 pmol of TrxR corresponds to a maximal activity of 1 enzyme unit.

GHOST Assay for Thioredoxin Reductase

A drawback of the TrxS$_2$ reduction assay is that it requires relatively large amounts of the substrate TrxS$_2$ in order to obtain a reliable and lasting oxidation rate of NADPH. This problem is circumvented in coupled assays, where reduced thioredoxin is reoxidized by a disulfide that is present in high concentrations in the assay mixture. Examples are the insulin reduction assay and the GHOST assay (or coupled GSSG reduction assay), which is described here. GHOST stands for *g*lutat*h*i*o*ne as *s*ubstrate of *t*hioredoxin.

This assay is based on the fact that glutathione disulfide is not a substrate of any known TrxR. In contrast, *P. falciparum* Trx(SH)$_2$ reduces glutathione disulfide, the rate constant for this reaction being 0.039 μM^{-1} min^{-1} at 25° (Kanzok *et al.*[8]).

TABLE II
KINETIC PARAMETERS OF *Plasmodium falciparum* THIOREDOXIN REDUCTASE
FOR VARIOUS SUBSTRATES[a]

Substrate	K_m (μM)	k_{cat} (min^{-1})	k_{cat}/K_m (μM^{-1} min^{-1})	Turnover at 10 μM substrate (μM min^{-1})
NADPH	2.8	2900	1035	780
TrxS$_2$	10.4	3100	300	490
GSNO	35	9.4	0.3	0.6 (3.4)[b]
GSSG	>20,000	<0.2	<0.00001	<0.000001 (3.9)[b]
DTNB	465	617	1.3	4.2

[a] Enzyme at 322 nM turns over 1000 μM min^{-1} TrxS$_2$, 206 μM min^{-1} DTNB, and 3 μM min^{-1} GSNO, respectively, under substrate saturation conditions (V_{max} conditions). The turnover at 10 μM substrate concentration was estimated by the Michaelis–Menten equation.

[b] Turnover in the presence of 10 μM *Pf* Trx(SH)$_2$, which is oxidized by the substrate nonenzymatically and then rereduced by *Pf* TrxR.

Example. The TrxR-containing sample is added to a cuvette containing 8.5 μM Trx and 100 μM NADPH in buffer T. The Trx-dependent rate of NADPH oxidation is determined to be 6.6 μM/min. When the reaction has come to an end, GSSG (90 μM final concentration) is added. The system resumes NADPH consumption at a rate (of approximately 6.1 μM/min in our example) that now represents a thioredoxin-catalyzed GSSG reduction. When using 1000 μM GSSG, the GHOST assay yields the full TrxR activity for a given Trx concentration (6.6 μM/min in this example).

This assay is well suited for screening samples for TrxR activity in the absence of GR. With respect to sensitivity, 0.3 nM TrxR can be measured routinely. Suspected contamination with GR in one of the additions is quantified by first adding GSSG and then Trx to the system.

It should be noted that the measured activity is not strictly proportional to the TrxR concentration because the steady state concentration of TrxS$_2$ varies both with TrxR activity and GSSG concentration.

5,5′-Dithiobis(2-nitrobenzoate) Reduction Assay for Thioredoxin Reductase

This assay is based on the TrxR-catalyzed reaction NADPH + H$^+$ + DTNB → NADP$^+$ + 2TNB. The enzyme is added to a solution of 3 mM DTNB in buffer T. After initiating the reaction with NADPH (200 μM final concentration), the increase in absorbance at 412 nm is monitored. One enzyme unit is defined as the NADPH-dependent production of 2 μmol of TNB (2-nitro-5-thiobenzoate; $\varepsilon_{412 nm} = 13.6$ mM^{-1} cm^{-1}) per minute.[16] To obtain the turnover number (k_{cat} in

Table II), the Michaelis–Menten equation with a K_m of 465 μM for DTNB is applied. It should be noted that mutants of TrxR that lack the redox center C535/C540 are able to reduce DTNB but not TrxS$_2$. This indicates that—at least in the mutants—DTNB can be reduced also at the flavin-near redox center C88/C93.[19]

Determination of Thioredoxin (*Plasmodium falciparum* Thioredoxin Peroxidase 1)

End-Point Determination Using Thioredoxin Reductase and NADPH

For quantification of *P. falciparum* thioredoxin in its oxidized form (TrxS$_2$), the thioredoxin reductase-catalyzed reaction can be employed (TrxR at 1 U/ml, 100 or 200 μM NADPH); the observed drop in NADPH concentration serves as a measure of the [TrxS$_2$] originally present. On the basis of the molarity, the absorbance at 280 nm is then determined; this yields an $\varepsilon_{280\ nm}$ of 11.7 mM^{-1} cm^{-1} (corresponding to an absorbance of 0.90 of a 1-mg/ml solution at 280 nm) for recombinant tagged *Pf*Trx-1 in oxidized form.

GHOST Assay Using NADPH, Thioredoxin Reductase, and Glutathione Disulfide

The principle is that Trx(SH)$_2$ reduces GSSG at a rate $v = k_2[\text{Trx(SH)}_2][\text{GSSG}]$ (Kanzok *et al.*[8]). Trx is kept in the reduced state by NADPH and TrxR. If TrxR is not rate limiting the observed rate of NADPH oxidation equals v. Thus [Trx(SH)$_2$] = [Trx$_{total}$] can be calculated[8] from the observed rate of

$$[\text{Trx}_{total}] = v(0.039^{-1})[\text{GSSG}]^{-1}\ \mu M$$

Example. To an assay mixture containing TrxR at >1 U/ml, 100 μM GSSG, and 100 μM NADPH at 25° in buffer T, an aliquot of a *Pf*Trx-containing column eluate is added. This leads to an absorbance decrease at 340 nm (0.012/min) that corresponds to a rate v of 2 μM/min. Using the above-described equation the Trx concentration in the assay mixture is calculated as 0.5 μM. This value is confirmed by repeating the assay with purified *Pf*Trx-1 (0.5 μM final concentration) instead of the unknown sample. In a third assay, TrxR is left out in order to measure and subtract glutathione reductase activity, which is negligible in this example.

The GHOST assay for Trx has the following advantages: (1) in the absence of major glutathione reductase activity it is robust, sensitive, and specific; (2) it accounts for both reduced and oxidized Trx; and (3) it measures biologically functional thioredoxin. It should be noted that the k_2 value depends on the source of the thioredoxin. The k_2 of PfTrx-1 is six times higher than the k_2 value of *E. coli* Trx.[8]

[19] M. Schirmer, M. Scheiwein, S. Gromer, K. Becker, and R. H. Schirmer, *Flavins Flavoprot.* **13**, 857 (1999).

Determination of Redox Potential of Thioredoxin 1

The reversibility of the reaction $NADPH + TrxS_2 + H^+ \rightleftharpoons Trx(SH)_2 + NADP^+$ can be employed for determining the equilibrium constant $K = [NADP^+][Trx(SH)_2]/([NADPH][TrxS_2])$. The absorbance change at 340 nm serves as a measure of NADPH oxidation with concomitant production of an equimolar amount of $Trx(SH)_2$.[16]

Example. In the presence of 50 nM TrxR, the initial concentrations of NADPH (between 20 and 150 μM), NADP$^+$ (1000 to 2000 μM), and TrxS$_2$ (10 to 30 μM) are varied systematically. These experiments yield an equilibrium constant K of 110 ± 10 at pH 7.4 and 25°, which corresponds to a difference in redox potential of 60.2 ± 2 mV. As the midpoint potential for the NADPH/NADP$^+$ system at pH 7.4, 25°, and 0.26 μ ionic strength can be estimated to be -330 mV, the *P. falciparum* Trx(SH)$_2$/TrxS$_2$ redox pair has a redox potential of -270 mV, which agrees well with the values reported for *E. coli* thioredoxin.[20,21]

Reduction of Peroxides by Thioredoxin/Thioredoxin Peroxidase System

Plasmodium falciparum thioredoxin peroxidase 1 efficiently catalyzes the reduction of H$_2$O$_2$, *tert*-butyl hydroperoxide (*t*BOOH), or cumene hydroperoxide in the presence of NADPH, TrxR, and *Pf*Trx-1.

$$Trx(SH)_2 + ROOH \rightarrow TrxS_2 + H_2O + ROH$$

As described for many other peroxiredoxins including glutathione peroxidases and tryparedoxin peroxidase, the reaction catalyzed by TPx-1 does not follow typical saturation kinetics at high substrate concentrations.[5,7,22] Under these conditions, definite V_{max} and K_m values cannot be determined. Furthermore, when approximating K_m values, different data are obtained for a given peroxide substrate when varying the Trx-1 concentration.

Example. TPx-1 assays can be carried out at 25° in 1 ml of assay mixture consisting of 50 mM HEPES (pH 7.2), 10 μM *Pf*Trx-1, 100 μM NADPH, *Pf*TrxR at $\sim$50 mU/ml (care should be taken that the *Pf*TrxR concentration is not rate limiting), and 200 μM H$_2$O$_2$, *t*BOOH, or cumene hydroperoxide. The assay is started with *Pf*TPx-1; NADPH consumption is then monitored spectrophotometrically at 340 nm. In this coupled assay system constantly high concentrations of reduced thioredoxin are maintained by the NADPH/TrxR system. When determining an apparent K_m in a Lineweaver–Burk plot in the presence of 4 to 200 μM H$_2$O$_2$, a K_m value of 10 μM is determined in the presence of 5 μM *Pf*Trx, a value of 17 μM is determined in the presence of 10 μM *Pf*Trx, and 41 μM in the presence

---

[20] H. Follmann and I. Häberlein, *Biofactors* **5,** 147 (1995).

[21] B. W. Lennon and C. H. Williams, Jr., *Biochemistry* **35,** 4704 (1996).

[22] E. Nogoceke, D. U. Gommel, M. Kiess, H. M. Kalisz, and L. Flohé, *Biol. Chem.* **378,** 827 (1997).

of 20 μM *Pf* Trx. The specific activities in the presence of 200 μM substrate and 10 μM *Pf* Trx are 3.1 U/mg (H_2O_2), 2.6 U/mg (*t*BOOH), and 1.9 U/mg (cumene peroxide), which corresponds to k_{cat} values of 67, 56, and 41 min^{-1}, respectively.

Nitrosoglutathione Reduction by Thioredoxin Reductase and Thioredoxin Reductase/Thioredoxin System

Nitrosoglutathione (GSNO) represents an important transport form of nitric oxide (NO) in biological systems, and NO is likely to be involved in the pathophysiology of cerebral malaria. As a scavenger of toxic radicals, GSNO is 100 times more effective than GSH.[23] Nitrosoglutathione has furthermore been shown to be an inhibitor of human glutathione reductase[24] and a substrate of the mammalian selenocysteine-containing thioredoxin system.[25]

GSNO was found to be a substrate of *Pf* TrxR but the k_{cat} was only 9.4 min^{-1} (Table II). When using *P. falciparum* glutathione reductase instead of TrxR, no NADPH-dependent GSNO reduction was observed. However, GSNO reacts with *P. falciparum* thioredoxin in a second-order reaction ($v = k_1[Trx(SH)_2][GSNO]$), the value for k_1 being 0.034 μM^{-1} min^{-1} at 25°.[8] This means that GSNO can be efficiently reduced by the thioredoxin system: It reacts with $Trx(SH)_2$ and the resulting $TrxS_2$ is rereduced by TrxR.[25]

Example of Nitrosoglutathione Reduction by Plasmodium falciparum Thioredoxin Reductase in Absence of Thioredoxin. *Pf* TrxR subunit (3 U/ml), (1 μM), in 50 mM Tris-HCl, 1 mM EDTA (pH 7.5), is incubated at 25° with various concentrations of GSNO ranging from 20 to 200 μM. The enzymatic reaction is started with 100 μM NADPH and is followed by an absorbance decrease at 340 nm. From these data the K_m (35 μM) and the specific activity (0.15 U/mg enzyme) for GSNO reduction are determined (see also Nikitovic and Holmgren[25]).

Example of Nitrosoglutathione Reduction by Thioredoxin Reductase/ Thioredoxin System. In the presence of thioredoxin GSNO is reduced more efficiently. In a typical assay (40 μM GSNO, 10 μM Trx, 16.4 nM TrxR, and 100 μM NADPH) the turnover of GSNO as measured by NADPH oxidation is 10 μM/min. The contribution of direct GSNO reduction by TrxR is only 0.08 μM under these conditions and thus negligible. Indeed, for the observed rate of 10 μM/min approximately 2000 nM enzyme would be needed in the absence of thioredoxin.

Calculations. Can we estimate the steady state concentrations of $TrxS_2$ and $Trx(SH)_2$ under assay conditions? One approach that neglects the possible inhibition of TrxR by GSNO is as follows. In the above-described example, $TrxS_2$ is

[23] C. C. Chiueh and P. Rauhala, *Free Radic. Res.* **31**, 641 (1999).

[24] K. Becker, S. N. Savvides, M. A. Keese, R. H. Schirmer, and P. A. Karplus, *Nat. Struct. Biol.* **5**, 267 (1998).

[25] D. Nikitovic and A. Holmgren, *J. Biol. Chem.* **271**, 19180 (1996).

reduced at a rate of 10 μM/min by TrxR; the Michaelis–Menten equation using $v = 10\ \mu M$/min, $V_{max} = 51\ \mu M$/min, and $K_m = 10.4\ \mu M$ yields a substrate concentration of $TrxS_2 = 2.6\ \mu M$. Accordingly, $[Trx(SH)_2]$ would be $10.0\ \mu M - 2.6\ \mu M = 7.4\ \mu M$. Looking from the other side, $TrxS_2$ is produced in the reaction $v = k_1[GSNO][Trx(SH)_2]$. With $v = 10\ \mu M$/min, $k_1 = 0.034\ \mu M^{-1}\ min^{-1}$, and 40 μM GSNO, the resulting $Trx(SH)_2$ concentration is 7.4 μM and the $TrxS_2$ concentration is, by difference, 2.6 μM. The self-consistency of these data suggests that our assumptions concerning the mechanism of GSNO reduction by the thioredoxin system are valid but this must be confirmed by further experiments.

The comparison given in Table II (3.4 vs. 0.6 μM/min) clearly shows that most GSNO is reduced by $Trx(SH)_2$. In the presence of 10 μM PfTrx-1, an apparent K_m of 15.2 μM was determined for GSNO. These data suggest that the *P. falciparum* thioredoxin system efficiently turns over *S*-nitrosoglutathione *in vitro* and probably also *in vivo* (see also Nikitovic and Holmgren[25]).

Glutathione Disulfide Reduction by Plasmodium falciparum Thioredoxin System

We have demonstrated that GSSG can be significantly reduced by the thioredoxin system even under quasiphysiologic conditions.[8]

Consider the reaction $Trx(SH)_2 + GSSG \rightleftharpoons TrxS_2 + 2GSH$, the rate constant for the forward reaction being k_2 and for the back reaction k_3. Applying the law of mass action yields

$$k_2/k_3 = K = [TrxS_2][GSH]^2/[Trx(SH)_2][GSSG]$$

The redox potentials at pH 7.4 and 25° are approximately -0.255 V for the GSSG/2GSH couple[26] and -0.270 V for $TrxS_2/Trx(SH)_2$.[8] Applying the Nernst equation, K is calculated to be 3.2 M. Using this value and $k_2 = 0.039\ \mu M^{-1}\ min^{-1}$ (Kanzok *et al.*[8]) we obtain $k_3 = k_2/K = 1.2 \times 10^{-8}\ (\mu M)^{-2}\ min^{-1}$.

First consider the forward reaction. With quasiphysiologic concentrations of 10 μM $Trx(SH)_2$ and 100 μM GSSG, the observed rate is $v = k_2[Trx(SH)_2][GSSG] = 0.039 \times 10 \times 100\ \mu M$/min $= 39\ \mu M$/min. Thus the rate of the forward reaction is significant *in vitro* and possibly also *in vivo* as shown in the following example.

Example. Under assay conditions at pH 7.4 and 25°, using 3.2 nM TrxR, 100 μM NADPH, 10 μM GSSG, and 21 μM $TrxS_2$, the turnover rate for GSSG reduction is measured to be 4.7 μM/min. For this turnover 6 nM GR would be required, that is, more GR than TrxR! These extravagant conditions prevail in cells where GR and TrxR have been largely inhibited by drugs such as carmustine.[27]

[26] C. H. Williams, Jr., *in* "Chemistry and Biochemistry of Flavoenzymes" (F. Müller, ed.), Vol. III, p. 121. CRC Press, Boca Raton, Florida, 1992.

[27] K. Becker, S. Kanzok, R. Iozef, I. Türbachova, and R. H. Schirmer, *Flavins Flavoprot.* **13,** 853 (1999).

Under other conditions, a high GSSG reduction rate ($>200 \ \mu M \ \text{min}^{-1}$ can be maintained in the absence of glutathione reductase by the thioredoxin system (1 μM TrxR and 20 μM Trx) when the concentration of GSSG is >1 mM.[8,11]

It is conceivable that even the back reaction is relevant, namely under the following conditions.[28] For instance, in a hitherto dormant cell that becomes reactivated, the starting conditions would be as follows: 25 μM TrxS$_2$, 7.5 mM glutathione, sufficient GR activity to keep glutathione in its reduced form, and no TrxR activity. The initial rate of TrxS$_2$ reduction would be

$$v = k_3[\text{TrxS}_2][\text{GSH}]^2 = 1.2 \times 10^{-8} \times 15 \times (7.5 \times 10^3)^2 = 10 \ \mu M/\text{min}$$

that is, most thioredoxin would be reduced within a few minutes after an external stimulus had roused the cell, and this Trx(SH)$_2$ could act as a metabolism-activating transcription factor or another signal.

Concluding Remarks

The characterization of *Pf*Trx, the natural substrate of *P. falciparum* thioredoxin reductase, contributes to our understanding of the redox metabolism in the parasites. This is also of practical importance because the proteins of the antioxidant defense system in *Plasmodium* represent promising targets for antimalarial drug design.[4,15,27,29]

GSSG reduction by the thioredoxin system may play a role in a variety of physiologic, pharmacologic, and pathologic conditions. This applies probably not only to different life stages of malaria parasites but also to cells and tissues from other organisms (like lymphocyte subtypes in acquired immunodeficiency syndrom (AIDS) patients and GR-deficient yeast mutants) even when the rate constant of the reaction between reduced thioredoxin and glutathione disulfide is not as high as in the case of *P. falciparum* thioredoxin.[8,25]

Note added in proof. The sketch shown in Fig. 1 is consistent with the recently published structure of the Sec498Cys mutant of rat TrxR.[30]

Acknowledgments

Our work is supported by the Deutsche Forschungsgemeinschaft [SFB 544 (R.H.S.) and SFB 535 (K.B.), and Be1540/4-1]. We thank Holger Bauer for providing Fig. 1 and Heiko Merkle for helpful discussions.

[28] R. C. Fahey, S. Brody, and S. D. Mikolajczyk, *J. Bacteriol.* **212,** 144 (1975).

[29] R. H. Schirmer, J. G. Müller, and R. L. Krauth-Siegel, *Angew. Chem. Ed. Engl.* **34,** 141 (1995).

[30] T. Sandalova, L. Zhong, Y. Lindqvist, A. Holmgren, and G. Schneider, *Proc. Natl. Acad. Sci. U.S.A.* **98,** 9533 (2001).

[36] Human Placenta Thioredoxin Reductase: Preparation and Inhibitor Studies

By Stephan Gromer, Heiko Merkle, R. Heiner Schirmer, and Katja Becker

Introduction

Mammalian thioredoxin reductases [EC 1.6.4.5; NADPH + H$^+$ + thioredoxin-S$_2$ ⇌ NADP$^+$ + thioredoxin-(SH)$_2$] is a dimeric (M$_r$ 55,000 per subunit) flavoenzyme. As a pyridine nucleotide–disulfide oxidoreductase it is closely related to glutathione reductase.[1-3] All members of this enzyme family have FAD and an adjacent redox active disulfide, but mammalian thioredoxin reductases contain, as a common feature, an additional redox-active center in the form of a penultimate Cys–Sec sequence.[4,5] This additional redox center is located on a flexible C-terminal extension, which explains another characteristic of large thioredoxin reductases—their unusual broad substrate specificity.[6] This aspect might explain a large variety of intracellular processes, such as electron delivery to ribonucleotide reductase but also extracellular processes, such as cytokine effects, that are based on the activities of thioredoxin reductase (TrxR) and its major substrate thioredoxin (Trx).[7-9]

Many studies on TrxR have been carried out with small TrxRs of bacterial origin. These enzymes, however, differ fundamentally from the large TrxRs[2,3] in size, structure, substrate spectrum, and catalytic mechanism. Furthermore, because of the broad substrate spectrum of large TrxRs, many in vitro findings may lead to functional misinterpretations, particularly because in vivo data are still rare. It is our goal to identify probes that selectively modify thioredoxin reductase and that can be employed as enzyme inhibitors both in vitro and in vivo. From the enzymologist's point of view many different inhibitors with clear cut effects on the intermediate

---

[1] A. Holmgren, J. Biol. Chem. 252, 4600 (1977).

[2] L. D. Arscott, S. Gromer, R. H. Schirmer, K. Becker, and C. H. Williams, Jr., Proc. Natl. Acad. Sci. U.S.A. 94, 3621 (1997).

[3] C. H. Williams, L. D. Arscott, S. Müller, B. W. Lennon, M. L. Ludwig, P. F. Wang, D. M. Veine, K. Becker, and R. H. Schirmer, Eur. J. Biochem. 267, 6110 (2000).

[4] V. N. Gladyshev, K. T. Jeang, and T. C. Stadtman, Proc. Natl. Acad. Sci. U.S.A. 93, 6146 (1996).

[5] S. Gromer, R. H. Schirmer, and K. Becker, FEBS Lett. 412, 318 (1997).

[6] S. Gromer, J. Wissing, D. Behne, K. Ashman, R. H. Schirmer, L. Flohé, and K. Becker, Biochem. J. 332, 591 (1998).

[7] S. Gromer, R. H. Schirmer, and K. Becker, Redox Rep. 4, 221 (1999).

[8] D. Mustacich and G. Powis, Biochem. J. 346, 1 (2000).

[9] E. S. J. Arnér and A. Holmgren, Eur. J. Biochem. 267, 6102 (2000).

states of the catalytic cycle are desirable. Physiological studies would benefit from highly selective, well-controllable inhibitors. Bearing in mind the medical implications of the enzyme (e.g., in tumor therapy[10]) such inhibitors should provide us with pharmacological lead substances—a prerequisite for rational drug design.

To assess the effects of potential inhibitors on enzyme activity a highly purified enzyme must be available not only for kinetic and X-ray crystallography studies but also as a reference for experiments on cell cultures. Enzyme preparations from native tissue are still necessary as the novel recombinant techniques for selenoenzymes[11,12] such as TrxR are promising but do not yet produce satisfying material.

Assay Procedures

Several different assays for the determination of thioredoxin reductase activity have been developed.[13] They differ in specificity, cost, and complexity. Here we present the most frequently applied spectrophotometric assays and comment on their specific properties. All assays are carried out at $25°$.

5,5′-Dithiobis(2-Nitrobenzoic Acid) Reduction Assay

The 5,5′-dithiobis(2-nitrobenzoic acid) (DTNB) reduction assay[13] is based on the broad substrate specificity of large thioredoxin reductases, which are capable of reducing DTNB (Ellman's reagent) to form two molecules of 2-nitro-5-thiobenzoate anion (TNB).

$DTNB^{2-}$ TNB^{2-} (dark yellow)

Because the thiolate of TNB is mainly responsible for the high absorbance at 412 nm the pH is of importance and should be above pH 7.0. Most publications utilize an $\varepsilon_{TNB, 412\,nm}$ of 13.6 mM^{-1} cm^{-1} for TNB. This value is not quite correct

[10] K. Becker, S. Gromer, R. H. Schirmer, and S. Müller, *Eur. J. Biochem.* **267**, 6118 (2000).
[11] E. S. J. Arnér, H. Sarioglu, F. Lottspeich, A. Holmgren, and A. Böck, *J. Mol. Biol.* **292**, 1003 (1999).
[12] S. Bar-Noy, S. N. Gorlatov, and T. C. Stadtman, *Free Radic. Biol. Med.* **30**, 51 (2001).
[13] E. S. J. Arnér, L. Zhong, and A. Holmgren, *Methods Enzymol.* **300**, 226 (1999).

(approximately ±1) under the conditions applied[14] but should be used in order to keep results comparable.

Unlike earlier protocols, we omit bovine serum albumin (BSA) in the buffer because BSA reacts with DTNB and might interfere with enzyme activity* and inhibitors. Although dilute human TrxR is much more stable in the presence of NADPH than other members of the disulfide reductase family, it should be noted that the loss of enzyme activity is approximately 50% after 1 hr when (pure) human TrxR is incubated at 1 nM concentration in assay buffer with 200 μM NADPH at 25°. Inhibitors that are not easily water soluble should be dissolved in dimethyl sulfoxide (DMSO) if possible. DMSO up to 10% (v/v) in the assay mixture does not alter the enzyme activity and is easier and more reliable to pipette than, for instance, ethanol.

Reagents

T-buffer: 100 mM Potassium phosphate, 2 mM EDTA; pH 7.4
DTNB: 100 mM DTNB in DMSO (store frozen, keep dark)
NADPH: 4 mM NADPH in T-buffer (prepare freshly every day, keep on ice)

In a total assay volume of 1 ml, 30 μl of DTNB (to 3 mM) is mixed with T-buffer, enzyme sample ($\sim$1–2 nM), and if applicable inhibitor solution. The enzymatic reaction is initiated by adding 50 μl of NADPH (to 200 μM). As the spectrophotometric signal of the reaction is usually not linear, particularly not at low enzyme concentrations, the TNB-dependent increase in absorbance at 412 nm should be monitored during the "most linear" phase within the first 3 min. For inhibition studies it is also helpful to state the time interval for which the ΔA min^{-1} is calculated (e.g., the range between 20 and 80 sec) in order to increase the reproducibility of the results. As 1 unit of enzymatic activity is defined as the consumption of 1 μmol of NADPH per minute (meaning the production of two TNB per NADPH), activity in the assay mixture (V $=$ 1 ml; d $=$ 1 cm) is calculated as follows:

$$\text{Volume activity} = \frac{\Delta A_{412\,\text{nm}}}{1\,\text{min} \cdot 2 \cdot 0.0136\,\mu M^{-1} \cdot 1}\left[\frac{\mu M}{\text{min}}\right] \quad \text{or} \quad \left[\frac{\text{mU}}{\text{ml}}\right] \quad (1)$$

The DTNB reduction assay is inexpensive, simple, and (because of the high ε_{412} value of TNB and the stoichiometry of the reaction) sensitive; it lacks, however, specificity. For instance, glutathione reductase can also reduce DTNB to give

---

* Increase up to 25%.

[14] P. W. Riddles, R. L. Blakeley, and B. Zerner, *Methods Enzymol.* **91**, 49 (1983).

TNB[15] (although with a low K_{cat} of $\sim$500 min^{-1}). Also, free SH groups (which are often present in crude extracts in relevant concentrations) react nonspecifically with DTNB to form mixed disulfides and TNB.[14] Furthermore, studies with non-selenium-containing large TrxRs[16] and selective proteolytic digests[6,17] of human TrxR indicate that DTNB may, unlike thioredoxin, be reduced not only at the C-terminal Cys–Sec redox center but also at the redox-active disulfide of the protein (Cys-57 and Cys-62).

Thioredoxin Reduction Assay

This is the most specific but also the most expensive assay system.

$$\text{Trx-S}_2 + \text{NADPH} + \text{H}^+ \rightarrow \text{Trx-(SH)}_2 + \text{NADP}^+$$

Native human thioredoxin (hTrx) causes technical difficulties due to dimer formation and because hTrx mutants lacking dimer formation are not yet affordable in the required amounts (IMCO, Stockholm, Sweden), *Escherichia coli* thioredoxin is often used for the reaction.[18] Thioredoxins of other species such as *Plasmodium falciparum* Trx exhibit low K_m values and protocols for rapid preparation are available. They should be considered as alternatives.[19,20]

Reagents

T-buffer: 100 mM Potassium phosphate, 2 mM EDTA; pH 7.4
Trx: 5 mM oxidized *E. coli* thioredoxin [$\varepsilon_{280\,nm}$ (*E. coli* Trx-S$_2$) = 13.7 mM^{-1} cm^{-1}] in T-buffer (store frozen, keep on ice)
NADPH: 4 mM NADPH in T-buffer (prepare freshly every day, keep on ice)

In a total assay volume of 1 ml, 25 μl of NADPH (to 100 μM) is mixed with T-buffer and enzyme sample (approximately 3 nM). The enzymatic reaction is initiated by adding 25 μl of (*E. coli*) Trx stock solution (to 125 μM) and the decrease in absorbance at 340 nm is monitored. One unit of enzyme activity is defined as the consumption of 1 μmol of NADPH per minute (V = 1 ml; d = 1 cm).

$$\text{Volume activity} = \frac{\Delta A_{340\,nm}}{1 \text{ min} \cdot 0.00622 \, \mu M^{-1} \cdot 1} \left[\frac{\mu M}{\text{min}}\right] \quad \text{or} \quad \left[\frac{\text{mU}}{\text{ml}}\right] \quad (2)$$

[15] K. K. Wong, M. A. Vanoni, and J. S. Blanchard, *Biochemistry* **27**, 7091 (1988).
[16] T. W. Gilberger, B. Bergmann, R. D. Walter, and S. Müller, *FEBS Lett.* **425**, 407 (1998).
[17] L. Zhong, E. S. J. Arnér, J. Ljung, F. Åslund, and A. Holmgren, *J. Biol. Chem.* **273**, 8581 (1998).
[18] S. B. Mulrooney, *Protein Expr. Purif.* **9**, 372 (1997).

As an alternative to *E. coli* Trx (K_m with human TrxR $= 25\ \mu M$), recombinant thioredoxin of the malarial parasite *P. falciparum*[19] (K_m with hTrxR $= 2\ \mu M$) can be used for assaying hTrxR activity. *Pf*Trx does not show dimer formation disturbing the reaction, it can be easily overexpressed in *E. coli,* purified by affinity chromatography, and—because of its low K_m with human TrxR—20 μM *Pf*Trx produces a stable spectrophotometric signal under substrate saturation over minutes.

GHOST Assay*

Despite its broad substrate specificity thioredoxin reductase is incapable of reducing glutathione disulfide (GSSG). Its product thioredoxin, in the reduced state, however, easily reduces GSSG to form reduced glutathione (GSH). Apart from its potential physiological importance, this novel assay system is useful to measure thioredoxin reductase activity in samples in which, for example, other assay constituents such as inhibitors interfere with the TNB determination at 412 nm. Furthermore, this assay yields a better linear signal than the DTNB reduction assay and it requires only catalytic concentrations of Trx.[19]

$$\text{Trx-S}_2 + \text{NADPH} + \text{H}^+ \rightarrow \text{Trx-(SH)}_2 + \text{NADP}^+$$
$$\text{GSSG} + \text{Trx-(SH)}_2 \rightarrow \text{Trx-S}_2 + 2\text{GSH}$$

$$\text{GSSG} + \text{NADPH} + \text{H}^+ \rightarrow 2\text{GSH} + \text{NADP}^+$$

As the overall reaction is also catalyzed by glutathione reductase, the presence of this enzyme can complicate the assay. Thus the functional absence of this enzyme is a prerequisite of this assay.

Reagents

T-buffer: 100 m*M* Potassium phosphate, 2 m*M* EDTA; pH 7.4
Trx: 1 m*M E. coli* thioredoxin in T-buffer (store frozen, keep on ice)
NADPH: 4 m*M* NADPH in T-buffer (prepare freshly every day, keep on ice)
GSSG: 20 m*M* Glutathione disulfide in T-buffer (store frozen, keep on ice)

Procedure. In a total assay volume of 1 ml, 25 μl of NADPH (to 100 μM) is mixed with T-buffer, 30 μl of Trx stock solution (to 30 μM), and enzyme sample (approximately 1–3 n*M*). The enzymatic reaction is initiated by adding 50 μl of GSSG (to 1 m*M*). If the assay mixture should still contain a small amount

---

* GHOST, Glutat*hi*one as *s*ubstrate of *t*hioredoxin.
[19] S. M. Kanzok, R. H. Schirmer, I. Turbachova, R. Iozef, and K. Becker, *J. Biol. Chem.* **275,** 40180 (2000).
[20] S. M. Kanzok, A. Fechner, H. Bauer, J. K. Ulschmid, H. M. Müller, J. Botella-Munoz, S. Schneuwly, R. H. Schirmer, and K. Becker, *Science* **291,** 643 (2001).

of glutathione reductase its share of the overall activity can be calculated and subtracted if GSSG is added before Trx; in other words, if Trx is used to start the assay.

One unit of enzymatic activity is defined as the consumption of 1 μmol of NADPH per minute. For activity calculations, see Eq. (2).

Instead of *E. coli* Trx, also thioredoxins of other species can be employed in this assay.[19]

Note that this assay system is influenced by the concentrations of GSSG, Trx, and TrxR. It should also be pointed out that thioredoxin stock solutions are often stored in the presence of reductants such as dithioerythritol (DTE). This "contamination" may lead to glutathione and thioredoxin disulfide consumption to relevant amounts. This would result in inaccurate enzyme activity measurements.

Auranofin Inhibition Assay of Biological Samples

Hill *et al.*[21] published a simple and effective method to determine thioredoxin reductase activity in biological samples by exploiting the inhibitory effects of gold thioglucose (aurothioglucose). In two parallel DTNB reduction assays DTNB reduction is determined, the first being a standard assay. The second, "non-TrxR" assay contains 10 μM aurothioglucose, leading to TrxR inhibition. The difference in activity between the two assays represents TrxR activity. We have modified this procedure by using auranofin instead of gold thioglucose. Auranofin is more stable in solution (dimethyl sulfoxide, DMSO) than aurothioglucose and lower concentrations are required for a specific and complete inhibition[22] of human TrxR. The mechanistically related enzyme glutathione reductase and the selenium-containing glutathione peroxidase are by orders of magnitude less susceptible to this inhibition. Because the Au(I)-induced TrxR inhibition is virtually not reversible by dialysis, an NADPH preincubation with and without auranofin for 5 min and consecutive dialysis can be performed. This could reduce the risk of inhibition of enzymes other than TrxR that contribute to the overall DTNB reduction. Otherwise an overestimation of TrxR activity is theoretically possible. However, the limitations of the DTNB reduction assay should be accounted for when interpreting the results.

Reagents

T-buffer: 100 mM Potassium phosphate, 2 mM EDTA; pH 7.4
DTNB: 100 mM DTNB in DMSO (store frozen, keep dark)
Auranofin: 1 mM in DMSO (store cold, keep dark)
NADPH: 4 mM NADPH in T-buffer (prepare freshly every day, keep on ice)

[21] K. E. Hill, G. W. McCollum, and R. F. Burk, *Anal. Biochem.* **253,** 123 (1997).
[22] S. Gromer, L. D. Arscott, C. H. Williams, Jr., R. H. Schirmer, and K. Becker, *J. Biol. Chem.* **273,** 20096 (1998).

Assay 1

In a total assay volume of 1 ml, 30 μl of DTNB is mixed with buffer and enzyme sample (approximately 1–2 nM). The enzymatic reaction is initiated by adding 50μl of NADPH (to 200 μM).

Assay 2: Non-TrxR

The procedure is the same as in assay 1, but 1 μl of buffer is replaced by 1 μl of auranofin solution (to 1 μM).

By using $\Delta A \, \mathrm{min}^{-1} = \Delta A \, \mathrm{min}^{-1}$ of assay 1 $- \Delta A \, \mathrm{min}^{-1}$ of assay 2 in Eq. (1), TrxR activity can be calculated. This assay can also be carried out with Trx-S$_2$ as oxidizing substrate instead of DTNB.

Enzyme Purification

The TE-buffer used throughout the preparation consists of 50 mM Tris-HCl, 1 mM EDTA, pH 7.6. In the method described here,[22] human placenta was used as starting material. As judged by SDS-PAGE, cellogel electrophoresis, kinetic studies and spectroscopic experiments this procedure yields an extremely pure and homogeneous enzyme. The procedure can be applied to other tissues and species, however, the elution conditions may however vary.

Chloroform–1-Butanol Extraction. Placenta tissue is weighed out and homogenized in a Waring blender with 0.6 ml of extraction solution [10 μM FAD and 40 μM phenylmethylsulfonyl fluoride (PMSF) in TE buffer] per gram of placenta. If the tissue is frozen it should be thawed in plastic bags together with the extraction buffer by placing the closed bags in a 40° water bath before homogenization. Immediately before treatment with chloroform–1-butanol[23] (which should be carried out in a hood), the homogenate is titrated to pH 8.3 with 5 M NH$_4$OH. The $-20°$ cold chloroform–1-butanol mixture (1 : 2.5, v/v; 120 μl/g of placenta) is added under vigorous stirring. It is important to rehomogenize the brownish suspension in the Waring blender. The resulting suspension is left for 30 min to 1 hr on ice and then centrifuged at $\geq$8000g for 90 min at 4°. The supernatant is filtered through glass wool and adjusted to pH 8.3 with 5 M NH$_4$OH. If the used tissue is limited the precipitate can be taken up in extraction solution (0.4 ml/g of placenta), homogenized, and centrifuged as described above and the supernatants are combined accordingly.

Acetone Precipitation. Per 1 ml of chloroform–butanol extract, 0.85 ml of ice-cold acetone is slowly added under stirring. The solution is left for approximately 1 hr on ice and then centrifuged at $\geq$3500g for 15 min at 4°. The clear, yellow supernatant is discarded. In bulk preparations when high-rpm centrifuges with large capacity become a limiting factor, the supernatant sometimes appears cloudy but hardly ever contains relevant amounts of activity. When in doubt a quick DTNB

[23] E. M. Scott, *Prep. Biochem.* **6,** 147 (1976).

reduction assay should be performed. The gluey brownish pellet is carefully taken up with a glass bar by adding small volumes of TE buffer to a final volume of approximately 100 ml. This sample is dialyzed exhaustively against 2-fold-diluted TE buffer (25 mM Tris-HCl, 0.5 mM EDTA, pH 7.6) overnight and centrifuged (30 min, $\geq$25,000g, 4°). The supernatant is set aside while the pellet is resuspended in TE buffer, mixed carefully, and centrifuged as described above. The combined supernatants are filtered through glass wool if gross impurities should persist and adjusted to pH 8.3 with 5 M NH$_4$OH.

DEAE-52 Cellulose Chromatography. The acetone-treated fraction is slowly applied at room temperature to a DEAE-52 cellulose column (Whatman, Clifton, NJ) equilibrated with TE buffer (approximately 0.6-ml final column volume per gram of placenta is required). After washing the column with 2 column volumes of TE buffer followed by 1 volume of 50 mM NaCl in TE buffer, thioredoxin reductase activity is eluted with 90 mM NaCl in TE buffer. In this step the enzyme comigrates with a deep red protein. The pool of active fractions is concentrated and washed with TE buffer in a Centriprep 30 (Amicon, Danvers, MA). The resulting solution is diluted 2-fold with TE buffer and the pH is adjusted to pH 7.6 with 100 mM HCl.

2′,5′-ADP-Sepharose 4B Affinity Chromatography. The above-described fraction is applied to a 30-ml (1.5 × 17 cm) 2′,5′-ADP-Sepharose 4B column (Pharmacia, Piscataway, NJ) in a jacketed chromatography tube. The tube is cooled to 6 ± 1°, the exact temperature being crucial for the purification success. The column is consecutively washed with 60 ml of TE buffer, 30 ml of 100 mM KCl in TE buffer, 20 ml of 200 mM KCl in TE, 30 ml of 100 mM KCl in TE, 60 ml of 2-fold-diluted TE, 60 ml of 500 μM NADH in TE, 60 ml of TE, 60 ml of 100 μM NADP$^+$ in TE, and 30 ml of 300 μM NADP$^+$ in TE. Finally, TrxR activity is eluted with 750 μM NADP$^+$ in TE buffer, concentrated, and washed with the buffer in a Centriprep 30. This solution, contains—on the basis of absorption spectra, specific activity, and sodium dodecyl sulfate–polyacrylamide gel electrophoresis (SDS–PAGE) analysis—more than 95% pure thioredoxin reductase. Because the remaining impurities contain small amounts of glutathione reductase a further purification step may be warranted.

Sephadex G-200 Gel Filtration. The above described fraction is applied to a Sephadex G-200 column equilibrated at 4° with TE buffer (Pharmacia; 1 × 100 cm) in a jacketed chromatography tube. Active and glutathione reductase-free fractions are pooled and concentrated. The enzyme solution is stored at 4°. (NH$_4$)$_2$SO$_4$ precipitation results in a tremendous loss of activity and is therefore not recommended.

Inhibitor Studies

A selection of TrxR inhibitors is shown in Table I.[24–32] Different methods can be applied to study a potential inhibitor. The easiest way is to add the inhibitor to

[24] K. U. Schallreuter, F. K. Gleason, and J. M. Wood, *Biochim. Biophys. Acta* **1054,** 14 (1990).

TABLE I
INHIBITORS OF MAMMALIAN THIOREDOXIN REDUCTASES[a]

Inhibitor	Inhibition			Comment	Refs.
	NADPH dependent	Reversible	Irreversible		
Carmustine (BCNU) and other nitrosoureas	+	–	+	Effective inhibitors; however, not selective because mechanistically related enzymes such as glutathione reductase are also inactivated and DNA is alkylated	2, 5, 24
Azelaic acid (dicarboxylic acids)	NA	NA	NA	Original data were obtained with *E. coli* TrxR. No inhibitory effects on hTrxR by up to 1 m*M* azelaic acid found	25
Adriamycin and other anthracyclines	(+)	NA	NA	Original work was done with rat enzyme. Results were not reproducible	26, 27
1-Chloro-2,4-dinitrobenzene (CDNB)	+	–	+	2,4-DNP-arylation of C-terminal Cys and Sec. Induces considerable oxidase activity	28–30, this report
Auranofin and aurothioglucose	+	–	+	Selective tight-binding inhibitors. Effect can be reversed by the reducing chelator BAL	21, 22
Alkyl-2-imidazolyl disulfide analogs	–	+	–	Mainly competitive inhibitors with K_i values in lower micromolar range. Some are metabolized by enzyme	31
(2,2':6',2''-Terpyridine)platinum(II) complexes	+	+	+	New platinum(II) derivatives with reversible and irreversible inhibitory component. IC_{50} values as low as 2 n*M*, that is in stoichiometric concentrations	32

[a] +, Yes; –, no; NA, not applicable.

the assay system and compare the activity with an untreated control. To further elucidate the mode of action but also to evaluate apparently weak inhibitors or covalent inhibitors that modify a protein time dependently, a preincubation protocol may be required. For members of the disulfide reductase family, the enzyme must often be in a reduced state before effective inhibition (e.g., by modification of the active site thiols) can take place. Because NADPH is an ingredient of all kinetic assay systems used, it is difficult to determine whether a specific substance does inhibit the oxidized, the reduced, or all forms of the enzyme. This problem can be overcome by incubating the enzyme in the absence and presence of NADPH with or without inhibitor; as stated above, native human TrxR is stable when exposed to NADPH. A sample of the preincubation solution is taken and assayed as usual (precaution must be taken to rule out that the sample carries over too much inhibitor into the assay vial, leading to additional inhibition). This procedure also allows the investigator to address the time and concentration dependence of the inhibition.

Example: Preincubation with Auranofin

In a final volume of 50 μl of T-buffer, 200 μM NADPH, 0.5 μM subunits TrxR, and 1 μM auranofin [1 μl of a 50-μl stock solution in dimethyl sulfoxide (DMSO)] are mixed at 25°. A corresponding control in which auranofin has been replaced by an equivalent volume of DMSO is also prepared.

Samples (3 μl) of each preincubation mixture are taken and activity is determined in separate DTNB reduction assays. This can be done at different time points if the inhibitior acts slower than auranofin. It is necessary to rule out that the transferred inhibitor concentration (here approximately 3 nM in the assay) has inhibitory effects.

Spectroscopy of Enzyme in Presence of Inhibitor. Mammalian TrxRs are flavoenzymes. Thus, during the catalytic cycle typical spectroscopic changes can be monitored.[2] Inhibitors can bind at different sites of the enzyme and may lead to a block in these spectroscopic events or even result in a modification of the signal. Because the spectroscopic changes of human TrxR closely resemble those of glutathione reductase it is justified to make use of the information gathered for the latter enzyme.[2]

[25] K. U. Schallreuter and J. M. Wood, *Cancer Lett.* **36,** 297 (1987).

[26] B. L. Mau and G. Powis, *Free Radic. Res. Commun.* **8,** 365 (1990).

[27] B. L. Mau and G. Powis, *Biochem. Pharmacol.* **43,** 1613 (1992).

[28] E. S. J. Arnér, *Biofactors* **10,** 219 (1999).

[29] J. Nordberg, L. Zhong, A. Holmgren, and E. S. Arnér, *J. Biol. Chem.* **273,** 10835 (1998).

[30] E. S. J. Arnér, M. Björnstedt, and A. Holmgren, *J. Biol. Chem.* **270,** 3479 (1995).

[31] J. E. Oblong, E. L. Chantler, A. Gallegos, D. L. Kirkpatrick, T. Chen, N. Marshall, and G. Powis, *Cancer Chemother. Pharmacol.* **34,** 434 (1994).

[32] K. Becker, C. Herold-Mende, J. J. Park, G. Lowe, and R. H. Schirmer, *J. Med. Chem.* **44,** 2784 (2001).

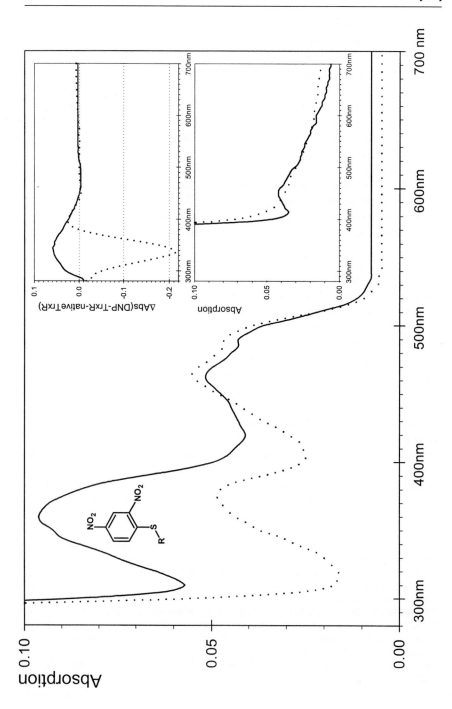

Modifying groups with their own typical absorption patterns may be studied per se. An example is shown in Fig. 1.[28,29,33] Changes in absorbance due to the microenvironment must be taken into consideration.

The inhibition of an enzyme can be reversible or irreversible. Thus the investigator should check the possible reversal of inhibition by dialysis or dilution, incubation with reducing agents [dithioerythritol (DTE), Cleland's reagent] or [in the case of metal–organic inhibitors such as gold(I) and platinum(II) complexes] by chelating agents[22] such as British-Anti-Lewisite (BAL; 2,3-dithiopropanol).

It is important to note, that reducing agents (DTE and BAL) interfere with the assay systems. The DTNB reduction assay, for instance, can be used only if these reducing agents are removed by gel filtration, dialysis, or chemical reactions in order to reduce the formation of nonenzymatically formed TNB. If the GHOST assay is employed the amount of transferred reducing equivalents can be calculated, and nonspecifically consumed GSSG can be replaced if necessary. GSH and GSSG (unlike higher concentrations of TNB) do not significantly interfere with the spectroscopic activity determination.

Apart from the spectroscopic methods to determine the site of action of a particular inhibitor, studies with mutant enzymes lacking the C-terminal Sec (e.g., Sec-496 → Cys) may provide additional information.[16,34,35]

Crystallization

The crystallization of mammalian thioredoxin reductases remains a challenge. We have screened a large variety of crystallization conditions, using the hanging drop method.[36] The most promising results have been obtained at 25° with 5 μl of protein solution human TrxR at (10 mg ml^{-1} in 2 mM EDTA, pH 7.4) mixed with 5 μl of 7% (w/v) polyethylene glycol (PEG) 8000 in 100 mM Tris-HCl, pH 7.4, in the drop. The reservoir contained 800 μl of 12% (w/v) PEG 8000 in 100 mM

FIG. 1. Spectral changes of human TrxR due to a covalently bound inhibitor. In this example, human TrxR in the presence of NADPH is modified by 1-chloro-2,4-dinitrobenzene (CDNB). Two enzyme samples were incubated with CDNB in the presence and absence of 200 μM NADPH for 1 hr. The samples were dialyzed thereafter and the spectra recorded. In the large diagram the absorption spectra of CDNB-modified (solid curve) and unmodified (dotted curve) human placenta TrxR are shown. The strong increase in absorbance between 300 and 400 nm is typical of an S-2,4-dinitrophenyl group [M. Bilzer, R. L. Krauth-Siegel, R. H. Schirmer, T. P. Akerboom, H. Sies, and G. E. Schulz, *Eur. J. Biochem.* **138**, 373 (1984)] (see *Inset* formula). *Lower inset:* Spectra for both enzyme forms in the presence of 200 μM NADPH. It indicates that the charge transfer complex between the proximal thiolate (Cys-62) and FAD is still generated in the modified enzyme. *Upper inset:* Differential spectra of native and CDNB-modified TrxR as calculated from the other spectra are shown. The solid curve represents the oxidized form of the enzyme (modified TrxR − native TrxR). Assuming an $\varepsilon_{340\ nm,\ Cys-2,4-DNP}$ of 9.6 mM^{-1} cm^{-1}, a binding of approximately 1.5 mol of dinitrophenyl groups per mole subunit can be calculated on the basis of this spectrum—a value that is in good agreement with other experimental data [E. S. J. Arnér, *Biofactors* **10**, 219 (1999); J. Nordberg, L. Zhong, A. Holmgren, and E. S. Arnér, *J. Biol. Chem.* **273**, 10835 (1998)]. The dotted curve in the upper inset shows the differential absorbance of the NADPH-reduced enzyme species. The prominent negative peak in the dashed curve is caused by NADPH consumption in the sample, which is due to the strong oxidase activity of the arylated enzyme.

Tris-HCl, pH 7.4. The large crystals harvested, however, do not yet diffract to a satisfactory resolution. For noncrystallographic applications, for example, biolistic *in vivo* studies of thioredoxin reductase, the crystals are satisfactory.[37]

---

[33] M. Bilzer, R. L. Krauth-Siegel, R. H. Schirmer, T. P. Akerboom, H. Sies, and G. E. Schulz, *Eur. J. Biochem.* **138**, 373 (1984).

[34] L. Zhong, E. S. J. Arnér, and A. Holmgren, *Proc. Natl. Acad. Sci. U.S.A.* **97**, 5854 (2000).

[35] L. Zhong, K. Persson, T. Sandalova, G. Schneider, and A. Holmgren, *Acta Crystallogr. D Biol. Crystallogr.* **56**, 1191 (2000).

[36] S. Gromer, *in* "Die Thioredoxinreduktase von Mensch und Maus—Ein Selenoenzym als Zielmolekül von Chemotherapeutika." MD Thesis. Ruprecht-Karls-Universität, Heidelberg, Germany, 1998.

[37] M. A. Keese, R. Saffrich, T. Dandekar, K. Becker, and R. H. Schirmer, *FEBS Lett.* **447**, 135 (1999).

[37] Classification of Plant Thioredoxins by Sequence Similarity and Intron Position

By YVES MEYER, FLORENCE VIGNOLS, and JEAN PHILIPPE REICHHELD

Thioredoxins are small proteins ($\sim$12 kDa) with a cysteine pair present in the almost conserved active site WCGPC, conferring a strong disulfide reductase activity. Our knowledge of plant thioredoxins has been reviewed.[1] One characteristic of plants is the presence of multiple thioredoxin types and of multiple genes for each thioredoxin type in a single genome. All genes are nuclear encoded. This article allows the reader to classify a newly sequenced plant cDNA (or genomic DNA fragment) within a particular type of thioredoxin. The method is based on the analysis of the thioredoxin gene family of the *Arabidopsis thaliana* genome, which is now completely sequenced. The validity of these characteristics to thioredoxin genes of other plants was tested with dicot and monocot sequences available in January 2001.

Arabidopsis Genes Encoding Thioredoxins or Proteins with Thioredoxin Domains

Classic Thioredoxins m, f, and h

The classic plant thioredoxins *m* and *f* were first isolated on a biochemical basis as *in vitro* activators of the chloroplastic enzymes malate dehydrogenase and fructose-1,6-bisphosphatase, respectively. The pea and spinach proteins

---

[1] P. Schurmann and J. P. Jacquot, *Annu. Rev. Plant Physiol. Plant Mol. Biol.* **51**, 371 (2000).

TABLE I
THIOREDOXIN GENE FAMILY IN *Arabidopsis thaliana*[a]

Name	Chromosome	BAC	Localization
Classical Thioredoxin Types			
Thioredoxins m			
*m*1	1	F21B7	Chloroplastic
*m*2	4–9	F9H3	Chloroplastic
*m*3	2–91	F9O13	Chloroplastic
*m*4	3	MJK13	Chloroplastic
Thioredoxins f			
*f*1	3	F13F9	Chloroplastic
*f*2	5	MQK4	Chloroplastic
Thioredoxins h			
*h*1	3	F24M12	Cytosolic
*h*2	5	MYH19	Cytosolic
*h*3	5	MBD2	Cytosolic
*h*4	1	F14P1–F6F9	Cytosolic
*h*5	1	F27F5	Cytosolic
*h*7	1	F23H11	Cytosolic
*h*8	1	T17F3	Cytosolic
*h*9	3	F17O14	Cytosolic
Additional Thioredoxin Types			
Thioredoxin x			
x	1	F14I3	?
Thioredoxins ch2			
ch2-1	1	F28O16	Chloroplastic
ch2-2	1	T10P12	Chloroplastic
Thioredoxins o			
o1	2–192	F19I13	?
o2	1	F17F8	?
Thioredoxins related to the h type			
CxxS1	1	T23J18	Cytosolic
CxxS2	2–220	T7D17	Cytosolic
TDX	3	MEB5	Cytosolic
Lilium type			
Lilium 1	1	F22O13	Cytosolic
Lilium 2	4–64	F20B18	?
Lilium 3	5	MFB13	?
Lilium 4	2–186	F4P9	?

[a] The localization of each gene on the chromosome, on the chromosome section (for chromosome 2 and 4) and on the bacterial artificial chromosome (BAC) is indicated. The potential cellular localization of the encoded protein was determined with Psort [K. Nakai and M. Kanehisa, *Proteins* **11**, 95 (1991)]. Question marks indicate no confident determination.

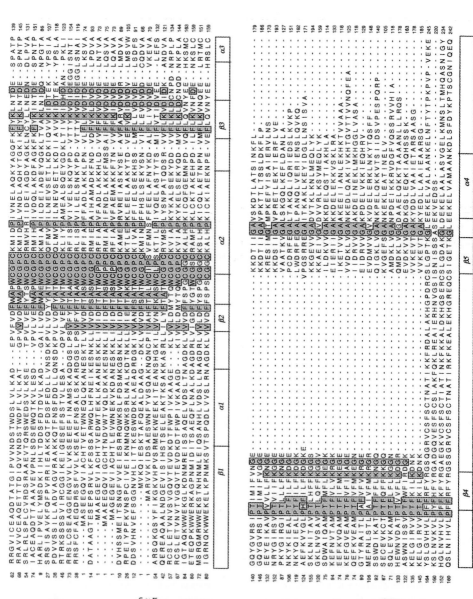

were sequenced and, later, cDNAs were cloned, giving access to the sequence of the transit peptide.[2,3] A third type of thioredoxin was isolated from nongreen spinach tissues. It was first called thioredoxin c for cytosolic and later renamed h (heterotrophic). This enzyme was partially sequenced and cDNAs were then isolated from *Chlamydomonas*[4] and tobacco.[5] Table I summarizes the different genes from A. *thaliana* encoding these three types of thioredoxins.

Four *Arabidopsis* genes[6] encode thioredoxin m precursor proteins ranging from 180 to 200 amino acids. Two other *Arabidopsis* genes encode thioredoxin f precursor proteins of 178 and 185 amino acids, f1 and f2, respectively. All thioredoxins m and f of *Arabidopsis* present the classic sequence WCGPC in the active site, like the pea and spinach sequences. Nevertheless, the sequence homologies between thioredoxins m and f are low. For both m and f types, putative transit peptides detected by Psort (Nakai and Kanehisa,[7] http://psort.nibb.ac.jp/form.html) address the mature proteins to the chloroplast. The putative cleavage sites between the transit peptide and the mature thioredoxin are in good agreement with the sequences of the mature thioredoxins m and f isolated from pea and spinach chloroplasts. The transit peptides addressing thioredoxins m and f to the chloroplast do not show obvious similarity between both types. This is a general rule for transit peptides addressing precursor proteins to organelles.

Eight *Arabidopsis* genes (Rivera-Madrid et al.,[8] and our unpublished data, 2001) encode thioredoxin h proteins, which have approximately the size of mature chloroplastic thioredoxins or of thioredoxins isolated from bacteria or animals, ranging from 114 to 148 amino acids. Within this group, five members present the classic sequence WCGPC in the active site, whereas the other three harbor a WCPPC site.

[2] N. Wedel, S. Clausmeyer, R. G. Herrmann, L. Gardet-Salvi, and P. Schurmann, *Plant Mol. Biol.* **18,** 527 (1992).

[3] M. Kamo, A. Tsugita, C. Wiessner, N. Wedel, D. Bartling, R. G. Herrmann, F. Aguilar, L. Gardet-Salvi, and P. Schurmann, *Eur. J. Biochem.* **182,** 315 (1989).

[4] M. Stein, J. P. Jacquot, E. Jeannette, P. Decottignies, M. Hodges, J. M. Lancelin, V. Mittard, J. M. Schmitter, and M. Miginiac-Maslow, *Plant Mol. Biol.* **28,** 487 (1995).

[5] I. Marty and Y. Meyer, *Plant Mol. Biol.* **17,** 143 (1991).

[6] D. Mestres-Ortega and Y. Meyer, *Gene* **240,** 307 (1999).

[7] K. Nakai and M. Kanehisa, *Proteins* **11,** 95 (1991).

[8] R. Rivera-Madrid, D. Mestres, P. Marinho, J. P. Jacquot, P. Decottignies, M. Miginiac-Maslow, and Y. Meyer, *Proc. Natl. Acad. Sci. U.S.A.* **92,** 5620 (1995).

FIG. 1. Multiple alignment of *Arabidopsis thaliana* thioredoxins. The transit peptides and a small part of the N-terminal part of some thioredoxins h cannot be aligned and are not shown. The secondary structure indicated under the alignment results from a comparison of the structures established for *Chlamydomonas h* [V. Mittard, M. J. Blackledge, M. Stein, J. P. Jacquot, D. Marion, and J. M. Lancelin, *Eur. J. Biochem.* **243,** 374 (1997)] and m [J. M. Lancelin, L. Guilhaudis, I. Krimm, M. J. Blackledge, D. Marion, and J. P. Jacquot, *Proteins* **41,** 334 (2000)] and for pea f and m [G. Capitani, Z. Marcovic-Housley, G. DelVal, M. Morris, J. N. Jansonius, and P. Schurmann, *J. Mol. Biol.* **302,** 135 (2000)].

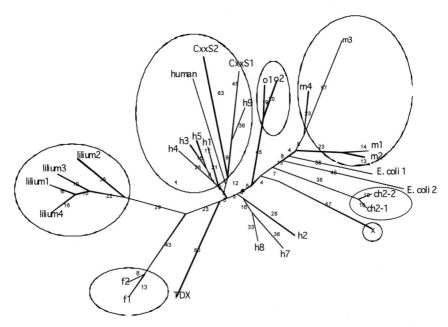

FIG. 2. Darwin tree of *Arabidopsis thaliana* thioredoxins. Each thioredoxin is named as in Table I. The PAM (point accepted mutations) distances between each node are indicated.

A multiple alignment performed with CLUSTAL W (Thompson *et al.*[9]; Fig. 1) shows conserved amino acids in all mature thioredoxins. This alignment based on sequence similarity fits with the thioredoxin structures established by nuclear magnetic resonance (NMR) for thioredoxins *m* and *h* from *Chlamydomonas*[10,11] and by X-ray diffraction for thioredoxins *m* and *f* from pea.[12] It shows that thioredoxins *h* present at the N-terminal part an insertion of three amino acids corresponding to a longer α_1 helix. Nevertheless, no obvious characteristic sequence pattern can be inferred for each type of thioredoxin.

In contrast, a phylogenetic tree (Fig. 2) constructed with the Darwin program (Hallett *et al.*,[13] http://cbrg.inf.ethz.ch/Darwin/index.html) associates all members of each type of thioredoxin in a particular group. As a comparison, the two *Escherichia coli* sequences and the cytosolic human sequence have been

[9] J. D. Thompson, F. Plewniak, and O. Poch, *Nucleic Acids Res.* **27,** 2682 (1999).

[10] V. Mittard, M. J. Blackledge, M. Stein, J. P. Jacquot, D. Marion, and J. M. Lancelin, *Eur. J. Biochem.* **243,** 374 (1997).

[11] J. M. Lancelin, L. Guilhaudis, I. Krimm, M. J. Blackledge, D. Marion, and J. P. Jacquot, *Proteins* **41,** 334 (2000).

[12] G. Capitani, Z. Markovic-Housley, G. DelVal, M. Morris, J. N. Jansonius, and P. Schurmann, *J. Mol. Biol.* **302,** 135 (2000).

[13] M. Hallett, G. Gonnet, C. Korostensky, and L. Bernardin, *Bioinformatics* **16,** 101 (2000).

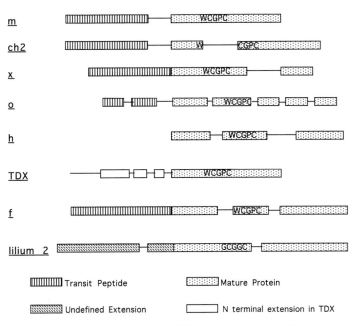

FIG. 3. Gene structures of *Arabidopsis thaliana* thioredoxins.

represented. Thioredoxins *m* are typical prokaryotic thioredoxins, whereas thioredoxins *f* and *h* are associated with vertebrate thioredoxins. Thioredoxins *h* form a complex and dispersed family with eight members.

An additional feature is the position of the introns in the genomic sequences.[14] As shown in Fig. 3, the four thioredoxin *m* genes present only one intron placed approximately at the cleavage site between the transit peptide and the mature protein. However, thioredoxin *f* genes have two introns, both located within the sequence encoding the mature protein. Four thioredoxins *h* present two introns, the second being located at the same position as the second intron of thioredoxins *f*. Thioredoxin *h5* presents only the first intron. The length of introns is variable and cannot be aligned even within a single type.

Arabidopsis Genes Encoding Proteins with One Thioredoxin Domain

In addition to these classic thioredoxins *m, f,* and *h,* the *Arabidopsis* genome presents several other genes encoding thioredoxin homologs. One gene encodes a prokaryotic-like thioredoxin that we have called thioredoxin *x.*[6] It presents an N-terminal extension with typical chloroplastic characteristics. Like prokaryotic thioredoxins, it has a short α_1 helix, but in contrast to thioredoxin *m* genes, it has

---

[14] M. Sahrawy, V. Hecht, J. Lopez-Jaramillo, A. Chueca A, Y. Chartier, and Y. Meyer, *J. Mol. Evol.* **42,** 422 (1996).

no intron between the transit peptide and the mature protein domain, but has an intron within the sequence encoding the mature protein.

Two genes encode prokaryotic-like thioredoxins distinct from thioredoxins m and x. They present an N-terminal extension with typical chloroplastic characteristics and have been named ch2-1 and ch2-2, respectively. Like thioredoxins m, they have an intron between the transit peptide and the mature protein domain but also an additional intron within the sequence encoding the mature protein. Like thioredoxins h, they have a long α_1 helix.

Two genes encode thioredoxin h-related proteins, with the nontypical active site CxxS. The first discovered gene was previously called AtTRX6,[14] but we now prefer to rename it CxxS1 because this modified active site is not compatible with typical thioredoxin activity. It may act as a disulfide isomerase and eventually as a reductase. The second variant is named CxxS2. Expressed sequence tags (ESTs) have been found for both genes, indicating that they are expressed. Both have a long α_1 helix and introns at the same position as typical thioredoxins h.

One gene that we named TDX encodes a protein with two domains: the N-terminal domain shows homology with a heat shock protein 70 (HSP70)-interacting protein, whereas the C-terminal domain is related to thioredoxins h but does not have any intron. This gene is also expressed. In addition, we have isolated a similar gene in *Nicotiana tabacum*.

Two genes form a family that we have called thioredoxin o (the first EST was sequenced in Orsay: ATTS5209 clone OBO83). Both encoded proteins have an N-terminal extension that may address the protein to organelles but Psort cannot discriminate between chloroplastic or mitochondrial adressing. As indicated in the tree in Fig. 2, the mature proteins form a clearly distinct group. The gene structure is also different from the other thioredoxins: four introns for o1 and five for o2, all located at positions different from all other thioredoxins.

First described in *Lilium grandiflorum* (GenPept accession AAA33400), another gene family with four members encodes proteins with homology to thioredoxins. It is not known whether these proteins have disulfide reductase activity. In the tree in Fig. 2, they form a separate group near thioredoxins f. They have an N-terminal extension. According to Psort, Lilium 1 is a cytosolic protein whereas Lilium 2, 3, and 4 extensions could have characteristics of a transit peptide. They present a long insertion in the loop between α_4 and α_5.

Arabidopsis Genes Encoding Proteins with Multiple Thioredoxin Domains

In addition to thioredoxins, *Arabidopsis,* like most eukaryotes, encodes proteins with multiple thioredoxin domains. One gene encodes a CDSP32 homolog, a protein with two thioredoxin domains first described in potato,[15] two genes of

[15] P. Rey, G. Pruvot, N. Becuwe, F. Eymery, D. Rumeau, and G. Peltier, *Plant J.* **13,** 97 (1998).

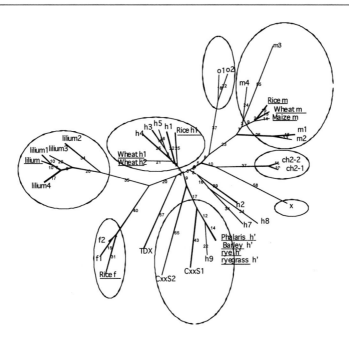

FIG. 4. Darwin tree of *Arabidopsis thaliana* thioredoxins with monocot thioredoxins extracted from Pir, SwissProt, or GenBank database. |pir||T50862, thioredoxin-like protein *Phalaris coerulescens;* |pir||T50864, thioredoxin-like protein *Hordeum bulbosum;* |pir||T50867, thioredoxin-like protein rye; |pir||T50865, thioredoxin-like protein perennial ryegrass; |pir||T03957, thioredoxin *m* maize; |pir||T04090, probable thioredoxin *h* rice; |sp|Q9ZP21|THIM_,wheat thioredoxin *m*-type; |sp|Q9ZP20|THIM_,*Orysa* thioredoxin *m*-type; |sp|O64394|THIH_,wheat thioredoxin *h*-type; |gb|AAF88067.1|, thioredoxin *h Triticum aestivum*; |gb|AAB82144.1|, thioredoxin *f* isoform *Oryza sativa* (an error in the translation phase was corrected before drawing the tree); |gb|AAA33400.1|, thioredoxin *Lilium longiflorum*. The monocot sequences are underlined.

nucleoredoxins,[16] and at least eight protein disulfide isomerases. These thioredoxins domains are divergent from the true thioredoxins and are not reviewed in this article.

Usability of *Arabidopsis* Data for Predicting Type of Plant Sequence

As stated above, two characteristics allow the classification of the various thioredoxins: a phylogenetic tree like that produced on the basis of sequence similarity by the Darwin program defines major classes. The presence of introns at the same sites within each class strongly reinforces the validity of this classification

[16] B. J. Laughner, P. C. Sehnke, and R. J. Ferl, *Plant Physiol.* **118**, 987 (1998).

and allows the reintroduction of some variants into a group (e.g., $h2, h7, h8$, CxxS1, and CxxS2 are considered as h on this basis) or the discrimination of two types (e.g., x cannot be associated with ch2 thioredoxins because they have different gene structure). It is questionable whether the groups defined for *Arabidopsis* thioredoxin genes represent a possible classification for all higher plants. As a test of validity, we extracted from GenPept all monocot sequences described as homologous to thioredoxins. We then eliminated redundant sequences and built a tree including the *Arabidopsis* set and the 12 remaining monocot sequences. As shown in Fig. 4, one rice protein is associated with *Arabidopsis* thioredoxins *f.* One rice, one maize, and one wheat protein are associated with *Arabidopsis* thioredoxins *m*. One *Lilium* thioredoxin is associated with *Arabidopsis* thioredoxins Lilium. Two wheat and one rice proteins are associated with the classic *Arabidopsis* thioredoxins *h* ($h1, h3, h4$, and $h5$). In addition, five sequences, one from *Phalaris,* one from *Hordeum,* two from rye, and one from ryegrass, form with $h9$ a new group rooting near from the typical *Arabidopsis* thioredoxins *h*. Further analysis of the genomic databases shows that two rice genes encode thioredoxins of this type. The two genomic sequences of rice as well as the genomic sequence of *Phalaris* each present two introns situated at the same position as in the typical *Arabidopsis* thioredoxins *h*. Thus this new group constitutes a subgroup of thioredoxins *h*. Thus, all monocot thioredoxin sequences available in January 2001 are members of the groups defined with *Arabidopsis* sequences. In addition, a survey of the ESTs from rice reveals the expression in this plant of a gene encoding a protein similar to CxxS2, having the redox site WCVTS. We have also found several rice ESTs encoding homologs of TDX. We have performed further analysis of all plant sequences in order to find homologs of the *Arabidopsis* thioredoxins x, o, and ch2 but we did not find any. It should be recalled that these genes are all expressed at a low level in *Arabidopsis*. Most plant sequences are EST sequences in which rare mRNAs are not represented.

A web site dedicated to plant thioredoxins is now in construction at http://lgdp. univ-perp.fr/redoxines in order to help annotation of the new plant sequences and provide plant physiologists with updated data.

[38] Ferredoxin-Dependent Thioredoxin Reductase: A Unique Iron–Sulfur Protein

By PETER SCHÜRMANN

Introduction

The ferredoxin : thioredoxin reductase (FTR) is the central enzyme of the ferredoxin/thioredoxin system of oxygenic photosynthetic cells.[1-3] This important system regulates carbon assimilation in response to light. It conveys a light signal originating in the illuminated thylakoids to key photosynthetic enzymes, thereby modulating their activity. The light signal in the form of electrons provided by the excitation of chlorophyll is transferred via ferredoxin to the FTR, which transforms it into a "thiol signal." This thiol signal is then transmitted through dithiol/disulfide interchanges via thioredoxins to target enzymes, activating or deactivating them by reduction. Photosynthetic eukaryotes contain in their chloroplasts two types of enzyme-specific thioredoxins, *f* and *m*, whereas in the photosynthetic prokaryotes the *f* type is missing. Reduced thioredoxin *f* activates (reduces) regulatory enzymes involved directly or indirectly in carbon assimilation [i.e., fructose-1,6-bisphosphatase (FBPase), sedoheptulose-1,7-bisphosphatase, phosphoribulokinase, ribulose-1,5-bisphosphate carboxylase (Rubisco) activase, coupling factor ATPase, and malate dehydrogenase]. The functions of thioredoxin *m* are less well defined. One demonstrated specific function of reduced thioredoxin *m* is the deactivation of NADP-dependent glucose-6-phosphate dehydrogenase of chloroplasts.[4] However, thioredoxin *m* might also be involved in the activation of transcription[5] and in the removal of reactive oxygen species (ROS).[6]

The FTR is a unique enzyme regarding its structure and mechanism. It reduces a disulfide bridge with the help of a 4Fe–4S cluster. The enzyme is composed of two dissimilar subunits of similar size. One, the catalytic subunit, contains a cubane [4Fe–4S] cluster and a redox-active disulfide bridge, functional in the reduction of thioredoxins. This subunit is highly conserved between cyanobacteria, algae, and higher plants. The other subunit, the variable subunit, is of variable size when

[1] P. Schürmann and B. B. Buchanan, *in* "Advances in Photosynthesis" (Govindjee, ed.), Vol. 11, p. 331. Kluwer Academic, Dordrecht, The Netherlands, 2001.

[2] P. Schürmann and J.-P. Jacquot, *Annu. Rev. Plant Physiol. Plant Mol. Biol.* **51,** 371 (2000).

[3] S. Dai, C. Schwendtmayer, K. Johansson, S. Ramaswamy, P. Schürmann, and H. Eklund, *Q. Rev. Biophys.* **33,** 67 (2000).

[4] I. Wenderoth, R. Scheibe, and A. von Schaewen, *J. Biol. Chem.* **272,** 26985 (1997).

[5] A. Danon and S. P. Mayfield, *Science* **266,** 1717 (1994).

[6] M. Baier and K. J. Dietz, *in* "Progress in Botany," Vol. 60, p. 282. Springer-Verlag, Berlin, Germany, 1999.

primary structures from different species are compared and appears to have only a structural function.

All the essential components of the ferredoxin/thioredoxin system have been cloned and overexpressed. This has considerably simplified their purification and provided ample material for functional and structural studies.[7-14] The FTR has been cloned and expressed from spinach[15] and *Synechocystis* sp. PCC 6803.[16] The cyanobacterial enzyme is functionally entirely comparable with the spinach enzyme, but is expressed in larger amounts and is more stable. The spinach enzyme, however, can be stabilized by truncation of the N terminus of the variable subunit.

In this article we describe the cloning, expression, and purification of the FTR from *Synechocystis* and of a truncation mutant of spinach thioredoxin *f*, which renders this protein more soluble and stable than the recombinant protein we had reported earlier.[17]

Ferredoxin/Thioredoxin Reductase

Assay Methods

Principle. There is no method allowing the direct measurement of FTR activity. FTR is usually assayed by measuring the light-dependent activation of a chloroplast target enzyme such as FBPase or NADP-dependent malate dehydrogenase in a two-stage assay. The FTR fraction to be tested is incubated in the light in the presence of thylakoids, ferredoxin, thioredoxin *f*, and FBPase and the resulting FBPase activity is measured after injection of an aliquot of the activation mixture into the reaction mixture for the spectrophotometric enzyme test.[17,18]

[7] C. Binda, A. Coda, A. Aliverti, G. Zanetti, and A. Mattevi, *Acta Crystallogr. D* **54**, 1353 (1998).

[8] S. Dai, C. Schwendtmayer, P. Schürmann, S. Ramaswamy, and H. Eklund, *Science* **287**, 655 (2000).

[9] G. Capitani, Z. Markovic-Housley, G. del Val, M. Morris, J. N. Jansonius, and P. Schürmann, *J. Mol. Biol.* **302**, 135 (2000).

[10] J. M. Lancelin, L. Guilhaudis, I. Krimm, M. J. Blackledge, D. Marion, and J. P. Jacquot, *Proteins* **41**, 334 (2000).

[11] M. Chiadmi, A. Navaza, M. Miginiac-Maslow, J. P. Jacquot, and J. Cherfils, *EMBO J.* **18**, 6809 (1999).

[12] P. D. Carr, D. Verger, A. R. Ashton, and D. L. Ollis, *Structure* **7**, 461 (1999).

[13] K. Johansson, S. Ramaswamy, M. Saarinen, M. Lemaire-Chamley, E. Issakidis-Bourguet, M. Miginiac-Maslow, and H. Eklund, *Biochemistry* **38**, 4319 (1999).

[14] C. R. Staples, E. Gaymard, A. L. Stritt-Etter, J. Telser, B. M. Hoffman, P. Schürmann, D. B. Knaff, and M. K. Johnson, *Biochemistry* **37**, 4612 (1998).

[15] E. Gaymard, L. Franchini, W. Manieri, E. Stutz, and P. Schürmann, *Plant Sci.* **158**, 107 (2000).

[16] C. Schwendtmayer, W. Manieri, M. Hirasawa, D. B. Knaff, and P. Schürmann, *in* "Photosynthesis: Mechanisms and Effects" (G. Garab, ed.), Vol. 3, p. 1927. Kluwer Academic, Dordrecht, The Netherlands, 1998.

[17] P. Schürmann, *Methods Enzymol.* **252**, 274 (1995).

[18] P. Schürmann and J.-P. Jacquot, *Biochim. Biophys. Acta Enzymol.* **569**, 309 (1979).

As a simplified alternative to the light-dependent reduction of FTR we have developed a chemical reduction system using either dithionite-reduced ferredoxin or viologens as electron donors.[19] Sodium dithionite alone cannot reduce FTR, but in the presence of either ferredoxin, the natural electron donor protein, or a mediator molecule such as methyl or benzyl viologen, it is efficient.

Although both assay methods have been developed for spinach FTR and involve spinach proteins as electron donors and acceptors, they work well with the *Synechocystis* FTR.

Procedure for Light-Dependent Activation. Activation is carried out in thick-walled conical glass tubes containing 100 μl of the following: 100 m*M* Tris-HCl (pH 7.9), 10 m*M* sodium ascorbate, 0.1 m*M* dichlorophenol indophenol, 14 m*M* 2-mercaptoethanol, 10 μ*M* ferredoxin, 1 μ*M* thioredoxin *f*, 0.5 unit of spinach FBPase,[20,21] heated thylakoids[22] equivalent to 10 μg of chlorophyll, and the FTR fraction to be assayed (1 to 5 pmol). The activation mixture is equilibrated for 5 min at 25° with argon, in the dark, and then illuminated for 5 to 15 min. After the activation a 5- or 10-μl aliquot is tested for FBPase activity by the spectrophotometric assay described below.

Procedure for Chemical Reduction of Ferredoxin: Thioredoxin Reductase. Activation is carried out in 1.5-ml conical plastic microtubes loosely covered with serum caps fitted with a hypodermic needle through which a gentle stream of argon is introduced. The activation mixture contains in a final volume of 100 μl the following: 100 m*M* Tris-HCl (pH 7.9), 5 m*M* sodium dithionite, 1 μ*M* benzyl or methyl viologen or 10 μ*M* ferredoxin, 14 m*M* 2-mercaptoethanol, 1 μ*M* thioredoxin *f*, 0.5 unit of FBPase, and the FTR fraction to be assayed.

The activation mixture, without the dithionite, is equilibrated for 5 min at room temperature with argon before 10 μl of 50 m*M* dithionite, dissolved in 100 m*M* Tris-HCl, pH 7.9, and kept under argon, is added. After an activation time of 5 to 15 min an aliquot of 5 or 10 μl of activation mixture is withdrawn and the FBPase activity is tested spectrophotometrically as described below.

Activations can be done in air, whereby only 60 to 90% of maximal activities are reached. However, to obtain reproducible results and complete activations the incubations should be performed under argon.

Spectrophotometric Fructose-1,6-bisphosphatase Assay. The activity of the FBPase is assayed in the spectrophotometer at 25°. The 1-ml reaction mixture contains 100 m*M* Tris-HCl (pH 7.9), 0.1 m*M* EGTA-Na, 1.5 m*M* MgSO₄, 1 m*M* fructose 1,6-bisphosphate, 0.3 m*M* NADP, 1.75 units of phosphoglucose isomerase, 0.7 unit of glucose-6-phosphate dehydrogenase, and 14 m*M* 2-mercaptoethanol. The reaction is monitored as the increase in absorbance at 340 nm.

[19] P. Schürmann, A.-L. Stritt-Etter, and J. Li, *Photosynth. Res.* **46,** 309 (1995).

[20] P. Schürmann and R. A. Wolosiuk, *Biochim. Biophys. Acta Enzymol.* **522,** 130 (1978).

[21] Y. Balmer and P. Schürmann, in "Photosynthesis: Mechanisms, Effects" (G. Garab, ed.), Vol. 3, p. 1935. Kluwer Academic, Dordrecht, The Netherlands, 1998.

[22] J.-P. Jacquot, M. Droux, M. Migniac-Maslow, C. Joly, and P. Gadal, *Plant Sci.* **35,** 181 (1984).

Cloning, Overexpression, and Purification of Recombinant Ferredoxin:
Thioredoxin Reductase from Synechocystis

Principle. FTR is composed of two nonidentical subunits, a catalytic subunit and a variable subunit, encoded by two different genes. As the entire genome of *Synechocystis* sp. PCC 6803 has been sequenced (see CyanoBase at *http://www.kazusa.or.jp/cyano/cyano.html*), the positions of the two genes are known and cosmids containing them can be obtained from the Kazusa DNA Research Institute (Chiba, Japan). The genes are easily isolated from the cosmids by polymerase chain reaction (PCR) amplification, using homologous oligonucleotides as elongation primers, containing, in addition, the information for restriction sites needed for subcloning of the subunits.

To obtain a functional enzyme both subunits must be expressed and assembled in a 1 : 1 ratio. To achieve this we have made a dicistronic construction with the genes for both subunits arranged in tandem, as already used successfully for the expression of spinach FTR.[15] This construct contains upstream ribosome-binding sites as well as start and stop codons for each subunit gene.

The purification is achieved through four consecutive chromatographic steps. A first hydrophobic interaction chromatography removes nucleic acid contaminants and some hydrophilic proteins. In the elution profile of the size-exclusion chromatography the FTR appears as a prominent peak and it can be detected by its spectral properties, and if only the main fractions containing FTR are pooled, the protein is pure after the following anion-exchange chromatography. A final purification is performed on either a hydroxyapatite or ferredoxin-Sepharose column.

Construction of Expression Vector. The cosmid clones Cs0377 and Cs0236 encode, respectively, the variable (SynA) and the catalytic (SynB) subunit genes of FTR in *Synechocystis* sp. PCC 6803. The coding sequences for the respective subunits are isolated by PCR, using the following sense (SynA1, SynB1) and antisense (SynA2, SynB2) primers:

M   N   V   G   D   R...
SynA1: 5′-GGTACC*CATATG*AATGTCGGCGATCGTG-3′
          *   D   E   I   L   T   V...
SynA2: 5′-AAGCTT*CTA*ATCTTCAATCAGGGTAAC-3′
                    M   T   S   S   D   T...
SynB1: 5′-AAGCTTGAAGAAGATATATAT*ATG*ACCAGCAGTGACACC-3′
          *   A   M   S   A...
SynB2: 5′-GGATCC*TA*GGCCATGCTGGCCT-3′

For the variable subunit the primer SynA1 introduces at the 5′ end two restriction sites, *Kpn*I followed by *Nde*I, where the *Nde*I site provides the start codon for this subunit. Primer SynA2 adds a *Hind*III site after the stop codon. For the catalytic subunit the primer SynB1 includes 5' of the start codon a *Hind*III restriction site and a ribosome-binding site (RBS), whereas the SynB2 primer provides a *Bam*HI site after the stop codon.

The isolated fragments are then sequentially subcloned in pBluescript SK+ (Stratagene, La Jolla, CA) in the same reading frame and on the same DNA, yielding a dicistronic construct. The correct presence of the *Synechocystis* fragments can be verified by sequencing. The *Nde*I–*Bam*HI fragment comprising the entire dicistronic construct is excised from Bluescript and ligated in the expression vector pET-3c (Stratagene) previously digested with *Nde*I and *Bam*HI.

Overexpression. *Escherichia coli* BL21 (DE3)pLysS is freshly transformed with the plasmid containing the dicistronic FTR construct. These cells are used to inoculate 500 ml of Luria–Bertani (LB) medium containing ampicillin (100 μg/ml) in a 1-liter Erlenmeyer flask. This preculture is grown overnight at 37° with vigorous shaking and used to inoculate 10 liters of LB medium supplemented with ampicillin (100 μg/ml) and 1 ml of antifoam A (Sigma, St. Louis, MO). The cells are grown in a 16-liter laboratory fermenter (Bioengineering, Wald, Switzerland) at 37°, sparged with air at 16 liters/min, and stirred at 500 rpm. When the culture reaches an $A_{600\,nm}$ of 0.8 to 1.0 the FTR expression is induced with 0.5 mM isopropyl thio-β-D-galactopyranoside (IPTG) and the culture is continued until the end of exponential growth. Cells are harvested by centrifugation (4000g for 10 min), resuspended in 200 ml of 50 mM Tris-HCl (pH 7.9), 0.1 mM phenylmethylsulfonyl fluoride (PMSF), 14 mM 2-mercaptoethanol, and frozen at $-20°$. All steps of the purification procedure are done at 4° unless otherwise indicated.

Cell Extract. Cells are thawed and sonicated 10 times (20 sec each) in a water–ice bath. Five units of Benzonase (Merck, Darmstadt, Germany) per milliliter of extract and MgCl$_2$ to 10 mM are added and the mixture is incubated at 25° until the viscosity due to the release of DNA has disappeared. The lysate is clarified by a 60-min centrifugation at 140,000g and 4°.

Hydrophobic Interaction Chromatography. Solid ammonium sulfate (226 mg/ml) is slowly added to the supernatant to reach 40% saturation and precipitated proteins are removed by centrifugation (48,000g, 15 min). After an additional filtration through a glass fiber filter (GF/C; Whatman, Clifton, NJ) to remove lipoproteins the clarified solution is loaded on a 5.0 × 5.0 cm Phenyl-Sepharose FF (Amersham Pharmacia, Piscataway, NJ) column equilibrated with 100 mM phosphate buffer (pH 7.3), 1.0 M (NH$_4$)$_2$SO$_4$, 14 mM 2-mercaptoethanol. The column is washed with 16 column volumes of equilibration buffer followed by elution of the retained proteins with 100 mM phosphate buffer (pH 7.3), 14 mM 2-mercaptoethanol. These proteins are concentrated by precipitation with ammonium sulfate at 90% saturation (603 mg/ml). After centrifugation (20 min at 32,000g) the pellets are dissolved in a minimal volume (≤40 ml) of 20 mM triethanolamine hydrochloride (pH 7.3), 400 mM NaCl, 14 mM 2-mercaptoethanol.

Gel Filtration. The concentrated protein solution is clarified by centrifugation and chromatographed at 90 ml/hr on a 5.0×90 cm Sephacryl S-100HR (Amersham Pharmacia) column equilibrated in 20 mM triethanolamine hydrochloride, (pH 7.3), 200 mM NaCl.

Anion-Exchange Chromatography. Active fractions are combined and diluted with an equal volume of 20 mM triethanolamine hydrochloride, pH 7.3, and loaded on a 5.0×11 cm Q-Sepharose FF (Amersham Pharmacia) column equilibrated with 20 mM triethanolamine hydrochloride (pH 7.3), 100 mM NaCl. The proteins are eluted with a 2000-ml gradient from 100 to 600 mM NaCl in 20 mM triethanolamine hydrochloride, pH 7.3. The active fractions are concentrated by ultrafiltration on a YM10 membrane (Amicon-Millipore).

Hydroxyapatite Chromatography. The concentrated fractions can be directly applied to a 1.5×14 cm hydroxyapatite column (40-μm Macro-Prep ceramic hydroxyapatite; Bio-Rad, Hercules, CA) equilibrated with 10 mM KP$_i$ buffer, pH 7.3. The column is washed with 100 ml of equilibration buffer before the FTR is eluted with 100 ml of 150 mM KP$_i$ buffer, pH 7.3.

Affinity Chromatography on Ferredoxin–Sepharose. As an alternative to the hydroxyapatite chromatography, the final purification can be achieved on a 1.5×6.5 cm ferredoxin-Sepharose column[23] equilibrated with 20 mM triethanolamine hydrochloride, pH 7.3. Before loading the concentrated fractions from the ion-exchange column must be diafiltered with the equilibration buffer. After adsorption of the FTR the column is washed with several column volumes of equilibration buffer until the absorption of the eluent is back to baseline. The FTR is desorbed with the same buffer containing 200 mM NaCl.

Concentration and Storage. The FTR fractions from the last purification step are combined, concentrated on a YM10 membrane, diafiltered with 20 mM triethanolamine hydrochloride, pH 7.3, and stored at liquid nitrogen temperature. Yields of 7 to 12 mg of pure FTR per liter of bacterial culture are obtained. The protein is stable over months when stored in the above described buffer at liquid nitrogen temperature.

Properties

Pure FTR is a yellow–brown protein, showing an absorbance spectrum typical for 4Fe–4S proteins with a flat peak in the visible range at about 410 nm and a molar absorbency of 17,400 M^{-1} cm^{-1} (Fig. 1). Its purity can be assessed by the 408 nm-to-278 nm absorbance ratio, which is 0.44 for the *Synechocystis* enzyme. This ratio is higher than for the spinach FTR because of the lower content of aromatic amino acids. In *Synechocystis* the variable subunit at 8000 Da is smaller than the catalytic subunit at 12,000 Da. The cyanobacterial FTRs lack a number of residues at the N-terminal end of the variable subunit that are present in the higher

[23] S. Ida, K. Kobayakawa, and Y. Morita, *FEBS Lett.* **65**, 305 (1976).

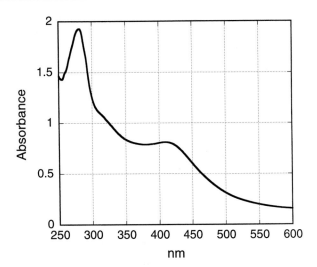

FIG. 1. Absorption spectrum of recombinant *Synechocystis* ferredoxin : thioredoxin reductase in 20 m*M* triethanolamine hydrochloride buffer, pH 7.3.

plant enzymes, where they account for the size variability of this subunit and its instability in the case of the spinach protein.[24]

The *Synechocystis* FTR has been crystallized by the hanging drop method and its three-dimensional structure determined by X-ray analysis.[8] It is a flat, disclike molecule. The variable subunit is a heart-shaped β-barrel structure on top of which sits the entirely α-helical catalytic subunit. In the center of the heterodimer, where the molecule is only 10 Å across, is the active site, composed of the 4Fe–4S cluster and the redox-active disulfide bridge. The Fe–S cluster is located on one side of the flat molecule close to the surface, containing three positive charges. The disulfide bridge is on the opposite surface, which has a more hydrophobic character. This arrangement allows the transfer of electrons across the FTR from ferredoxin to thioredoxin with a positively charged docking site for the negatively charged ferredoxin on one side and a more hydrophobic docking site for thioredoxins on the other side of the flat molecule. These properties make the FTR a versatile thioredoxin reductase, capable of accepting electrons from diverse ferredoxins and reducing the disulfides of various thioredoxins.

Spinach Thioredoxin *f*

Thioredoxin *f* is a necessary component for the activation of certain target enzymes. Its overexpression in the form of inclusion bodies and its solubilization and purification have been described in an earlier volume of this series.[17] Truncation

---

[24] A. Tsugita, K. Yano, L. Gardet-Salvi, and P. Schürmann, *Protein Seq. Data Anal.* **4**, 9 (1991).

of the N terminus of our original construct and insertion in the pET expression vector provide a thioredoxin f that is rapidly purified to yield a soluble, stable protein whose crystal structure has been solved.[9]

Assay Methods

Principle. The presence of thioredoxin f is detected by its capacity to catalyze FBPase activation in the presence of dithiothreitol (DTT). The resulting FBPase activity is measured, after dilution into a reaction mixture, either by a colorimetric determination of the P_i released or by the spectrophotometric test described for the FTR.

Activation. FBPase (0.25 units) is activated for 5 to 15 min at 25° in 50 μl containing 100 mM Tris-HCl (pH 7.9), 5 mM dithiothreitol, and thioredoxin f to be assayed.

Colorimetric Assay. After activation 0.45 ml of reaction mixture containing 100 mM Tris-HCl (pH 7.9), 0.1 mM EDTA-Na, 1.5 mM MgSO$_4$, and 1.5 mM sodium-fructose 1,6-bisphosphate is added and incubated at 25°. The reaction is stopped after 5 to 15 min by the addition of 2 ml of reagent for the P_i analysis. The samples are allowed to stand for 5 min before the absorbance at 730 nm is determined spectrophotometrically. P_i (0.5 μmol) yields an absorbance of about 0.67 in a cuvette with a 1-cm light path.

Preparation of P_i Analysis Reagent. H$_2$O (150 ml) is combined with 25 ml of 7 M H$_2$SO$_4$ and 25 ml of ammonium molybdate [(NH$_4$)$_6$Mo$_7$O$_{24} \cdot$4H$_2$O, 6.6 g/ 100 ml of H$_2$O] and well mixed, and then 3.1 g of FeSO$_4$ is added and dissolved. The reagent is stable for 1 to 2 days.

Production of Recombinant Truncated Thioredoxin f

Construction of Expression Vector. From a spinach leaf cDNA library the coding region for an N terminal-truncated thioredoxin f is isolated and amplified by PCR with the following primers:

M   E   A   I   V   G...

Sense primer TF1: 5'-<u>CC*ATGG*</u>AAGCCATTGTAGGG-3'

*   S   S   R   A   A...

Antisense primer TF2: 5'-<u>GGATCC</u>TCAACTACTTCGAGCAGC-3'

With the primer TF1 the protein starts at Met-10, thus removing the first nine residues. This primer also introduces an *Nco*I restriction site with the start codon. The primer TF2 introduces a *Bam*HI site after the stop codon. The amplified fragment is purified by electrophoresis, recovered from the agarose gel, and ligated into the expression vector pET3d previously digested with *Nco*I and *Bam*HI.

Principle. This thioredoxin f is expressed in *E. coli* as soluble protein and can be quantitatively released from the bacteria by osmotic shock.[25] This method has the advantage that relatively few nucleic acids and contaminating proteins are present in the extract and are easily removed by cation-exchange and size-exclusion chromatography.

Overexpression. *Escherichia coli* BL21 (DE3) is freshly transformed with the pET3d plasmid. The preculture and culture in LB medium are done at 37° as described for the FTR, except that the protein production is induced with 0.1 mM IPTG and the cells are harvested before reaching the end of exponential growth.

Release by Osmotic Shock. The harvested bacterial cells are carefully resuspended with a glass rod in one-twentieth the original volume of cold hypertonic buffer [20% (w/v) sucrose, 20 mM morpholinoethanesulfonic acid (MES)–NaOH, 2.5 mM EDTA-Na, pH 6.1] and left on ice for 10 minutes. The cells are collected by centrifugation for 5 min at 15,000g. The supernatant is decanted and the cells are resuspended as described previously in an equivalent volume of cold hypotonic buffer [20 mM MES–NaOH, 2.5 mM EDTA-Na (pH 6.1), 100 μM PMSF]. After 10 min of incubation on ice the cells are removed by centrifugation for 10 min at 15,000g.

Purification of Recombinant Truncated Thioredoxin f

The following purification steps are all done at 4°.

Cation-Exchange Chromatography. The supernatant from the osmotic shock, containing thioredoxin f, is centrifuged (48,000g for 30 min, 4°) and applied on an SP-Sepharose column (5 × 5.5 cm; Amersham Pharmacia) equilibrated in 50 mM malonate-HCl, pH 6.0. Thioredoxin f is eluted with a 2000-ml gradient of 0–200 mM NaCl in the same buffer and emerges from the column as a unique peak at about 120 mM NaCl. The peak fractions are concentrated on a YM10 membrane (Amicon).

Size-Exclusion Chromatography. The concentrated protein solution ($\leq$15 ml) is clarified by centrifugation and separated on a Sephadex G-50F column (2.6 × 95 cm) equilibrated and developed with 20 mM MES–NaOH (pH 6.1), 200 mM NaCl at 25 ml/hr. This step removes some minor, higher molecular weight contaminants.

Concentration and Storage. Thioredoxin f is concentrated on a YM10 membrane and diafiltrated with 20 mM MES–NaOH, pH 6.1, before storage at $-20°$. Yields of pure protein of 15 mg/liter bacterial culture are routinely obtained.

Properties

Truncated thioredoxin f has a molecular mass of 12,579, a pI of 7.15, and a molar absorbency at 278 nm of 14,200 M^{-1} cm^{-1}. It is essentially indistinguishable from the native protein in its capacity to activate FBPase.[26] The protein is stable when stored frozen and can be lyophilized.

Acknowledgment

Work done in the author's laboratory has been supported by the Schweizerischer Nationalfonds (31-47107.96 and 31-56761.99).

[25] N. G. Nossal and L. A. Heppel, *J. Biol. Chem.* **241,** 3055 (1966).
[26] G. del Val, F. Maurer, E. Stutz, and P. Schürmann, *Plant Sci.* **149,** 183 (1999).

[39] Plant Thioredoxin Gene Expression: Control by Light, Circadian Clock, and Heavy Metals

By STEPHANE D. LEMAIRE, MYROSLAWA MIGINIAC-MASLOW, and JEAN-PIERRE JACQUOT

Introduction

Thioredoxins (TRX) are small ubiquitous proteins ($\cong$12 kDa) found in all living organisms.[1,2] In plants, two isoforms called *m* and *f* are located in the chloroplast and are involved in the control of key carbon fixation enzymes by light. Under illumination the photosynthetic electron transfer chain generates reduced ferredoxin (Fd), which transfers its electrons to many acceptors including TRXs *m* and *f* via ferredoxin–thioredoxin reductase. In turn, thioredoxins are able to reduce several key enzymes of carbon fixation metabolism such as fructose-1,6-bisphosphatase and NADP malate dehydrogenase,[3,4] which are converted from an inactive to an active form. Another thioredoxin isoform, the *h*-type, is located in the cytosol and is reduced by NADPH through an NADPH thioredoxin reductase.[5] The exact function of the cytoplasmic isoform(s) is not fully understood in green leaves and algae.[2]

The posttranslational mechanisms involving TRXs have been extensively studied but little is known about the regulation of the expression of the different TRX genes. We have shown that TRX genes are regulated by light, the circadian clock, and heavy metals in the unicellular green alga *Chlamydomonas reinhardtii*.[6,7] In this article, we describe the experimental procedures used to study their expression.

Experimental Procedures

Strains, Media, and Culture Conditions

Strains and Media. Chlamydomonas reinhardtii cell wall-less strain CW15 (137c, mt⁺, cw15) can be obtained from the *Chlamydomonas* Genetics Center at

[1] J. P. Jacquot, J. M. Lancelin, and Y. Meyer, *New Phytol.* **136,** 543 (1997).

[2] P. Schürmann and J. P. Jacquot, *Annu. Rev. Plant Physiol. Plant Mol. Biol.* **51,** 371 (2000).

[3] M. Miginiac-Maslow, K. Johansson, E. Ruelland, E. Issakidis-Bourguet, I. Schepens, A. Goyer, M. Lemaire-Chamley, J.-P. Jacquot, P. Le Maréchal, and P. Decottignies, *Physiol. Plant.* **110,** 323 (2000).

[4] J. P. Jacquot, J. Lopez-Jaramillo, M. Miginiac-Maslow, S. Lemaire, J. Cherfils, A. Chueca, and J. Lopez-Gorge, *FEBS Lett.* **401,** 143 (1997).

[5] F. J. Florencio, P. Gadal, and B. B. Buchanan, *Plant Physiol. Biochem.* **31,** 649 (1993).

[6] S. D. Lemaire, M. Stein, E. Issakidis-Bourguet, E. Keryer, V. Benoit, B. Pineau, C. Gerard-Hirne, and J. P. Jacquot, *Planta* **209,** 221 (1999).

Duke University (Durham, NC). This strain is widely used because the absence of the cell wall facilitates cell fractionation, RNA extraction, and cell transformation. Cells are grown in HSM photoautotrophic minimal medium and pregrowth cultures are performed under continuous illumination in TAP medium. Basic procedures for the preparation of culture media are described in Harris.[8] The cultures are continuously stirred, bubbled with 5% CO_2, and maintained at 28°. Light intensity is 300 $\mu E/m^2 \cdot$ sec at the level of the flask culture. Cell growth can be monitored by counting the cells in a hemocytometer as described in Harris.[8]

Circadian Clock Setting. By definition, a circadian rhythm is entrained by environmental time cues such as alternating light/dark cycles. This rhythm has a period of approximately 24 hr, it persists under continuous conditions, and the period is temperature compensated.[9] In fact, when the circadian clock is reset by a 12-hr dark period, the phase of the rhythm is shifted so that the rhythm begins at the onset of illumination.[10] Once started, the rhythm persists under continuous conditions of illumination, that is, in continuous light or continuous darkness.

To demonstrate that the expression of a gene is controlled by the circadian clock, all three conditions must be fulfilled (period of approximately 24 hr, persistence under continuous conditions, and same period at different temperatures). When analyzing circadian oscillations the term "circadian time" (CT) is used to describe the phase of the rhythm under continuous conditions. A circadian cycle is divided in 24 circadian hours so that the subjective day spans CT 0 to CT 12 and the subjective night spans CT 12 to CT 24 (CT 24 is similar to CT 0).

For each experiment cell samples are collected every 4 hr from CT 20 to CT 20 over 48 hr. This corresponds to two cycles and should allow observation of two consecutive peaks. Shorter sampling times might be necessary to determine precisely the period of the rhythm but also at the dark-to-light and light-to-dark transitions, where rapid variations of expression are observed.

The cells are first cultured under light/dark cycles (12 hr/12 hr) for three cycles and the experiment starts at the fourth cycle. This allows determination of the period of the rhythm, which should be close to 24 hr to fulfill the first criterion. It is interesting to note that under these conditions the cells are synchronized. This is important because it has been reported that some plant TRX genes are highly expressed in dividing cells.[11] Thus, this experiment allows the investigator to test whether the expression is increased during cell divisions. Cell divisions occur

[7] S. Lemaire, E. Keryer, M. Stein, I. Schepens, E. Issakidis-Bourguet, C. Gerard-Hirne, M. Miginiac-Maslow, and J. P. Jacquot, *Plant Physiol.* **120**, 773 (1999).

[8] E. Harris, *in* "The *Chlamydomonas* Sourcebook." Academic Press, San Diego, California, 1989.

[9] L. N. Edmunds, *in* "Cellular and Molecular Bases of Biological Clocks." Springer-Verlag, New York, 1988.

[10] T. Kondo, C. H. Johnson, and J. W. Hastings, *Plant Physiol.* **95**, 197 (1991).

[11] F. Regad, C. Hervé, O. Marinx, C. Bergounioux, D. Tremousaygue, and B. Lescure, *Mol. Gen. Genet.* **248**, 703 (1995).

mainly after transfer of the cells into darkness.[12] Consequently, the level of an mRNA encoding a protein whose function is necessary for cell division would be expected to increase during this period. For example, in the case of *Chlamydomonas* TRXs a decrease in the mRNA level is observed during this period. This indicates that these TRXs are not likely to be involved in cell division processes.

In a second step, the expression is monitored in cells shifted to continuous conditions. The cells are grown in the light at 28°, shifted for 12 hr to the dark in order to reset the clock, transferred for 12 hr to the light in order to initiate the rhythm, and either maintained in continuous light or transferred to continuous darkness. To fulfill the second criterion, the oscillations should persist under these continuous conditions with a period of approximately 24 hr.

Finally, the period is also measured with cells grown at 21° to check that it is temperature compensated. The period should remain close to 24 hr to fulfill the third criterion. If the three criteria are met it can be concluded definitely that the expression is controlled by the circadian clock.

DCMU and DBMIB Treatments. In plant cells, regulation of gene expression by light may implicate various signaling pathways including classic photoreceptors and photosynthesis. To distinguish between these two possibilities, the effect of inhibitors of the photosynthetic electron transfer chain on the expression of light-regulated genes can be analyzed.

The expression patterns are monitored in cultures exposed to light in the presence or absence of DCMU [3-(3,4-dichlorophenyl)-1,1-dimethylurea] or DBMIB (2,5-dibromo-3-methyl-6-isopropyl-*p*-benzoquinone). The cells are first grown in the light and subjected to a resetting 12-hr dark period. DCMU and DBMIB are added 20 min before illumination at a final concentration of 50 μM and 1 μM, respectively. Because these compounds are dissolved in 2-propanol, control cultures are supplemented with an equal volume of 2-propanol at the same moment. Cell samples are collected at the dark-to-light transition (0 hr) and after 6 hr of illumination (6 hr), which correspond to the maximum RNA level for *C. reinhardtii* TRX *m* and TRX *h*. A control experiment is done on cultures maintained in continuous darkness.

Heavy Metal Treatments. To avoid interference with the light-dependent induction of thioredoxin gene expression,[6] the heavy metal treatments are performed on cultures maintained in darkness. Cells are grown under continuous illumination and the cultures are transferred to the dark 5 hr before the addition of heavy metals. The culture is separated into three bottles before transfer to the dark. The cultures are supplemented with either sterile water (control), $HgCl_2$, or $CdCl_2$ at final concentrations of 1 and 100 μM, respectively. These concentrations correspond to a sublethal dose for each cation. Cell samples are collected each hour from -2 to 6 hr, 0 hr being the time of heavy metal addition.

---

[12] S. D. Lemaire, M. Hours, C. Gérard-Hirne, A. Trouabal, O. Roche, and J. P. Jacquot, *Eur. J. Phycol.* **34**, 279 (1999).

Northern Blots

RNA Extraction. Approximately 30 million cells are collected in a sterile 50-ml tube (Falcon type) for each extraction and pelleted by centrifugation (3000g, 5 min, 4°). The pellet is immediately resuspended in 1 ml of TRIzol reagent (GIBCO-BRL, Gaithersburg, MD) or TRI reagent (Sigma, St. Louis, MO). Polysaccharides, membranes, and unlysed cells are eliminated by centrifugation (12,000g, 10 min, 4°). At this step the supernatant is either stored at −80° or immediately treated according to the supplier protocol. Briefly, add 200 μl of chloroform, incubate at room temperature for 2 min, and centrifuge (12,000g, 15 min, 4°). The supernatant (600 μl) is supplemented with 500 μl of 2-propanol and centrifuged (12,000g, 10 min, 4°) after a 10-min incubation at room temperature. The RNA pellet is washed [1 ml of 70% (v/v) ethanol and centrifugation at 8000g for 10 min at 4°], dried at room temperature for 30 min, and resuspended in 20 μl of Milli-Q (Millipore, Bedford, MA)-treated sterile water. Total RNA extracts can be stored at −20° for at least 1 month.

Gel Electrophoresis and Blotting. Before use all electrophoretic equipment is treated with 0.1 N NaOH to eliminate RNases. A 1.5% (w/v) agarose gel is prepared in 1× MOPS buffer [20 mM morpholine propane sulfonic acid (MOPS), 30 mM sodium acetate, 10 mM Na$_2$EDTA, pH 7]. After melting of the agarose in a microwave oven, formaldehyde is added (20%, v/v) and the gel is poured. For each sample the RNA concentration is determined by spectrophotometry and a volume corresponding to 10 μg is complemented with Milli-Q sterile water to a final volume of 8 μl. Each sample is supplemented with 14 μl of formamide, 5 μl of formaldehyde, and 3 μl of 10× MOPS buffer, denatured at 70° for 10 min, and supplemented with 3 μl of sterile loading buffer [0.25% (w/v) bromphenol blue, 30% (v/v) glycerol]. The samples are then loaded onto the gel. When necessary, 8 μl of RNA ladder (RNA Ladder; GIBCO-BRL) is also loaded. Migration is performed in 1× MOPS buffer at a constant 100 V for 3 hr.

After migration, the gel can be stained with ethidium bromide (1.25 μg/ml, 5 min). The gel is then destained by several 30-min washes in sterile water. Washes are stopped when ribosomal RNAs and ladder bands can be visualized under UV. The RNAs are blotted onto a positive membrane (Appligene; Qbiogene, Carlsbad, CA) by capillarity in 2× SSC (1× SSC is 0.15 M NaCl, 0.015 M sodium citrate) for 16 hr. The membrane is rinsed in distilled water for 5 min and the RNAs are covalently fixed onto the membrane by UV cross-linking (UV Stratalinker 1800; Stratagene, La Jolla, CA).

Labeling of Probes. The probes are synthesized by random priming (Nonaprimer kit II; Appligene, Qbiogene). The constitutive probe is a fragment of the coding region of a G protein β subunit-like polypeptide.[13] Because TRXs belong to a multigene family it is important to avoid cross-hybridization between different

---

[13] J. A. Schloss, *Mol. Gen. Genet.* **221,** 443 (1990).

TRXs. For this purpose, the TRX probes correspond to the 3' untranslated region of thioredoxin h and m isoforms,[14] obtained by polymerase chain reaction (PCR). The Fd probe, used as a control, also corresponds to the 3' untranslated region of the gene.

Hybridization and Washing. The membranes are prehybridized for 2 hr and hybridized overnight in 0.5 M sodium phosphate buffer (pH 7.2), 1 mM Na$_2$EDTA, 7% (w/v) sodium dodecyl sulfate (SDS), and 1% (w/v) bovine serum albumin (BSA). Hybridization is performed at 62° for all probes. Washes are performed in 40 mM sodium phosphate buffer (pH 7.2), 1 mM Na$_2$EDTA, 1% (w/v) SDS at ambient temperature (twice, 5 min each) and the final wash is performed at 62° for 20 min in the same buffer. Several exposure times are used in order to be in the linear response range of the film (X-Omat; Kodak, Rochester, NY).

Membrane Stripping. After hybridization the blot can be stripped by two rounds of boiling in 0.01× SSC, 0.5% (w/v) SDS and then hybridized with another probe. This stripping can be performed four or five times without altering the signals.

Western Blots

Protein Extraction. Total protein extracts are prepared by resuspension of a pellet of 30 million cells in 100 to 200 μl of extraction buffer [30 mM Tris-HCl (pH 7.5), 100 μM phenylmethylsulfonyl fluoride (PMSF)] followed by two rounds of freezing in liquid nitrogen. The extraction is facilitated by the fact that the strain is cell wall-less. The extracts are centrifuged at 12,000g for 15 min at 4° and the protein concentration of the supernatant is determined by the Bradford dye-binding assay (Bio-Rad, Hercules, CA).

Sodium Dodecyl Sulfate–Polyacrylamide Gel Electrophoresis and Blotting. Proteins are separated as a function of their molecular mass by electrophoresis on an SDS-containing polyacrylamide gel.[15] The procedure for SDS–polyacrylamide gel electrophoresis (PAGE) is standard and is described by Sambrook *et al.*[16] Protein samples (30 to 60 μg) are denatured for 3 min at 100° in loading buffer [2% (w/v) SDS, 10% (v/v) glycerol, 50 mM Tris-HCl (pH 6.8), 2% (v/v) 2-mercaptoethanol, 0.1% (w/v) bromphenol blue] and loaded on the gel with molecular mass markers. Migration is performed at a constant 25 mA per gel until the blue dye reaches the bottom of the gel (approximately 1 hr and 30 min).

After electrophoresis, the proteins are blotted onto a nitrocellulose membrane (0.45 μm, Hybond-C extra; Amersham, Piscataway, NJ). The blotting is performed

[14] M. Stein, J. P. Jacquot, E. Jeannette, P. Decottignies, M. Hodges, J. M. Lancelin, V. Mittard, J. M. Schmitter, and M. Miginiac-Maslow, *Plant Mol. Biol.* **28**, 487 (1995).

[15] U. K. Laemmli, *Nature (London)* **227**, 680 (1970).

[16] J. Sambrook, E. F. Fritsch, and T. Maniatis, *in* "Molecular Cloning: A Laboratory Manual." Cold Spring Harbor Laboratory Press, Cold Spring Harbor, New York, 1989.

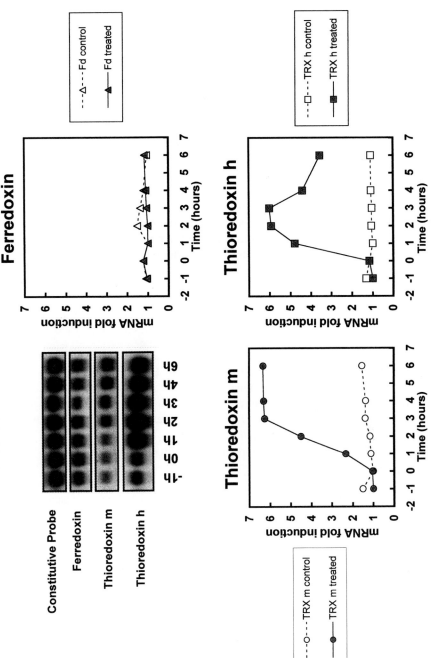

FIG. 1. Induction of thioredoxin gene expression by mercury. Cultures were maintained in the dark for 5 hr before the addition of mercury ($HgCl_2$, 1 μM final concentration). Open symbols and dashed lines: control culture; gray symbols and continuous line: culture supplemented with mercury. Triangles, Fd; squares, TRX h; circles, TRX m.

for 16 hr at 30 mA and 4° in transfer buffer [25 mM Tris-HCl (pH 8.3), 150 mM glycine, 20% (v/v) ethanol].[17]

Immunodetection. The blots are blocked with dry powdered milk (5%, w/v) in TBS buffer [20 mM Tris-HCl (pH 7.9), 150 mM NaCl] for 30 min. The antibodies used as a first-layer reagent are rabbit polyclonal monospecific anti-TRX h or anti-TRX m IgG [1000-fold dilution in 5% (w/v) milk–TBS for 1 hr]. The membranes are rinsed in 5% (w/v) milk–TBS (twice, 15 min each) and exposed to secondary antibodies (goat anti-rabbit IgG conjugated with horseradish peroxidase; Bio-Rad) [2500-fold dilution in 5% (w/v) milk–TBS for 1 hr]. After thorough washing [15 min in 5% (w/v) milk–TBS and twice for 15 min each in TBS], the immunoreactive proteins are detected by the color produced on reaction of horseradish peroxidase with H_2O_2 and 4-chloro-1-naphthol in TBS buffer. To prepare the revealing solution dissolve 30 mg of 4-chloro-1-naphthol in 10 ml of ethanol, add 60 μl of H_2O_2 (8.8 M), and complete to 60 ml with TBS. The reaction is stopped after 5 to 30 min by rinsing the membrane in distilled water. All incubations are performed at room temperature. Alternatively, the immunoreactive proteins can be revealed by the enhanced chemiluminescence (ECL) technique instead of with 4-chloro-1-naphthol.[18]

Analysis of Results

The autoradiograms and Western blots can be quantified by densitometric scans (Masterscan; Scanalytics, CSPI, Billerica, MA). This allows normalization of the signals to a constitutive probe signal to take into account loading variations. Results can be expressed as induction fold or as relative levels. In the former case the lowest value is considered as the 1.0 reference whereas in the latter the highest value is considered as the 100% reference.

Heavy Metal Control. A classic Northern blot result for heavy metal induction of TRX gene expression is presented in Fig. 1. The results are also presented as induction fold. A rapid induction of both TRX mRNAs is observed after addition of mercury. The Fd mRNA level is an excellent control because it does not increase after addition of the metal ion whereas it is circadian clock controlled like both TRX genes.[6]

Circadian Clock Control. The results presented in Fig. 2 demonstrate that *Chlamydomonas* TRX m and h genes are controlled by the circadian clock. The three criteria are fulfilled because the period of the rhythm is close to 24 hr, the rhythm persists under continuous conditions, and the period is temperature compensated because it is unchanged at 21° compared with 28°. However, when

---

[17] H. Towbin, T. Staehelin, and J. Gordon, *Proc. Natl. Acad. Sci. U.S.A.* **76,** 4350 (1979).
[18] I. Durrant, *Nature (London)* **346,** 297 (1990).

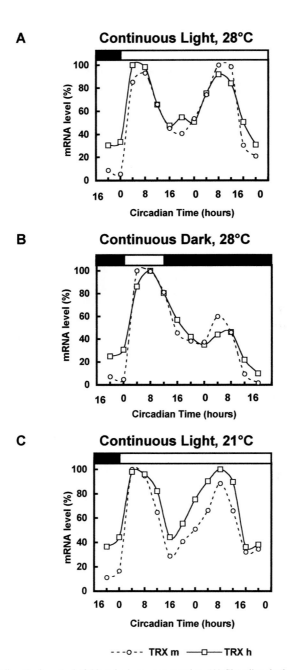

FIG. 2. Circadian clock control of thioredoxin gene expression. (A) Circadian rhythm of expression in cells shifted to continuous light at 28°. (B) Circadian rhythm of expression in cells shifted to continuous darkness at 28°. (C) Circadian rhythm of expression in cells shifted to continuous light at 21°. Relative mRNA levels for TRX *m* (circles) and TRX *h* (squares). The solid and open boxes represent the dark and light periods, respectively.

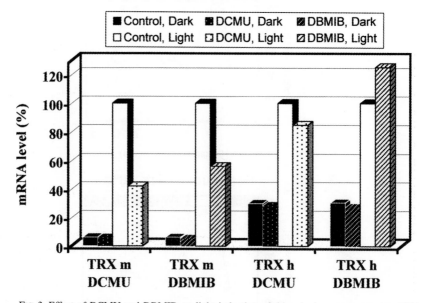

Fig. 3. Effect of DCMU and DBMIB on light induction of thioredoxin gene expression. RNA samples were prepared at the dark-to-light transition (0 hr) and 6 hr later. Dark columns, control, 0 hr; open columns, control, 6 hr; dark dotted columns, DCMU treated, 0 hr; open dotted columns, DCMU treated, 6 hr; dark dashed columns, DBMIB treated, 0 hr; open dashed columns, DBMIB treated, 6 hr.

the experiment is performed in the dark, although the rhythm persists, the second peak is always lower than the first. This phenomenon, known as dampening of the rhythm, is often observed when studying circadian rhythms.[10,19]

DCMU and DBMIB Treatments. Among the two inhibitors used, DCMU blocks the electron transfer chain just before plastoquinone (PQ), between quinones Q_A and Q_B of photosystem II. On the other hand, DBMIB blocks the chain after PQ, between PQ and the cytochrome $b_6 f$ complex. Thus, in the presence of DCMU the PQ pool is oxidized whereas it is reduced in the presence of DBMIB. It has been shown that the redox state of the PQ pool can control the expression of nuclear genes.[20,21] Thus, if the expression is controlled by the redox state of the PQ pool, we expect an opposite effect of DCMU and DBMIB.

[19] W. C. Taylor, *Plant Cell.* **1,** 259 (1989).
[20] S. Karpinski, C. Escobar, B. Karpinska, G. Creissen, and P. M. Mullineaux, *Plant Cell.* **9,** 627 (1997).
[21] J. M. Escoubas, M. Lomas, J. Laroche, and P. G. Falkowski, *Proc. Natl. Acad. Sci. U.S.A.* **92,** 10237 (1995).

An example of such an analysis in the case of *Chlamydomonas* TRX *m* and *h* treatments is presented in Fig. 3. Two distinct results are observed. In the case of TRX *m*, the light induction is decreased in the presence of the inhibitors but not completely abolished. This suggests that several light transduction pathways might be involved in the control of TRX *m* gene expression by light, one of them being dependent on the photosynthetic electron transfer chain. However, DCMU and DBMIB affect similarly the expression of this gene. Thus, the redox state of the plastoquinone pool is clearly not involved in this control. On the other hand, the inhibitors have little or no effect on the light induction of TRX *h* mRNA. This indicates that the TRX *h* mRNA level is likely to be regulated by a different light transduction pathway, independent of the photosynthetic electron transfer chain.

[40] Thioredoxin Genes in Lens: Regulation by Oxidative Stress

By KAILASH C. BHUYAN, PABBATHI G. REDDY, and DURGA K. BHUYAN

Introduction

Thioredoxin (Trx), along with glutaredoxin (Grx) and protein disulfide isomerases (PDIs), belongs to the class of thiol/disulfide oxidoreductases shown to catalyze oxidation–reduction of cysteine residues, which play a major role in the three-dimensional structure and function of several proteins.[1–3] The redox-active disulfide/dithiol moiety at the conserved catalytic site sequence, -Cys-Gly-Pro-Cys- in Trx, and -Cys-Pro-Tyr-Cys- in Grx, and conserved domains homologous to Trx in PDI, is the characteristic feature of these protein molecules.[2–4] The Grx system requires the reduced form of glutathione (GSH) to catalyze GSH-disulfide oxidoreduction. The resulting product, the oxidized form of glutathione (GSSG), is a potent inhibitor of this reaction, showing that the overall cellular concentrations of GSH and GSSG are important determinants of the functional efficacy of this system.[5] Trx, being independent of GSH, operates by the NADPH-dependent flavin adenine dinucleotide-containing enzyme, thioredoxin reductase (TR) (NADPH: oxidized thioredoxin oxidoreductase; EC 1.6.4.5). The conserved catalytic site of

[1] J. C. A. Bardwell and J. Beckwith, *Cell* **74**, 769 (1993).
[2] B. B. Buchanan, *in* "Thioredoxin and Glutaredoxin Systems" (A. Holmgren, C. I. Branden, H. Jornvall, and B. M. Sjoberg, eds.), p. 233. Raven Press, New York, 1986.
[3] A. Holmgren and M. Björnstedt, *Methods Enzymol.* **252**, 199 (1995).
[4] G. Powis, D. Mustacich, and A. Coon, *Free Radic. Biol. Med.* **29**, 312 (2000).
[5] A. Holmgren, *J. Biol. Chem.* **254**, 3672 (1979).

TR, -Cys-Val-Asn-Val-Gly-Cys-, undergoes reversible oxidation reduction similar to Trx. The TR catalyzes reduction of Trx-S_2 to Trx-$(SH)_2$, which spontaneously reduces protein disulfide (protein-S_2) to protein-$(SH)_2$ [Eqs. (1) and (2)].[3,4]

$$\text{Trx-}S_2 + \text{NADPH} \cdot H^+ \underset{}{\overset{TR}{\rightleftharpoons}} \text{Trx-}(SH)_2 + \text{NADP}^+ \qquad (1)$$

$$\text{Trx-}(SH)_2 + \text{protein-}S_2 \underset{}{\overset{Spontaneous}{\rightleftharpoons}} \text{Trx-}S_2 + \text{protein-}(SH)_2 \qquad (2)$$

The redox potential (E_0') of NADPH $\cdot$ H^+/NADP$^+$ is -315 mV at pH 7.0, and 25°, and that of *Escherichia coli* Trx-$(SH)_2$/Trx-S_2 is -270 mV.[2–4] Thioredoxin-like domains in PDI have been shown to be the catalytic site for native disulfide formation during protein folding in the endoplasmic reticulum of cells actively synthesizing disulfide-linked cell surface receptors and secretory proteins.[6]

Trx was originally found in *E. coli* as a hydrogen donor for ribonucleotide reductase, which is essential for DNA synthesis.[2,3] Subsequently, this redox-active low molecular weight protein of about 12 kDa has been isolated from a wide variety of prokaryotic and eukaryotic cells.[2–4] The Trx gene family is widely distributed in several mammalian tissues, and the multifunctional roles of Trx and TR have been reviewed.[7–10] Trx participates in the repair of damaged sulfhydryl groups of several proteins including major transcription factors. The significance of the Trx protein in the regulation of transcription factors such as nuclear factor-κB (NF-κB) and activator protein 1 (AP-1) became apparent by its translocation from cytoplasm to nucleus on oxidative stress. Its association with apoptosis signal-regulating kinase 1 (ASK-1) indicated its participation in protection against cell death or apoptosis. This was further supported by the higher susceptibility of Trx-1 mutant yeast cells to H_2O_2 toxicity. Trx-1 null mutant mice survived only 10 days after birth, showing its involvement in cellular development and differentiation during embryogenesis. Trx was linked to cell signaling via p38, and was identified as an activator of the endogenous glucocorticoid or interleukin 2 receptors. Trx and TR function as an antioxidant in the removal of cytotoxic reactive oxygen species (ROS) at the membrane site.[7–10] The toxicity of ROS generated by chemical agents in the retinal pigment epithelial cells,[11] and of H_2O_2 to lens epithelial cells,[12] was decreased

[6] R. B. Freedman, R. Hirst, and M. F. Tuite, *Trends Biochem. Sci.* **19**, 331 (1994).

[7] J. J. Mieyal, S. A. Gavvina, P. A. Mieyal, U. Srinivasan, and D. W. Stark, in "Biothiol in Health and Disease" (L. Packer and E. Cadenas, eds.), p. 305. Marcel Dekker, New York, 1995.

[8] H. Nakamura, K. Nakamura, and J. Yodoi, *Annu. Rev. Immunol.* **15**, 351 (1997).

[9] C. K. Sen, *Biochem.Pharmacol.* **55**, 1747 (1998).

[10] K. U. Schallreuter and J. M. Wood, *Free Radic. Biol. Med.* **6**, 519 (1989).

[11] H. Shibuki, N. Katai, S. Kuroiwa, T. Kurokara, J. Yodoi, and N. Yoshimura, *Invest. Ophthalmol. Vis. Sci.* **39**, 1470 (1998).

[12] A. Spector, G. H. Yan, R. R. Huang, M. J. McDermitt, P. R. Gascoyne, and V. Pigiet, *J. Biol. Chem.* **263**, 4984 (1988).

when Trx was added to culture media, showing its protective ability. Antioxidants as sensors of oxidative stress and their redox regulatory role in gene transcription have been discussed.[9,13]

Superoxide dismutase (SOD), catalase, and glutathione peroxidase (GSHPx) systems are primary defenses against cellular damage by ROS.[14,15] In the ocular lens under physiological conditions, activities of antioxidant enzymes, catalase,[16] GSHPx,[16,17] SOD,[18] and an excess of GSH over GSSG[19] are responsible for maintenance of the reducing milieu of cytosol and plasma membrane integral protein–lipid components. Other contributors are ascorbic acid[18] and a high ratio of NAD(P)H to NAD(P).[20] Thiol:disulfide oxidoreductases can act as redox sensors and participate in the redox regulation of transmembrane signal transduction and gene transcription for cellular functions at normal levels of ROS.[8,9] Studies of experimental cataracts *in vivo*[16,18,21–23] and *in vitro*[24,25] provided convincing evidence that oxidative stress, generated by ROS, shifts the lenticular redox potential to the oxidizing state and plays a significant role in the formation of lens opacity known as cataract. In elderly human normal lenses and in early cataracts, there is depletion of the GSH pool with protein S-thiolation forming protein-glutathione (PSSG) and protein-cysteine (PSSC) mixed disulfides.[26,27] Formation of high molecular weight aggregates by covalent disulfide cross-linking of proteins unfolded by oxidative stress contributes to loss of lens transparency, leading to age-related human cataract.[26,27] It has been proposed that thiol:disulfide oxidoreductases, including Trx, provide a mechanism to protect and repair the sulfhydryl groups of proteins after oxidative stress.[2–4,7,9]

[13] C. K. Sen and L. Packer, *FASEB J.* **10,** 709 (1996).

[14] I. Fridovich, *in* "Biochemical and Medical Aspects of Oxygen" (O. Hayaishi and K. Asada, eds.) p. 171. University Park Press, Baltimore, Maryland, 1977.

[15] B. Chance, H. Sies, and A. Boveris, *Physiol. Rev.* **59,** 527 (1979).

[16] K. C. Bhuyan and D. K. Bhuyan, *Biochim. Biophys. Acta* **497,** 641 (1977).

[17] A. Pirie, *Biochem. J.* **96,** 244 (1965).

[18] K. C. Bhuyan and D. K. Bhuyan, *Biochim. Biophys. Acta* **542,** 28 (1978).

[19] J. G. Bellows, *in* "Cataract and Anomalies of the Lens," p. 212. C.V. Mosby, St. Louis, Missouri, 1944.

[20] F. J. Giblin and V. N. Reddy, *Exp. Eye Res.* **31,** 601 (1980).

[21] K. C. Bhuyan and D. K. Bhuyan, *Curr. Eye Res.* **3,** 67 (1984).

[22] D. K. Bhuyan and K. C. Bhuyan, *Methods Enzymol.* **233,** 630 (1994).

[23] D. K. Bhuyan, X. Huang, G. Kuriakose, W. H. Garner, and K. C. Bhuyan, *Curr. Eye Res.* **16,** 519 (1997).

[24] S. D. Varma, S. Kumar, and R. D. Richards, *Proc. Natl. Acad. Sci. U.S.A.* **76,** 3504 (1979).

[25] J. S. Zigler, Jr., H. M. Jernigan, D. Garland, and V. N. Reddy, *Arch. Biochem. Biophys.* **241,** 163 (1985).

[26] J. J. Harding and K. J. Dilley, *Exp. Eye Res.* **22,** 1 (1976).

[27] R. C. Augusteyn, *in* "Mechanisms of Cataract Formation in the Human Lens" (G. Duncan, ed.), p. 71. Academic Press, New York, 1981.

In this article, experimental procedures with data are described to demonstrate Trx-1 and Trx-2 gene and protein expression in the ocular lens, activities of cytosolic and membrane-bound TR in normal and cataractous lenses of the human, and Trx gene and protein regulation in the Emory mouse lens in the presence of oxidative stress *in vivo*.

Experimental Procedures

Human Donor Lenses

Eyes from human donors, ages 19–96 years, received at the Eye-Bank for Sight Restoration (New York, NY), were transported on ice to the laboratory within 12–48 hr postmortem. The lenses were immediately dissected from the eyes and examined for clarity with a slit-lamp biomicroscope (slit-lamp). Intact normal or decapsulated lenses, and lens capsule epithelium, were collected in plastic vials in liquid nitrogen. The tissues were either processed on the same day or kept frozen at −85° until use. A single lens from an individual donor eye was taken for isolation of total RNA, and the lens from the contralateral eye was used for analysis of Trx protein or cytoplasmic and membrane-bound TR activities.

Cataractous lenses used in this study were the few, among Eye-Bank donor eyes, that had lens opacities as observed with a slit-lamp. The capsule epithelium (capsulorhexis) samples, provided by A. M. Cotliar and C. B. Camras, were obtained during the routine surgical extraction of cataracts from patients at the Edward S. Harkness Eye Institute of Columbia University (New York, NY) and Department of Ophthalmology at the University of Nebraska Medical Center (Omaha, NE), respectively.

It is important to perform correlative studies with fresh animal lenses to confirm the data obtained from human lenses because of variable sampling conditions such as different pathological states, dietary habbits, drug intake, diverse genetic background, and environmental impact. Reliable samples are those received 12 to 24 hr postmortem.

Test Animals

The animal investigations were conducted according to the *Guide for the Care and Use of Laboratory Animals* [DHEW Publication No. (NIH) 86-23, Office of Science and Health Reports, DRR/NIH, Bethesda, MD], and Recommendations from the Declaration of Helsinki. The Emory mouse develops cortical cataract at old age (9 to 11 months), and has been considered to be a model for age-related human cataract.[28] This substrain of mice, which was a special strain derived from Carworth Farm Webster (CFW) mice,[28] is from our inbred colony maintained in the

[28] J. F. R. Kuck, Jr., *Exp. Eye Res.* **50,** 659 (1990).

animal facility at the Institute of Comparative Medicine of Columbia University. For the *in vivo* studies, 4-week-old healthy mouse pups of both sexes were selected after confirming by examination with a slit-lamp that their eyes were normal. For *in vivo* photochemical oxidative stress to mice, feeding of a dietary supplement of riboflavin along with whole-body fluorescent light exposure was considered ideal because it appears to simulate the human condition of exposure to sunlight in the presence of an endogenous photosensitizer such as riboflavin, present at a level of 12 μM in the lens epithelium[29] and 30 nM in the aqueous humor.[30]

An experimental group of mouse pups (50) was fed a special diet prepared by supplementing 0.05% (w/w) riboflavin in fresh ground Purina Rodent Lab Chow 5001 (obtained from Ralston Purina, St. Louis, MO), along with whole-body exposure to fluorescent light of 5812-lux illumination intensity daily according to a 12-hr : 12-hr light and dark cycle. They were placed, at five per cage, on a stainless steel rack equipped with fluorescent tube lights. Ultraviolet (UV) radiation-cutoff Plexiglas filters were installed above the cages to avoid UV from the incident visible light. A control group of mouse pups (25), maintained on the regular diet, received a similar daily exposure according to a 12-hr : 12-hr light and dark cycle, and a second control group of pups (25) was raised on the special diet without additional light exposure. All the mice received tap water for drinking, had access to the respective diets *ad libitum,* and were housed in an environmentally controlled (25° and 45–55% relative humidity) animal facility.

The eyes of the mice were examined periodically with a Haag–Streit slit-lamp for lens opacities. Before each examination, the pupils were fully dilated by topical application of 20 μl of 0.1% (v/v) Mydriacyl (tropicamide ophthalmic solution; Alcon, Fort Worth, TX). The lenses of the mice exposed to photochemical oxidative stress for 1 to 12 weeks remained clear. At varying times, the mice were killed by CO_2 euthanasia and their eyes were enucleated. The dissected lenses, liver, kidney, and tail were collected in liquid nitrogen and stored at −85° until processed for analysis of redox-related gene and protein expression.

Reactive Oxygen Species in Eye

By applying techniques described previously[22,23] to lenses of Emory mice subjected to *in vivo* photochemical oxidative stress, the cumulative level of O_2^- was determined on the basis of the SOD-inhibitable reduction of ferricytochrome c at 550 nm.[31] OH was estimated at 510 nm by quantitating 2,3-dihydroxybenzoate produced by hydroxylation of salicylate,[32] and H_2O_2 was assayed at 480 nm by

[29] V. E. Kinsey and C. E. Frohman, *AMA Arch. Ophthalmol.* **46,** 536 (1951).
[30] F. J. Philpot and A. Pirie, *Biochem. J.* **37,** 250 (1943).
[31] I. Fridovich, *in* "Handbook of Methods for Oxygen Radical Research" (R. A. Greenwald, ed.), p. 121. CRC Press, Boca Raton, Florida, (1986).

TABLE I
REACTIVE SPECIES OF OXYGEN IN MOUSE LENS: EFFECT OF *in vivo* OXIDATIVE STRESS

Mouse	Duration of oxidative stress	Reactive oxygen species $[\mu M]^a$		
		H_2O_2	O_2^-	$\cdot OH$
Control[b]	None	9 ± 1	18 ± 2	9 ± 2
Experimental[b]	4.5 weeks	83 ± 6	195 ± 2	157 ± 3

[a] Mean $\pm$ SEM, $n = 4$ groups of 4 to 10 lenses of each. Lens water is taken as 65% of the wet weight.
[b] Details regarding the control and experimental groups of mice are given in the section Test Animals in text.

estimating pink ferrithiocyanate produced by oxidation of $Fe(NH_4)_2(SO_4)_2$ by H_2O_2 in the sample in the presence of potassium thiocyanate.[33]

From the results given in Table I, it is seen that, in the lenses of mice fed a riboflavin-supplemented diet along with light exposure for 4.5 weeks, generation of oxidative stress was indicated by a rise in O_2^- and H_2O_2 by 9- to 10-fold, and $\cdot OH$ by about 17-fold compared with the respective levels of ROS in the lenses of control mice. It is important to note that production of oxidative stress in human eyes with cataracts has been suggested by observations of significantly higher than normal levels of H_2O_2, about 3-fold in the aqueous humor, $120 \pm 10 \ \mu M$ (mean $\pm$ SEM, $n = 10$ eyes) of cataract patients,[34] about 2-fold in the vitreous humor, $160 \pm 8 \ \mu M$ ($n = 14$ eyes) in cataract,[35] and 5- to 6-fold in the cataractous lenses, $186 \pm 16 \ \mu M$ ($n = 6$ lenses).[36]

mRNAs Encoding Thioredoxin Genes in Lens

Isolation of Total RNA

Total RNA from lenses is isolated and purified by using TRIzol reagent according to the manufacturer instructions. The TRIzol reagent (GIBCO-BRL, Life Technologies, Grand Island, NY) is a monophasic solution of phenol and guanidine isothiocyanate, which improves the single-step isolation method[37] of total RNA from cells and tissues. To prevent RNase contamination, disposable gloves and

---

[32] B. Halliwell and J. M. C. Gutteridge, *in* "Handbook of Methods for Oxygen Radical Research" (R. A. Greenwald, ed.), p. 177. CRC Press, Boca Raton, Florida, 1986.

[33] A. G. Hildebrandt, I. Root, M. Tjoe, and G. Heinemeyer, *Methods Enzymol.* **52**, 342 (1978).

[34] K. C. Bhuyan, D. K. Bhuyan, and S. M. Podos, *Life Sci.* **38**, 1463 (1986).

[35] K. C. Bhuyan, D. K. Bhuyan, and S. M. Podos, *IRCS Med. Sci.* **9**, 126 (1981).

[36] D. K. Bhuyan, C. B. Camras, H. K. Lakhani, and K. C. Bhuyan, *Invest. Ophthalmol. Vis. Sci.* **33**, 798 (1992).

[37] P. Chomczynski and N. Sacchi, *Anal. Biochem.* **162**, 156 (1987).

sterile plasticware are used. A homogenate of 8 to 10 mouse lenses prepared in 0.5 ml of TRIzol, or of a single human lens prepared in 1 ml of TRIzol, using a Tekmar tissue homogenizer, is held at room temperature ($\sim$25°) for 5 min to achieve complete dissociation of nucleoprotein complexes. This is followed by addition of 0.2 ml of chloroform per milliliter of TRIzol reagent used. The sample tube is capped and contents are shaken vigorously for 15 sec, incubated at 25° for 2 to 3 min, and centrifuged at 12,000g for 15 min at 5°. After centrifugation the colorless upper aqueous phase is transferred to a fresh tube, and RNA is precipitated by mixing it with 2-propanol (0.5 ml/ml of TRIzol reagent used initially), incubated at 25° for 10 min, and centrifuged at 12,000g for 10 min at 5°. The gel-like pellet of RNA obtained is washed with 75% (v/v) ethanol and centrifuged at 7500g for 5 min at 5°. The purified RNA is dissolved in 25 μl of water treated with diethyl pyrocarbonate (DEPC) to eliminate any RNase or DNase activity, and the solution is stored at −85° until use. An absorbance ratio >1.8 at 260/280 nm indicates a pure RNA preparation. About 25 μg of total RNA from 10 lenses of mice and about 20 μg of total RNA from a single human lens can be obtained.

Thioredoxin Gene: Primer Design and Synthesis

Already known murine and human cDNA sequences of Trx-1 and Trx-2 genes are taken from GenBank (available at *www.ncbi.nlm.nih.gov*) for primer design. The primers are chosen so that the polymerase chain reaction (PCR) products amplified by a given cDNA primer pair are specific for the genes under investigation, and show little possible sequence homology to other known genes. The primers are synthesized by the DNA facility at Columbia University and Applied BioSystems (Foster City, CA). The PCR conditions are tested for each primer set for each tissue and optimized. The primer sequences of murine and human Trx genes are given in Table II.

Semiquantitative Reverse Transcriptase-Polymerase Chain Reaction

For analyses of expression of the Trx gene family by reverse transcriptase (RT)-PCR, a Gene Amp RNA PCR core kit from Perkin-Elmer Roche Molecular Systems (Foster City, CA), designed for analyses of gene expression at the RNA level, is used according to the instructions of the manufacturer on the basis of the published technique.[38] A reaction mixture containing 5 mM MgCl$_2$, PCR buffer [50 mM KCl, 10 mM Tris-HCl (pH 8.3)], dGTP, dATP, dTTP, and dCTP (1 mM each), 20 U of RNase inhibitor, 50 U of cloned murine leukemia virus (MuLV) reverse transcriptase, 2 μg of total RNA, isolated from lens, is added in a final volume of 20 μl, adjusted with autoclaved deionized,

[38] R. K. Saiki, D. H. Gelfand, S. Stoffel, S. J. Scharf, R. Higuchi, G. T. Horn, K. B. Mullis, and H. A. Erlich, *Science* **239,** 487 (1988).

TABLE II
THIOREDOXIN 1 AND THIOREDOXIN 2 cDNA SEQUENCES USED FOR SYNTHESIS OF PRIMER PAIRS

Gene	GenBank accession number	Primer Sequences (sense and antisense)	Position	PCR product size (bp)
Trx-1, human	AF276919	5'-ATGGTGAAGCAGATCGAGAG-3'	1–20	420
		5'-GTCACGCAGATGGCAACTGGTT-3'	400–421	
Trx-2, human	AF276920	5'-ATGGCTCAGCGACTTCTTCT-3'	1–20	590
		5'-GAGGGAGGCAGCAGGAAGGG-3'	571–590	
β-Actin, human	AB004047	5'-ACCCCGTGCTGCTGACCGAGG-3'	312–332	731
		5'-TGGAGCCGCCGATCCACAC-3'	1025–1043	
Trx-1, murine	X77585	5'-GTCGTGGTGGACTTCTCTG-3'	141–159	325
		5'-GGGTATAGACTCTCCACAC-3'	468–486	
Trx-2, murine	U85089	5'-GATGGCTCAGCGGCTCCTCC-3'	51–70	589
		5'-ATGACAGCTGAGGGCTGGGA-3'	621–640	
GAPDH, murine	M32599	5'-CAAATTCAACGGCACAGTC-3'	202–220	498
		5'-GACCTTGCCCACAGCCT-3'	684–700	

ultrafiltered water (purified water), and incubated at 25° for 10 min to reverse transcribe the RNA template into cDNA.

For the PCR, in a final volume of 100 μl assay mixture including 20 μl of reverse transcriptase reaction mixture; 78 μl of solution containing 2 mM MgCl$_2$, PCR buffer, purified water, and 2.5 U of AmpliTaq DNA polymerase; and 2 μl containing 10 pmol each of the upstream and downstream primers is added. PCR amplification is done in a thermal cycler (Omn-E; Hybaid, Middlesex, UK). The amplification conditions are 95° for 4 min for denaturation of RNA/DNA hybrids; 55° for 1 min and 72° for 1 min for the first cycle; and 94° for 30 sec, 55° for 1 min, 72° for 1 min for 40 cycles of annealing. The last cycle includes extension at 72° for 10 min.

Glyceraldehyde-phosphate dehydrogenase (GAPDH) primers for murine RNA and β-actin primers for human RNA are used to monitor equal amounts of RNA used in the RT-PCR. The PCR products are separated by ethidium bromide-stained 1.5% (w/v) agarose gel electrophoresis, visualized under UV light, and photographed. The PCR products are extracted from the gel by using a DNA extraction kit (Qiagen, Santa Clarita, CA) and quantitated for graphic analyses. The gel-purified PCR products are cloned into pGEM-T cloning vector system II Promega, Madison, WI). The plasmid DNA is isolated from an overnight-grown single colony culture, using QIA prep plasmid kit (Qiagen). The plasmid DNA is sequenced by an ABI Prism 377 automated DNA sequencer (Perkin-Elmer Roche Molecular Systems) at a concentration of 40 fmol in 12 μl of purified water with 3.2 pmol of either upstream or downstream primer. Negative controls with no cDNA in the PCR mixture are included in all experiments. All the RT-PCR in a single experiment are done with the same stock reagents.

Western Blot Analysis of Thioredoxin

In a typical experiment, a homogenate of two mouse lenses or of the capsule epithelium of a single human lens is prepared in 200 μl of buffer A [2 mM Tris (pH 7.5), 1 mM EDTA, 10 mM 2-mercaptoethanol, and 20 mM phenylmethylsulfonyl fluoride (PMSF)]. Protein concentrations in the lens extracts are measured,[39] using crystallized bovine serum albumin as standard. Total protein, 150 μg of the mouse lens or 100 μg of the human lens epithelium, is resolved into polypeptide components by 10% (w/v) sodium dodecyl sulfate–polyacrylamide gel electrophoresis (SDS–PAGE),[40] and electroblotted onto nitrocellulose membrane. Nonspecific binding is blocked by soaking the Trans-Blot (Bio-Rad, Hercules, CA) overnight with 5% (w/v) nonfat dry milk in TTBS buffer [20 mM Tris-HCl buffer, pH 7.6, containing 137 mM NaCl and 0.05% (v/v) Tween 20], followed by washing with TTBS for 1 hr, incubation with polyclonal antibody to murine thioredoxin 1 reactive to human and murine Trx-1 (kindly provided by J. Yodoi, Kyoto University, Kyoto, Japan) at 1 : 5000 dilution in TTBS, and a wash for 1 hr with TTBS. The immunoreactive protein bands are visualized by chemiluminescence after labeling with horseradish peroxidase-labeled anti-rabbit IgG at 1 : 1000 dilution. The protocol for enhanced chemiluminescence (ECL) Western blot analysis is as described by Amersham Life Sciences (Arlington Heights, II). The chemiluminescent signal from the immunoreactive band of Trx protein recorded on Hyperfilm by autoradiography is scanned with a densitometer (model 300A; Molecular Dynamics, Sunnyvale, CA) at the DNA Facility, Columbia-Presbyterian Cancer Center (New York, NY).

Using semiquantitative RT-PCR, cDNA transcripts of mRNAs encoded by Trx-1 and Trx-2 genes in normal lenses of humans, ages 19, 41, 45, 60, and 74 years, are identified (Fig. 1). The experiments are done in duplicate. The expression of Trx-1 and Trx-2 genes in the lens appears to decrease with advanced age. However, to confirm this observation, more human lenses of different ages need to be studied. The nucleotide sequences of the cDNA representing mRNA bp 1–421 of Trx-1, and of the cDNA representing mRNA bp 1–590 of Trx-2 of the human lens, have been deposited in GenBank under accession numbers AF276919 and AF276920, respectively. The amino acid sequences deduced

[39] M. M. Bradford, *Anal. Biochem.* **72,** 258 (1976).
[40] U. K. Laemmli, *Nature* (*London*) **227,** 680 (1970).

Age (yr)   19   41   45   60   74

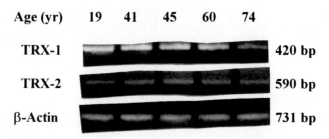

TRX-1   420 bp

TRX-2   590 bp

β-Actin   731 bp

FIG. 1. RT-PCR-characterized expression of mRNAs encoded by Trx-1 and Trx-2 genes in normal lens of humans of varying ages. PCR-amplified cDNA transcripts of Trx-1 (420 bp) and Trx-2 (590 bp), generated from the mRNAs from a single lens, were resolved by 1.5% (w/v) agarose gel electrophoresis, stained with ethidium bromide, and visualized under UV light. A 731-bp β-actin cDNA transcript of the lens mRNA was analyzed with each sample as a control to ensure that equal amounts of RNA were used in the RT-PCR.

from the nucleotide sequences of human lens cytoplasmic thioredoxin (Trx-1) and mitochondrial thioredoxin (Trx-2) are as follows.

Trx-1 :   MVKQIESKTAFQEALDAAGDKLVVVDFSATWCGPCKMINPFF
          HSLSEKYSNVIFLEVDVDDCQDVASECEVKCTPTFQFFKKGQK
          VGEFSGANKEKLEATINELV

Trx-2 :   MAQRLLLRRFLASVISRKPSQGQWPPLTSKALQTPQCSPGGLT
          VTPNPARTIYTTRISLTTFNIQDGPDFQDRVVNSETPVVVDFHA
          QWCGPCKILGPRLEKMVAKQHGKVVMAKVDIDDHTDLAIEY
          EVSAVPTVLAMKNGDVVDKFVGIKDEDQLEAFLKKLIG

Representative Trx-1 protein expression detected by Western blot analysis of human normal lenses of varyious ages, is shown in Fig. 2. Trx protein in the lens is high at younger age (19 years) compared with lenses of older individuals. Studies using more lenses are needed to confirm this finding.

Age (yr)   19   41   45   60   74

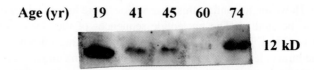

12 kD

FIG. 2. Protein levels of Trx-1 in epithelia of normal lenses of humans aged 19, 41, 45, 60, and 74 years. Lens epithelial protein, 100 μg, was resolved into polypeptide components by 10% (w/v) SDS–PAGE, electroblotted on nitrocellulose membrane, probed with polyclonal antibody to Trx-1, and visualized by enhanced chemiluminescence assay. The immunoreactive band of Trx-1 protein as identified is seen in the photograph.

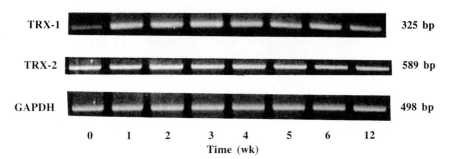

TRX-1   325 bp

TRX-2   589 bp

GAPDH   498 bp

0   1   2   3   4   5   6   12
Time (wk)

Fig. 3. RT-PCR-characterized expression of mRNAs encoded by Trx-1 and Trx-2 genes in the normal lens of the Emory mouse subjected to *in vivo* photochemical oxidative stress (described in Experimental Procedures) for a period varying from 1 to 12 weeks. PCR-amplified cDNA transcripts, 325 bp for Trx-1 and 589 bp for Trx-2, generated from the mRNAs from the lens, were resolved by 1.5% (w/v) agarose gel electrophoresis, stained with ethidium bromide, and visualized under UV light. A GAPDH cDNA transcript (498 bp) of the lens mRNA, was analyzed with each sample as a control to ensure that equal amounts of RNA were used in the RT-PCR. [Reprinted with permission from P. G. Reddy, D. K. Bhuyan, and K. C. Bhuyan, *Biochem. Biophys. Res. Commun.* **265**, 345 (1999).]

Trx-1 is a cytosolic and membrane-bound protein, and Trx-2 is a mitochondrial protein encoded by a nuclear gene. Both Trx-1 and Trx-2 are redox regulatory genes, and their protective functions against damage by oxidants in many tissues such as retina, brain, liver, kidney, keratinocytes, and lymphocytes are documented.[8–10] Because such information for lens is limited, an investigation of the status of expression of these genes in lenses under *in vivo* oxidative stress was attempted.[41] In Fig. 3, RT-PCR characterized cDNA transcripts of mRNA expression of Trx-1 and Trx-2 genes in lenses of Emory mice subjected to *in vivo* photochemical oxidative stress for 1 to 12 weeks are shown. The expression of the Trx-1 gene, but not of the Trx-2 gene, in the lens (Fig. 3) increased steadily to 5-fold after 3 weeks of oxidative stress, and gradually declined thereafter. This increase in Trx-1 gene expression in the lens was confirmed by DNA quantitation as reported elsewhere.[41] The expression of Trx-2 gene decreased about 50% in lenses of mice exposed to oxidative stress for 6 weeks (Fig. 3). However, the expression of Trx-1 gene in the liver, kidney, and tail of these mice under oxidative stress was unaltered, as reported earlier.[41] The possibility of enhanced oxidative stress in the lens compared with other body tissues is likely because it receives direct exposure to sunlight. It is important to note that, in the lenses of Emory mice under oxidative stress for 3 weeks, there was an ~4-fold increase in the level of Trx-1 protein as detected by Western blot analysis (Fig. 4, lane 4). The cDNA transcripts of mRNAs encoding Trx-1 and Trx-2 of mouse lens, isolated by RT-PCR, cloned,

[41] P. G. Reddy, D. K. Bhuyan, and K. C. Bhuyan, *Biochem. Biophys. Res. Commun.* **265**, 345 (1999).

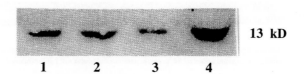

FIG. 4. Protein levels of Trx-1 in lenses of Emory mice. Lens protein, 150 μg, was resolved into polypeptide components by 10% (w/v) SDS–PAGE, and the Trx-1 protein was identified by immunoblot analysis with polyclonal antibody for murine Trx-1. Lanes 1–4 represent Trx-1 protein in lenses of 4-week-old mouse pups raised for 3 weeks on: (1) regular diet and ambient light exposure, (2) special diet [regular diet supplemented with 0.05% (w/v) riboflavin] and ambient light exposure, (3) regular diet and fluorescent light exposure, and (4) special diet and fluorescent light exposure, respectively. The details are given in the section Test Animals in text. [Reprinted with permission from P. G. Reddy, D. K. Bhuyan, and K. C. Bhuyan, *Biochem. Biophys. Res. Commun.* **265**, 345 (1999).]

and sequenced revealed their identity with the reported sequences of murine Trx-1 and Trx-2.[42]

It is reported that riboflavin or light alone induces the expression of several genes in tissues.[43] To verify this, the expression of Trx-1 and Trx-2 genes[41] and of Trx-1 protein (Fig. 4, lanes 2 and 3) in the lenses of control mice fed a riboflavin-supplemented diet or exposed to light alone for 1–4 weeks is analyzed. The results show no effect of riboflavin or light alone. Thus, it is apparent that the transient upregulation of Trx-1 gene and protein in mouse lens under *in vivo* oxidative stress is a tissue-specific adaptation. It shows an early protective response of the normal lens to the altered cellular redox status.

Thioredoxin Reductase Activity in Human Lens

Preparation of Enzyme Extracts

From a single lens, capsule epithelium, 30–40 mg, and fiber cells (decapsulated lens), 180–220 mg, are isolated, and the homogenates of epithelium in 1 ml and of fiber cells in 4 ml obtained with the extractant [25 mM KH$_2$PO$_4$–K$_2$HPO$_4$ buffer (pH 7.4), containing 1 mM EDTA], are centrifuged at 48,200g for 20 min at 0–4°. The supernatants, dialyzed for 18 hr at 5° against 5 mM phosphate buffer, pH 7.4, containing 0.1 mM EDTA, are used as the source of cytosolic TR.

The crude membrane pellets (48,200g) are washed three times with the extractant buffer, 1 ml for lens epithelium and 4 ml for decapsulated lens, and suspended in lysis buffer [10 mM phosphate buffer (pH 7.4), 150 mM NaCl, 1 mM phenylmethyl-sulfonyl fluoride (PMSF), 0.5% (v/v) octylphenol ethylene oxide condensate

[42] Y. Tagaya, Y. Maeda, A. Mitsui, N. Kondo, H. Natsui, J. Hamuro, N. Brown, K. I. Arai, T. Yokota, H. Waksugi, and J. Yodoi, *EMBO J.* **8**, 757 (1994).

[43] M. Nagao and K. Tanaka, *J. Biol. Chem.* **267**, 17925 (1992).

(Nonidet P-40), a nonionic surfactant, and 2.5 mM EDTA]. The solubilized membrane extracts are used as the source of TR. TR, using NADPH $\cdot$ H$^+$, specifically reduces Trx-S$_2$ to Trx-(SH)$_2$ [Eq. (1)]. In the technique[44] used, NADPH $\cdot$ H$^+$-dependent reduction of 5,5′-dithiobis (2-nitrobenzoic acid) (DTNB) by enzyme extract, in the presence or absence of aurothioglucose, is determined at 412 nm and aurothioglucose-inhibited DTNB reduction becomes the measure of TR activity.

The assay system, at 25°, consists of 100 mM potassium phosphate buffer (pH 7.4), 10 mM EDTA, 0.2 mg of bovine serum albumin (BSA), 0.2 mM NADPH, 5 mM DTNB, and an appropriate amount of enzyme extract (0.1 to 0.2 ml) with or without 50 μM aurothioglucose in a final volume of 1 ml. Before addition to the assay system, the enzyme extract is incubated with 0.5 mM aurothioglucose for 10 min at 25°, and the appropriate amount along with NADPH is added to the assay system to achieve a final concentration of 50 μM gold thioglucose. It is important to run simultaneous control reactions, with or without aurothioglucose in the absence of enzyme extract or in the presence of heat-denatured enzyme extract, to apply correction to the observed absorbance at 412 nm due to reduction of DTNB to yellow 2-nitro-5-mercaptobenzoate dianion ($\varepsilon_{412\,nm} = 1.36 \times 10^4 M^{-1}$ cm^{-1}). The activity of lens TR is expressed as units (U) per milligram of protein. One unit of TR represents 1 nmol of DTNB reduced per minute at 25° under the assay conditions described.

TR activities in the human lens, determined by the aforementioned technique, are given in Table III. In normal human lenses, age 30–40 years, cytosolic and membrane-bound TR activities (U/mg protein) in the lens epithelial cells are 5–8 and 2–3 U, respectively. The TR activity in lens fiber cells is 8–10 U in the cytosolic fraction and 13–21 U in the membrane fraction. The cytosolic and membrane-associated TR activities in the lens epithelial and fiber cells appear to decrease with age, and the activities are low in the epithelia and fiber cells of cataractous lenses. Decreased activity of TR in lens with age or in cataract raises an important question that is not unfamiliar—is the decreased TR activity a cause of cataract or a postcataract change? The evidence that forskolin (FK506) decreases Trx expression in tissues,[45] and is also cataractogenic in rats,[46] suggests that impairment of the TR–Trx system could be a causative factor in cataract formation. There are several reports showing Trx to be a sensor of oxidative stress, and indicating a role for the TR–Trx system in transducing signals for cell growth and differentiation by redox regulation.[4,8,9] This enzyme system, a potent protein/disulfide reductase,[2–4] could be of significance in relation to thiol/disulfide

[44] K. E. Hill, G. W. McCollum, and R. F. Burk, *Anal. Biochem.* **253**, 123 (1997).

[45] K. Furuke, H. Nakamura, T. Hori, S. Iwata, N. Maekawn, T. Inemoto, Y. Yamoka, and J. Yodoi, *Int. Immunol.* **7**, 985 (1995).

[46] H. Ishida, T. Mitamura, Y. Takahashi, A. Hisatomi, Y. Fukuhara, K. Murato, and K. Ohara, *Toxicology* **123**, 167 (1997).

TABLE III

THIOREDOXIN REDUCTASE ACTIVITY IN HUMAN LENS CYTOSOL AND CELL MEMBRANES

		TR activity (U/mg protein)[a]			
		Lens capsule-epithelium		Lens fiber cells	
Lens	Age (years)	Cytosol	Membranes	Cytosol	Membranes
Normal[b]	34	4.62	1.52	7.75	16.41
	39	8.03	1.27	8.58	20.53
	40	4.71	1.39	9.15	12.95
	40	7.91	2.53	7.56	12.62
	53	6.33	2.46	9.98	6.54
	76	6.20	2.33	6.68	5.39
	76	5.51	2.29	—	—
Cataract[c]	82	4.38	1.11	3.01	5.48
	82	2.76	1.16	5.05	4.13
	84	1.03	0.83	2.12	1.77
	84	0.92	0.90	2.43	1.73
	96	2.76	1.00	0.24	0.71
	96	2.21	1.03	0.19	0.60
Cataract[d]	64–80	1.93	0.91	—	—

[a] Nanomoles of 5,5′-dithiobis(2-nitrobenzoic acid) reduced per minute under the assay conditions, represent units of TR activity. The assay system is described under Experimental Procedures: Estimation of TR Activity.
[b] Human lenses were obtained from Eye-Bank donor eyes received at the Eye-Bank for Sight Restoration (New York, NY).
[c] Eye-Bank donor eyes with lens opacities (Early cataract).
[d] Lens capsule epithelium (capsulorhexis) samples were removed from patients during cataract surgery by A. M. Cotliar and C. B. Camras.

redox regulation for the maintenance of structural and functional integrity of lens proteins. In the plasma membrane, Trx and TR have been observed in keratinocytes, melanoma,[10] and in several human cell lines.[47] It was suggested that this enzyme exerts its action by reducing soluble mediators or cellular receptors involved in cell-to-cell communications.[47] Hormone- and growth factor-stimulated thiol/disulfide oxidoreductase activities in plasma membranes of both plants and animals have been observed and reviewed,[48] and their participation during growth factor- and receptor interaction-initiated cell growth by plasma membrane expansion has been proposed.[48] It is likely that the TR–Trx system in lenticular plasma membranes is involved in the growth and differentiation of lens epithelial cells into fiber cells at the equator (germinative zone), where extensive cellular elongation with several

[47] E. E. Wollman, A. Kahan, and D. Fradelizi, *Biochem. Biophys. Res. Commun.* **230,** 602 (1997).
[48] D. J. Morre, *J. Bioenerget. Biomembr.* **26,** 421 (1994).

hundred-fold expansion of the plasma membrane occurs during differentiation of lens epithelial cells into fiber cells.[49]

Concluding Remarks

The mammalian lens epithelial cells in the germinative zone are in a process of continuous proliferation and differentiation into new fiber cells throughout life. Therefore, it is conceivable that there is a redox regulatory role of Trx in transmembrane cell signaling controlling this process. A shift in the cellular redox state of the lens under oxidative stress, resulting in altered signal transduction at advanced age, might be a factor that initiates age-related human cataract. Considering the decline of GSH and formation of protein disulfides in the lens with age and in cataracts, the TR–Trx system could be of importance in maintenance of the cysteinyl residues of lens proteins, which are prone to oxidative damage.

The Trx-1 and Trx-2 genes are expressed in human and mouse lenses, and *in vivo* oxidative stress upregulates lens-specific Trx-1 gene and protein expression in the Emory mouse. Thioredoxin reductase activities are present in cytosolic and membrane fractions of the human lens epithelial and fiber cells, and this enzyme activity is found to decrease in cataracts. Thus, thioredoxin might have a protective role in lens against oxidative stress in addition to its participation as a redox sensor in the cell for DNA synthesis. This suggested therapeutic role of thioredoxin could also lead to modeling of anticataract drugs.

Acknowledgments

The authors thank Stanley Chang, M.D., Anthony Donn, M.D., Abraham Spector, Ph.D., and Harold L. Kern, Sc.D., for their help. We are greateful to the Department of Ophthalmology (College of Physicians and Surgeons, Columbia University, New York, NY) for supporting the Membrane Biochemistry Laboratory for cataract research by an unrestricted grant from Research to Prevent Blindness, Inc., NY.

---

[49] R. M. Broekhuyse, *in* "The Human Lens in Relation to Cataract" (K. Elliot and D. Fitzsimons, eds.), p. 135. Elsevier, Amsterdam, 1973.

[41] Thioredoxin Overexpression in Transgenic Mice

By HAJIME NAKAMURA, AKIRA MITSUI, and JUNJI YODOI

Introduction

There is accumulating evidence that thioredoxin (TRX) plays a crucial role in intracellular and extracellular various biological functions including signal transduction and cytoprotection.[1] TRX is induced by various oxidative stresses including viral infection and translocated from the cytoplasm into the nucleus on oxidative stress. In the nucleus, TRX interacts with redox factor 1 (Ref-1), an endoexonuclease, to enhance the DNA-binding activity of transcription factors such as AP-1 and p53.[2,3] TRX shows cytoprotective functions against oxidative stress by scavenging reactive oxygen species (ROS) in cooperation with peroxiredoxin/TRX-dependent peroxidase. In addition, TRX is secreted from the cells through a leaderless pathway. Extracellularly, TRX shows cytokine-like and chemokine-like functions. To analyze the biological functions of TRX, we first tried to develop TRX knockout mice. However, TRX knockout mice were embryonic lethal.[4] We then developed thioredoxin-overexpressing mice (TRX–Tg mice), in which human TRX are systemically expressed in C57/BL6 strain mice under the control of the β-actin promoter.[5] Human TRX cDNA was inserted between the human β-actin promoter and human β-actin terminator and used to generate the transgenic mice. The phenotype and behavior of TRX–Tg mice are apparently normal. TRX-Tg mice contain severalfold larger amounts of human TRX protein in most organs compared with endogenous mouse TRX protein levels. TRX–Tg mice are more resistant to focal cerebral ischemic injury[5] and excitotoxic hippocampal injury[6] compared with wild-type C57BL/6 mice. Moreover, our observations indicate that TRX-Tg mice are more resistant to a variety of oxidative stresses and survive longer than wild-type mice.[7] In addition, pancreatic β-cell-specific overexpression

[1] H. Nakamura, K. Nakamura, and J. Yodoi, *Annu. Rev. Immunol.* **15,** 351 (1997).

[2] K. Hirota, M. Matsui, S. Iwata, A. Nishiyama, K. Mori, and J. Yodoi, *Proc. Natl. Acad. Sci. U.S.A.* **94,** 3633 (1997).

[3] M. Ueno, H. Masutani, R. J. Arai, A. Yamauchi, K. Hirota, T. Sakai, T. Inamoto, Y. Yamaoka, J. Yodoi, and T. Nikaido, *J. Biol. Chem.* **274,** 35809 (1999).

[4] M. Matsui, M. Oshima, H. Oshima, K. Takaku, T. Maruyama, J. Yodoi, and M. M. Taketo, *Dev. Biol.* **178,** 179 (1996).

[5] Y. Takagi, A. Mitsui, A. Nishiyama, K. Nozaki, H. Sono, Y. Gon, N. Hashimoto, and J. Yodoi, *Proc. Natl. Acad. Sci. U.S.A.* **96,** 4131 (1999).

[6] Y. Takagi, I. Hattori, K. Nozaki, A. Mitsui, M. Ishikawa, N. Hashimoto, and J. Yodoi, *J. Cereb. Blood Flow Metab.* **20,** 829 (2000).

[7] A. Mitsui, H. Nakamura, N. Kondo, Y. Hirabayashi, S. Ishizaki-Koizumi, T. Hirakawa, T. Inoue, and J. Yodoi, unpublished observation (2001).

of TRX in nonobese diabetic (NOD) mice by using the insulin promoter prevents autoimmune and streptozotocin-induced diabetes.[8] Thus, TRX-Tg mice are good models with which to investigate the biological functions and roles of TRX *in vivo*.

Thioredoxin Knockout Mice

The mouse genome contains one active TRX gene on chromosome 1 and one processed pseudogene on chromosome 4.[9] The TRX gene extends over 12 kb and contains five exons separated by four introns.[10] To develop TRX knockout mice, a part of the mouse TRX gene including the translation start codon was deleted by homologous recombination in embryonic stem (ES) cells. Heterozygotes are viable, fertile, and appear normal. TRX heteroknockout mice are now available at the Institute for Virus Research (Kyoto University, Kyoto, Japan) and under investigation for stress sensitivity. In contrast, homozygous mutants die shortly after implantation at the egg cylinder formation stage.[4] By day 6.5 of gestation, the embryos proper were not found in the deciduas and the tissues could not be recovered for genotyping. One possible explanation for the early lethality of TRX homoknockout embryos is impaired DNA replication after maternal TRX is lost in the embryo. Interestingly, Ref-1-deficient mice also die shortly after implantation, at day 5.5 of gestation.[11] Because Ref-1 and TRX operate coordinately in the redox-sensitive activation of transcription factors such as AP-1 or in the DNA repair/replication, Ref-1 and TRX may be essential for a critical stage of early embryonic development.

Characteristics of Thioredoxin-Transgenic Mice

Resistance to Focal Cerebral Ischemic Injury

There is no gross anatomical difference in the brain between TRX–Tg mice and wild-type C57BL/6 mice. Immunohistochemical staining with anti-human TRX antibody shows that human TRX is overexpressed in the vascular endothelial cells and glia cells in the brain of TRX-Tg mice. There are no differences in the expression of antioxidant enzymes such as Cu/Zn-SOD, Mn-SOD, and glutathione peroxidase between TRX-Tg mice and wild-type mice. To examine the effect of TRX overexpression against ischemic injury, a focal cerebral ischemia model is applied in TRX-Tg mice and wild-type mice.[5] Focal cerebral ischemia is

[8] M. Hotta, F. Tashiro, H. Ikegami, H. Niwa, T. Ogihara, J. Yodoi, and J.-I. Miyazaki, *J. Exp. Med.* **188**, 1445 (1998).

[9] M. Matsui, Y. Taniguchi, K. Hirota, M. Taketo, and J. Yodoi, *Gene* **152**, 165 (1995).

[10] M. Taketo, M. Matsui, J. M. Rochelle, J. Yodoi, and M. F. Seldin, *Genomics* **21**, 251 (1994).

[11] S. Xanthoudakis, R. J. Smeyne, J. D. Wallace, and T. Curran, *Proc. Natl. Acad. Sci. U.S.A.* **93**, 8919 (1996).

induced by occlusion of the middle cerebral artery, using the intralumenal filament technique,[12] in male mice weighing 30–35 g under general anesthesia. Twenty-four hours later, the animals are killed and the brain sections are analyzed. The infarcted areas and volume in TRX-Tg mice are significantly smaller than in wild-type C57BL/6 mice. Because oxidative modification of proteins is accompanied by the generation of protein carbonyl derivatives, the protein carbonyl contents of the soluble fraction of crude brain cortical extract preparations are analyzed 24 hr after ischemia. The protein carbonyl contents in TRX-Tg mice are significantly less than in wild-type mice. Taken together, TRX overexpression reduces ischemia-induced oxidative stress and protects the brain tissues from ischemic injury.

Resistance to Excitotoxic Hippocampal Injury

Excessive activations of excitatory neurotransmitter receptors are involved in several neurodegenerative diseases and ischemic brain injury. Kainic acid binds to such an excitatory amino acid receptor and induces neuronal damage by generation of reactive oxygen species (ROS). To examine the effect of TRX overexpression against kainic acid-induced neuronal damage, kainic acid is injected into TRX–Tg mice and wild-type C57BL/6 mice.[6] Kainic acid is diluted in phosphate-buffered saline (PBS) at a concentration of 5 mg/ml. Mice weighing 30–40 g are injected intraperitoneally with kainic acid (20 mg/kg). The mice are observed for seizure incidence for 1 hr after the injection. Seven of 10 TRX-Tg mice and all 10 wild-type mice exhibit severe seizure. The mean seizure score, described elsewhere,[13] is significantly lower in TRX-Tg mice than in wild-type mice at 30 min and 45 min after treatment. Seven days after kainic acid treatment, hippocampal neuronal damage is assessed by cresyl violet staining. TRX-Tg mice show significantly more intact neurons in the hippocampal CA1 and CA3 regions than wild-type mice. These results indicate that TRX overexpression attenuates kainic acid-induced neuronal damage.

Resistance to Oxidative Stress and Prolonged Survival

Data show that TRX-Tg mice are more resistant to oxidative stress than wild-type C57BL/6 mice. For example, fasting-induced lipid peroxidation in the liver measured by estimation of thiobarbituric acid-reactive substance (TBARS) formation is reduced in TRX-Tg mice compared with wild-type mice. Bone marrow cells from TRX-Tg mice are more resistant to ultraviolet light exposure-induced cytocide than those from wild-type mice. Moreover, TRX-Tg mice survive longer than wild-type C57BL/6 mice[7] (Fig. 1). We also have found that TRX-Tg mice are more resistant to bleomycin-induced lung fibrosis and cytokine-induced lethal interstitial

[12] E. G. Longa, P. R. Weinstein, S. Carlson, and R. Cummins, *Stroke* **20,** 84 (1989).
[13] D. D. Yang, C. Y. Kuan, A. J. Whitmarsh, M. Rincon, T. S. Zheng, R. J. Davis, P. Rakic, and R. A. Flavell, *Nature (London)* **389,** 865 (1997).

Wilcoxon p= 0.027
Log-Rank p= 0.036

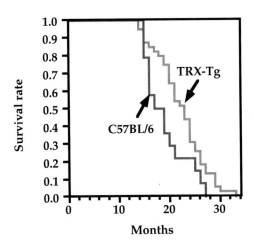

FIG. 1. Survival curve of thioredoxin–transgenic mice (TRX–Tg) and wild-type C57BL/6 mice.

pneumonia than wild-type mice.[14] Leukocyte migration in the inflammatory site is inhibited by circulating TRX when recombinant TRX is intravenously injected in mice or when TRX-Tg mice are stimulated by lipopolysaccharide.[15,16] Moreover, when wild-type E7.5 mouse embryos are cultivated *in vitro* under a high oxygen concentration, their growth is retarded and developmental abnormalities are frequently produced.[17] In contrast, most of the embryos from TRX-Tg mice grow normally. These results suggest that TRX-Tg mice are resistant to a variety of oxidative stresses.

Pancreatic β Cell-Specific Overexpression of Thioredoxin

Several reports have shown that locally produced ROS are involved in the destruction of pancreatic β cells, resulting in the development of diabetes mellitus (DM). NOD mice spontaneously develop autoimmune diabetes with remarkable similarity to human insulin-dependent diabetes mellitus. NOD transgenic mice

[14] T. Hoshino, H. Nakamura, M. Okamoto, S. Araya, O. Shimozato, H. A. Young, K. Oizumi, and J. Yodoi, unpublished observation (2001).

[15] H. Nakamura, S. De Rosa, J. Yodoi, A. Holmgren, P. Ghezzi, L. A. Herzenberg, and L. A. Herzenberg, *Proc. Natl. Acad. Sci. U.S.A.* **98**, 2688 (2001).

[16] H. Nakamura, L. A. Herzenberg, J. Bai, S. Araya, N. Kondo, Y. Nishinaka, L. A. Herzenberg, and J. Yodoi, *Proc. Natl. Acad. Sci. U.S.A.,* in press (2001).

[17] M. Kobayashi, H. Nakamura, J. Yodoi, and K. Shiota, unpublished observation (2001).

that overexpress TRX in pancreatic β cells by using the human insulin promoter were produced to elucidate the effect of TRX on the ROS-induced development of DM.[8] The Ins-TRX transgene, containing human TRX cDNA under the control of the human insulin promoter, was microinjected into fertilized eggs of NOD mice to generate NOD transgenic mice. Human TRX mRNA was exclusively detected in the pancreas by RT-PCR. Western blotting with anti-human TRX antibody showed a high level of human TRX expression in the lysate of pancreatic islets from transgenic mice. There was no difference in insulin secretion capacity between NOD TRX-transgenic mice and TRX-negative littermates. To examine the effect of TRX expression on the development of DM, female NOD TRX-transgenic mice and TRX-negative littermates were monitored for glucosuria up to 32 weeks of age. The onset of glucouria was significantly retarded and the cumulative incidence of diabetes was significantly reduced in NOD TRX-transgenic mice compared with TRX-negative littermates.

Next, the protective role of TRX was examined in another diabetes model induced by streptozotocin, an ROS-inducing agent. Male NOD TRX-transgenic mice were mated with female C57BL/6J mice to produce (NOD $\times$ B6) F_1 mice. There was no difference in fasting blood glucose levels between (NOD $\times$ B6) F_1 TRX-transgenic mice and their nontransgenic littermates. Streptozotocin (250 mg/kg) was injected intraperitoneally into 8-week-old (NOD $\times$ B6) F_1 TRX-transgenic mice and their nontransgenic littermates. Seven days after streptozotocin injection, the blood glucose levels were elevated more than 500 mg/dl in seven of nine nontransgenic mice and only two of nine TRX-transgenic mice. Pancreatic insulin content was significantly higher in TRX-transgenic mice than in TRX-negative littermates.

These results clearly indicate that TRX plays a crucial role in preventing the oxidative stress-induced development of diabetes mellitus.

Concluding Remarks

TRX-Tg mice provide us with direct evidence that TRX plays a protective role against oxidative stress-associated disorders *in vivo*. *In vitro* studies have demonstrated that TRX shows a variety of intracellular and extracellular biological functions including redox regulation of transcription factors and cytokine/chemokine-like fuctions. TRX-Tg mice are good tools for further analysis of the unknown biological *in vivo* functions of TRX.

Acknowledgment

This work was supported by a grant-in-aid of Research for the Future from the Japan Society for the Promotion of Science.

[42] Multiplex Reverse Transcription-Polymerase Chain Reaction for Determining Transcriptional Regulation of Thioredoxin and Glutaredoxin Pathways

By CARMEN PUEYO, JUAN JURADO, MARÍA-JOSÉ PRIETO-ÁLAMO, FERNANDO MONJE-CASAS, and JUAN LÓPEZ-BAREA

Introduction

Thioredoxins (Trxs) and glutaredoxins (Grxs) are small redox-active proteins that have been isolated as hydrogen donors for ribonucleotide reductase (RRase), the key enzyme for deoxyribonucleotide and DNA biosynthesis. Trxs and Grxs contain two redox-active cysteine thiols, which by dithiol–disulfide interchange reduce an acceptor disulfide in the active center of RRase.[1,2]

Escherichia coli has two Trxs (Trx1 and Trx2), three Grxs (Grx1, Grx2, and Grx3), a novel redoxin with Trx-like activity but Grx-like amino acid sequence (NrdH), and two aerobic class I RRases (NrdAB and NrdEF). Grx1 and Trx1 are the two main reductants of the NrdAB enzyme and NrdH is the specific hydrogen donor of the NrdEF system. Trx1, Trx2, Grx1, Grx2, and Grx3 are encoded by the *trxA, trxC, grxA, grxB,* and *grxC* genes, respectively.[3] The *nrdA* and *nrdB* genes that encode the NrdAB class Ia reductase constitute a transcription unit that does not include the genes for either Trxs or Grxs. However, the *nrdE* and *nrdF* genes that encode the NrdEF class Ib reductase form a conserved *nrdHIEF* operon that includes the gene encoding the NrdH redoxin.[2]

Trxs and Grxs differ in their reductive pathways, although ultimately reducing equivalents come from NADPH. The thioredoxin system is composed of NADPH, the flavoprotein thioredoxin reductase (the product of *trxB*), and Trx; the glutaredoxin system consists of NADPH, the flavoprotein glutathione reductase (the product of *gorA*), the ubiquitous tripeptide glutathione (GSH), and Grx.[1]

Different inducible responses are critical in protecting *E. coli* from oxidative damage. The two best characterized oxidative stress responses are controlled by the OxyR and SoxR transcriptional activators of *E. coli*.[4] In addition to the classic function of acting as reductants for RRase, several components of the Trx and Grx/GSH pathways are involved in the oxidative response of the OxyR and SoxR transcription factors during their regulatory cycles.[5,6] Therefore, (1) OxyR is activated

[1] A. Holmgren, *J. Biol. Chem.* **264,** 13963 (1989).

[2] A. Jordan and P. Reichard, *Annu. Rev. Biochem.* **67,** 71 (1998).

[3] A. Rietsch and J. Beckwith, *Annu. Rev. Genet.* **32,** 163 (1998).

[4] G. Storz and J. A. Imlay, *Curr. Opin. Microbiol.* **2,** 188 (1999).

[5] M. Zheng, F. Åslund, and G. Storz, *Science* **279,** 1718 (1998).

[6] H. Ding and B. Demple, *Biochemistry* **37,** 17280 (1998).

(in the absence of added H_2O_2) by oxidation of the cellular thiol-disulfide state, caused by the simultaneous impairment of Trx and Grx/GSH systems[7,8]; (2) this OxyR activation, involving the formation of an intramolecular disulfide bond, is reversed by the cellular disulfide-reducing machinery, with particular dependence on Grx1[5]; and (3) genes that encode components of both Trx and Grx/GSH systems are upregulated by OxyR (see data in Table II), thus ensuring that this transcription factor is active for a defined period of time.

Here we described a method to quantitate the *in vivo* expression of most known components of the *E. coli* thioredoxin (*trx* genes) and glutaredoxin/GSH (*grx* and *gorA* genes) pathways, of related enzymes (*nrd* genes), and of key members of the OxyR (*oxyS*) and SoxR/S (*soxS*) regulons.

RNA

RNA Isolation

Escherichia coli cultures (25 ml), harvested at the indicated time or OD_{600} value, are frozen in liquid nitrogen. Immediately after, total RNA is prepared by the hot phenol extraction method of Emory and Belasco[9] with slight modifications. Pelleted cells are suspended in 0.125 ml (culture at OD_{600} of 0.2) or 0.250 ml (culture at OD_{600} of 0.7) of ice-cold 0.3 M sucrose, 0.01 M sodium acetate buffer (pH 4.5). After addition of 0.125 or 0.250 ml of 2% (w/v) sodium dodecyl sulfate (SDS) in the same buffer, the cell suspension is heated for 3 min at 70° and extracted for 3 min at 70° with 0.25 or 0.50 ml of hot phenol at pH 4.5 (Amresco, Solon, OH). The RNA is then ethanol precipitated and treated for 60 min at 37° with RNase-free DNase I ($\geq$50 U; Roche Diagnostics, Mannheim, Germany) in 0.1 M sodium acetate, 5 mM $MgSO_4$ (pH 5.0), to remove contaminating genomic DNA. After that, the RNA is phenol extracted, ethanol precipitated, quantified spectrophotometrically, and stored at $-80°$ in diethyl pyrocarbonate (DEPC) treated water until use. We have observed no appreciable deterioration of RNA samples stored for up to 1 year. The quality of the preparation can be checked by electrophoresis of 5 μg of RNA in a 1% agarose gel containing 2.2 M formaldehyde. Lack of DNA contamination must be checked by polymerase chain reaction (PCR) amplification of RNA samples. No fluorescent bands should be visible from these reactions. At least two independent RNA isolations should be accomplished for each experimental condition.

In Vitro RNA Synthesis

The external standard RNA (*CYP1A* transcript in this work) is synthesized *in vitro* from an insert cloned into a vector containing a T7 polymerase-binding

[7] F. Åslund, M. Zheng, J. Beckwith, and G. Storz, *Proc. Natl. Acad. Sci. U.S.A.* **96**, 6161 (1999).

[8] M. J. Prieto-Álamo, J. Jurado, R. Gallardo-Madueño, F. Monje-Casas, A. Holmgren, and C. Pueyo, *J. Biol. Chem.* **275**, 13398 (2000).

site, by means of a commercial RNA transcription kit (Stratagene, La Jolla, CA). The DNA template, still present after the transcription reaction, is removed with RNase-free DNase I (3 U/μg DNA, 15 min at 37°). The synthesized RNA is phenol extracted, ethanol precipitated, quantified spectrophotometrically, and stored at $-80°$ until use. The external standard RNA is mixed with total RNA before the reverse transcription step.

cDNA

Synthesis of cDNA is carried out with the GeneAmp RNA PCR kit (Applied Biosystems, Foster City, CA). In short, 1 μg of total RNA plus 16 pg of external standard RNA are incubated for 10 min at 25° and then retrotranscribed for 15 min at 42° with 50 U of murine leukemia virus (MuLV) reverse transcriptase, using random hexamers. The enzyme is inactivated by heating for 5 min at 99°. Each RNA sample is retrotranscribed on an average of three separate occasions. We synthesize cDNA in a 20-μl reaction and then use 1 μl of the cDNA solution in each multiplex PCR amplification tube.

Primers

Primer Design

Primers (20- to 30-mers) must be exactly complementary to the desired template location while preventing the formation of internal secondary structures. Primers were designed with Oligo 6.1.1/98 software (Molecular Biology Insights, Plymouth, MN). To obtain the highest specificity and acceptability for use in multiplex PCR reactions, primers are chosen to have high T_m and optimal 3′ ΔG values (Table I).

Primer Label

One primer of each pair is labeled at the 5′ end with a fluorescent dye. Up to three different fluorescent dyes can be used (a fourth dye color is reserved for labeling the size standard). By making use of differences in color, fragments of identical sizes can be distinguished.[10] Nevertheless, on the basis of our experience we recommend labeling the primers with the same dye label and, thus, distinguishing the fragments by differences in size. Fragments differing by about 6 bases in length (Fig. 1) are easily separated and quantified, under both standard and highly induced conditions (Table II).

---

[9] S. A. Emory and J. G. Belasco, *J. Bacteriol.* **172,** 4472 (1990).

[10] R. Gallardo-Madueño, J. F. M. Leal, G. Dorado, A. Holmgren, J. López-Barea, and C. Pueyo, *J. Biol. Chem.* **273,** 18382 (1998).

TABLE I
PCR PRIMER CHARACTERISTICS

Primers[a]	Fragment size (bp)	Sequence	$T_m{}^b$ (°C)	$3' \Delta G^c$ (kcal/mol)
gorA*	87	5'-GGCACCTGCGTAAATGTTGGCTGTG	80.9	−3.0
gorA		5'-CGGGCCGTACATATGGATCGCTTCA	81.8	−3.0
CYPA1*	93	5'-TCCTTCAACCCAGACCGTTTCCTCA	79.5	—
CYPA1		5'-CCGCTTTCCCAAGCCAAAAACCATC	81.3	−1.0
oxyS*	99	5'-GGAGCGGCACCTCTTTTAACCCTTG	79.2	−3.0
oxyS		5'-TCCTGGAGATCCGCAAAAGTTCACG	80.0	−6.9
grxB*	105	5'-ATCCCCGTCGAATTACATGTTCTGCTCA	80.7	−6.4
grxB		5'-ATAGCGGCTGTCATCTTTTTGCAGAATGG	80.5	−3.7
grxC*	112	5'-AAGAGATGATCAAACGCAGCGGTCG	80.1	−3.4
grxC		5'-TCCAGTCCACCACGTGCATCCAAT	80.5	−3.3
grxA*	118	5'-GCTGAGAAATTGAGCAATGAACGCGATG	81.8	−3.3
grxA		5'-GCACGGTTCTACGGGTTTACCTGCCTTT	81.6	−3.7
nrdA*	124	5'-GACGCCTATGAGCTGCTGTGGGAAA	80.1	−8.0
nrdA		5'-GGAAGCGTGACGGATCGTAGTTGG	78.6	−4.2
trxA*	130	5'-TCGTCGATTTCTGGGCAGAGTGGTG	79.0	−5.2
trxA		5'-GCAGTGCCAGGGTTTTGATCGATG	79.5	−5.2
soxS*	136	5'-GACGCATCAGACGCTTGGCGATT	80.3	−3.4
soxS		5'-GCGGGAGAAGGTCTGCTGCGAGA	81.8	−3.4
gapA*	143	5'-CGTTCTGGGCTACACCGAAGATGACG	81.1	−3.0
gapA		5'-AACCGGTTTCGTTGTCGTACCAGGA	79.1	−3.6
trxC*	152	5'-CCAATTTTTGAAGATGTCGCGCAAGAGC	82.0	−4.2
trxC		5'-AGCATGTCGACAACCTGACCGTTTTTGA	81.6	−4.9
trxB*	157	5'-CTGGAAGAAGTGACCGGCGATCAAA	80.0	−8.0
trxB		5'-CCAGTTCCAGCTGCCCTTCGAAA	79.4	−5.2
nrdE*	164	5'-CCTCGGGGCATGGAAGTTTTACACC	79.8	−2.5
nrdE		5'-AGCGTCCTGACAGCATTTCATCGGT	79.7	−5.9

[a] Forward primers, labeled with 4,7,2',4', 5',7' -hexachloro-6-carboxyfluorescein, are marked with asterisks.

[b] T_m (melting temperature) values (calculated by the nearest-neighbor method) give the dissociation temperature of each primer/template duplex.

[c] $3' \Delta G$ values (calculated by multiplex analysis) report the free energy of cross-dimerization between each primer and the primer forming the most stable dimer. A $3'$ dimer ΔG value greater than −1.0 (meaning no likelihood of dimerization) is indicated by a dash.

Primer Specificity

A highly specific multiplex PCR generates only PCR products of the predicted sizes. Nevertheless, the PCR products can be further verified by nucleotide sequencing. Primer design and PCR conditions are adjusted for maximum specificity. To eliminate competing side reactions and unspecific PCR products, the use of a DNA polymerase that is supplied in an inactive form and becomes activated by heating the reaction mixture is highly recommended.

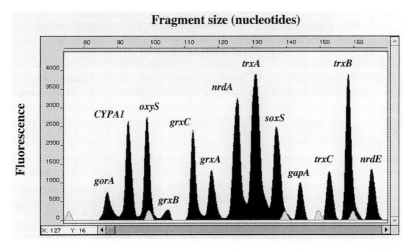

Fragment size (nucleotides)

FIG. 1. GeneScan electropherogram display of one representative sample. Peak height is shown on the y axis and fragment size is shown on the x axis. Solid peaks are identified by the name of each gene. Size standards are open peaks. The sample loaded corresponds to untreated wild-type bacteria (Table II). Primers were at the following concentrations: 0.02 μM (*grxB*), 0.03 μM (*CYP1A*), 0.04 μM (*gorA, nrdA*), 0.05 μM (*gapA*), 0.06 μM (*grxC*), 0.08 μM (*soxS*), 0.12 μM (*grxA, trxB*), 0.24 μM (*oxyS, trxA*), and 0.32 μM (*trxC, nrdE*). Forward and reverse primers were at identical concentrations. Coamplification took place for 27 cycles.

Multiplex Polymerase Chain Reaction

Standard Conditions

The standard reaction mixture (25-μl final volume) contains MPCR buffer 3 (Maxim Biotech, San Francisco, CA) supplemented with 1 mM MgCl$_2$, a 250 μM concentration of each dNTP, a 0.08 μM concentration of each primer, 1 μl of cDNA solution, and 1.25 units of AmpliTaq Gold (Applied Biosystems). The so-called standard reaction mixture is regarded as the point of departure to explore modifications and potential improvements. In particular, the primer concentrations must be adjusted (see legend to Fig. 1) so that all genes yield quantifiable fluorescent signals over a wide range of expression levels (Table II). Before running the PCR amplification, the AmpliTaq Gold is activated by heating for 10 min at 95°. Each PCR cycle consists of 1 min of denaturation at 94°, and 45 sec of annealing and extension at 70°. The number of PCR cycles is fixed by means of a C-type multiplex PCR (Fig. 2).

Quantification of Polymerase Chain Reaction Products

After amplification, 0.25 μl of the multiplex PCR solution is mixed with 0.5 μl of deionized formamide, 0.125 μl of Prism Genescan-350 Tamra ladder (fragment

TABLE II

INDUCTION BY OXIDATIVE STRESS OF EXPRESSION OF GENES ENCODING KNOWN
COMPONENTS OF THIOREDOXIN AND GLUTAREDOXIN/GSH PATHWAYS AND
RELATED ENZYMES

Gene	Wild-type basal levels[a]	Wild-type induced levels[a]	OxyR[c] mutant basal levels[b]
grxA	1.54 ± 0.04^{c}	$65.59 \pm 2.92\ (42.6)^{*}$	$41.84 \pm 2.22\ (27.2)^{*}$
grxB	0.27 ± 0.02	$0.34 \pm 0.04\ (1.3)$	$0.27 \pm 0.02\ (1.0)$
grxC	2.12 ± 0.06	$2.68 \pm 0.14\ (1.3)$	$1.46 \pm 0.05\ (0.7)$
gorA	0.91 ± 0.02	$6.31 \pm 0.32\ (6.9)^{*}$	$2.76 \pm 0.18\ (3.0)^{*}$
trxA	7.95 ± 0.47	$9.31 \pm 0.55\ (1.2)$	$10.67 \pm 0.45\ (1.3)$
trxC	1.44 ± 0.07	$44.69 \pm 1.25\ (31.0)^{*}$	$38.64 \pm 1.37\ (26.8)^{*}$
trxB	4.08 ± 0.07	$12.61 \pm 0.89\ (3.1)^{*}$	$7.15 \pm 0.41\ (1.8)^{*}$
nrdA	4.21 ± 0.17	$2.57 \pm 0.20\ (0.6)$	$5.03 \pm 0.21\ (1.2)$
nrdE	1.55 ± 0.03	$5.29 \pm 0.44\ (3.4)^{*}$	$0.54 \pm 0.03\ (0.3)^{*}$
oxyS	2.42 ± 0.07	$239.95 \pm 8.10\ (99.2)^{*}$	$87.91 \pm 5.21\ (36.3)^{*}$
soxS	3.25 ± 0.09	$67.46 \pm 3.83\ (20.8)^{*}$	$1.64 \pm 0.06\ (0.5)^{*}$
CYP1A	2.45 ± 0.06	$2.71 \pm 0.22\ (1.1)$	$3.91 \pm 0.31\ (1.6)$

[a] Wild-type bacteria (UC5710) grown in M9 minimal medium to reach an OD_{600} of 0.2 were treated (induced levels) or not treated (basal levels) with 30 μM H_2O_2 for 1 min.
[b] Untreated OxyR[c] mutant (UC1394).
[c] Data represent the mean $\pm$ SEM ($n = 6$) of the fluorescence signal of each PCR product relative to that of the control gapA gene. Values relative to those of untreated wild-type bacteria are indicated in parentheses. Statistical significance ($p \le 0.01$) is marked with an asterisk. Coamplification took place for 27 cycles.

size standard), and 0.125 μl of loading buffer (blue dextran, 50 mg/ml; EDTA, 25 mM). Samples are denatured at 92° for 3 min and run on a denaturing 4.25% (w/v) polyacrylamide gel at 750 V in an ABI 377 DNA sequencer (Applied Biosystems). Data are collected and analyzed with ABI Prism 377 Collection 2.1/97 and GeneScan Analysis 2.0.2/95 software (Applied Biosystems), respectively. The GeneScan system sizes and quantitates DNA fragments by automated fluorescence detection. The results are displayed as electropherograms, as tabular data, or as a combination of both. Electropherograms show fluorescence as a function of fragment size (Fig. 1) and tabular data provide detailed sizing and quantitative information. The relative concentrations of different-sized fragments in a sample can be determined by comparing either the peak heights or peak areas given in the tabular data. We normalize putative differences among multiplex PCR outcomes by dividing the fluorescent signal strength (peak area) of each target fragment by the fluorescent signal strength of the standard (see below) in the same reaction.

Standard

In semiquantitative or relative reverse transcription (RT)-PCR, the amounts of the target gene transcripts are quantified in relation to the level of a coamplified

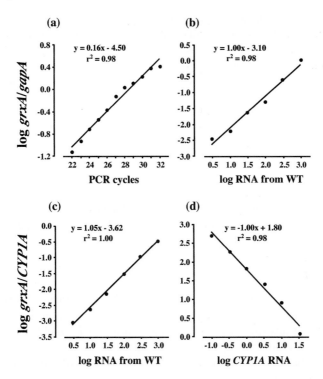

Fig. 2. C-type (a) T-type (b–d) multiplex PCR. In (a), a single cDNA sysnthesis reaction was undertaken with a fixed amount (1 μg) of total RNA from wild-type bacteria (UC5710) and a fixed amount (16 pg) of the external standard (*CYP1A*). Synthesized cDNA was then amplified for 22 to 32 cycles. To analyze the results, the log values of the ratios of the fluorescent signal of each sequence to that of *gapA* (internal standard) were plotted against the number of PCR cycles. Data are illustrated by representing the regression line of the *grxA/gapA* ratio plot. The slope values ranged from 0.07 (*grxC/gapA*) to 0.22 (*oxyS/gapA*); correlation coefficient values were $r^2 \geq 0.97$. In (b) and (c), six separate cDNA synthesis reactions were undertaken with increasing amounts (ng) of total RNA from wild-type bacteria (UC5710). The final amount of RNA was kept equal to 1 μg by adding decreasing quantities of RNA from a $\Delta grxA$ mutant (UC848) (b) or decreasing quantities of human lung RNA (c). In (b), the log values of the *grxA/gapA* ratio were plotted against the log values of the amount of total RNA from UC5710. In (c), the log values of the ratios of the fluorescent signal of each sequence to that of *CYP1A* (external standard) were plotted against the log values of the amount of total RNA from UC5710. Data are illustrated by representing the regression line of the *grxA/CYP1A* ratio plot. The slope values ranged from +1.04 (*grxC/CYP1A*) to +1.15 (*nrdA/CYP1A*); correlation coefficient values were $r^2 \leq 0.97$. In (d), six separate cDNA synthesis reactions were undertaken with increasing amounts (pg) of the external standard (*CYP1A*) and a fixed amount of total RNA (1 μg) from wild-type bacteria (UC5710). The log values of the ratios of the fluorescent signal of each sequence to that of *CYP1A* were plotted against the log values of the amount of *CYP1A* RNA. Data are illustrated by representing the regression line of the *grxA/CYP1A* ratio plot. The slope values ranged from −0.89 (*nrdA/CYP1A*) to −1.04 (*trxC/CYP1A*); correlation coefficient values were $r^2 \geq 0.96$. Coamplification took place for 27 cycles in (b), (c), and (d).

standard. The best standard is a gene whose expression level remains constant under the experimental conditions studied. We routinely use as internal standard (also named reference, control, or housekeeping gene) the *gapA* gene, which encodes D-glyceraldehyde-3-phosphate dehydrogenase, a key enzyme of both the glycolytic and gluconeogenic pathways. Nevertheless, it cannot be taken for granted that the control gene will maintain a steady level of expression under all circumstances.[11] Hence, an *in vitro*-synthesized RNA must also be included as external standard to control, by means of either competitive or noncompetitive RT-PCR strategies, the potential variability of the reference gene. As external standard we have used a fragment of a gene (*CYP1A* which encodes cytochrome P4501A from *Liza aurata*) with no homology with the bacterial genome and/or a laboratory-engineered fragment of the *gapA* gene.[8,11] Although the *CYP1A* noncompetitor heterologous standard is as useful as the *gapA* competitor to monitor changes in the expression level of the reference gene, the competitor has the additional advantage of sharing with the reference gene idential PCR primer-binding sites.

Statistical Analysis

Samples for comparison of different experimental conditions or different bacterial strains must be handled in parallel. Data are presented as the means $\pm$ standard errors of the means (SEM) from $n \geq 6$ multiplexed PCR amplifications. Statistical comparisons are done by a hierarchical analysis of variance test that consider treatment (or bacterial mutant strain) and RNA as sources of variations.

Optimization of Multiplex Reverse Transcription-Polymerase Chain Reaction

Any attempt to quantitate mRNA levels by PCR must be limited to the analysis of products generated only during the exponential phase of amplification. Theoretically, within the exponential phase, PCR products accumulate according to Eqs. (1) and (2),

$$T_n = T_0(1 + E^{\mathrm{T}})^n \tag{1}$$

$$S_n = S_0(1 + E^{\mathrm{S}})^n \tag{2}$$

where the initial amounts (before PCR) of T (target) and S (standard) sequences are designated as T_0 and S_0, the amounts of products following a number n of PCR cycles as T_n and S_n, and the mean efficiencies as E^{T} and E^{S}, respectively. Making a ratio of both equations and taking the logarithm, we obtain

$$\log(T_n/S_n) = \log T_0 - \log S_0 + n \log[(1 + E^{\mathrm{T}})/(1 + E^{\mathrm{S}})] \tag{3}$$

[11] M. Manchado, C. Michán, and C. Pueyo, *J. Bacteriol.* **182**, 6842 (2000).

The exponential phase of the multiplex RT-PCR is defined by using two variants (C- and T-type) of RT-PCR, which were initially developed for absolute quantification of mRNA level.[12]

C-Type and T-Type Multiplex Polymerase Chain Reaction

In the C-type RT-PCR variant, the amounts of target RNA and standard RNA are kept constant and the number of PCR cycles is altered. According to Eq. (3), straight lines having slope values within the range $-\log 2$ ($E^T = 0$ and $E^S = 1$) to $+\log 2$ ($E^T = 1$ and $E^S = 0$) are expected if E^T and E^S values remain constant in the course of the PCR amplification process. As exemplified in Fig. 2a for the *grxA/gapA* ratio, data from the C-type variant must be consistent with these theoretical predictions, hence demonstrating that PCR amplification efficiencies for all targets and standards remain unchanged in the reactions along the 10 PCR cycles that were undertaken.

In the T-type RT-PCR variant, the amount of target RNA (or external standard RNA) in the RT-PCR is altered, keeping the amount of standard RNA (or target RNA) and the number of PCR cycles constant. According to Eq. (3), straight lines having a slope of $+1$ in case of serial dilutions for T_0 or a slope of -1 in the case of serial dilutions for S_0, are expected if E^T and E^S values remain constant (or change in such a way that the ratio $[(1 + E^T)/(1 + E^S)]$ remains constant). As exemplified in Fig. 2b for the *grxA/gapA* ratio, and in Fig. 2c and d for the *grxA/CYP1A* ratio, data from the T-type variant must fulfill these theoretical predictions. Of particular note is that, in contrast to C-type RT-PCR, T-type reactions detect changes in the efficiency of both cDNA synthesis and PCR amplification.

Both C- and T-type variants must be undertaken to test the fundamental assumption of RT-PCR: the initial ratio of target RNA copies to standard RNA copies is maintained throughout the RT-PCR process.

Application

Once the multiplex RT-PCR is optimized and the exponential amplification phase is precisely defined, the methodology can be used with confidence to detect from small variations to an at least 100-fold difference in the *in vivo* expression of the selected target genes. The examples in Table II serve to illustrate the potency of this technique.

Table II analyzes the expression profile of the genes in response to oxidative stress by H_2O_2. As shown, treatment with H_2O_2 (30 μM, 1 min) did not have the same effect on all RNA species. Hence, while no significant changes in mRNA levels were found for 4 (*grxB, grxC, trxA,* and *nrdA*) of the 11 target genes, highly

---

[12] J. Zhang, M. Desai, S. E. Ozanne, C. Doherty, C. N. Hales, and C. D. Byrne, *Biochem. J.* **321,** 769 (1997).

significant ($p \leq 0.01$) variations (from 3.1- to 99.2-fold) were quantified for the transcripts of the remaining 7 (*grxA, gorA, trxC, trxB, nrdE, oxyS,* and *soxS*) loci.

The role of OxyR in the upregulation of *grxA, gorA, trxC, trxB,* and *nrdE* genes on exposure to H_2O_2 was further explored by studying the effect of the constitutive *oxyR2* mutant allele on basal levels of expression. As shown in Table II, transcription of *grxA, gorA, trxC,* and *trxB* is upregulated in the constitutive strain, thus indicating an OxyR-dependent regulation. In contrast, no upregulation of *nrdE* gene is observed in the strain with the constitutive *oxyR2* mutation, indicating that OxyR is not involved in its response to H_2O_2. Transcription of *oxyS* was as expected from an OxyR-regulated gene, and transcription of *soxS* was in agreement with our finding that the SoxR/S response, which protects cells against superoxide toxicity, is triggered also by H_2O_2, in an OxyR-independent, yet SoxR-dependent, manner.[11]

It is important to note that the ratio of the fluorescent signal of the *CYP1A* RNA (external standard) to that of the *gapA* transcript (internal standard) remained constant in the experiments outlined in Table II, hence ensuring that changes detected with reference to the housekeeping *gapA* gene are properly attributed to variations in the expression levels of the target genes under analysis.

Final Remarks

The main goals in gene expression studies are to compare the steady state levels of a group of specific transcripts among cell types or strains, and to quantify the effect of particular stimuli on the expression of these genes. As shown here, our multiplex RT-PCR procedure is in practice a quick, reliable, and robust methodology for precise relative quantification of *in vivo* gene expression. In addition, because of the PCR amplification step, our method displays a much higher sensitivity than that of current techniques for mRNA quantitation, such as Northern blotting or primer extension analyses.

By employing the multiplex RT-PCR technique, as detailed above, we have obtained evidences that transcription of genes that encode several components of both thioredoxin (*trxC* and *trxB*) and glutaredoxin/GSH (*grxA* and *gorA*) systems are induced by oxidative stress, in an OxyR-dependent manner.[8] In contrast, transcription of the *nrdAB* operon is not induced by oxidative stress, yet it is induced by hydroxyurea (an inhibitor of RRase) or when the NrdAB enzyme must operate in the absence of its two main reductants (Trx1 and Grx1).[8,10] Comparatively, transcription of the *nrdHIEF* operon is readily induced by oxidative stress, although in both an OxyR- and SoxR/S-independent manner.[13] Our multiplex RT-PCR protocol is a versatile technique that should be easily applicable to many other situations and cell types in which it would be desirable to measure accurately and

[13] F. Monje-Casas, J. Jurado, M. J. Prieto-Álamo, A. Holmgren, and C. Pueyo, *J. Biol. Chem.* **276,** 18031 (2001).

reproducibly differences in the transcript levels of genes encoding proteins that protect cells against reactive oxygen species toxicity and for related enzymes.

Acknowledgments

This work was subsidized by Grant PB98-1627 (DGES). J. Jurado and M.-J. Prieto-Álamo are recipients of postdoctoral contracts from the Ministerio de Educación y Cultura, and F. Monje-Casas is a predoctoral fellow from the Junta de Andalucía. We thank M. Manchado and G. Dorado for helpful discussions regarding the method.

[43] Redox Regulation of Cell Signaling by Thioredoxin Reductases

By QI-AN SUN and VADIM N. GLADYSHEV

Introduction

Mammalian thioredoxin reductases (TRs) are central enzymes in the thioredoxin (Trx) redox pathway. These proteins are FAD-containing pyridine nucleotide disulfide oxidoreductases that utilize NADPH for reduction of active-site disulfide of Trx.[1] Three mammalian TR isozymes and their alternative splicing forms have been identified,[2,3] with TR1[4] (also called TRα[5] and TrxR1[6]) being a cytosolic enzyme, TR3 (also called TRβ[5] and TrxR2[6]) being a mitochondrial enzyme, and TR2 (also called Trx and glutathione reductase, TGR[7]) being an isozyme that has been detected only in mouse testes.[2,7] The three mammalian TRs contain conserved selenocysteine (Sec), encoded by TGA, at the C-terminal penultimate position, which is essential for catalytic activities of the enzymes (Fig. 1). The reasons why this rare amino acid residue rather than cysteine (Cys) is utilized for TR function are beginning to emerge. Sec is a better nucleophile than sulfur and, under

[1] A. Holmgren, *J. Biol. Chem.* **264**, 13963 (1989).

[2] Q.-A. Sun, Y. Wu, F. Zappacosta, K.-T. Jeang, B. J. Lee, D. L. Hatfield, and V. N. Gladyshev, *J. Biol. Chem.* **274**, 24522 (1999).

[3] Q.-A. Sun, F. Zappacosta, V. M. Factor, P. J. Wirth, D. L. Hatfield, and V. N. Gladyshev, *J. Biol. Chem.* **276**, 3106 (2001).

[4] V. N. Gladyshev, K.-T. Jeang, and T. C. Stadtman, *Proc. Natl. Acad. Sci. U.S.A.* **93**, 6146 (1996).

[5] P. Y. Gasdaska, M. M. Berggren, M. J. Berry, and G. Powis, *FEBS Lett.* **442**, 105 (1999).

[6] S.-R. Lee, J.-R. Kim, K.-S. Kwon, H W. Yoon, R. L. Levine, A. Ginsburg, and S. G. Rhee, *J. Biol. Chem.* **274**, 4722 (1999).

[7] Q.-A. Sun, L. Kirnarsky, S. Sherman, and V. N. Gladyshev, *Proc. Natl. Acad. Sci. U.S.A.* **98**, 3673 (2001).

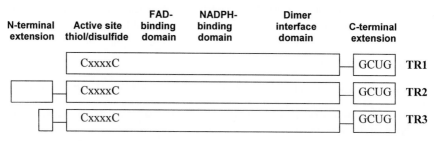

FIG. 1. Domain organization of mammalian TR isozymes. The three TRs display 50–75% identity in amino acid sequences and possess an active center thiol/disulfide, FAD- and NADPH-binding domains, and a dimer interface domain whose locations within the TR isozymes are indicated above the large rectangular boxes. TRs also contain a conserved C-terminal GCUG tetrapeptide that is indicated by the small rectangular boxes on the right. In addition, TR3 contains a mitochondrial-targeting signal peptide that is removed in the mature protein, and TR2 contains a larger N-terminal domain, which is homologous to glutaredoxin, and these are indicated by the small rectangular boxes on the left. Homology analyses indicate that TR1 and TR2 are evolutionarily related enzymes, whereas TR3 is a more distantly related homolog.

physiological conditions, Sec is ionized whereas Cys is protonated.[8] These properties of Sec result in mammalian TRs playing two important functions. First, Sec is directly involved in reactions catalyzed by TRs.[9,10] It is a component of the C-terminal Gly-Cys-Sec-Gly tetrapeptide, which is analogous to glutathione (GSH, the oxidized form is GSSG), which is fused to a pyridine nucleotide disulfide oxidoreductase protein domain.[7] Second, the unusual location of Sec in a C-terminal, conformationally flexible sequence allows this amino acid to function as a cellular redox sensor and thus to support a role for mammalian TR in redox-regulated cell signaling.[2] In contrast, TRs in bacteria, archaea, yeast, and plants show limited homology to animal TRs and lack the C-terminal Sec. Therefore, the role of TR in redox regulation of cellular processes may be different in organisms other than animals.[11]

At low concentrations, reactive oxygen species (ROS), such as hydrogen peroxide (H_2O_2), superoxide anion radical, and hydroxyl radical, function as essential participants in cell signaling,[12] although excessive production of ROS or insufficiency of antioxidant defenses has also been implicated in apoptosis, aging,

[8] T. C. Stadtman, *Annu. Rev. Biochem.* **65**, 83 (1996).

[9] S.-R. Lee, S. Bar-Noy, J. Kwon, R. L. Levine, T. C. Stadtman, and S. G. Rhee, *Proc. Natl. Acad. Sci. U.S.A.* **97**, 2521 (2000).

[10] L. Zhong and A. Holmgren, *J. Biol. Chem.* **275**, 18121 (2000).

[11] L. D. Arscott, S. Gromer, R. H. Schirmer, K. Becker, and C. H. Williams, Jr., *Proc. Natl. Acad. Sci. U.S.A.* **94**, 3621 (1997).

[12] M. Sundaresan, Z. X. Yu, V. J. Ferrans, K. Irani, and T. Finkel, *Science* **270**, 296 (1995).

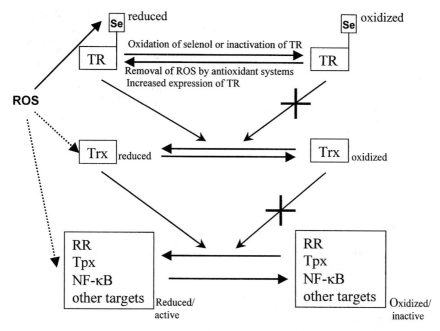

FIG. 2. Proposed role of thioredoxin reductase in redox regulation of cell signaling. ROS can directly target selenol of Sec, which results in decreased activity of the enzyme. Because the redox state of Trx is dependent on that of TR, oxidized Trx accumulates in cells, which affects activities of various redox-regulated pathways, including biosynthesis of deoxyribonucleotides by ribonucleotide reductase (RR), antioxidant defense by thioredoxin peroxidase (Tpx), gene expression by transcription factors such as NF-κB, cell signaling by kinases and phosphatases and other cellular pathways/targets. It is not known whether ROS can regulate these pathways by directly targeting indicated proteins (shown by dashed arrows), but available data suggest this possibility. The redox signaling cycle involving TR is completed by increased expression of this enzyme and degradation of ROS by antioxidant systems. See text for further discussion.

and cancer.[13] Redox signaling is thought to be achieved through the coupling of ROS with oxidation–reduction processes, which involve thiol groups in proteins. The redox state of thiol groups is controlled by two cellular redox systems: the Trx (TR, Trx, and Trx peroxidase) and GSH (GSSG reductase, GSH, and GSH peroxidase) systems.[14] Data have revealed that Sec in TR1 is the direct target of ROS[2] (Fig. 2). The intracellular generation of ROS results in oxidation of the selenol group of TR1, which transiently upsets catalytic activity of the enzyme. The decrease in TR1 activity may result in the oxidation of Trx, which affects activities

[13] E. R. Stadtman, *Science* **257**, 1220 (1992).
[14] A. Holmgren and M. Bjornstedt, *Methods Enzymol.* **252**, 199 (1995).

of signaling and regulatory proteins, such as protein tyrosine phosphatases, Trx peroxidase, and various redox-regulated transcription factors. It was also suggested that ROS directly target these proteins.[15] The subsequent elimination of ROS by antioxidant systems and induction of TR1 expression complete the redox signaling cycle involving Sec in TR1[2] (Fig. 2). In this article, we discuss approaches for characterization of a redox sensor role of Sec in TRs *in vivo* and *in vitro*. Procedures for isolation of the three TR isozymes are presented. TR1, the best characterized of the three isozymes, is used as a model enzyme in most redox signaling experiments. However, other mammalian TR isozymes likely share these same functions.

Isolation of Mammalian Thioredoxin Reductase Isozymes

Three mammalian TR isozymes (TR1, TR2, and TR3) were previously identified.[2] TR1 and TR3 were isolated from mouse and rat liver, and TR2 was isolated from mouse or rat testes.[2,7] One procedure for isolation of TR1 has been described in this series.[16] In this article, we describe a modified method, which is sufficient to isolate different TR isozymes. A three-step procedure is utilized that includes anion-exchange chromatography on a DEAE-Sepharose column, chromatography on an ADP-Sepharose column, and hydrophobic interaction chromatography on a phenyl-Sepharose or phenyl-HPLC column. The first chromatographic step allows partial separation of TR isozymes along with significant enrichment of their specific activities. The second step utilizes an affinity column and provides a dramatic enrichment of isozymes. The third step further separates TR isozymes as well as completes their separation from other proteins. To monitor isozymes during isolation, extracts from ^{75}Se-labeled cells are used. Chromatographic fractions are analyzed by immunoblot assays with isozyme-specific antibodies and by sodium dodecyl sulfate (SDS)–polyacrylamide gels followed by detection of radioactivity with a PhosphorImager (Molecular Dynamics, Sunnyvale, CA). To obtain mouse tissues that are metabolically labeled with ^{75}Se, 1 mCi of freshly neutralized [^{75}Se]selenite (University of Missouri Research Reactor) is injected into a mouse 2 days before tissue removal. To label cells in culture, [^{75}Se]selenite is added directly to cell culture medium and cells are grown for 2 days. Before protein isolation, ^{75}Se-labeled cells or tissues are mixed with the corresponding unlabeled cells or tissues.

Materials

[75]Se-labeled and unlabeled mouse or rat tissues (liver or testes)
DEAE-Sepharose (Amersham Pharmacia, Piscataway, NJ)

[15] S. G. Rhee, *Exp. Mol. Med.* **31,** 53 (1999).
[16] E. S. Arner, L. Zhong, and A. Holmgren, *Methods Enzymol.* **300,** 226 (1999).

ADP-Sepharose (Amersham Pharmacia)
Phenyl-HPLC column
Antibodies specific for TR1, TR2, or TR3
Buffer A: 25 mM Tris-HCl, 1 mM EDTA; pH 8.0
Buffer B: 25 mM Tris-HCl, 1 mM EDTA; pH 7.5
Other reagents: Phenylmethylsulfonyl fluoride (PMSF), aprotinin, leupetin, pepstatin A

Method

1. To isolate TR1 and TR3, liver is used, and testes are used for isolation of TR1 and TR2. Sixty grams of mouse or rat tissue (liver or testes) is homogenized and sonicated in 5 volumes of buffer A containing 1 mM EDTA, 1 mM PMSF, aprotinin (5 μg/ml), leupeptin (5 μg/ml), and pepstatin A (5 μg/ml).
2. The homogenate is centrifuged at 18,000g for 40 min at 4°.
3. The supernatant is applied onto a DEAE-Sepharose column equilibrated in buffer A. A gradient of 0–0.4 M NaCl in buffer A is applied, and TR1, TR2, and TR3 are detected in fractions from the column by immunoblot assays with antibodies specific to different TR isozymes. TR2 is eluted at ~150 mM NaCl, before TR1, which is eluted at ~200 mM NaCl. TR3 is eluted at ~100 mM NaCl or is located in flowthrough fractions.
4. Fractions containing different isozymes of TR are collected separately and diluted in 2.5 volumes of buffer B (this step is not necessary for TR3 fractions). Each TR pool is applied separately onto an ADP-Sepharose column equilibrated in buffer B. The column is sequentially washed with buffer B and buffer B containing 200 mM NaCl, and TRs are eluted with buffer B containing 1 M NaCl.
5. The fractions containing TR are made in 0.8 M (NH$_4$)$_2$SO$_4$ by addition of 4 M ammonium sulfate and applied onto a phenyl-HPLC column equilibrated in 0.8 M (NH$_4$)$_2$SO$_4$ in buffer B. A gradient of 0.8 M (NH$_4$)$_2$SO$_4$ in buffer B to buffer B is applied. Elution of TRs is analyzed by immunoblot assays with isozyme-specific antibodies and by PhosphorImager analyses of SDS–polyacrylamide gels. TR1 is typically eluted from the phenyl column at ~500 mM (NH$_4$)$_2$SO$_4$, TR2 at ~100 mM (NH$_4$)$_2$SO$_4$, and TR3 at ~700 mM (NH$_4$)$_2$SO$_4$.

Using this procedure, TR1 and TR2 are not completely separated after the DEAE-Sepharose column. These isozymes may be combined at this step. The last isolation step, the phenyl-HPLC chromatography, is usually sufficient to separate TR1 and TR2 completely. TRs can also be analyzed in chromatographic fractions by determining their catalytic activities by procedures described elsewhere in this volume. All three isozymes catalyze NADPH-dependent reduction of DTNB and mammalian and *Escherichia coli* thioredoxins.[16] In addition, TR2 has GSSG reductase and glutaredoxin activities.[7] Finally, these isozymes may be distinguished by SDS–polyacrylamide gels, on the basis of their size. A major form of mouse TR2 migrates on SDS–polyacrylamide gels as a 65-kDa polypeptide, and those of TR1 and TR3, as 55-to 57-kDa polypeptides. Purification of the three TR isozymes is summarized in Scheme 1.

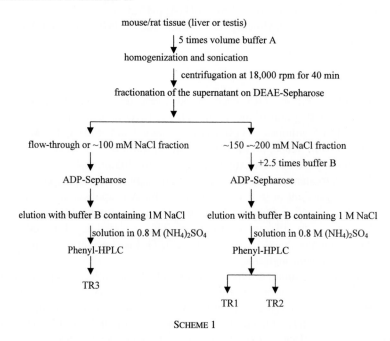

mouse/rat tissue (liver or testis)

↓ 5 times volume buffer A

homogenization and sonication

↓ centrifugation at 18,000 rpm for 40 min

fractionation of the supernatant on DEAE-Sepharose

flow-through or ~100 mM NaCl fraction   ~150 -~200 mM NaCl fraction

↓ +2.5 times buffer B

ADP-Sepharose   ADP-Sepharose

elution with buffer B containing 1M NaCl   elution with buffer B containing 1 M NaCl

↓solution in 0.8 M (NH₄)₂SO₄   ↓solution in 0.8 M (NH₄)₂SO₄

Phenyl-HPLC   Phenyl-HPLC

TR3   TR1   TR2

Scheme 1

Analysis of Thioredoxin Reductase 1 Redox State, Using 5-Iodoacetamidofluorescein

Changes in the redox state of TR1 may be monitored by alkylation of this enzyme with 5-iodoacetamidofluorescein (5-IAF; Molecular Probes, Eugene, OR). Only the reduced from of TR1 can be modified by this compound. The presence of fluorescein in the modified enzyme is determined by immunoblot assays using antibodies specific for fluorescein (Molecular Probes). The procedure involves incubation of the purified mouse liver TR1 with or without NADPH, followed by incubation in the presence of 5-IAF. The use of ^{75}Se-labeled enzyme is advantageous in that it allows monitoring of the selenium content of the enzyme. Selenium may be lost from TR1 and possibly other isozymes on incubation with ROS or growth of cells in a high oxygen concentration.[17] 5-IAF has previously been used to detect proteins containing reactive thiols.[18]

[17] S. N. Gorlatov and T. C. Stadtman, *Proc. Natl. Acad. Sci. U.S.A.* **95,** 8520 (1998).
[18] Y. Wu, K. S. Kwon, and S. G. Rhee, *FEBS Lett.* **440,** 111 (1998).

Materials

[75]Se-labeled mouse liver TR1
NADPH, 0.5 mM
5-IAF, 0.1 mM
Tris-HCl buffer, 0.25 mM, with 1 mM EDTA; pH 7.5
Antibodies specific for TR1
Antibodies specific for fluorescein

Method

An aliquot of mouse liver TR1 (1 μl, ~0.5 μg) is added to each of four Eppendorf tubes followed by addition of 2 μl of NADPH to tubes 3 and 4. After 10 min, 2-μl aliquots of 0.1 mM 5-IAF are added to tubes 2 and 4, buffer is supplied to make the total volume of each sample 10 μl, and the four samples are further incubated at 25° for 10 min. Subsequently, unreacted 5-IAF is eliminated by the addition of excess dithiothreitol (DTT), and the samples are analyzed by SDS–polyacrylamide gel electrophoresis (PAGE) and immunoblot assays with antibodies specific for TR1, and separately, for fluorescein. The selenium content in TR1 samples is determined by PhosphorImager analyses on SDS–polyacrylamide gels. Normally, anti-TR1 immunoblot assays give identical signals in the four samples, whereas anti-fluorescein signals indicate TR1 redox state (high anti-fluorescein signal indicates the presence of the reduced form of TR1).

Analysis of Changes in Redox State of Isolated Mouse Liver Thioredoxin Reductase 1 *in Vitro*

Using the above-described approach, it is possible to analyze changes in the redox state of TR1 on addition of oxidants. For example, the addition of H_2O_2 leads to oxidation of the NADPH-reduced TR1.[2] The increase in H_2O_2 concentration results in lowering the fluorescein signal, while not affecting the protein level or the amount of ^{75}Se in the protein. Thus, hydrogen peroxide oxidizes thiol and/or selenol groups in TR1.

Materials

[75]Se-labeled purified TR1
NADPH, 0.5 mM
H_2O_2: 0.1, 1, and 5 mM
5-IAF, 0.12 mM
Tris-HCl buffer, 0.25 mM, containing 1 mM EDTA; pH 7.5
Antibodies specific for TR1
Antibodies specific for fluorescein

Method

Aliquots of TR1 (1 μl, ~0.5 μg) are distributed into each of five tubes and 2 μl of NADPH is added to each of tubes 1–4. After a 10-min incubation, 1-μl volumes of 0.1, 1, and 5 mM H_2O_2 are added to tubes 2, 3, and 4, respectively, giving final H_2O_2 concentrations of 10, 100, and 500 μM, respectively. Buffer is also added to make the final volume of each sample 10 μl. After incubation at 25° for 10 min, 2 μl of 5-IAF is added to each tube and samples are incubated for an additional 10 min. The samples are analyzed as described above.

Analysis of Thioredoxin Reductase 1 Redox State in Cell Culture System

Human epidermoid carcinoma A431 cells can be used as a model cell culture system because these cells are known to generate ROS (predominantly, hydrogen peroxide) in response to epidermal growth factor (EGF) stimulation.[19] In these cells, generation of ROS results in initial oxidation followed by Trx-dependent reduction of protein tyrosine phosphatase 1B, which increases tyrosine phosphorylation of proteins.[20] In addition, ROS generation results in reversible changes in the protein disulfide isomerase redox state.[21] Likewise, it was found that stimulation of A431 cells with EGF results in a time-dependent oxidation of TR1 determined as a decrease in the fluorescein signal in TR1 modified with 5-IAF in extracts of A431 cells. Thus, oxidation of thiol or selenol groups in TR1 occurred in response to the generation of ROS in EGF-treated cells. Reactivity of Sec in TR1, along with accessibility of this residue and a general affinity of selenols for hydrogen peroxide, suggested that this effect was due to direct interaction of Sec in TR1 with ROS.[2]

Materials

A431 cells
Dulbecco's modified Eagle's medium (DMEM)
Fetal bovine serum (FBS)
EGF
5-IAF, 2 mM
Lysis buffer: 50 mM Tris-HCl, 1 mM EDTA, PMSF, aprotinin (5 μg/ml), leupeptin (5 μg/ml), pepstatin A (5 μg/ml), 1% (v/v) Igepal CA-630, 0.04% (w/v) NaN$_3$; pH 7.5
ADP-Sepharose
Antibodies specific for TR1
Antibodies specific for fluorescein

[19] Y. S. Bae, S. W. Kang, M. S. Seo, I. C. Baines, E. Tekle, P. B. Chock, and S. G. Rhee, *J. Biol. Chem.* **272,** 217 (1997).

[20] S. R. Lee, K. S. Kwon, S. R. Kim, and S. G. Rhee, *J. Biol. Chem.* **273,** 15366 (1998).

[21] Y. L. Wu, S. W. Kang, R. L. Levine, M. Bourdi, L. Pohl, and S. G. Rhee, *FASEB J.* **12,** A1480 (1998).

Method

A431 cells are grown on DMEM containing 10% (v/v) FBS until near confluency and then incubated in DMEM containing 0.1% (v/v) FBS overnight. EGF (final concentration, 500 ng/ml) is added directly to the medium and cells are collected at various time points within 1 hr of EGF stimulation. The collected cells are lysed at 37° in an anaerobic lysis buffer containing 20 μM 5-IAF. The unreacted 5-IAF is then eliminated by the addition of 1 mM DTT. To separate the TR1 signal from that of other cellular proteins that are modified with 5-IAF, TR1 is enriched using ADP-Sepharose. The ADP-Sepharose-bound TR1 is eluted with buffer B containing 1 M NaCl and is further analyzed by immunoblot assays with antibodies specific for TR1, and separately, for fluorescein. The fluorescein signal divided by the TR1 signal reflects the redox state of this enzyme, similarly to the test tube analyses described above.

Specific Alkylation of Selenocysteine in Thioredoxin Reductase 1 by 5-Iodoacetamidofluorescein

To identify which amino acid is oxidized in TR1 when ROS are generated intracellularly, TR1 that is modified with 5-IAF in cell extracts may be isolated to near homogeneity and analyzed by tryptic digest and mass spectrometry. The use of ^{75}Se-labeled TR1 is advantageous as it provides additional means of TR1 and Sec detection in chromatographic fractions. For this experiment, A431 cells are labeled with ^{75}Se and the cell lysate is incubated with 5-IAF as described above. The 5-IAF-modified ^{75}Se-labeled TR1 is isolated from A431 cell extracts, using the procedure described above for isolation of TR isozymes. If the amount of TR biomass is limiting, then only the first two chromatographic steps in the isolation procedure are used. The purified 5-IAF-modified TR1 is further alkylated with iodoacetamide under denaturing conditions to protect previously unmodified thiol and selenol groups, digested by trypsin, and peptides are fractionated by reversed-phase high-performance liquid chromatography (HPLC). The Sec-containing C-terminal peptide is detected by the presence of ^{75}Se in fractions from a C_{18} column and the presence of 5-IAF or iodoacetamide-modified groups is determined by tandem mass spectrometry. Using this procedure, we found that the selenol group in TR1 is the only site of modification by 5-IAF and, therefore, Sec is the only residue in TR1 that is oxidized by ROS in A431 cell.[2]

Materials

A431 cells
Dulbecco's modified Eagle's medium (DMEM)
Fetal bovine serum (FBS)
[75Se] selenite
PBS: 10 mM phosphate buffer with 137 mM NaCl, 2.7 mM KCl; pH 7.2

Lysis buffer: 50 mM Tris-HCl, 1 mM EDTA, PMSF, aprotinin (5 μg/ml),
    leupeptin (5 μg/ml), pepstatin A (5 μg/ml), 1% (v/v) Igepal CA-630,
    0.04% (w/v) NaN$_3$; pH 7.5
5-IAF, 2 mM
DEAE-Sepharose
ADP-Sepharose
C$_{18}$ reversed-phase HPLC column

Method

A431 cells are grown in the presence of ^{75}Se until near confluency, incubated
in DMEM containing 0.1% (v/v) FBS, washed four times with PBS, harvested
by scraping in an anaerobic lysis buffer, and alkylated with 20 μM 5-IAF as
described above. The alkylation reaction is quenched by the addition of 1 mM
DTT and the solution is dialyzed against 25 mM Tris-HCl, pH 7.5, containing
leupeptin (2 μg/ml) and aprotinin (2 μg/ml). TR1 is isolated by chromatography
on DEAE-Sepharose and ADP-Sepharose as described above for the isolation of
the enzyme from mouse and rat tissues. Purified 5-IAF-modified TR1 ($\sim$50 μg) is
further incubated with 2 mM DTT in the presence of 8 M guanidine hydrochloride
and reacted with 10 mM iodoacetamide for 30 min to alkylate unmodified thiol
and selenol groups.

The protein is then digested with trypsin and the resulting peptides are separated
on a C$_{18}$ reversed-phased HPLC column with a gradient of acetonitrile. A sample
without TR1 is used as control. The eluted peptides are detected by measurement
of abosorbance at 214 nm to detect peptides and at 440 nm to detect fluorescein-
containing peptides. The eluted peptides are also analyzed with a γ counter for the
presence of [75]Se. The fraction that contains both ^{75}Se and fluorescein is subjected
to mass spectrometry analysis to determine the mass of the peptide, followed by
amino acid sequencing by tandem mass spectrometry.

Using this procedure, we found that the only detected 5-IAF-labeled peptide is
the C-terminal peptide of TR1 containing sequence SGASILQAGCUG (U is Sec).
The experimentally determined mass for this peptide is $[M + 2H]^{2+} = 779.73$,
which matched the predicted mass for the C-terminal tryptic peptide of TR1
that is modified with a single 5-IAF group and a single iodoacetamide group
($[M + 2H]^{2+} = 779.48$). Subsequent sequencing of the peptide revealed that the 5-
IAF-modified group was Sec, and the iodoacetamide-modified group was cysteine.

Effect of Stimulation of A431 Cells with Epidermal Growth Factor
or H$_2$O$_2$ on Expression of Thioredoxin Reductase 1

Changes in the redox state of TR1 in a cell culture model may be compared with
changes in its expression induced by stimulation of cells with growth factors or

treated with oxidants or antioxidants. For example, we investigated how prolonged incubation of A431 cells with EGF and H_2O_2 affects TR1 expression and found the amount of TR1 was increased as a result of incubation of A431 cells with EGF and H_2O_2. The use of ^{75}Se labeling allows monitoring of changes in expression of other major selenoproteins, such as glutathione peroxidases 1 and 4 and the 15-kDa selenoprotein.[2] We found that expression of these proteins changed little on incubation of cells with hydrogen peroxide or EGF. Thus, radioactive labeling provides an additional control of protein expression.

Materials

A431 cells
Dulbecco's modified Eagle's medium (DMEM)
Fetal bovine serum (FBS)
[^{75}Se]Selenite
Lysis buffer: 50 mM Tris-HCl, 1 mM EDTA, PMSF, aprotinin (5 μg/ml), leupeptin (5 μg/ml), pepstatin A (5 μg/ml), 1% (v/v) Igepal CA-630, 0.04% (w/v) NaN$_3$; pH 7.5
EGF
H_2O_2, 20 mM

Method

A431 cells are grown and labeled with ^{75}Se as described above. The cells are incubated for various time periods (0–8 hr) with EGF (500 ng/ml), 0.2 mM H_2O_2 or water, lysed, and subjected to SDS–PAGE and PhosphorImager analysis to detect ^{75}Se-labeled proteins and to immunoblot analysis with antibodies specific for TR1 or other proteins.

Concluding Remarks

The presence of Sec and its C-terminal penultimate location are conserved in mammalian TR isozymes. Studies suggest that in addition to a role in Trx reduction, Sec in TR1 may sense the presence of ROS, and thus TR1 participates in redox signaling pathways. Removal of either Sec tRNA or Trx genes from the mouse genome causes embryonic death,[22,23] suggesting the importance of this redox system. In contrast to the role of TR1 and Trx in redox signaling, a role of the GSH pathway in this process remains to be demonstrated.

[22] M. Matsui, M. Oshima, H. Oshima, K. Takaku, T. Maruyama, J. Yodoi, and M. M. Taketo, *Dev. Biol.* **178**, 179 (1995).
[23] M. R. Bosl, K. Takaku, M. Oshima, S. Nishimura, and M. M. Taketo, *Proc. Natl. Acad. Sci. U.S.A.* **94**, 5531 (1997).

Author Index

Subject Index

A

AP-1, thioredoxin redox regulation, 281
Ascorbic acid, *see* Vitamin C
Atlas cDNA expression array, *see* DNA
  microarray
Auranofin, thioredoxin reductase inhibition
  assay, 387–388, 391
Aurothioglucose, deiodinase inhibition,
  135, 153

B

Bone
  Kashin–Beck disease and selenium
    deficiency, 177
  monocyte-derived cell expression of
    glutathione peroxidase and thioredoxin
    reductase, 176
  osteoblast
    bone remodeling, 168
    differentiation, 168
    glutathione peroxidase expression and
      regulation, 171, 178
    selenoprotein P expression, 178
    thioredoxin reductase expression and
      regulation
      assay sensitivity and specificity, 176
      serum response, 175–176
      vitamin D response, 171–172, 175
  osteoclast
    bone remodeling, 169
    differentiation, 169
  redox signaling and reactive oxygen species,
    177
  selenium-75 labeling of selenoprotein
    expression and cell distribution,
    169–171
  thioredoxin reductase
    functions, 178–179
    inhibition in rheumatoid arthritis treatment,
      177

D

D1, *see* Deiodinase type 1
D2, *see* Deiodinase type 2
D3, *see* Deiodinase type 3
DBMIB, *see* 2,5- Dibromo-3-methyl-6-
  isopropyl-*p*-benzoquinone
DCMU, *see* 3-(3,4-Dichlorophenyl)-1,1-
  dimethylurea
Deiodinase type 1
  assay
    animal assays, 166–167
    cell assays, 164–165
    chromatography techniques, 157–159
    immunosequestration, 163–164
    organ assays, 165–166
    radioimmunoassay, 160–161
    radioiodide release assay, 161–163
    radiolabeled substrate purification, 159–160
  essential residues, 137–138
  functional overview, 125–127, 131
  gene structure and function, 130
  inhibitors
    alkylating agents, 134–135
    aurothioglucose, 135
    modeling, 140–141
    thiouracils, 135–137
  kinetic parameters, 138–140
  pH optimum, 139
  reaction mechanism, 130, 132–134
  SECIS element activity assay
    expression constructs, 20
    principles, 19–20
    radioactivity assay, 20–21
    transient transfection, 20
  selenium effects on messenger RNA levels, 50
  solubilization, 129–130
  stereospecificity, 141
  structure, 127, 129
  subcellular localization, 129
  substrate specficity, 138–141
  turnover, 134

F

Ferredoxin : thioredoxin reductase
assay
  activation by chemical reduction, 405
  light-dependent activation, 405
  principles, 404–405
  spectrophotometric
    fructose-1,6-bisphosphatase assay, 405
function, 403
iron–sulfur cluster, 403, 409
sources, 404
Synechocystis enzyme
  absorption spectrum, 408
  crystal structure, 409
  gene cloning, 406
  purification from recombinant *Escherichia coli*
    anion-exchange chromatography, 408
    cell growth and induction, 407
    extraction, 407
    ferredoxin affinity chromatography, 408
    gel filtration, 408
    hydrophobic interaction
      chromatography, 407
    hydroxyapatite chromatography, 408
    storage, 408
    vector, 406–407
  subunits, 406, 408
FTR, *see* Ferredoxin : thioredoxin reductase

G

GHOST assay
  placental thioredoxin reductase,
    386–387
  Plasmodium falciparum enzymes
    thioredoxin, 377
    thioredoxin reductase, 375–376
Glutaredoxin
  assays
    dehydroascorbic acid reduction, 296
    high-molecular-weight disulfide reduction,
      296
    high-molecular-weight mixed disulfide
      reduction, 296
    low-molecular-weight disulfide reduction,
      296
  catalytic mechanism, 293
  classification, 295–296

Escherichia coli types and functions,
    293–294, 361, 441
functional overview, 286, 291
human functions, 294–295
oxidative stress induction in *Escherichia coli*
  multiplex reverse transcriptase–polymerase
    chain reaction
    amplification reaction, 445
    complementary DNA synthesis, 443
    C-type, 447
    expression profile, 449–450
    internal standard, 445–446
    optimization, 447, 449
    primer design, labeling, and specificity,
      443–444
    quantitative analysis, 445
    RNA isolation, 442
    statistical analysis, 446
    transcription reaction, 442–443
    T-type, 447, 449
  OxyR activation, 441–442, 450
  SoxR/S regulons, 441–442, 450
sequence homology between species,
    291–292
yeast types and functions, 294
Glutathione peroxidase, *see also* Glutathione
    peroxidase 1; Phospholipid hydroperoxide
    glutathione peroxidase
  bone
    monocyte-derived cell expression, 176
    osteoblast expression and regulation, 171,
      178
  classification, 101
  functions, 102–103
  high-throughput microtiter plate assay for
    specific activity determination
    bicinchoninic acid assay for protein
      materials, 115
    overview, 113
    reaction and detection, 115–116
    sensitivity, 118
  calculations, 118
  coupled assay with glutathione reductase,
    113–114
  data analysis, 121
  linearity, 118, 120
  materials, 116
  path length considerations, 120–121
  reaction conditions, 116–117
  sample preparation, 114

ISBN 0-12-182248-6